INTRODUCTION TO ADVANCED ASTROPHYSICS

GEOPHYSICS AND ASTROPHYSICS MONOGRAPHS

AN INTERNATIONAL SERIES OF FUNDAMENTAL TEXTBOOKS

VOLUME 17

INTRODUCTION TO ADVANCED ASTROPHYSICS

by

V. KOURGANOFF

Professor Emeritus, University of Paris

D. REIDEL PUBLISHING COMPANY

DORDRECHT : HOLLAND / BOSTON : U.S.A.

LONDON : ENGLAND

Library of Congress Cataloging in Publication Data

CIP

Kourganoff, Vladimir, 1912–
Introduction to advanced astrophysics.

(Geophysics and astrophysics monographs ; v. 17)
Includes bibliographical references and index.
1. Astrophysics. I. Title. II. Series.
QB461.K69 523.01 79-20463
ISBN 90–277–1002–3
ISBN 90–277–1003–1 pbk.

Published by D. Reidel Publishing Company,
P.O. Box 17, Dordrecht, Holland

Sold and distributed in the U.S.A., Canada and Mexico
by D. Reidel Publishing Company, Inc.
Lincoln Building, 160 Old Derby Street, Hingham,
Mass. 02043, U.S.A.

Printed in The Netherlands

TABLE OF CONTENTS

CHAPTER 6. NEUTRON STARS

CHAPTER 7. PULSARS

Part III. Newton's Law, Binary Systems and Galactic X-ray Sources

CHAPTER 8. THEORY OF SPECTROSCOPIC AND ECLIPSING BINARIES. STELLAR MASSES

CHAPTER 9. GALACTIC X-RAY SOURCES

Part IV. Cosmology: Elementary Theory and Basic Observational Data

CHAPTER 10. ELEMENTARY THEORETICAL COSMOLOGY: THE NEWTONIAN APPROACH

CHAPTER 11. BASIC CONCEPTS OF RELATIVISTIC COSMOLOGY

CHAPTER 12. RELATIVISTIC EFFECTS IN OBSERVATIONAL COSMOLOGY. THE COSMOLOGICAL REDSHIFT IN EXPANDING UNIVERSE

CHAPTER 13. BASIC DATA IN OBSERVATIONAL COSMOLOGY: ACTIVE GALACTIC NUCLEI AND CLUSTERS OF GALAXIES

PREFACE

The purpose of this textbook is to provide a basic knowledge of the main parts of modern astrophysics for all those starting their studies in this field at the undergraduate level. The reader is supposed to have only a high school training in physics and mathematics. In many respects this *Introduction to Advanced Astrophysics* could represent a volume of the Berkeley Physics Course. Thus, the primary audience for this work is composed of students in astronomy, physics, mathematics, physical chemistry and engineering. It also includes high school teachers of physics and mathematics. Many amateur astronomers will find it quite accessible.

In the frame of approximations proper to an introductory textbook, the treatment is quite rigorous. Therefore, it is also expected to provide a firm background for a study of advanced astrophysics on a postgraduate level.

A rather severe selection is made here among various aspects of the Universe accessible to modern astronomy. This allows us to go beyond simple *information* on astronomical phenomena – to be found in popular books – and to insist upon *explanations* based on modern general physical theories.

More precisely, our selection of topics is determined by the following considerations:

The study of the solar system (the Moon and the planets) has recently progressed at a tremendous rate. However, the very rich harvest of observations provided by space research is mainly purely descriptive and is perfectly presented in review papers of *Scientific American*, *Science*, *Physics Today* and similar magazines.

A detailed description of our own galaxy, and of external galaxies, is given in dozens of well known books, illustrated with excellent reproductions of photographs obtained with giant telescopes.

On the other hand, a systematic use of electronic computers in stellar dynamics has recently made it possible to attempt an explanation of the spiral structure of galaxies. However, a clear account of this problem would oblige us to expand too much the scope of our book.

The description of solar activity (sunspots, prominences, solar flares, etc.) and of geophysical phenomena associated with this activity (aurorae, magnetic storms, etc.) can also be found in many textbooks. Moreover, the corresponding explanations involve theories (plasma physics, magneto-hydrodynamics, . . .) probably too advanced for our prospective readers.

Similarly, the study of interstellar medium relies more on advanced physics than on advanced astrophysics.

Finally, Otto Struve's fundamental *Elementary Astronomy* can give all basic astrophysics omitted from our textbook.

After all these exclusions, we are left with several important and interesting problems:

the internal constitution of ordinary stars, of white dwarfs and of neutron stars – elementary cosmology – recent observations of pulsars, quasars, and X-ray sources. All these topics are discussed very thoroughly, and the physical meaning of the necessary formalism is strongly stressed everywhere.

Four chapters (1, 4, 8 and 11) deal with the theory of radiative transfer, some elementary properties of the degenerate Fermi gas, celestial mechanics applied to binary systems (stellar masses), and relativistic concepts, respectively. All these preliminary chapters represent a necessary and useful introduction to different astrophysical theories discussed in other chapters.

Some further comments on this programme may be appropriate. In our opinion, scientific textbooks often summarize all necessary preliminary *specialized* knowledge in a very condensed way, or refer the reader – from the start – to other, more or less specialized, treatises: such is the case, e.g. for the tensor calculus, in very many textbooks on cosmology.

We prefer a more pedagogical method, which consists either in avoiding any reference to this more specialized knowledge – see our Chapter on the Newtonian approach to cosmology – or in a detailed presentation of the necessary preliminaries like the ones mentioned above.

A second tendency of many textbooks is to leave the major part of the derivations to the reader, often without sufficient guidance as to how the results are obtained. Moreover, the phrase 'it easily follows' too often conceals an obscure argument. This over-condensation usually aims at a reduction of printing expenses and sometimes expresses the fear of the author to be accused of pedantry. And yet already Boltzmann recommended that one leave the care for elegance – in scientific publications – to tailors, whereas Niels Bohr and Werner Heisenberg used to say that 'only abundance leads to clarity' and that 'the interest for details is sound and necessary'.*

We have paid special attention to both difficulties. Those formulae which form the basis of the main argument are always reproduced for the convenience of the reader, unless they have just been given. Moreover, by quoting – as S. Chandrasekhar – the equation numbers of all the formulae involved, we give full guidance for the derivation of each result.

Finally, a last word of warning is perhaps necessary before we leave these introductory remarks concerning the 'elegance' in textbooks. As the matter becomes clear, thanks to abundant and detailed explanations, some readers tend to underestimate the difficulties and the mystery which elsewhere surround the problem. What in other books seems strange, and even incomprehensible, becomes almost evident and too simple to need an explanation or genuine research. If we risk to irritate a few readers, we hope that many others will appreciate a textbook perhaps less entertaining but just as easy to read as a detective story.

It is almost impossible to be entirely original in the domains of astrophysics as classical as the radiative transfer, the internal constitution of normal stars and white dwarfs, the elementary statistical physics, the binary stars or the elementary cosmology. However, from the pedagogical point of view these parts of the book are perhaps the most original.

Chapter 1 represents a digest of our *Introduction to the General Theory of Particle*

* Heisenberg, W.: 1969, *Der Teil und das Ganze*, Piper, München, p. 333.

Transfer (full references are given at the end of each chapter). Chapters 2 and 3 summarize our *Introduction to the Physics of Stellar Interiors*. Chapter 4 is strongly inspired by C. Kittel's excellent *Thermal Physics*. Chapter 5 owes very much to S. Chandrasekhar's fundamental *Introduction to the Study of Stellar Structure* (the part dealing with white dwarfs). Chapter 8, on stellar masses, is based partly on our *Astronomie Fondamentale Elémentaire* and partly on J. A. Hynek's contribution to the *Topical Symposium* of the Yerkes Observatory.

The sources of the other Chapters are more diffuse and represent a more typical 'didactic synthesis': each of the Chapters 9 and 13 represents a digest of more than 100 original articles. We also must acknowledge our indebtness to R. N. Manchester and J. H. Taylor whose excellent recent monograph on '*Pulsars*' is very often quoted in Chapter 7.

Our presentation of the theoretical cosmology owes very much to all those who developed – after E. A. Milne and W. H. McCrea – the Newtonian approach to cosmology: O. Heckmann, H. Bondi, Ya. B. Zel'dovich, E. R. Harrison, P. J. E. Peebles, and all those who clarified and simplified the approach to relativistic cosmology: R. C. Tolman, H. P. Robertson, G. C. McVittie, S. Weinberg, and many others.

Such as it stands, our textbook is divided in four almost independent parts:

I. *Radiative transfer and internal structure of normal stars*;
II. *White dwarfs, neutron stars and pulsars*;
III. *Newton's law, binary systems and galactic X-ray sources*;
IV. *Cosmology: elementary theory and basic observational data.*

It represents our translation, from French into English, of our lectures on astrophysics made at the university of Paris from 1961 to 1977, brought up to date until November 1978.

Many colleagues of British, Canadian or American origin, contributed – one or two hours each – in polishing our English: Miss A. Underhill, MM. D. Thomas, B. Kandel, J. Piret, B. Carter, S. Leach, D. Flower, L. Celnikier, S. Frautshi, W. P. Allis, C. Moser, and, for the whole of the book, the editor B. McCormac. To all of them we express our sincere gratitude. In addition, one could mention that, as stated by Zel'dovich: "Specific broken English (not Shakespearian or slang) plays the role played by Latin in the time of Copernicus".

PART I

RADIATIVE TRANSFER AND INTERNAL STRUCTURE OF NORMAL STARS

CHAPTER 1

INTRODUCTION TO THE THEORY OF RADIATIVE TRANSFER

1. Basic Concepts in the Description of a Radiation Field

1.1. THE DIRECTIONAL ENERGY DENSITY OF A RADIATION FIELD

Consider a point P in a radiation field (a field of photons) at time t. Let (dV) be an element of volume dV surrounding P, and let $\boldsymbol{\omega}$ denote a unit vector originating at P (see Figure 1.1).

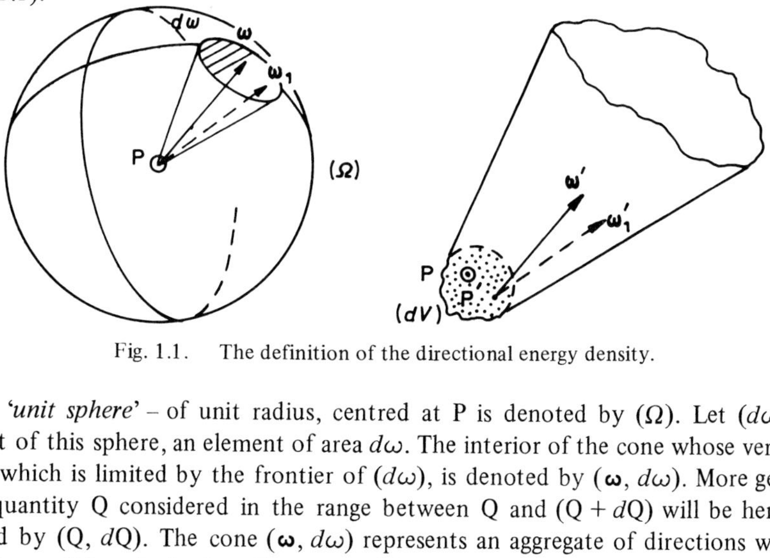

Fig. 1.1. The definition of the directional energy density.

The *'unit sphere'* – of unit radius, centred at P is denoted by (Ω). Let $(d\omega)$ be an element of this sphere, an element of area $d\omega$. The interior of the cone whose vertex is at P, and which is limited by the frontier of $(d\omega)$, is denoted by $(\boldsymbol{\omega}, d\omega)$. More generally, every quantity Q considered in the range between Q and $(Q + dQ)$ will be henceforth denoted by (Q, dQ). The cone $(\boldsymbol{\omega}, d\omega)$ represents an aggregate of directions within an elementary solid angle $d\omega$ (usually measured in *steradians*) around $\boldsymbol{\omega}$.

The left part of the figure shows the cone $(\boldsymbol{\omega}, d\omega)$ and the unit sphere (Ω). The dotted unit vector $\boldsymbol{\omega}_1$ defines *one* of the directions within the cone.

The right part of the figure shows, in two dimensions, the volume element (dV) around the point P and *one* of the very numerous points P′ inside (dV). The unit vector $\boldsymbol{\omega}'_1$ is equivalent to the unit vector $\boldsymbol{\omega}_1$ in the left part of the figure.

Let us consider the total *energy* of all photons of frequency $(\nu, d\nu)$ which at time t are, at different points P′, within (dV), and whose direction is one of the various directions $\boldsymbol{\omega}'_1$ inside $(\boldsymbol{\omega}, d\omega)$. Let $dE_\nu(P, \boldsymbol{\omega}, t, dV, d\omega, d\nu)$ denote this energy.

The theoretical analysis of the properties of a radiation field is based on the following fundamental principle: It is assumed that the ratio

$$u_\nu(P, \boldsymbol{\omega}, t) = \frac{dE_\nu(P, \boldsymbol{\omega}, t, dV, d\omega, d\nu)}{dV\,d\omega\,d\nu}, \tag{1}$$

is independent of the form and of the dimensions of (dV) and $(d\omega)$, provided that these dimensions become sufficiently small but remain sufficiently far from zero.

In these conditions, the ratio written above takes what is generally called *a plateau value* (before the start of erratic fluctuations for *very* small values of dV, $d\omega$ or $d\nu$, due to the discrete nature of the particles considered in such statistics).

In the specific case of a radiation field, the plateau value of the ratio written above is called the *directional energy density* of the field at P for the direction $\boldsymbol{\omega}$ at time t for frequency ν. It is denoted by $u_\nu(\mathrm{P}, \boldsymbol{\omega}, t)$.

When the variables ν and ω are *explicitly indicated*, as in $u_\nu(\mathrm{P}, \boldsymbol{\omega}, t)$, one can speak more briefly of *the energy density* $u_\nu(\mathrm{P}, \boldsymbol{\omega}, t)$.

Thus we have the fundamental relation

$$\boxed{dE_\nu(\mathrm{P}, \boldsymbol{\omega}, t, dV, d\omega, d\nu) = u_\nu(\mathrm{P}, \boldsymbol{\omega}, t)\, dV\, d\omega\, d\nu,} \tag{1'}$$

which, of course, is valid only for those values of dV, $d\omega$, and $d\nu$, for which the concept of density has a meaning.

1.1.1. *Remarks*

1.1.1.1. It is obvious that if we drop the differentiation in direction introduced in the definition of $u_\nu(\mathrm{P}, \boldsymbol{\omega}, t)$, in a similar way we can define a *non-directional energy* density $u_\nu(\mathrm{P}, t)$ related to $u_\nu(\mathrm{P}, \boldsymbol{\omega}, t)$ by

$$u_\nu(\mathrm{P}, t) = \int_\Omega u_\nu(\mathrm{P}, \boldsymbol{\omega}, t)\, d\omega, \tag{2}$$

where the integration is extended over the whole of the unit sphere (Ω), i.e. over all directions of space.

1.1.1.2. Similarly, if we drop the differentiation in frequency we can define the 'integrated' directional energy density $u(\mathrm{P}, \boldsymbol{\omega}, t)$ related to $u_\nu(\mathrm{P}, \boldsymbol{\omega}, t)$ by

$$u(\mathrm{P}, \boldsymbol{\omega}, t) = \int_0^\infty u_\nu(\mathrm{P}, \boldsymbol{\omega}, t)\, d\nu. \tag{3}$$

Note that, generally, quantities qualified as 'integrated' always refer in the theory of radiative transfer to an integration in *frequency* and are denoted by the same symbol as the corresponding monochromatic quantities, but with the suffix ν omitted.

1.1.1.3. If we drop the differentiation in direction and in frequency, we obtain an energy density which can be denoted by $u(\mathrm{P}, t)$ and which will be related to the other densities by

$$u(\mathrm{P}, t) = \int_0^\infty u_\nu(\mathrm{P}, t)\, d\nu = \int_\Omega u(\mathrm{P}, \boldsymbol{\omega}, t)\, d\omega. \tag{4}$$

It is obvious that $u(\mathrm{P}, t)$ represents the total energy of all photons contained at time t in a unit volume (of adequate dimensions) near the point P.

1.1.1.4. Note finally that it is very important to avoid confusions between the different densities defined above. This is performed by the use of a specific notation derived from that of the directional energy density $u_\nu(\mathrm{P}, \boldsymbol{\omega}, t)$ by suppression of appropriate suffixes or arguments.

1.2. The Intensity of a Radiation Field

It is sometimes more convenient to describe the properties of a radiation field by a concept other than $u_\nu(\mathrm{P}, \boldsymbol{\omega}, t)$, $u(\mathrm{P}, \boldsymbol{\omega}, t)$, $u_\nu(\mathrm{P}, t)$ or $u(\mathrm{P}, t)$. Indeed, all these concepts describe the situation at a given instant t and concern an element of volume of the field.

However, *near* the same instant t we can also consider the flow of energy carried during some interval of time (dt) through a *geometrical* infinitesimal ring (dA_ω), of area dA_ω, normal to the unit vector $\boldsymbol{\omega}$, by the photons of direction $(\boldsymbol{\omega}, d\omega)$ and of frequency $(\nu, d\nu)$ (see Figure 1.2).

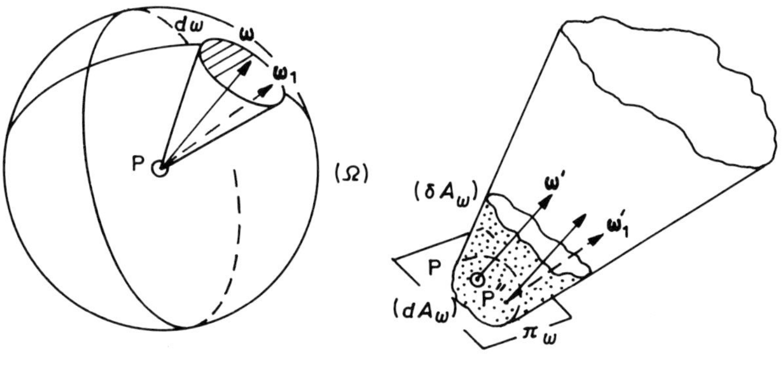

Fig. 1.2. The definition of the intensity of a radiation field.

The left part of the figure is similar to the left part of Figure 1.1. It shows the cone $(\boldsymbol{\omega}, d\omega)$, the unit sphere (Ω), and the dotted unit vector $\boldsymbol{\omega}_1$ defining one of the directions within the cone $(\boldsymbol{\omega}, d\omega)$.

The right part of the figure shows the plane (π_ω), through the point P, normal to $\boldsymbol{\omega}$; the point P″ is one of the very numerous points inside the ring (dA_ω), of arbitrary shape, traced in the plane (π_ω).

An obvious notation for the energy considered above is $dE_\nu(\mathrm{P}, \boldsymbol{\omega}, t, dA_\omega, d\omega, d\nu, dt)$.

We can now introduce *the plateau value* $I_\nu(\mathrm{P}, \boldsymbol{\omega}, t)$, called *Intensity*, of the ratio

$$I_\nu(\mathrm{P}, \boldsymbol{\omega}, t) = (\text{Plateau value of}) \frac{dE_\nu(\mathrm{P}, \boldsymbol{\omega}, t, dA_\omega, d\omega, d\nu, dt)}{dA_\omega\, d\omega\, d\nu\, dt}. \tag{5}$$

Thus we have the fundamental relation

$$\boxed{dE_\nu(\mathrm{P}, \boldsymbol{\omega}, t, dA_\omega, d\omega, d\nu, dt) = I_\nu(\mathrm{P}, \boldsymbol{\omega}, t) dA_\omega d\omega \, d\nu \, dt} \quad , \tag{6}$$

which, of course, is valid only for those values of $dA_\omega, d\omega, d\nu, dt$ for which the concept of plateau value (explained in the previous subsection) has a meaning.

1.2.1. Remarks

1.2.1.1. As in Subsection 1.1, we can introduce the concept of integrated intensity $I(\mathrm{P}, \boldsymbol{\omega}, t)$ connected with $I_\nu(\mathrm{P}, \boldsymbol{\omega}, t)$ by

$$I(\mathrm{P}, \boldsymbol{\omega}, t) = \int_0^\infty I_\nu(\mathrm{P}, \boldsymbol{\omega}, t) \, d\nu. \tag{7}$$

1.2.1.2. However, one easily realizes that symbols such as $I_\nu(\mathrm{P}, t)$ or $I(\mathrm{P}, t)$ have no physical meaning.

On the contrary, integrals such as

$$\int_\Omega I_\nu(\mathrm{P}, \boldsymbol{\omega}, t) \, d\omega \quad \text{or} \quad \int_\Omega I(\mathrm{P}, \boldsymbol{\omega}, t) \, d\omega, \tag{8}$$

do have a meaning, which will be given later by Equations (11) and (12).

1.2.1.3. The quantities $I_\nu(\mathrm{P}, \boldsymbol{\omega}, t)$ and $u_\nu(\mathrm{P}, \boldsymbol{\omega}, t)$ are not independent, but are connected by the following simple relation:

$$I_\nu(\mathrm{P}, \boldsymbol{\omega}, t) = c u_\nu(\mathrm{P}, \boldsymbol{\omega}, t), \tag{9}$$

where c is the speed of light in vacuum.

Let us consider, indeed, the volume element (dV), in the right part of Figure 1.2, which is a frustum of a cone. Its first base is (dA_ω) and its second base is (δA_ω), at a distance $c\,dt$ from the first. Quantities of second order can be neglected and (dV) can be considered as a cylinder if dt is sufficiently small. Then $dV = dA_\omega(c\,dt)$.

The energy dE_ν of the photons $(\nu, d\nu)$ within (dV) will take, according to Equation (1′), the value $u_\nu(\mathrm{P}, \boldsymbol{\omega}, t) c\,dt\,dA_\omega d\omega \, d\nu$. But according to Equation (6) the same energy will be given by $I_\nu(\mathrm{P}, \boldsymbol{\omega}, t) dA_\omega d\omega \, d\nu \, dt$, equal to the energy which entered (dV) during the time dt. Hence Equation (9).

On integrating in frequency this relation (9) we obtain, by Equations (7) and (3):

$$I(\mathrm{P}, \boldsymbol{\omega}, t) = c\,u(\mathrm{P}, \boldsymbol{\omega}, t). \tag{10}$$

1.2.2.4. Equations (9) and (10) disclose the physical meaning of the integrals (8).

Indeed, on using Equations (10) and (4), we find:

$$\int_\Omega I(\mathrm{P}, \boldsymbol{\omega}, t) \, d\omega = cu(\mathrm{P}, t). \tag{11}$$

Similarly, on using Equations (9) and (2), we find:

$$\int_\Omega I_\nu(P, \boldsymbol{\omega}, t)\, d\omega = c u_\nu(\mathrm{P}, t). \tag{12}$$

Thus, for instance, the integral at the left-hand side of Equation (11), divided by c, represents the energy density $u(\mathrm{P}, t)$ of photons of all frequencies at P at time t.

1.3. MEAN ENERGY AND MEAN INTENSITY

Generally every real radiation field is anisotropic: the directional densities $u_\nu(\mathrm{P}, \boldsymbol{\omega}, t)$ do not have the same value for different directions $\boldsymbol{\omega}$.

Nevertheless, even in the presence of an anisotropic real field, we can introduce the concept of another field (F) equivalent, in a sense defined below, to the actual field.

By definition (F) has to be isotropic (its directional energy density $\bar{u}_\nu(\mathrm{P}, \boldsymbol{\omega}, t)$ must have the same value for all directions $\boldsymbol{\omega}$), but the non-directional energy density $\bar{u}_\nu(\mathrm{P}, t)$ of (F) must be that of the actual field.

Note that in order to indicate the physical dimensions of $\bar{u}_\nu(\mathrm{P}, \boldsymbol{\omega}, t)$ it is necessary to keep the parameter $\boldsymbol{\omega}$ in the parentheses, in spite of the fact that, by definition, the average isotropic field (F) described by $\bar{u}_\nu(\mathrm{P}, \boldsymbol{\omega}, t)$ does *not* depend on $\boldsymbol{\omega}$.

In Equation (2) applied to the field (F) we can take the function $\bar{u}_\nu(\mathrm{P}, \boldsymbol{\omega}, t)$, which is physically independent of $\boldsymbol{\omega}$, out of the integral sign, and replace $\int_\Omega d\omega$, equal to the area of the unit sphere, by 4π.

On equating the values of $u_\nu(\mathrm{P}, t)$ given by Equation (2) for the real and for the average field (F) we find:

$$\bar{u}_\nu(\mathrm{P}, \boldsymbol{\omega}, t) = \frac{1}{4\pi} \int_\Omega u_\nu(\mathrm{P}, \boldsymbol{\omega}, t)\, d\omega = \frac{1}{4\pi} u_\nu(\mathrm{P}, t). \tag{13}$$

Thus, the directional energy density $\bar{u}_\nu(\mathrm{P}, \boldsymbol{\omega}, t)$ of the isotropic theoretical field (F) associated with the real field appears *mathematically* as a *directional average* of the actual directional energy density $u_\nu(\mathrm{P}, \boldsymbol{\omega}, t)$.

On integrating Equation (13) with respect to frequency, we obtain:

$$\bar{u}(\mathrm{P}, \boldsymbol{\omega}, t) = \frac{1}{4\pi} \int_\Omega u(\mathrm{P}, \boldsymbol{\omega}, t)\, d\omega = \frac{1}{4\pi} u(\mathrm{P}, t). \tag{14}$$

The quantity $\bar{u}_\nu(\mathrm{P}, \boldsymbol{\omega}, t)$ is called *the mean directional energy density* and $\bar{u}(\mathrm{P}, \boldsymbol{\omega}, t)$ is called *the mean integrated directional energy density*.

In a similar way, if we denote by $\bar{I}_\nu(\mathrm{P}, \boldsymbol{\omega}, t)$ *the mean intensity* of the theoretical isotropic average field (F) defined above, Equation (9), which can be applied to the average field as to any field, will give:

$$\bar{I}_\nu(\mathrm{P}, \boldsymbol{\omega}, t) = c\bar{u}_\nu(\mathrm{P}, \boldsymbol{\omega}, t), \tag{15}$$

hence, according to Equation (13):

$$\bar{I}_\nu(\mathrm{P}, \boldsymbol{\omega}, t) = \frac{c}{4\pi} u_\nu(\mathrm{P}, t). \tag{16}$$

This Equation (16) is usually written with $u_\nu(\mathrm{P}, t)$ at the left-hand side:

$$u_\nu(\mathrm{P}, t) = \frac{4\pi}{c} \bar{I}_\nu(\mathrm{P}, \boldsymbol{\omega}, t). \tag{17}$$

On comparing Equations (16) and (12) we find:

$$\bar{I}_\nu(\mathrm{P}, \boldsymbol{\omega}, t) = \frac{1}{4\pi} \int_\Omega I_\nu(\mathrm{P}, \boldsymbol{\omega}, t)\, d\omega. \tag{18}$$

And on integrating with respect to frequency we obtain:

$$\bar{I}(\mathrm{P}, \boldsymbol{\omega}, t) = \frac{1}{4\pi} \int_\Omega I(\mathrm{P}, \boldsymbol{\omega}, t)\, d\omega. \tag{19}$$

On the other hand, on integrating Eqution (17) with respect to frequency, and taking into account Equations (4) and (7), applied respectively to the real and to the average field, we find:

$$u(\mathrm{P}, t) = \frac{4\pi}{c} \bar{I}(\mathrm{P}, \boldsymbol{\omega}, t). \tag{20}$$

According to the previous definitions the quantity $\bar{I}(\mathrm{P}, \boldsymbol{\omega}, t)$ represents *the integrated mean intensity*, i.e., the intensity of the theoretical isotropic field whose integrated non-directional energy density $u(\mathrm{P}, t)$ is identical to that of the real field.

Equation (7) applied to the average field yields:

$$\bar{I}(\mathrm{P}, \boldsymbol{\omega}, t) = \int_0^\infty \bar{I}_\nu(\mathrm{P}, \boldsymbol{\omega}, t)\, d\nu. \tag{21}$$

1.4. The Flux and the Flux Density in Radioastronomy

In radioastronomy one uses the term '*flux*' for the description of the radiation received by a radiotelescope per second – in the form of radio waves – over the entire effective area of the telescope, integrated over all radio frequencies accessible to the receivers of the telescope.

In Subsection 3.3 we shall introduce a quite different concept of *net flux* with respect to a given direction.

On the other hand, discrete radio-sources are often characterized in radioastronomy by their *flux density*, defined as the energy received per second and per unit frequency interval over the entire effective area of the radiotelescope normal to the direction of the source. It is measured in *flux units* equal to $10^{-26}\,\mathrm{W\,m^{-2}\,Hz^{-1}}$. Thus the flux density of the radioastronomers is a concept very analogous to the concept of *intensity* I_ν defined in Subsection 1.2.

2. Relations between Macroscopic and Microscopic Parameters Describing the Interactions between Matter and Radiation

2.1. A Brief Summary of Some Elementary Fundamental Relations

It is well known that a neutral or an ionized atom (A) can normally only have a set of discrete quantized stationary energy states ($E_1, E_2, E_3, \ldots E_n, \ldots E_{\mathrm{lim}}$).

The values ($E_1, E_2, E_3, \ldots E_n, \ldots E_{\text{lim}}$) also characterize the energy levels of (A) on *a level diagram*.

These levels are not infinitely sharp: they have a certain width, which can be neglected in a first approximation but must be taken into account in a more advanced study.

The lowest level E_1 is called *the ground level*. The highest level of the diagram corresponds to *the limit energy* E_{lim} of the atom (A). Beyond this limit possible energy states form a continuum (no quantization!).

One says that (A) is excited when its energy takes on one of the values $E_2, E_3, \ldots$ other than E_1. Consequently the levels $E_2, E_3, \ldots$ are called *the excited levels*.

Consider an atom (A) which is initially in an energy state, 'on the level', E_j. We shall label such an atom (A_j).

The existence of material media which are more or less transparent to incident radiation, and other reasons discussed later, lead to the following principles.

Statistically, incident photons can be divided into two groups: the group of those which, passing in the vicinity of atoms (A_j), are simply '*transmitted*', i.e., pass without interaction with atoms (A_j), and the group of those which undergo an interaction with atoms (A_j).

From the point of view of the transfer theory, this interaction always appears – concerning the photons present in the incident beam – as a total annihilation of photons by the interaction, These photons give up the totality of their energy (and of their momentum) to the corresponding atoms (A_j). They cease to exist as photons.

One says in the theory of radiative transfer that atoms (A_j) operate a (positive) '*removal*' of the corresponding photons from the incident beam. Each microscopic interaction with one atom representing microscopically the loss of one photon.

However, if photons initially present in the incident beam, and undergoing the process of a removal, can only disappear, the passage of the beam through the matter containing atoms (A_j) can also lead to another type of interaction.

In this latter process the presence in the incident beam of photons of a certain energy has as consequence the loss of energy by some of the atoms of the material medium to the benefit of a creation of photons, the result being an enrichment of the beam by the photons so created.

From the point of view of the transfer theory, such an enrichment of the beam represents the opposite of a removal and it is appropriate to describe it as a '*negative removal*'.

This terminology – proper to the theory of radiative transfer – is not the same as that used in atomic physics and in quantum mechanics, where one uses respectively the terms of '*absorption*' (for removal) and '*stimulated emission*' (for negative removal).

However, our terminology presents the advantage of using the same name (removal), qualified as positive or negative, for all interactions corresponding to the existence of an incident beam (or – what is equivalent – to the existence of a certain radiation field of non-zero density).

Nevertheless, the composition of the emergent beam depends not only upon the composition of the incident beam and upon positive or negative removals suffered by the incident beam, but depends also upon the photons 'injected' into the beam during its passage through the medium.

This injection is due to processes of '*spontaneous emission*', i.e. to creation of photons at the expense of the energy of the atoms of the medium (and not at the expense of the energy of the radiation field). This creation is due simply to the instability of the energy states of excited atoms whose lifetimes, generally of the order of 10^{-8} s, are always finite. (For some 'metastable states' these lifetimes can become much longer.) One might say that any excited atom must sooner or later 'give birth to a photon'.

Another fundamental difference between the three physical processes mentioned above, also suggests to group positive and negative *removals* rather than to group stimulated and spontaneous *emissions*.

As a matter of fact, in the presence of a narrow monodirectional beam (a light ray), all photons created by stimulated emission (negative removal) *move in the direction of the incident beam* – a property which plays a fundamental role in a laser – whereas photons created by spontaneous emission are (statistically) distributed in all directions. Only a part of these last photons goes in the direction of the incident beam.

After this replacement in the transfer theory of the term of stimulated emission by that of negative removal, it becomes unnecessary to keep the word 'stimulated' in connexion with the stimulated emission.

Henceforth, we simply use the term of *emission* whenever there is no ambiguity.

2.1. Remark

One should note that the concept of *transmission* (passage of the light through matter without interaction) very clear in a conception of the light as formed by discrete particles (photons), and justified experimentally by the existence of transparent media, is more difficult to understand when the light is conceived as an electromagnetic wave. This difficulty is connected with the 'duality' between waves and particles: some phenomena can be better explained using the concept of particles, others being better understood in the frame of the wave theory.

A very fundamental reason, however, commands to admit the existence of photons simply transmitted through the material medium.

Indeed, interaction is only possible when a kind of resonance occurs between the energy $h\nu$ of an incident photon and the discrete quantized values of energy states permitted to the atom (A). For a removal (positive or negative) of a photon $h\nu$ to take place, atom (A) must necessarily have the levels E_j and E_k separated by an energy difference equal to $h\nu$.

Certainly, the existence of levels which are not infinitely narrow attenuates the rigour of this rule. Reciprocally the very narrow width of most levels with respect to the separation between successive levels, evidenced by the existence of spectral lines, suggests a very low probability of interaction for photons whose energy $h\nu$ is not closely equal to a certain difference $(E_k - E_j)$.

2.2. The Directional Density of Positive Removal and the Coefficient of Positive Removal Σ_ν^+

For different historical reasons, a removal for a given beam usually is not described in terms of the intensity $I_\nu(\mathrm{P}, \boldsymbol{\omega}, t)$ of the radiation field associated with the beam, but in terms of the corresponding energy density. Relation (9) shows that this is rather unimportant.

More troublesome is the fact that *microscopic* parameters of interaction are traditionally defined as a function of the non-directional density $u_\nu(\mathrm{P}, t)$.

Even this is still not too bad, as we show below, since usually one deals with isotropic material media, i.e. media which interact with photons independently of their direction: this allows us to go very easily from relations involving $u_\nu(\mathrm{P}, t)$ to relations involving $u_\nu(\mathrm{P}, \boldsymbol{\omega}, t)$.

The only serious difficulty, introduced by the weight of traditions, comes from the fact that usually one does not take into account the width of the spectral lines, and consequently the width of the energy levels, in the definition of the 'coefficients' describing microscopic interactions. The result is that instead of starting from refined, well differentiated, concepts (as we did in Section 1) one is forced here to start from global concepts. To refine these concepts progressively is never easy and never entirely intellectually satisfying.

Anyhow, let us consider a material medium containing, at time t in the vicinity of a point P, $N^j(\mathrm{P}, t)$ atoms (A_j) per cm^3.

Let then $u_r(\mathrm{P}, t)$ represent the non-directional energy density of the radiation field at P at time t for the frequency ν_r defined by

$$h\nu_r = E_k - E_j. \tag{22}$$

The subscript r denotes here a given spectral line (r) supposed to be of negligible width (just as the levels E_k and E_j). In the real case, where the width of a spectral line is not negligible, ν_r would represent the central frequency of the line, i.e. the frequency for which interaction parameters, in the domain of the line (r), are maximum.

Let finally $N^{jk}(\mathrm{P}, t)$ represent the number of atoms (A_j) which, per cm^3 and per second, pass from the state E_j to a more energetic state E_k as a consequence of the removal of photons of frequency ν_r. Henceforth we shall usually write simply N^j instead of $N^j(\mathrm{P}, t)$ and N^{jk} instead of $N^{jk}(\mathrm{P}, t)$.

One can also say that N^{jk} represents the number of transitions from the lower level j to an upper level k, per cm^3 and per second, for atoms (A_j).

In 1917, Einstein adopted as principle the idea that N^{jk} is proportional to N^j and to $u_r(\mathrm{P}, t)$, *the transition probability* B_{jk} defined by

$$B_{jk} = \frac{N^{jk}}{N^j u_r(\mathrm{P}, t)}, \tag{23}$$

being a *constant* characteristic of the atomic structure of the atoms (A). This *Einstein's coefficient* B_{jk} is thus supposed to be independent of N^j and of $u_r(\mathrm{P}, t)$.

Today, quantum mechanics allow, for sufficiently simple atoms, *to compute* the values of B_{jk}. For more complex atoms, these values are determined by application of the theory of radiative transfer to an interpretation of laboratory *experiments*.

Thus we have the fundamental global relation:

$$N^{jk} = B_{jk} N^j u_r(\mathrm{P}, t). \tag{23$'$}$$

In this relation the dependence of N^j and of N^{jk} on P and t is implicit.

This relation allows different important interpretations, generalizations and transformations.

One can note first that Equation (23′) describes the interaction in terms of the number of atoms undergoing some modification of their energy state, i.e. directs one's attention towards the material *medium* participating in the interaction.

In the theory of radiative transfer, it is more useful to express the interaction in terms of its influence on the *radiation field*, and to consider that the physical process described by Equation (23′) also corresponds to a positive removal of photons $h\nu_r$ from the radiation field.

Let then $p_r^+(\mathrm{P}, t)$ be the *decrease* suffered *per second* by the energy density $u_r(\mathrm{P}, t)$ of the field under the influence of removals associated with transitions from the state j to the state k. The superscript (+) recalls that we deal with real positive removals; the negative ones will be considered later. The quantity $p_r^+(\mathrm{P}, t)$ is called the *non-directional density of positive removal*, at P near time t, for the frequency ν_r.

Since each transition of an atom (A_j) from the state j to the state k demands an 'annihilation' of one photon of energy $h\nu_r$, we have evidently:

$$p_r^+(\mathrm{P}, t) = h\nu_r N^{jk} = h\nu_r B_{jk} N^j(\mathrm{P}, t) u_r(\mathrm{P}, t). \tag{24}$$

In order to take into account the finite width of the spectral lines, let us consider that the coefficient B_{jk} relates to the *whole* of the line (r), and introduce the function $\psi_r(\nu)$ defined by

$$\psi_r(\nu) = \frac{p_{\nu,r}^+(\mathrm{P}, t)}{p_r^+(\mathrm{P}, t)}. \tag{25}$$

This function $\psi_r(\nu)$ describes the variations of the non-directional density of removals as a function of the frequency ν, in the neighbourhood of the center ν_r of the line (r). The function $\psi_r(\nu)$ is normalized to unity:

$$\int_{(\text{Domain of the line } r)} \psi_r(\nu)\, d\nu = 1. \tag{26}$$

In Equation (25), and in Equation (26′) given below, $p_{\nu,r}^+(\mathrm{P}, t)$ represents the non-directional density of positive removal, in the neighbourhood of a frequency (ν, r) near the central frequency ν_r of the line (r). Of course, (ν, r) is different from the central frequency ν_r and can also be written simply as ν.

All this can be summarized by Equation (26′).

$$p_{\nu,r}^+(\mathrm{P}, t) = h\nu_r B_{jk} N^j(\mathrm{P}, t) u_r(\mathrm{P}, t) \psi_r(\nu). \tag{26'}$$

A determination of the function $\psi_r(\nu)$ represents a difficult problem of atomic physics and quantum mechanics. Its expression depends essentially on the physical processes responsible in each case for the broadening of atomic levels and spectral lines. In the theory of radiative transfer this function $\psi_r(\nu)$ *is supposed to be known*. Its main general property is to become negligible as soon as the frequency ν differs significantly from the central frequency ν_r.

It is usual to assume that in Equation (26′) $u_r(\mathrm{P}, t)$ can be replaced by $u_{\nu,r}(\mathrm{P}, t)$. The justification for this substitution is that in the domain of the line (r) the variations of the energy density of the radiation field are in general *continuous* and relatively small. This is certainly true for what can be called the 'first contact' between the incident beam and the material medium. It is less sure after a succession of interactions as the beam progresses through the matter.

However, it can be argued that the error introduced by the replacement of $u_r(\mathrm{P}, t)$ by $u_{\nu,r}(\mathrm{P}, t)$ just compensates the error introduced by the use of $u_r(\mathrm{P}, t)$ in the definition of B_{jk}, since it is obvious that $p^+_{\nu,r}(\mathrm{P}, t)$ must physically depend on $u_{\nu,r}(\mathrm{P}, t)$ and not on $u_r(\mathrm{P}, t)$.

Thus we can write:

$$p^+_{\nu,r}(\mathrm{P}, t) = h\nu B_{jk}(\nu) N^{j'}(\mathrm{P}, t) u_\nu(\mathrm{P}, t), \tag{26''}$$

where we have introduced – instead of the global B_{jk} – a more refined coefficeint $B_{jk}(\nu)$, incorporating $\psi_r(\nu)$, relative to the frequency ν.

Actually we can consider Equation (26″) as a *definition* of $B_{jk}(\nu)$. Note also that we have written $N^{j'}$ instead of N^j. Indeed, the departure level for the transition of frequency $(\nu, r) = \nu$ is not the central level j but some near level j'.

Note finally that $h\nu_r$ has been replaced in Equation (26″) by $h\nu$.

In other words Equation (26″) is simply a *generalization* of Equation (24) for the case where one takes into account the finite width of the levels. The notation $B_{jk}(\nu)$ presents the advantage of recalling the historical origin of the formula, in spite of the fact that the levels rigorously implied in its definition are j' and k', corresponding to the frequency ν, and no longer j and k.

A last generalization, which does not present any difficulty, consists in an introduction of *directional* densities instead of non-directional ones, by writing:

$$p^+_{\nu,r}(\mathrm{P}, \boldsymbol{\omega}, t) = h\nu B_{jk}(\nu) N^{j'}(\mathrm{P}, t) u_\nu(\mathrm{P}, \boldsymbol{\omega}, t). \tag{27}$$

Taking into account Equation (9) we finally find:

$$p^+_\nu(\mathrm{P}, \boldsymbol{\omega}, t) = \frac{h\nu}{c} B_{jk}(\nu) N^{j'}(\mathrm{P}, t) I_\nu(\mathrm{P}, \boldsymbol{\omega}, t) = \Sigma^+_\nu I_\nu(\mathrm{P}, \boldsymbol{\omega}, t). \tag{28}$$

The last Equation represents both *a definition* of the *coefficient of positive removal* Σ^+_ν and gives its expression as a function of $B_{jk}(\nu)$.

This fundamental relation shows that the energy removed from the radiation field as a result of the interaction – per cm^3, steradian, second and per unit frequency interval (near point P, direction $\boldsymbol{\omega}$, frequency ν and time t) – is *proportional* to the intensity $I_\nu(\mathrm{P}, \boldsymbol{\omega}, t)$ of the radiation field.

The coefficient of proportionality Σ^+_ν, i.e. the coefficient of positive removal, depends on: (1) The *nature* of the medium, through $B_{jk}(\nu)$; (2) The *state* of the medium, through $N^{j'}(\mathrm{P}, t)$; and (3) The *frequency* ν, through $B_{jk}(\nu)\, h\nu$.

Generally, Σ^+_ν is a function of ν, P and t. This could be made more explicit by writing $\Sigma^+_\nu(\mathrm{P}, t)$. However, in usual *homogeneous* and *stationary* media the dependence on P and t becomes irrelevant and one can simply write Σ^+_ν.

It is important to note that Σ^+_ν does *not* depend on *the state of the radiation field* – i.e. on the properties of the incident beam – near the point P and time t.

Thus, describing the interaction by the relation

$$p^+_\nu(\mathrm{P}, \boldsymbol{\omega}, t) = \Sigma^+_\nu I_\nu(\mathrm{P}, \boldsymbol{\omega}, t), \tag{28'}$$

we obtain a separation between quantities – such as $I_\nu(\mathrm{P}, \boldsymbol{\omega}, t)$ – which depend on the radiation field and quantities – such as Σ^+_ν – independent of the radiation field.

In the traditional presentation of the transfer theory, the coefficient of removal Σ_ν^+ is introduced directly by defining it as the ratio p_ν^+/I_ν. Our presentation has the advantage, at the expense of some complexity, to explicit the dependence of Σ_ν^+ on the physical parameters involved in the interaction.

2.3. THE DIRECTIONAL DENSITY OF NEGATIVE REMOVAL AND THE COEFFICIENT OF NEGATIVE REMOVAL Σ_ν^-

A genial idea of A. Einstein (leading later to the invention of the laser) was to assume that the presence in the radiation field of the photons of requency ν_r could induce, or 'stimulate', a deexcitation of atoms initially in an excited state of energy E_k (when a *lower* state of energy E_j was possible) provided that $h\nu_r$ were equal to $(E_k - E_j)$.

This effect is rather surprising from the point of view of a corpuscular conception of the light. From the 'classical' point of view it can be easily understood as an emission of electromagnetic waves by electrons of the medium undergoing *forced vibrations* imposed by electrodynamical forces generated by the incident electromagnetic waves.

Einstein described this phenomenon of *stimulated emission*, considered in the transfer theory as *a negative removal*, by a relation similar to Equation (23′), viz.

$$N^{kj} = B_{kj}N^k u_r(\mathrm{P}, t), \tag{29}$$

In Equation (29) the quantities N^{kj}, B_{kj} and N^k, respectively, have the same *physical dimensions* as N^{jk}, B_{jk} and N^j. Note however the inversion of the order of subscripts and superscripts between N^{kj} and N^{jk} or between B_{kj} and B_{jk}.

On introducing the same type of generalization as in Subsection 2.2, we can write, by analogy with Equation (28), the following expression for the *directional density of negative removal* $p_\nu^-(\mathrm{P}, \boldsymbol{\omega}, t)$:

$$p_\nu^-(\mathrm{P}, \boldsymbol{\omega}, t) = \frac{h\nu}{c} B_{kj}(\nu)N^{k'}(\mathrm{P}, t)I_\nu(\mathrm{P}, \boldsymbol{\omega}, t) = \Sigma_\nu^- I_\nu(\mathrm{P}, \boldsymbol{\omega}, t), \tag{30}$$

where Σ_ν^- represents the *coefficient of negative removal.*

2.4. THE TOTAL DIRECTIONAL DENSITY OF REMOVAL AND THE COEFFICIENT OF REMOVAL Σ_ν

On combining Equations (28) and (30), we obtain the expression for the balance between positive and negative removals, i.e. the (total) directional density of removal: $p_\nu(\mathrm{P}, \boldsymbol{\omega}, t) = p_\nu^+(\mathrm{P}, \boldsymbol{\omega}, t) - p_\nu^-(\mathrm{P}, \boldsymbol{\omega}, t)$.

On introducing the (total) coefficient of removal $\Sigma_\nu = \Sigma_\nu^+ - \Sigma_\nu^-$, $p_\nu(\mathrm{P}, \boldsymbol{\omega}, t)$ can be written:

$$p_\nu(\mathrm{P}, \boldsymbol{\omega}, t) = \Sigma_\nu I_\nu(\mathrm{P}, \boldsymbol{\omega}, t), \tag{31}$$

where

$$\Sigma_\nu = \frac{h\nu}{c} [B_{jk}(\nu)N^{j'} - B_{kj}(\nu)N^{k'}]. \tag{31′}$$

It is well known (see for instance Feynman, 1964, or Kittel, 1969, Ch. 6) that in a given enclosure, containing a given number of particles in a state of thermodynamic equilibrium, the proportion of particles in energy states E_k and E_j is given by the Boltzmann law. Considering the relation $h\nu_r = E_k - E_j$, this law can be expressed by:

$$\frac{N^k/g_k}{N^j/g_j} = \frac{e^{-E_k/kT}}{e^{-E_j/kT}} = e^{-(E_k-E_j)/kT} = e^{-h\nu_r/kT}. \tag{32}$$

In Equation (32) g_k and g_j are the statistical weights of the levels k and j; T is the temperature of the medium; and k is the *Boltzmann constant*: $k = 1.38 \times 10^{-16}$ c.g.s.

If the system formed by the atoms (A) is *not* in an equilibrium state we cannot speak of *the* temperature of the medium. However, we can still associate with the numbers N^k and N^j, *such as they are* at a given time at a given point of the system, an *'excitation temperature'* T_{kj}, relative to the levels k and j, *defined* by

$$\frac{N^k/g_k}{N^j/g_j} = e^{-h\nu_r/kT_{kj}}. \tag{33}$$

(Obviously, no confusion is possible between the subscript k and the Boltzmann constant k.)

On introducing the excitation temperature T_{kj} we preserve the *form* of the relation (32). However it must be clearly understood that T_{kj} – for a given value of ν_r – is deduced, by Equation (33), from a known ratio N^k/N^j, whereas by Equation (32) the ratio N^k/N^j is deduced from a known value of the temperature T.

When the left-hand side of Equation (33) is greater than unity, which occurs in a laser, the excitation temperature T_{kj} becomes *negative*.

In the absence of an equilibrium there can exist many different excitation temperatures, for different pairs of levels k and j.

The elementary theory recalled in the Appendix to this Chapter shows that the coefficients B_{jk} and B_{kj} are not independent. They are connected by the relation:

$$B_{kj}g_k = B_{jk}g_j. \tag{34}$$

On eliminating the ratio g_k/g_j from Equation (33) by means of this relation, we obtain:

$$B_{kj}N^k = B_{jk}N^j e^{-h\nu_r/kT_{kj}}. \tag{35}$$

A generalization of this Equation for the case where the levels are of finite width gives:

$$B_{kj}(\nu)N^{k'} = B_{jk}(\nu)N^{j'} e^{-h\nu/kT_{k'j'}}. \tag{36}$$

Thus we can put the expression (31′) for Σ_ν into the form:

$$\Sigma_\nu = \frac{h\nu}{c} B_{jk}(\nu)N^{j'}(\mathrm{P}, t)(1 - e^{-h\nu/kT_{k'j'}}). \tag{37}$$

The comparison with the definition (28) of Σ_ν^+ shows that the contribution of the negative removals to the value of the total coefficient Σ_ν is negligible every time the ratio $h\nu/kT_{k'j'}$ is sufficiently great. This occurs in the domain of the optical frequencies. The opposite takes place in the domain of low frequencies encountered in radioastronomy and in the case of a laser.

2.4.1. *Remarks*

2.4.1.1. We have considered hitherto only the removals of photons of energy $h\nu_r = E_k - E_j$, (or of energy $h\nu$ near $h\nu_r$), by the atoms (A) of the medium, corresponding to transitions between *discrete* levels j and k below the limit of quantization E_{lim}. However, the removals of photons of the same energy $h\nu_r$ can also correspond to the *'bound–free' transitions* between a discrete level n higher than the level j and a non-quantized level c (a level of the continuum).

Of course, when one takes into account the finite width of the levels these bound–free transitions are associated with the corresponding *sublevels* n' and c'.

Finally, a removal of photons of energy $h\nu_r$ (or of nearby energy $h\nu$) can also correspond to 'free–free' transitions between two non-quantized levels c' and c''.

Note that the concept of freedom implied here in the word 'free' is related only to the constraints imposed by *quantization*. It must not be confused with the freedom of a free material particle which conserve, in classical mechanics, its kinetic energy when not acted upon by any force.

The old quantum theory gives a convenient image of the 'free' (non-quantized) states. In the presence of electrodynamical forces exerted on the 'optical electron' by the rest of the atom, this electron describes an *elliptical* orbit for discrete states and a *hyperbolic orbit* for the non-quantized states (continuum).

Similarly a spacecraft launched with a sufficient velocity '*escapes*' along a hyperbolic orbit and can therefore be considered, in some way, as '*free*'. However, this spacecraft is still acted upon by the gravitational attraction of the Earth or of the Sun, and is not free from the mechanical point of view. Launched with a smaller initial velocity it would describe a closed elliptical trajectory – a '*bound*' one.

In this model a 'free–free' transition would correspond to a jump from one hyperbolic orbit to another hyperbolic orbit, this jump implying a modification in the total energy of the system formed by the optical electron and the rest of the atom.

Let us call $p_{\nu,r}$ the density of removal corresponding to transitions between discrete levels – the subscript r recalling that such transitions usually correspond to a spectral line (r). And let us call $p_{\nu,c}$ the density of removal which, for the same frequency ν, corresponds to a transition (bound–free or free–free) which ends in the continuum. The total density of removal p_ν is given by the sum $(p_{\nu,r} + p_{\nu,c})$.

Just as indicated by Equation (31) for $p_{\nu,r}$, it can be shown (see for instance Kourganoff, 1969) that $p_{\nu,c}$ is also proportional to I_ν.

In order to obtain the *complete* coefficient of removal, corresponding to $(p_{\nu,r} + p_{\nu,c})$, one must add to the right-hand side of Equation (37), which corresponds only to $p_{\nu,r}$, the part $\Sigma_{\nu,c}$ corresponding to $p_{\nu,c}$. This part is given by Kourganoff (1969).

2.4.1.2. Some authors use the symbol k_ν instead of our Σ_ν for the (total) coefficient of removal, and call it '*the absorption coefficient*'.

However, in quantum mechanics the term *absorption* corresponds (as already pointed out above) to the *positive* removals alone. Thus, such a practice can lead to many errors.

Some more careful authors use again the symbol k_ν instead of our Σ_ν^+, which really corresponds to the absorption coefficient of quantum mechanics. And they use for our total coefficient Σ_ν the notation k'_ν.

The use of the symbol Σ has the major advantage of providing a bridge between the theory of radiative transfer and the theory of neutron diffusion (nuclear reactors), as explained in Kourganoff (1969).

2.4.1.3. In the *stationary case*, when both I_ν and Σ_ν keep the same value all the time, one can omit the argument t in Equation (31) and write

$$\text{(Stationary case)} \quad p_\nu(\mathrm{P}, \boldsymbol{\omega}) = \Sigma_\nu I_\nu(\mathrm{P}, \boldsymbol{\omega}). \tag{37'}$$

However, one must remember that the definitions of $I_\nu(\mathrm{P}, \boldsymbol{\omega})$ and $p_\nu(\mathrm{P}, \boldsymbol{\omega})$ still imply the *unity* of the *interval of time*.

2.5. THE DIRECTIONAL EMISSION DENSITY

The result of *all* physical processes contributing to *inject* photons of energy $h\nu$ in the beam of direction $\boldsymbol{\omega}$ – near the point P and near time t – is described by *the directional emission density* $e_\nu(\mathrm{P}, \boldsymbol{\omega}, t)$.

Let us now take into account the well known property of all material media to emit, at non-zero absolute temperature, more or less photons. This *thermal emission* obeys a law which is particularly simple in the case of a thermally insulated enclosure at constant temperature, i.e. in the case of *thermodynamic equilibrium*.

In this case the stationary radiation field generated by the matter *of* – or *inside* – the enclosure is of the kind known as *black body radiation*, or as *a radiation field inside a black body*, (the term 'black radiation' is elegant but somewhat misleading!)

This radiation field is *homogeneous*, *isotropic*, *stationary* and its intensity depends only on the frequency ν of photons and on the temperature T in the enclosure. (This is proved in all current treatises on thermodynamics.) Contrary to the intensity $I_\nu(\mathrm{P}, \boldsymbol{\omega}, t)$ of any general radiation field, the intensity of the black body radiation does *not* depend on the situation of the point P inside the enclosure, on the direction $\boldsymbol{\omega}$ or on time t.

Moreover, the intensity of the radiation field inside a black body is independent from the *'chemical nature'* of the matter *of* – or *inside* – the enclosure.

In order to express all these particular properties of the black body radiation its *intensity* is denoted by $B_\nu(T)$. This particular intensity is a *universal function*, given below by Equation (43), of the temperature T inside the enclosure and of the frequency ν. It does not depend on anything else, and has the same physical dimensions as $I_\nu(\mathrm{P}, \boldsymbol{\omega}, t)$. Care must be taken to avoid the widely spread confusion between $B_\nu(T)$ and the corresponding *non-directional energy density* of the field which we shall denote by $u_\nu^*(T)$. The asterisk in $u_\nu^*(T)$ recalls that this quantity refers to the *black body* radiation. Like the corresponding intensity $B_\nu(T)$, it does not depend on P or t.

In an *isotropic* field – such as that of a black body – of intensity $B_\nu(T)$, the *mean* intensity $\bar{I}_\nu$ (defined in Subsection 1.3) of the average field coincides in all directions with the intensity $B_\nu(T)$ of the real field. Therefore, according to Equation (17), $u_\nu^*(T)$ will be given as a function of $B_\nu(T)$, by

$$u_\nu^*(T) = \frac{4\pi}{c} B_\nu(T). \tag{38}$$

On applying Equation (31) to the particular case of the black body radiation we obtain:

$$p_\nu^*(\mathrm{P}, \boldsymbol{\omega}, t) = \Sigma_\nu B_\nu(T), \tag{39}$$

written more usually as

$$B_\nu(T) = p_\nu^*(\mathrm{P}, \boldsymbol{\omega}. t)/\Sigma_\nu. \tag{40}$$

We keep the arguments P, $\boldsymbol{\omega}$, t in spite of the homogeneity, isotropy, and stationarity of the field in order to preserve and to recall the *physical dimensions* of $p_\nu^*(\mathrm{P}, \boldsymbol{\omega}, t)$. We also avoid in this way a possible confusion with $p_\nu^*(\mathrm{P}, t)$.

Equation (39) shows that, beside the obvious dependence on ν, the directional density of removal $p_\nu^*(\mathrm{P}, \boldsymbol{\omega}, t)$ depends not only, through $B_\nu(T)$ and Σ_ν, on the *state* of the medium, but also, through the coefficient Σ_ν, on the *nature* of the material medium.

Since the radiation field inside a black body is both stationary and homogeneous, its intensity $B_\nu(T)$ remains permanently *the same* at two neighbouring points of any radiation beam. This is possible if, and only if, the directional density of removal $p_\nu^*(\mathrm{P}, \boldsymbol{\omega}, t)$ is *fully compensated* by the directional density of emission $e_\nu^*(\mathrm{P}, \boldsymbol{\omega}, t)$. One sometimes says that this compensation expresses the '*microreversibility*' of radiative processes in the state of thermodynamic equilibrium.

Thus for the radiation field inside a black body we have, according to Equation (39):

$$e_\nu^*(\mathrm{P}, \boldsymbol{\omega}, t) = \Sigma_\nu B_\nu(T). \tag{41}$$

Equation (41) represents one of the forms of *Kirchhoff's law*.

Through integration over all directions, the isotropy of the right-hand side of Equation (41) yields another form of the same law.

$$e_\nu^*(\mathrm{P}, t) = 4\pi \Sigma_\nu B_\nu(T). \tag{42}$$

According to this form of Kirchhoff's law, the ratio of the non-directional emission density $e_\nu^*(\mathrm{P}, t)$ to the coefficient of removal Σ_ν does *not* depend on the *nature* of the matter inside a black body. This ratio depends only on the temperature T of the medium and on the frequency ν – through the universal function $B_\nu(T)$.

The explicit form for the function $B_\nu(T)$ is given by quantum statistics. The quantum theory predicts for $B_\nu(T)$ the expression:

$$B_\nu(T) = \frac{2h\nu^3}{c^2}(e^{h\nu/kT} - 1)^{-1}, \tag{43}$$

where h is Planck's constant, k is the Boltzmann constant, and c is the speed of light in vacuum. This expression for $B_\nu(T)$ is called Planck's law or better *Planck's function*.

The difference between the *physical definition* of $B_\nu(T)$ – as intensity of the radiation field inside a black body – and *the mathematical expression* of its dependence on T and ν, given by Equation (43), must be emphasized since it is very often overlooked.

It results from Equations (38) and (43) that $u_\nu^*(T)$, sometimes erroneously called $B_\nu(T)$, is given explicitly by

$$u_\nu^*(T) = \frac{8\pi h\nu^3}{c^3}(e^{h\nu/kT} - 1)^{-1}. \tag{44}$$

Kirchhoff's law is a *local* relation between $e_\nu^*(\mathrm{P}, \boldsymbol{\omega}, t)$ and $\Sigma_\nu(\mathrm{P}, t)$ at the point P. One can therefore admit *as a first approximation* that it remains valid when in an enclosure the temperature, instead of being the same everywhere, varies sufficiently slowly as a function of the distance between different points.

In such a situation (like the situation in the stars far from the surface), each small volume element can be considered to be in '*local thermodynamic equilibrium*' (LTE), i.e. can be assimilated to a *black body* at a given uniform temperature proper to each element.

This approximation is very useful and very usual in elementary astrophysics since it permits the use of $e_\nu^*(\mathrm{P}, \boldsymbol{\omega}, t)$ instead of $e(\mathrm{P}, \boldsymbol{\omega}, t)$. This assumption leads, by Equation (41), to

$$e_\nu(\mathrm{P}, \boldsymbol{\omega}, t)_{\mathrm{LTE}} = \Sigma_\nu B_\nu(T), \tag{45}$$

where T is now the *local* temperature near the point P, and the subscript LTE has an obvious meaning.

When Σ_ν does not depend on t Equation (45) takes the form:

$$\text{(Stationary case)} \quad e_\nu(\mathrm{P}, \boldsymbol{\omega})_{\mathrm{LTE}} = \Sigma_\nu B_\nu(T). \tag{45'}$$

Another particular type of interaction between the matter and a radiation field, which gives a simple expression for $e_\nu(\mathrm{P}, \boldsymbol{\omega}, t)$ is the one called the '*perfect isotropic scattering*'. It corresponds to the case when the whole of the emission is due to coherent isotropic scattering supposed to be '*perfect*', i.e. such that the *non-directional* densities $e_\nu(\mathrm{P}, t)$ and $p_\nu(\mathrm{P}, t)$ are equal.

In this case all photons removed by the interaction are 're-created' *with the same frequency*, since the scattering is coherent, and scattered isotropically in all directions.

The *directional* emission density $e_\nu(\mathrm{P}, \boldsymbol{\omega}, t)$, corresponding to the *unit of solid angle*, will be given in this case, since the scattering is supposed to be isotropic, by

$$e_\nu(\mathrm{P}, \boldsymbol{\omega}, t) = \frac{1}{4\pi} e_\nu(\mathrm{P}, t) = \frac{1}{4\pi} p_\nu(\mathrm{P}, t). \tag{46}$$

On integrating Equation (31) over all directions and on applying Equations (18) and (46) we obtain:

$$e_\nu(\mathrm{P}, \boldsymbol{\omega}, t)_{\mathrm{PIS}} = \Sigma_\nu \bar{I}_\nu(\mathrm{P}, \boldsymbol{\omega}, t), \tag{47}$$

where the subscript PIS obviously indicates the perfect isotropic scattering.

This relation is *mathematically* very like the relation (45). Nevertheless, it must be emphasized that *physically* the perfect isotropic scattering is very different from the local thermodynamic equilibrium.

When a gas is sufficiently *cold* to minimize the thermal emission, and sufficiently *diluted* to avoid collisional excitation, the perfect isotropic scattering can represent a rather satisfactory approximation in the case of *resonance scattering*, i.e. when the interaction concerns only the fundamental and the *first* excited level (though in this case the scattering is not rigourously isotropic).

Finally, when neither LTE nor PIS represent an acceptable approximation, electronic computers can be applied if the state is assumed to be *steady* (*stationary*). Then different '*populations*' N^j, N^k, N^c etc. and different *intensities* I_ν of the radiation field being considered as unknown quantities, one can write a system of equations expressing the conservation of the population of each level. The interactions to be considered in these equations are those between the atoms of the medium and the photons of the radiation field, but also between the material particles alone (collisional excitations and deexcitations).

Unfortunately, when all real levels are taken into account, the complexity of the corresponding system of equations is so great, that even with electronic computers one is obliged to operate with more or less artificial models. In these models several real levels are *grouped* and replaced by a finite number of *average levels*.

In general, one must also take into account the *spontaneous emissions* due to the *finite lifetime* of excited levels. These emissions are described, according to A. Einstein, by the relations of the form

$$N_{\mathrm{sp}}^{kj} = A_{kj} N^k. \tag{48}$$

Here the subscript sp means *spontaneous* and N_{sp}^{kj} represents the number of atoms (A_k) undergoing the transition from the state E_k to a lower state E_j, per cm^3 and per s.

Equation (48) is simply a *definition* of *the transition probability for spontaneous emission* (or the 'Einstein's coefficient') A_{kj} for discrete infinitely narrow levels.

When the finite width of levels is taken into account and when emission is supposed to be isotropic, the generalizations analogous to those made previously give for the part $\delta e_\nu(\mathrm{P}, \boldsymbol{\omega}, t)$ of $e_\nu(\mathrm{P}, \boldsymbol{\omega}, t)$ corresponding to spontaneous emissions:

$$\delta e_\nu(\mathrm{P}, \boldsymbol{\omega}, t) = \frac{h\nu}{4\pi} A_{kj}(\nu) N^{k'}(\mathrm{P}, t). \tag{49}$$

3. Equation of Transfer and the Corresponding Equation of Continuity

3.1. The Purely Kinetic Variation of the Directional Energy Density at a Fixed Point as a Function of Time

Let us consider the physically frequent case of a *non-homogeneous* radiation field, i.e. a field where – at a given time t and in a given arbitrary direction $\boldsymbol{\omega}$ – the directional energy density $u_\nu(\mathrm{P}, \boldsymbol{\omega}, t)$ takes on different values at *different points* P.

Then, the *mobility* of the photons associated with the *non-homogeneity* of the field will generate a purely *kinetic* variation of the density $u_\nu(\mathrm{P}, \boldsymbol{\omega}, t)$ *as a function of time*, independently of any interaction with a material medium and in particular for photons travelling in the vacuum.

Figure 1.3 is a diagram where the successive situations relative to times $(t - \Delta t)$ and t are schematically indicated by a 'vertical' displacement along an axis of times Ot directed downwards.

Different squares represent *geometrical* volume elements of identical volumic capacity. They are located at *fixed* points P′, P and P_1, along some given fixed direction $\boldsymbol{\omega}$, at a distance $\Delta s_\omega = c\Delta t$ from each other.

At different times and at different points, these volume elements are more or less filled with photons.

Inclined arrows indicate that after Δt seconds the photons initially contained in the element called (A') go into the element called (A); and, similarly, those initially contained in (A) go into (A_1). The *vertical arrow* at the level of the point P (i.e. of the element A) corresponds to the evolution of the photon density in a given volume element – such as (A) – between times $(t - \Delta t)$ and t.

The photon density, in different volume elements at different times, is symbolized by the 'density of cross-hatching'. Of course, the dotted squares are never empty, but are not cross-hatched in order to make the figure clearer. In Figure 1.3 it is assumed that at time t the photon density is *greater* at P in the element (A) than at P′ in the element (A').

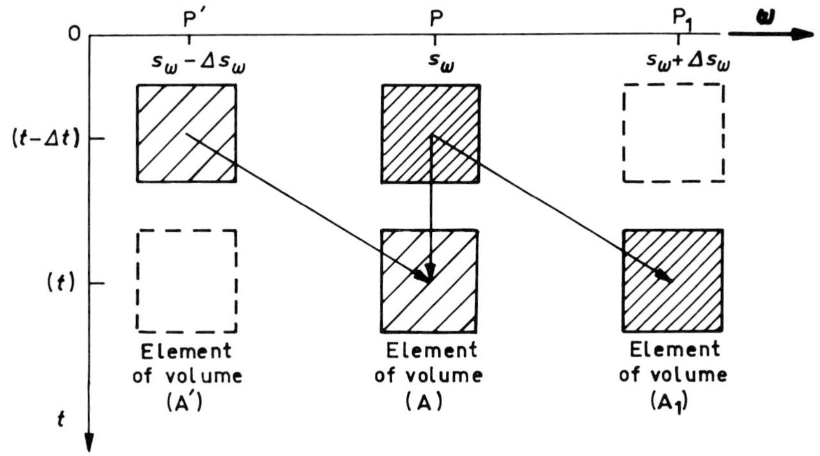

Fig. 1.3. The concept of the purely kinetic variation of the directional energy density at a given point as a function of time.

The distance Δs_ω between P′ and P, and between P and P_1, is completely determined by the choice of the time interval Δt. Indeed, Δs_ω must be equal to $c\Delta t$ in order that the photons present in the element (A') at time $(t - \Delta t)$ – and travelling in vacuum with speed c – could fill the element (A) at time t; those initially contained in the element (A) going during the same time into the element (A_1).

Figure 1.3 shows how the *increase* of the density in the direction $\boldsymbol{\omega}$ at a given time – i.e., *a positive density gradient*, leads to a decrease of the density in a given element in the course of time – i.e. a *negative time derivative*.

More precisely, between the time $(t - \Delta t)$ and the time t, the density u_ν at P changes from the value $u_\nu(\mathrm{P}, \boldsymbol{\omega}, t - \Delta t)$ to the value $u_\nu(\mathrm{P}, \boldsymbol{\omega}, t)$. But, because of the *substitution* due to the mobility of photons, $u_\nu(\mathrm{P}, \boldsymbol{\omega}, t)$ is equal to $u_\nu(\mathrm{P}', \boldsymbol{\omega}, t - \Delta t)$ – as indicated by the first inclined arrow in Figure 1.3.

Hence, a variation of the density u_ν at P during the time interval Δt: the photons of density $u_\nu(\mathrm{P}, \boldsymbol{\omega}, t - \Delta t)$ have been evacuated from the element (A) and replaced by the photons of a *smaller* density $u_\nu(\mathrm{P}', \boldsymbol{\omega}, t - \Delta t)$.

Therefore, indicating our assumption of an absence of interaction by the subscript '*kin*', we can write:

$$\left[\frac{\partial u_\nu(\mathrm{P},\boldsymbol{\omega},t)}{\partial t}\right]_{\mathrm{kin}} - \lim_{\Delta t\to 0}\frac{u_\nu(\mathrm{P},\boldsymbol{\omega},t)-u_\nu(\mathrm{P},\boldsymbol{\omega},t-\Delta t)}{\Delta t}$$

$$= \lim_{\substack{\Delta t\to 0\\ \Delta s_\omega\to 0}}\left[\frac{u_\nu(\mathrm{P}',\boldsymbol{\omega},t-\Delta t)-u_\nu(\mathrm{P},\boldsymbol{\omega},t-\Delta t)}{\Delta s_\omega}\right]\frac{\Delta s_\omega}{\Delta t}. \tag{50}$$

If t and s_ω were independent the fraction between the square brackets would give – as Δs_ω tends to zero – the derivative (with a minus sign) of the scalar function $u_\nu(\mathrm{P}, \boldsymbol{\omega}, t-\Delta t)$ of the point P along the direction $\boldsymbol{\omega}$. However, since Δs_ω is equal to $c\Delta t$, the interval of time Δt tends to zero as Δs_ω tends to zero. Thus the actual limit of the fraction between the square brackets will be $[-\partial u_\nu(\mathrm{P}, \boldsymbol{\omega}, t)/\partial s_\omega]$. As to the fraction $\Delta s_\omega/\Delta t$ it is obviously always equal to c.

We find finally:

$$\left[\frac{\partial u_\nu(\mathrm{P},\boldsymbol{\omega},t)}{\partial t}\right]_{\mathrm{kin}} = -c\,\frac{\partial u_\nu(\mathrm{P},\boldsymbol{\omega},t)}{\partial s_\omega}. \tag{51}$$

In the derivative of u_ν with respect to s_ω the vector $\boldsymbol{\omega}$ simply plays the role of a constant parameter.

On applying the relation (9) between u_ν and I_ν we can put Equation (51) in a more elegant and usual form:

$$\left[\frac{\partial u_\nu(\mathrm{P},\boldsymbol{\omega},t)}{\partial t}\right]_{\mathrm{kin}} = -\frac{\partial I_\nu(\mathrm{P},\boldsymbol{\omega},t)}{\partial s_\omega}. \tag{52}$$

3.1.1. Remark

We can express the consequences of the non-homogeneity of the field and of the mobility of the photons in another way.

In Figure 1.4, let us consider a quasi-cylindrical volume element (dV), whose 'axis' is in the direction $\boldsymbol{\omega}$ and whose 'bases', of area dA_ω, are centred on points P and P′ respectively. This element is actually a frustum of a cone but the difference between the areas of the two bases tends to zero when the 'height' of the element tends to zero.

The distance between P and P′ is again Δs_ω, but here we call it ds_ω in order to be near the notation of Section 1.

Thus, the displacement from P′ to P is described by $(ds_\omega)\boldsymbol{\omega}$.

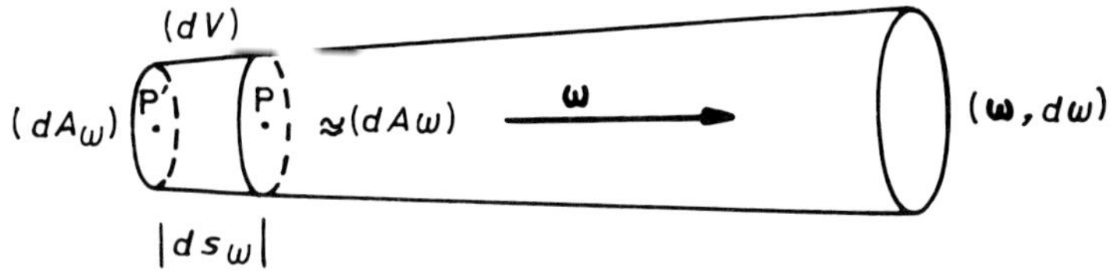

Fig. 1.4. An other introduction of the concept of purely kinetic variation of the directional energy density.

According to the definition of $u_\nu(\mathrm{P}, \boldsymbol{\omega}, t)$, the energy $dE_\nu(t)$ of the photons ($\boldsymbol{\omega}, d\omega$; $\nu, d\nu$) present at time t in the element (dV) is given by Equation (1), viz.

$$dE_\nu(t) = u_\nu(\mathrm{P}, \boldsymbol{\omega}, t)\, dV\, d\omega\, d\nu. \tag{53}$$

According to the same relation (53), the corresponding energy $dE_\nu(t-dt)$ contained in (dV) at time $(t-dt)$ is given by the limited Taylor expansion:

$$\begin{aligned} dE_\nu(t-dt) &= u_\nu(\mathrm{P}, \boldsymbol{\omega}, t-dt)\, dV\, d\omega\, d\nu \\ &= dE_\nu(t) - \left[\frac{\partial u_\nu(\mathrm{P}, \boldsymbol{\omega}, t)}{\partial t}\right]_{\text{kin}} dt\, dV\, d\omega\, d\nu. \end{aligned} \tag{54}$$

Physically the excess of $dE_\nu(t)$ over $dE_\nu(t-dt)$ must be equal to the excess of the energy $I_\nu(\mathrm{P}', \boldsymbol{\omega}, t-dt)\, dA_\omega\, d\omega\, d\nu\, dt$, entering during the interval of time dt through the base centered on P$'$, over the energy $I_\nu(\mathrm{P}, \boldsymbol{\omega}, t)\, dA_\omega\, d\omega\, d\nu\, dt$, leaving (dV) during the same time dt through the base centered on P.

Thus, neglecting the difference of second order in (dt) between $I_\nu(\mathrm{P}', \boldsymbol{\omega}, t-dt)\, dt$ and $I_\nu(\mathrm{P}', \boldsymbol{\omega}, t)\, dt$, we can write:

$$\begin{aligned} dE_\nu(t) - dE_\nu(t-dt) &= I_\nu(\mathrm{P}', \boldsymbol{\omega}, t)\, dA_\omega\, d\omega\, d\nu\, dt - \\ &\quad - I_\nu(\mathrm{P}, \boldsymbol{\omega}, t)\, dA_\omega\, d\omega\, d\nu\, dt. \end{aligned} \tag{55}$$

Let us now substitute Equation (54) into Equation (55) and divide by $dt\, dV\, d\omega\, d\nu$. We obtain, on replacing dV/dA_ω by ds_ω:

$$\left[\frac{\partial u_\nu(\mathrm{P}, \boldsymbol{\omega}, t)}{\partial t}\right]_{\text{kin}} = -\frac{I_\nu(\mathrm{P}, \boldsymbol{\omega}, t) - I_\nu(\mathrm{P}', \boldsymbol{\omega}, t)}{ds_\omega}. \tag{56}$$

When (dt) tends to zero, $ds_\omega = c\,dt$ tends also to zero and Equation (56) becomes identical with Equation (52), viz.

$$\left[\frac{\partial u_\nu(\mathrm{P}, \boldsymbol{\omega}, t)}{\partial t}\right]_{\text{kin}} = -\frac{\partial I_\nu(\mathrm{P}, \boldsymbol{\omega}, t)}{\partial s_\omega}. \tag{56$'$}$$

If the method used in this remark is not fully as clear as the other one, this is due to the following circumstances.

In Figure 1.4 the points P$'$ and P are not defined in the same way as in Figure 1.3. In the method of the remark one uses implicitly the principle, introduced in Subsection 1.1, according to which the element of volume (dV) *surrounding* the point P in Equation (53) can have *any* position around P. However, in our proof, the point P is situated *on* the very *frontier* of the element (dV). Nevertheless, this procedure gives an exact result because here, when dt tends to zero, the element (dV) becomes infinitely *flat*. This effaces the difference between the frontier and the interior.

3.2. The Different Forms of the Equation of Transfer

Let us now consider a more general case of a non-homogeneous and non-stationary radiation field whose photons can be removed or 'enriched', by various emissive processes, through *interaction with a material medium*.

The *total* variation of the directional energy density $u_\nu(\mathrm{P}, \boldsymbol{\omega}, t)$, at a given point P, in a given direction $\boldsymbol{\omega}$, *per unit time* in the neighbourhood of time t, will be denoted by $\partial u_\nu(\mathrm{P}, \boldsymbol{\omega}, t)/\partial t$ *without the subscript 'kin'*.

To obtain $\partial u_\nu(\mathrm{P}, \boldsymbol{\omega}, t)/\partial t$ we must *add* the directional emission density $e_\nu(\mathrm{P}, \boldsymbol{\omega}, t)$ to the *purely kinetic* variation $[\partial u_\nu/\partial t]_{\mathrm{kin}}$ of u_ν and *subtract* the corresponding directional density of removal $p_\nu(\mathrm{P}, \boldsymbol{\omega}, t)$. Thus $\partial u_\nu/\partial t$ will be given by

$$\frac{\partial u_\nu(\mathrm{P}, \boldsymbol{\omega}, t)}{\partial t} = \left[\frac{\partial u_\nu(\mathrm{P}, \boldsymbol{\omega}, t)}{\partial t}\right]_{\mathrm{kin}} + e_\nu(\mathrm{P}, \boldsymbol{\omega}, t) - p_\nu(\mathrm{P}, \boldsymbol{\omega}, t). \tag{57}$$

This Equation can be transformed, by application of Equations (9), (52) and (31), into:

$$\boxed{\frac{1}{c}\frac{\partial I_\nu(\mathrm{P}, \boldsymbol{\omega}, t)}{\partial t} = -\frac{\partial I_\nu(\mathrm{P}, \boldsymbol{\omega}, t)}{\partial s_\omega} + e_\nu(\mathrm{P}, \boldsymbol{\omega}, t) - \Sigma_\nu I_\nu(\mathrm{P}, \boldsymbol{\omega}, t)} \; . \tag{58}$$

On expressing the derivative of I_ν along the direction $\boldsymbol{\omega}$ as a function of **grad** I_ν, the same equation can be written

$$\frac{1}{c}\frac{\partial I_\nu(\mathrm{P}, \boldsymbol{\omega}, t)}{\partial t} = -\boldsymbol{\omega}.\mathbf{grad}\, I_\nu(\mathrm{P}, \boldsymbol{\omega}, t) - \Sigma_\nu I_\nu(\mathrm{P}, \boldsymbol{\omega}, t) + e_\nu(\mathrm{P}, \boldsymbol{\omega}, t). \tag{59}$$

This is the most general form of *the equation of transfer*, when effects of refraction are neglected.

In the very important particular case of a *stationary situation*, the left-hand side of Equations (58) or (59) is permanently equal to zero, and the equation of transfer takes the form

$$\text{(Stationary case)} \quad \boxed{\begin{aligned} \frac{\partial I_\nu(\mathrm{P}, \boldsymbol{\omega})}{\partial s_\omega} &= \boldsymbol{\omega}.\mathbf{grad}\, I_\nu(\mathrm{P}, \boldsymbol{\omega}) \\ &= -\Sigma_\nu I_\nu(\mathrm{P}, \boldsymbol{\omega}) + e_\nu(\mathrm{P}, \boldsymbol{\omega}) \end{aligned}} \; . \tag{60}$$

In equation (60) we omit the argument t because of the assumption of stationary. However, one must remember that $I_\nu(\mathrm{P}, \boldsymbol{\omega})$ and $e_\nu(\mathrm{P}, \boldsymbol{\omega})$ still imply the unity of the *interval* of time in their definition.

In the stationary case, the variable s_ω is often replaced by a new variable $\tau_{\omega,\nu}$ defined by the element of *physical* – or 'optical' – *displacement*:

$$d\tau_{\omega,\nu} = \Sigma_\nu\, ds_\omega . \tag{61}$$

The dependence of τ on $\boldsymbol{\omega}$ is sometimes considered as implicit, and one writes simply τ_ν. However, this can lead to confusion.

The division of both sides of Equation (60) by Σ_ν introduces also the function $S_\nu(\mathrm{P}, \boldsymbol{\omega}, t)$ – called *the source function* – defined by

$$S_\nu(\mathrm{P}, \boldsymbol{\omega}, t) = \frac{e_\nu(\mathrm{P}, \boldsymbol{\omega}, t)}{\Sigma_\nu}. \tag{62}$$

The use of the definitions (61) and (62) allows us to write Equation (60) in a more compact form:

$$(\text{Stationary case}) \quad \frac{\partial I_\nu(\mathrm{P}, \boldsymbol{\omega})}{\partial \tau_{\omega,\nu}} = -I_\nu(\mathrm{P}, \boldsymbol{\omega}) + S_\nu(\mathrm{P}, \boldsymbol{\omega}). \tag{63}$$

The argument t is omitted again, for the same reason as above.

3.2.1. *Remark*

According to the definition (62) of the source function S_ν, Equations (45) and (47), respectively characteristic of the local thermodynamic equilibrium and of the perfect isotropic scattering, give the following expressions for S_ν:

$$S_\nu(\mathrm{P}, \boldsymbol{\omega}, t)_{\mathrm{LTE}} = B_\nu(T), \tag{64}$$

and

$$S_\nu(\mathrm{P}, \boldsymbol{\omega}, t)_{\mathrm{PIS}} = \bar{I}_\nu(\mathrm{P}, \boldsymbol{\omega}, t). \tag{65}$$

3.3. THE NET FLUX WITH RESPECT TO A DIRECTION n

It is sometimes useful to abandon the differentiation with respect to direction and to consider simply (see Figure 1.5) the 'flow of energy' of the photons which pass first through the *unit* area normal to a unit vector **n** and then through the hemisphere (Ω_n^+), of axis **n**, of unit radius. This hemisphere is a kind of a dome supported by a central pillar **n**.

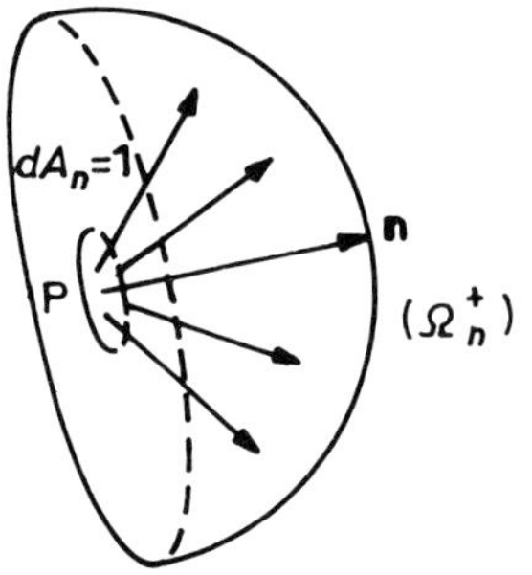

Fig. 1.5. The definition of the positive flux with respect to a given direction.

This defines the quantity $\pi F_\nu^+(\mathrm{P}, \mathbf{n}, t)$ called *the positive flux with respect to the direction* **n** at the point P of a radiation field near the time t.

Similarly, we can consider the 'flow of energy' of the photons which pass first through the unit area normal to **n** and then through the hemisphere (Ω_n^-) of axis $(-\mathbf{n})$. This defines the quantity $\pi F_\nu^-(\mathrm{P}, \mathbf{n}, t)$ called *the negative flux with respect to the direction* $\mathbf{n}$, at the point P near the time t.

Both πF_ν^+ and πF_ν^- are in some respects analogous to $I_\nu(\mathrm{P}, \boldsymbol{\omega}, t)$, since they correspond to the energy of photons *in the unit of the frequency interval* near the frequency ν (this trivial point was omitted, for brievity's sake, in the defintion given above), to a *unit area* and to the *unit of interval of time*.

However, they are very different from $I_\nu(\mathrm{P}, \boldsymbol{\omega}, t)$ by the fact that they correspond to a unit area which is no longer normal to $\boldsymbol{\omega}$ but is normal to $\mathbf{n}$.

Moreover, they correspond to quantities which have not entirely the same physical dimensions as I_ν since they are *no longer defined per unit solid angle* (about $\boldsymbol{\omega}$), but concern *all directions* defined either by the hemisphere (Ω_n^+) or by the hemisphere (Ω_n^-).

Finally, one must avoid any assimilation of πF_ν^+ with a simple integral of $I_\nu(\mathrm{P}, \boldsymbol{\omega}, t)$ over (Ω_n^+) and of πF_ν^- with a simple integral of I_ν over (Ω_n^-), as one is often tempted to do. The actual relation between πF_ν^+ or πF_ν^- and I_ν will be given below by Equations (69) and (70)

Let us consider, for the moment, *the difference*

$$\pi F_\nu(\mathrm{P}, \mathbf{n}, t) = \pi F_\nu^+(\mathrm{P}, \mathbf{n}, t) - \pi F_\nu^-(\mathrm{P}, \mathbf{n}, t). \tag{66}$$

This difference is called the *net* flux with respect to the direction $\mathbf{n}$, at the point P and in the neighbourhood of time t.

The relation between the net flux and intensity I_ν can be obtained in the following way.

Let us consider first a surface element (dA_n) normal to $\mathbf{n}$ but of small *arbitrary* area dA_n. Then, let us fix our attention (see Figure 1.6) on the photons $(\boldsymbol{\omega}, d\omega; \nu, d\nu)$ crossing (dA_n) within the solid angle $(d\omega)$ about $\boldsymbol{\omega}$.

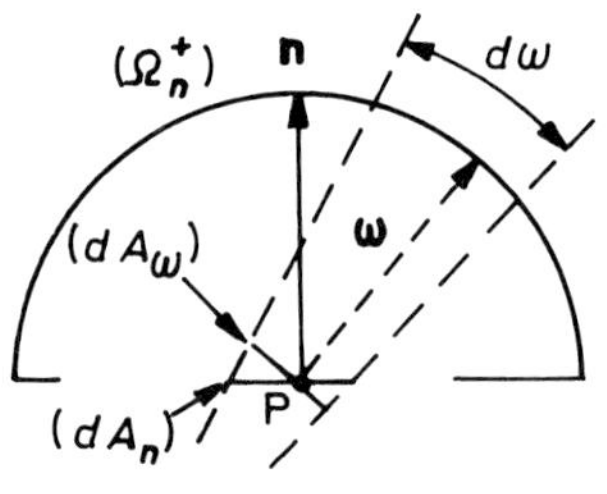

Fig. 1.6. An illustration of the identity $dE_\nu(\ldots, dA_n, \ldots) \equiv dE_\nu(\ldots, dA_\omega, \ldots)$.

In Figure 1.6 the section (dA_n) and the section (dA_ω) of the corresponding beam by a plane normal to $\boldsymbol{\omega}$, of area dA_ω, are represented in two dimensions and are symbolized by their traces.

Obviously,

$$dA_\omega = (\boldsymbol{\omega} \cdot \mathbf{n})\, dA_n. \tag{67}$$

The energy $dE_\nu(\mathrm{P}, \boldsymbol{\omega}, t, dA_n, d\omega, d\nu, dt)$ of the photons $(\boldsymbol{\omega}, d\omega; \nu, d\nu)$ crossing the element (dA_n) during the interval of time dt must be equal to the energy $dE_\nu(\mathrm{P}, \boldsymbol{\omega}, t, dA_\omega, d\omega, d\nu, dt)$ crossing the element (dA_ω). Indeed, the *same* photons $(\boldsymbol{\omega}, d\omega; \nu, d\nu)$

cross the section (dA_ω) and the section (dA_n) during the interval of time dt. Therefore, according to Equation (6), we have

$$dE_\nu(\mathrm{P}, \boldsymbol{\omega}, t, dA_n, d\omega, d\nu, dt) = I_\nu(\mathrm{P}, \boldsymbol{\omega}, t)\, dA_\omega\, d\omega\, d\nu\, dt$$

$$= I_\nu(\mathrm{P}, \boldsymbol{\omega}, t)(\boldsymbol{\omega} \cdot \mathrm{n})\, dA_n\, d\omega\, d\nu\, dt. \qquad (68)$$

In the definition of πF_ν^+ and in the definition of πF_ν^- we took dA_n, $d\nu$ and dt respectively equal to unity. Thus we can write:

$$\pi F_\nu^+(\mathrm{P}, \mathrm{n}, t) = \int_{\Omega_n^+} I_\nu(\mathrm{P}, \boldsymbol{\omega}, t)(\boldsymbol{\omega} \cdot \mathrm{n})\, d\omega, \qquad (69)$$

and

$$\pi F_\nu^-(\mathrm{P}, \mathrm{n}, t) = + \int_{\Omega_n^-} I_\nu(\mathrm{P}, \boldsymbol{\omega}, t)(-\mathrm{n} \cdot \boldsymbol{\omega})\, d\omega$$

$$= - \int_{\Omega_n^-} I_\nu(\mathrm{P}, \boldsymbol{\omega}, t)(+\mathrm{n} \cdot \boldsymbol{\omega})\, d\omega. \qquad (70)$$

The fact that $[-(-\mathrm{n} \cdot \boldsymbol{\omega})] = +(\boldsymbol{\omega} \cdot \mathrm{n})$ transforms the *physical difference* (66) into an *algebraic sum* of a positive and a negative quantity. Thus, the complete unit sphere centred on P being denoted by (Ω), we obtain:

$$\boxed{\pi F_\nu(\mathrm{P}, \mathrm{n}, t) = \int_\Omega I_\nu(\mathrm{P}, \boldsymbol{\omega}, t)(\boldsymbol{\omega} \cdot \mathrm{n})\, d\omega} \qquad (71)$$

Let (α, β, γ) denote the cartesian components of $\boldsymbol{\omega}$ with respect to some cartesian axes (Px, Py, Pz) and let $\pi \mathbf{F}_\nu(\mathrm{P}, t)$ denote the *vector* whose cartesian components are respectively

$$\int_\Omega I_\nu(\mathrm{P}, \boldsymbol{\omega}, t)\alpha\, d\omega; \quad \int_\Omega I_\nu(\mathrm{P}, \boldsymbol{\omega}, t)\beta\, d\omega; \quad \int_\Omega I_\nu(\mathrm{P}, \boldsymbol{\omega}, t)\gamma\, d\omega. \qquad (72)$$

This definition can be symbolized by

$$\pi \mathbf{F}_\nu(\mathrm{P}, t) = \int_\Omega I_\nu(\mathrm{P}, \boldsymbol{\omega}, t)\, \boldsymbol{\omega}\, d\omega. \qquad (73)$$

The left-hand side of Equation (73) is called *the net flux vector* at the point P at time t. For each given radiation field there exists only *one* such vector at each point P at time t.

On expressing $(\boldsymbol{\omega} \cdot \mathrm{n})$ as a function of the cartesian components of $\boldsymbol{\omega}$ and of n, one finds that Equation (71) can be rewritten in the form:

$$\boxed{\pi F_\nu(\mathrm{P}, \mathrm{n}, t) = \pi(\mathbf{F}_\nu \cdot \mathrm{n})} \qquad (74)$$

Thus, the *single* vector $\pi \mathbf{F}_\nu(\mathrm{P}, t)$ is able to represent the net flux $\pi F_\nu(\mathrm{P}, \mathrm{n}, t)$ corresponding to *all* possible *orientations* of n in a given radiation field. One just has to take the algebraic measure of the projection of the net flux vector upon the corresponding n for obtaining $\pi F_\nu(\mathrm{P}, \mathrm{n}, t)$.

If we consider, in a given field, the value of $\pi F_\nu(\mathrm{P}, \mathbf{n}, t)$ as a function of $\mathbf{n}$, this net flux will go through a *maximum* when the vector $\mathbf{n}$ is aligned along the net flux vector.

3.3.1. *Remarks*

3.3.1.1. The concept of *flux* introduced above is not the usual one. In vector analysis and in its applications to physics, given a vector field $V(\mathrm{P})$ – function of point P – and the outward normal $\mathbf{n}$ to a closed convex surface (S), the usual outward flux of $\mathbf{V}$ through a surface element (dA_n) of (S) is defined by $(\mathbf{V} \cdot \mathbf{n})\, dA_n$.

This element of outward flux brings into play the area (dA_n) and not, as in the theory of radiation transfer, a *unit* area. Moreover, the usual definition of the outward flux, recalled above, implies that the vector field $V(\mathrm{P})$ varies over the domain (dA_n) *sufficiently slowly* (in modulus and in direction) to experience practically no variation as P moves over (dA_n). One the contrary, in the theory of radiation transfer one takes into account particles crossing (dA_n) in *all directions* of a half-space for the positive or the negative flux (and in all directions of space for the net flux).

3.3.1.2. For a photon of frequency ν moving in vacuum in the direction $\boldsymbol{\omega}$, the *individual* velocity vector is $c\boldsymbol{\omega}$. Let us then apply the definition of the *average* vector velocity of particles at a given point P at time t, to the case when the particles are photons. In this particular case the usual definition of the average vector velocity, such as it is given in all textbooks on the *Kinetic Theory of Fluids*, will give the average vector velocity $\mathbf{v}_\nu(\mathrm{P}, t)$ of photons, in a general radiation field, by the (self-explanatory) relation:

$$\mathbf{v}_\nu(\mathrm{P}, t) = \int_\Omega (c\boldsymbol{\omega}) \quad [\text{proportion of photons } \nu \text{ of type } (\boldsymbol{\omega}, d\omega)]. \tag{75}$$

According to the definition of the directional energy density $u_\nu(\mathrm{P}, \boldsymbol{\omega}, t)$ introduced in Subsection 1.1 Equation (75) is equivalent to

$$\begin{aligned} \mathbf{v}_\nu(\mathrm{P}, t) &= \int_\Omega c\boldsymbol{\omega} \left[\frac{u_\nu(\mathrm{P}, \boldsymbol{\omega}, t)\, d\omega / h\nu}{u_\nu(\mathrm{P}, t)/h\nu} \right] = \frac{1}{u_\nu(\mathrm{P}, t)} \int_\Omega I_\nu \boldsymbol{\omega}\, d\omega \\ &= \frac{\pi \mathbf{F}_\nu(\mathrm{P}, t)}{u_\nu(\mathrm{P}, t)}, \end{aligned} \tag{76}$$

where we have used Equations (73) and (9).

Thus the true analogue of the net flux *vector* of the theory of radiative transfer is this *average vector velocity* of the kinetic theory of fluids applied to photons.

3.4. THE EQUATION OF CONTINUITY OF TRANSFER THEORY

Equation (57) describes the transfer of photons of a single given direction. Considering now the photons of *all directions*, let us look for an expression for the variation of *the non-directional* density $u_\nu(\mathrm{P}, t)$ at a given point P as a function of time t.

The terms of Equation (57) such as $e_\nu(\mathrm{P}, \boldsymbol{\omega}, t)$ and $p_\nu(\mathrm{P}, \boldsymbol{\omega}, t)$ in general will be replaced by

$$e_\nu(\mathrm{P}, t) = \int_\Omega e_\nu(\mathrm{P}, \boldsymbol{\omega}, t)\, d\omega \quad \text{and} \quad p_\nu(\mathrm{P}, t) = \int_\Omega p_\nu(\mathrm{P}, \boldsymbol{\omega}, t)\, d\omega. \tag{77}$$

In the stationary case one should write:

$$\text{(Stationary case)} \quad e_\nu(\mathrm{P}) = \int_\Omega e_\nu(\mathrm{P}, \boldsymbol{\omega})\, d\omega; \; p_\nu(\mathrm{P}) = \int_\Omega p_\nu(\mathrm{P}, \boldsymbol{\omega})\, d\omega. \tag{77'}$$

The computation of the term $[\partial u_\nu(\mathrm{P}, t)/\partial t]_{\mathrm{kin}}$ is less immediate. It cannot be done, as in Subsection 3.1, by considering the photons of a *single* direction $\boldsymbol{\omega}$. Instead, let us consider the same volume element (dV) as the one used, in Subsection 1.1, for the definition of the non-directional density $u_\nu(\mathrm{P}, t)$, and let Q denote a point *on* the frontier (S) of (dV). It is thus assumed that we are able to distinguish, as through a microscope, the details of the surface (S), in spite of the smallness of (dV). The unit outward normal to (S) at Q is denoted by $\mathbf{n}$ and a surface element of (S), surrounding Q, is denoted by (dA_n).

From the point of view of the theory of radiative transfer (see Subsection 3.3) the net flux $\pi F_\nu(\mathrm{Q}, \mathbf{n}, t)$ at the point Q represents, according to the role played by $\mathbf{n}$ with respect to (S), *the excess* of the outward flow over the inward flow of energy per *unit* area of (S) near Q (and, more trivially, per unit frequency interval, and per unit time).

For *the whole* of the element (dA_n) the corresponding excess is therefore equal to $\pi F_\nu(\mathrm{Q}, \mathbf{n}, t)\, dA_n$, which – on replacing $\pi F_\nu(\mathrm{Q}, \mathbf{n}, t)$ by its expression (74) – finally is given by $[\pi \mathbf{F}_\nu(\mathrm{Q}, t) \cdot \mathbf{n}]\, dA_n$.

For *the whole of the surface* (S) the excess of the inward flow of energy over the outward flow of energy per unit frequency interval and per *unit time* (note the reversal of the sign!) will be given by $\int_{(\mathrm{S})} - [\pi \mathbf{F}_\nu(\mathrm{Q}, t) \cdot \mathbf{n}]\, dA_n$.

Let us now take into account the definition of $u_\nu(\mathrm{P}, t)$ – in Subsection 1.1 – as the plateau value when (dV) tends to zero (a process we shall denote here by $\overline{\overline{\mathrm{p.v.}}}$) of the energy of photons of *all directions* (in unit frequency interval near ν) contained in the *unit of volume* of (dV) at time t.

The kinetic variation of $u_\nu(\mathrm{P}, t)$ per unit time, caused by the excess of the inward flow over the outward flow through (S) computed above, will thus be given by

$$\left[\frac{\partial u_\nu(\mathrm{P}, t)}{\partial t}\right]_{\mathrm{kin}} \overline{\overline{\mathrm{p.v.}}} -\frac{1}{dV} \int_{(\mathrm{S})} [\pi \mathbf{F}_\nu(\mathrm{Q}, t) \cdot \mathbf{n}]\, dA_n. \tag{78}$$

From the point of view of the usual vector analysis, as pointed out in Subsection 3.3.1.1, the integral at the right-hand side of Equation (78) represents the *outward flux* of the vector field $\pi \boldsymbol{F}_\nu(\mathrm{Q}, t)$ through the surface (S). We can thus rewrite Equation (78) as

$$\left[\frac{\partial u_\nu(\mathrm{P}, t)}{\partial t}\right]_{\mathrm{kin}} \overline{\overline{\mathrm{p.v.}}} -\frac{\text{outward flux of } \pi \mathbf{F}_\nu \text{ through (S)}}{dV}. \tag{78'}$$

This shows, according to the usual definition of *the divergence* of a vector field (applied in the Stocke's theorem), that $[\partial u_\nu(\mathrm{P}, t)/\partial t]_{\mathrm{kin}}$ is simply given by

$$\left[\frac{\partial u_\nu(\mathrm{P}, t)}{\partial t}\right]_{\mathrm{kin}} = -\operatorname{div} [\pi \mathbf{F}_\nu(\mathrm{P}, t)]. \tag{78''}$$

On totalizing this kinetic effect and the effects of interactions we obtain:

$$\boxed{\frac{\partial u_\nu(\mathrm{P},t)}{\partial t} = -\operatorname{div}[\pi \mathbf{F}_\nu(\mathrm{P},t)] - p_\nu(\mathrm{P},t) + e_\nu(\mathrm{P},t)}\;. \tag{79}$$

Equation (79) is called the '*equation of continuity of transfer theory*', since it represents a generalization of the usual *equation of continuity*:

$$\frac{\partial \rho}{\partial t} = -\operatorname{div}(\rho \mathbf{v}), \tag{79'}$$

This statement is explained in Subsection 3.4.1.1. For the moment note that in the stationary case Equation (79) – with the same notation as in Equations (60) and (63) – takes the form:

$$\text{(Stationary case)} \quad \boxed{\operatorname{div}[\pi \mathbf{F}_\nu(\mathrm{P})] = e_\nu(\mathrm{P}) - p_\nu(\mathrm{P})}\;, \tag{80}$$

3.4.1. Remarks

3.4.1.1. The *number density* $n_\nu(\mathrm{P}, t)$ of the photons corresponding to the unit of frequency interval near the frequency ν is associated with the non-directional energy density $u_\nu(\mathrm{P}, t)$ by the obvious relation $u_\nu(\mathrm{P}, t) = h\nu\, n_\nu(\mathrm{P}, t)$.

On dividing both sides of Equation (79) by $h\nu$ and on taking into account Equation (76) we obtain the equivalent equation:

$$\frac{\partial n_\nu(\mathrm{P},t)}{\partial t} = -\operatorname{div}[n_\nu(\mathrm{P},t)\,\mathbf{v}_\nu(\mathrm{P},t)] - \frac{p_\nu(\mathrm{P},t)}{h\nu} + \frac{e_\nu(\mathrm{P},t)}{h\nu}. \tag{81}$$

Let us now return to the equation of continuity (79′) applied, as usual in fluid mechanics and in electrodynamics, to a system of identical particles either of individual mass m or of individual charge e, and let $n(\mathrm{P}, t)$ represent the *number density* of the particles of the system.

In fluid mechanics ρ usually represents the mass density $m\,n(\mathrm{P}, t)$ and in electrodynamics ρ usually represents the charge density $e\,n(\mathrm{P}, t)$ of the system.

On dividing both sides of Equation (79′) either by m or by e we obtain a little more general equation of continuity

$$\frac{\partial n(\mathrm{P},t)}{\partial t} = -\operatorname{div}[n(\mathrm{P},t)\,\mathbf{v}], \tag{81'}$$

analogous to Equation (81), but *less general* than Equation (81), since it neglects the interaction terms p_ν and e_ν.

Of course, Equation (79′) can be obtained more directly by conventional methods. However, our treatment emphasizes the relation between the concept of the net flux vector in the theory of the radiative transfer and the concept of the average velocity in statistical mechanics.

3.4.1.2. Advanced readers, familiar with general statistical mechanics, will easily recognize, after considering Equation (76), that Equation (79) represents from the point of view of the kinetic theory of fluids the *macroscopic transport equation of energy* for the photons of the radiation field.

4. Applications to the Physics of Stellar Interiors

4.1. Fick's Law

The radius of a star is always so great that one can neglect, in a first approximation, the *local* curvature of each concentric layer.

In general, the intensity $I_\nu(\mathrm{P}, \boldsymbol{\omega})$ of a stationary radiation field depends both on the position of the point P in space and on the direction $\boldsymbol{\omega}$ of the beam.

In a star it is natural to specify the direction $\boldsymbol{\omega}$ by its orientation with respect to the unit vector **N** normal to the locally plane–parallel layers. Let then denote by Θ the angle between **N** (*outward* directed) and $\boldsymbol{\omega}$

It is usually assumed that in a star the intensity $I_\nu(\mathrm{P}, \boldsymbol{\omega})$ of the radiation field, for a given value of Θ, is the same for all different values of the 'azimuthal' angle φ of $\boldsymbol{\omega}$ about N, (see Figure 1.7). Such a field is called *'axially symmetric'*.

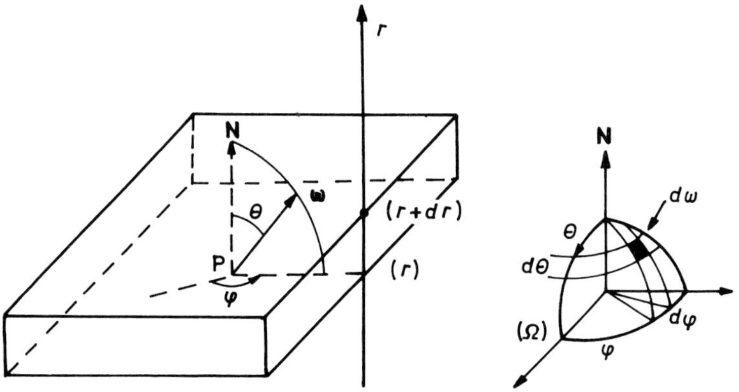

Fig. 1.7. The definition of the axial symmetry and of the corresponding $d\omega$.

The left part of Figure 1.7 represents a portion of the layer situated between the distances r and $(r + dr)$ from the center of the star. The angle Θ can take all values from 0 to π. The azimuthal angle φ is reckoned from an arbitrary fixed direction in the plane normal to N. The right part of the figure illustrates the computation of the element of solid angle $d\omega$ in the presence of an axial symmetry. On introducing the usual notation

$$\mu = \cos\Theta = (\boldsymbol{\omega}\cdot\mathbf{N}), \tag{82}$$

we find:

$$d\omega = |(\sin\Theta\, d\varphi)\, d\Theta| = |(d\mu)\, d\varphi|. \tag{83}$$

Again because of the axial symmetry, the integrations with respect to $\boldsymbol{\omega}$ over the unit sphere (Ω) centred at P can now be performed replacing $\int_0^{2\pi} \ldots d\varphi$ by 2π and integrating with respect to μ from $\mu = -1$ to $\mu = +1$. This takes into account the fact that $d\omega$ is essentially positive, whereas $d\mu$ is negative when Θ varies from 0 to π.

Thus, we can indicate the integrations with respect to $\boldsymbol{\omega}$ by the following *integral operators*:

$$\int_{\Omega} \ldots d\omega = 2\pi \int_{-1}^{+1} \ldots d\mu, \tag{84}$$

and

$$\int_{\Omega} \ldots (\boldsymbol{\omega} \cdot \mathrm{N}) d\omega = 2\pi \int_{-1}^{+1} \ldots \mu \, d\mu. \tag{85}$$

Note finally that in a star the intensity I_ν of the radiation field depends on the point P only through a single parameter: the 'level' of P, i.e. the distance r of P from the center of the star.

We consider below only *stationary* situations. Therefore we can omit any explicit mention of the variable t. However, as already pointed out above, one must remember that all intensities and fluxes are defined per *unit of the interval of time* (i.e. per second).

Thus the *physical quantity* $I_\nu(\mathrm{P}, \boldsymbol{\omega})$ can be described in a star by some essentially positive *mathematical function* $I_\nu(r, \mu)$ of the variables r and μ. Similarly, the physical quantity $\pi F_\nu(\mathrm{P}, \mathrm{N})$ can be described by some mathematical function $\pi F_\nu(r)$ of a single variable r. In both cases, ν plays the role of a parameter.

On applying the operator (85) to Equation (71) we obtain:

$$\pi F_\nu(r) = 2\pi \int_{-1}^{+1} I_\nu(r, \mu) \mu \, d\mu. \tag{86}$$

This shows that we can divide both sides of Equation (86) by π and obtain a more simple expression for $F_\nu(r)$. The factor π was introduced in the definition of the *net flux* (in Subsection 3.3) precisely in view of this simplification.

According to the definition (Section 3) of the elementary displacement ds_ω along the direction $\boldsymbol{\omega}$, this displacement is always positive. On the other hand, dr and μ are both positive when μ is positive, and are both negative when μ is negative. Thus, quite generally:

$$ds_\omega = \frac{dr}{\mu}. \tag{87}$$

Therefore we can put Equation (60), i.e. the equation of transfer in the stationary case, into the form:

$$\mu \frac{\partial I_\nu(r, \mu)}{\partial r} = -\Sigma_\nu I_\nu(r, \mu) + e_\nu(\mathrm{P}, \boldsymbol{\omega}). \tag{88}$$

In normal stars the temperature gradient is very small. Thus, for instance, the radius $R_\odot$ of the Sun is equal to 7×10^{10} cm, whereas the temperature varies from ≈ 6000 K at the surface up to $\approx 15 \times 10^6$ K at the center; hence an average gradient of about 2×10^{-3} K cm^{-1}.

Therefore, the hypothesis of local thermodynamic equilibrium (LTE) is acceptable as a first approximation, and we can apply Equation (45). This gives:

$$\boxed{\mu \frac{\partial I_\nu(r,\mu)}{\partial r} = -\Sigma_\nu I_\nu(r,\mu) + \Sigma_\nu B_\nu(T)} \quad . \tag{89}$$

Note that here the coefficient of removal Σ_ν can depend on r, since the nature and the state of the layer (r, dr) can depend on r. On the other hand, in a spherically symmetrical star the temperature T depends only on r and is described by the mathematical function $T(r)$.

Let us apply the integral operator (85) to Equation (89). Obviously, the derivation with respect to r and the integration with respect to μ commute. Therefore, if we apply Equation (86) we obtain:

$$2\pi \frac{\partial}{\partial r} \int_{-1}^{+1} I_\nu(r,\mu)\mu^2 \, d\mu = -\pi F_\nu(r)\Sigma_\nu, \tag{90}$$

(the isotropic term proportional to $B_\nu(T)$ gives an integral equal to zero!).

By analogy with the notations introduced above we can write:

$$u_\nu(\mathrm{P}, \boldsymbol{\omega}) = u_\nu(r,\mu) \quad \text{and} \quad u_\nu(\mathrm{P}) = u_\nu(r). \tag{91}$$

Thus Equation (9) takes the form:

$$I_\nu(r,\mu) = c u_\nu(r,\mu), \tag{92}$$

and the integral at the left-hand side of Equation (90) is transformed into $c\int_{-1}^{+1} u_\nu(r,\mu)\mu^2 \, d\mu$.

In a first approximation, we can replace in this last integral $u_\nu(r,\mu)$ by its *average* value $(1/4\pi)u_\nu(r)$. The errors introduced by this substitution are reduced by the presence, under the integration sign, of the factor μ^2, which is very small within the most part of the integration limits $(-1, +1)$.

Finally, on solving Equation (90) for $\pi F_\nu(r)$ we obtain:

$$\boxed{\pi F_\nu(r) = -\frac{c}{3\Sigma_\nu} \frac{\partial u_\nu(r)}{\partial r}} \quad . \tag{93}$$

This relation represents the transposition, for the astrophysical case, on the well known *Fick's law* in the theory of gaseous diffusion.

4.2. THE EQUATION OF CONTINUITY FOR STELLAR INTERIORS

In the particular case of stellar interiors, the equation of continuity (79) can be considerably simplified. Indeed, we are dealing here with locally *plane-parallel layers*; our radiation field is *axially symmetric*; and we assume, as usual, a *stationary state*.

However, from the pedagogical point of view, it is never advisable to use a suitable simplification of a complicated general relation, when the result can be obtained directly from the corresponding assumptions.

Therefore, let us consider (Subsection 4.1) a cylindrical volume element (dV), (see the left part of Figure 1.7). Its axis is parallel to the line going from the center O to some point P of the star. The point P within (dV) is at a distance r from the center. Both bases of the cylinder (dV), at the levels r and $(r + dr)$ respectively, are supposed to be of *unit area*.

Let us take into account the photons of *all directions*, in the unit of frequency interval near the frequency ν, and consider, for (dV), the energy balance between the 'import' and the 'export' of these photons per second. Let us call '*the net import*' the difference (import–export).

The *net flux* at the level r with respect to the direction **N** being denoted again by $\pi F_\nu(r)$, we see that:

The *net import* through the base at the level $(r) = \pi F_\nu(r)$;
The *net import* through the base . . . $(r + dr) = -\pi F_\nu(r + dr)$.

Therefore the *total net import* for (dV) – on taking into account the fact that because of the axial symmetry of the field the balance through the frontiers parallel to OP is zero – will be given by $[\pi F_\nu(r) - \pi F_\nu(r + dr)]$, i.e. by $-\pi(\partial F_\nu(r)/\partial r)\,dr$.

Since, 'by construction' $dV = 1\,dr$, we finally find:

$$\left[\frac{\partial u_\nu(r)}{\partial t}\right]_{\text{kin}} = \text{'net import' per cm}^3\text{, per second} = -\pi\frac{\partial F_\nu(r)}{\partial r}. \tag{94}$$

This very simple relation represents the particular form of the general equation (78″) for the case of normal stellar interiors. The corresponding particular form of the general equation of continuity of transfer theory (80) is now obviously:

$$\pi\frac{\partial F_\nu(r)}{\partial r} = e_\nu(r) - p_\nu(r). \tag{95}$$

The assumption of local thermodynamic equilibrium (LTE), and the application of Equations (77′) and (45′) corresponding to the stationary case, yield:

$$e_\nu(r) = e_\nu(\text{P}) = \Sigma_\nu\,B_\nu(T)\int_\Omega d\omega = 4\pi B_\nu(T)\Sigma_\nu\,. \tag{96}$$

On the other hand, Equations (77′) and (37′) together with the relations corresponding to Equations (9) and (2) in the stationary case, yield:

$$p_\nu(r) = p_\nu(\text{P}) = \Sigma_\nu\int_\Omega I_\nu(\text{P}, \boldsymbol{\omega})\,d\omega = \Sigma_\nu\int_\Omega cu_\nu(r, \mu)\,d\omega$$

$$= \Sigma_\nu\,cu_\nu(r). \tag{97}$$

On solving Equation (95) for $u_\nu(r)$ we obtain:

$$u_\nu(r) = \frac{4\pi}{c}B_\nu(T) - \frac{\pi}{c\Sigma_\nu}\frac{\partial F_\nu(r)}{\partial r}. \tag{98}$$

On comparing with Equations (38) and (96) we find that we can write:

$$u_\nu(r) = u_\nu^*(T)(1-\epsilon) \quad \text{where} \quad \epsilon = \frac{\pi}{e_\nu(r)} \frac{\partial F_\nu(r)}{\partial r}. \tag{99}$$

This shows, rather unexpectedly, that the assumption of LTE is *not* strictly equivalent to an equality between the local non-directional energy density $u_\nu(r)$ and the local non-directional energy density $u_\nu^*(T)$ of a radiation field inside a black body at the corresponding temperature $T = T(r)$.

Nevertheless, according to Equation (99), when ϵ is sufficiently small one can, by *an additional approximation*, assimilate $u_\nu(r)$ with $u_\nu^*(T)$. It is usually admitted that in a real star we actually can neglect ϵ in (99), and this assumption is confirmed by a numerical analysis of stellar models. We can then write:

$$u_\nu(r) = u_\nu^*(T) = \frac{4\pi}{c} B_\nu(T). \tag{100}$$

On integrating in frequency, and on taking into account the classical expression for the non-directional *integrated* energy *density* of the radiation of a black body at temperature T (see for instance Kittel, 1969, p. 256) we obtain:

$$u(r) = u^*(T) = aT^4 = \frac{4\pi}{c} B(T) = \frac{4\pi}{c}\frac{\sigma}{\pi} T^4, \tag{101}$$

where

$$a = 7.56 \times 10^{-15}\,\text{c.g.s.} \quad \text{and} \quad \sigma = 5.67 \times 10^{-5}\,\text{c.g.s.}. \tag{101'}$$

4.2.1. Remarks

4.2.1.1. A part of Equation (101) is equivalent to $B(T) = (\sigma/\pi)T^4$. The strange presence of the divisor π in this relation has the following origin.

The constant σ is introduced, historically, by the Stefan's law:

$$\pi F^{++}(T) = \sigma T^4. \tag{101''}$$

This law states that the *positive* integrated flux $\pi F^{++}(T)$ *emerging* from a plane hole opened in *a black body* at the temperature T is proportional to T^4 and the constant of proportionality is usually called σ.

Let us then assimilate in Equation (69) the normal **n** with a normal to the plane of the hole, and introduce the usual notations: μ for $(\boldsymbol{\omega} \cdot \mathbf{n})$ and $B_\nu(T)$ for the intensity of the black body radiation field (see Subsection 2.5).

After integration in frequency of both sides of Equation (69) applied to a black body radiation, we obtain:

$$\pi F^{++} = \int_\Omega B(T)\mu\, d\omega = \int_0^1 B(T)\mu 2\pi\, d\mu = 2\pi B(T)[\mu^2/2]_0^1 = \pi B(T). \tag{102}$$

If we compare this result with Stefan's law (101″) we find indeed:

$$B(T) = (\sigma/\pi)T^4. \tag{102'}$$

4.2.1.2. The fundamental Equation (98) can also be obtained by an application of the integral operator (84) to the equation of transfer (89). This last procedure emphasizes the difference between *the Fick's law* (93) and *the equation of continuity* (98), which are often confused.

4.3. THE OPACITY DEFINED BY ROSSELAND'S METHOD. THE NET OUTPUT OF INTEGRATED RADIATION ACROSS A SPHERE OF RADIUS r

Let L_r represent the *net* output of *integrated* radiation – in ergs per second when c.g.s. units are used – across a sphere of radius r inside a star. In other words, let L_r represent *the difference*, per second, between the energy of photons of *all frequencies* emerging from this sphere and the energy of photons entering from outside into this sphere.

Let us take into account the definition of the net flux (with respect to the direction of the outward normal) at the level r of a star and the notation $\pi F_\nu(r)$ introduced in Subsections 3.3 and 4.1. Then, $\pi F(r)$ denoting $\pi F_\nu(r)$ integrated in frequency, we obtain:

$$L_r = 4\pi r^2 \int_0^\infty \pi F_\nu(r)\, d\nu = 4\pi r^2 \pi F(r). \tag{103}$$

On replacing in Equation (93) $\partial/\partial r$ by $(dT/dr)(\partial/\partial T)$ and $u_\nu(r)$ by its approximate value (100), we find after integration in frequency:

$$\pi F(r) = -\frac{4\pi}{3}\frac{dT}{dr}\int_0^\infty \frac{1}{\Sigma_\nu}\frac{\partial B_\nu(T)}{\partial T}\, d\nu. \tag{104}$$

Following the Norwegian astrophysicist S. Rosseland, we can write:

$$\int_0^\infty \frac{1}{\Sigma_\nu}\frac{\partial B_\nu(T)}{\partial T}\, d\nu = \frac{1}{\bar{k}_R}\int_0^\infty \frac{\partial B_\nu}{\partial T}\, d\nu = \frac{1}{\bar{k}_R}\frac{\partial}{\partial T}\int_0^\infty B_\nu(T)\, d\nu = \frac{1}{\bar{k}_R}\frac{dB(T)}{dT}. \tag{105}$$

The *Rosseland's opacity* (also called the *Rosseland's mean*) $\bar{k}_R$ is thus defined as the 'harmonic' mean of Σ_ν, i.e. as an average value of $1/\Sigma_\nu$.

On using Equation (101) we obtain:

$$L_r = 4\pi r^2 \frac{4\pi}{3}\left(-\frac{dT}{dr}\right)\left(\frac{4\sigma}{\pi}T^3\right)\frac{1}{\bar{k}_R}. \tag{106}$$

It will be shown in Subsection 4.5 (Equation (147)), that $\bar{k}_R$ is given for the current models of the Sun (and for similar normal stars) by

$$\bar{k}_R = (2.40\times 10^{23})\rho^{1.75}(1+X)^{0.75}T^{-3.5}. \tag{107}$$

Here $\rho = \rho(r)$ represents the mass density (in grams per cm^{-3}) and T represents the temperature of the stellar mixture. X is the mass in grams of hydrogen contained, before ionization, in each gram of the stellar mixture, (X is always less than or equal to 1).

4.4. THE RADIATION PRESSURE AS A FUNCTION OF THE TEMPERATURE

In most textbooks on electrodynamics and in most textbooks on thermodynamics (see for instance Kittel, 1969, p. 257) it is shown that the radiation pressure P_{rad} exerted by an isotropic stationary radiation field is equal to one third of the non-directional integrated energy density $u(Q)$ of the field at the point Q.

However, most of the proofs of this relation are rather abstract and indirect. Therefore, it can be worthwhile to use a different approach based on an application of the concepts introduced in Section 1.

Let us consider (see Figure 1.8) a volume element (dV) of a *material medium* placed in a stationary radiation field of intensity $I_\nu(Q, \boldsymbol{\omega})$, where Q denotes (as in Subsection 3.4) a point *on* the frontier (S) of (dV).

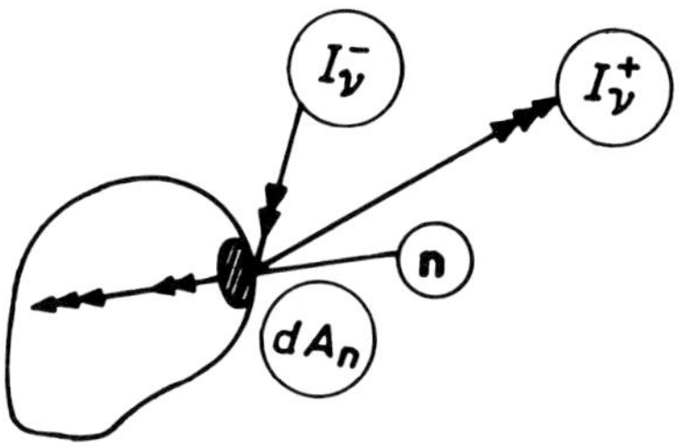

Fig. 1.8. The active pressure created by the incident photons, and the reactive pressure created by the emergent photons.

The *unit* outward normal to (S) at Q is denoted by **n** and the surface element of (S) surrounding Q is denoted by (dA_n). As usual, $\boldsymbol{\omega}$ denotes the *unit* vector describing the direction of a beam $(\boldsymbol{\omega}, d\omega)$ of photons $(\nu, d\nu)$, and μ is defined by

$$\mu = (\boldsymbol{\omega} \cdot \mathbf{n}). \tag{108}$$

In order to emphasize that μ is *positive* for *emergent* photons – i.e. for photons which leave (dV) through (dA_n) – we denote the corresponding intensity by I_ν^+ or more precisely by $I_\nu^+(Q, \boldsymbol{\omega})$.

Similarly, since μ is *negative* for the *incident* photons – i.e. for the photons which strike (dV) from outside – we denote the corresponding intensity by I_ν^- or by $I_\nu^-(Q, \boldsymbol{\omega})$.

From a dynamical point of view, the emergent photons act upon (dV) as a gas ejected from the nozzle of a rocket: they produce *a 'recoil'*. The *normal* component (normal to dA_n) of this *'reactive pressure'* is directed towards the interior of (dV), i.e. along the direction $(-\mathbf{n})$.

Similarly, the incident photons act upon (dV) as projectiles bombarding (dV) from outside. The normal component of this *'active pressure'* is *also* directed towards the interior of (dV), and therefore *adds up* to the normal component of the reactive pressure.

It is well known (see for instance Kourganoff, 1969, p. 184) that the active dynamical pressure produced by incident particles of any nature – atoms, molecules or photons – and the reactive dynamical pressure produced by emergent particles are given by

$$P_a = \text{'active pressure'} = p_i D_i; \quad P_r = \text{'reactive pressure'} = p_e D_e. \tag{109}$$

In these relations D_i and D_e denote respectively the *number* of particles incident or emergent *per cm² of* (S) *per second*; and p_i and p_e denote respectively *the modulus of the normal component* of the *momentum* of incident or emergent particles.

According to the discussion leading to Equation (68) (Subsection 3.3) – the energy brought per cm² of (dA_n) per second by the incident photons $(\boldsymbol{\omega}, d\omega; \nu, d\nu)$ is equal to $I_\nu^-(\mathrm{Q}, \boldsymbol{\omega})(\boldsymbol{\omega}\cdot\mathbf{n})\,d\omega\,d\nu$. The energy of each photon $(\nu, d\nu)$ being equal to $h\nu$, the corresponding *number* of incident photons, i.e. the corresponding value of D_i, will be given by

$$D_i = I_\nu^-|\mu|\,d\omega\,d\nu/h\nu. \tag{110}$$

On the other hand, the modulus of the momentum of each photon $(\nu, d\nu)$ being given by

$$\frac{h\nu}{c}, \tag{111}$$

the modulus of the normal component p_i of their momentum will be given by

$$p_i = \frac{h\nu}{c}|\mu|. \tag{112}$$

Whatever the sign of μ we have: $|\mu|^2 = \mu^2$. Therefore, on combining Equations (110) and (112), the element $P_{\mathrm{rad},i}$ $(\boldsymbol{\omega}, d\omega; \nu, d\nu)$ of the radiation pressure produced by the *incident* photons $(\boldsymbol{\omega}, d\omega; \nu, d\nu)$ can be written as:

$$P_{\mathrm{rad},i}(\boldsymbol{\omega}, d\omega; \nu, d\nu) = \frac{1}{c} I_\nu^- \mu^2\,d\omega\,d\nu. \tag{113}$$

Obviously, a similar treatment will give, for the *emergent* photons: the element $P_{\mathrm{rad},e}$:

$$P_{\mathrm{rad},e}(\boldsymbol{\omega}, d\omega; \nu, d\nu) = \frac{1}{c} I_\nu^+ \mu^2\,d\omega\,d\nu. \tag{114}$$

Since the pressures produced by the incident and the emergent photons are *additive*, the element $P_{\mathrm{rad}}(\nu, d\nu)$ of the radiation pressure produced by photons $(\nu, d\nu)$ of *all directions*, will be given by an integration of $P_{\mathrm{rad}}(\boldsymbol{\omega}, d\omega; \nu, d\nu)$ – given by Equations (113) and (114) – over the whole of the unit sphere (Ω), i.e. by

$$P_{\mathrm{rad}}(\nu, d\nu) = \frac{1}{c}\int_\Omega I_\nu(\mathrm{Q}, \boldsymbol{\omega})\mu^2\,d\omega\,d\nu = d\nu \int_\Omega u_\nu(\mathrm{Q}, \boldsymbol{\omega})\mu^2\,d\omega, \tag{115}$$

Similarly, the element $P_{\mathrm{rad}}(\boldsymbol{\omega}, d\omega)$ of the radiation pressure produced by photons $(\boldsymbol{\omega}, d\omega)$ of *all frequencies* will be given – on taking into account Equation (3) applied to the stationary case – by

$$\begin{aligned} P_{\mathrm{rad}}(\boldsymbol{\omega}, d\omega) &= \frac{1}{c}\int_0^\infty I_\nu(\mathrm{Q}, \boldsymbol{\omega})\mu^2\,d\omega\,d\nu = \left[\int_0^\infty u_\nu(\mathrm{Q}, \boldsymbol{\omega})\,d\nu\right]\mu^2\,d\omega \\ &= u(\mathrm{Q}, \boldsymbol{\omega})\mu^2\,d\omega, \end{aligned} \tag{116}$$

Finally, the radiation pressure P_{rad} produced at Q by the photons of *all frequencies* and *all directions* of a *stationary* field will be given by the *rigorous* relation:

$$\boxed{P_{\text{rad}} = \int_{\Omega} u(\mathrm{Q}, \boldsymbol{\omega})\mu^2 d\omega} \quad . \tag{117}$$

In this relation $u(\mathrm{Q}, \boldsymbol{\omega})$ represents the *integrated directional energy* density of the radiation field at the point Q and in direction $\boldsymbol{\omega}$.

For an *isotropic* stationary radiation field, as inside a black body, the integrated directional density $u(\mathrm{Q}, \boldsymbol{\omega})_{\text{iso}}$ can be expressed as a function of the integrated *non-directional* energy density $u(\mathrm{Q})$ by

$$u(\mathrm{Q}, \boldsymbol{\omega})_{\text{iso}} = \frac{u(\mathrm{Q})}{4\pi} . \tag{118}$$

On introducing the integral operator (84), we obtain for the radiation pressure created by an isotropic stationary field, the rigorous relation:

$$\text{(Isotropic field)} \quad P_{\text{rad}} = \frac{u(\mathrm{Q})}{4\pi} 2\pi \int_{-1}^{+1} \mu^2 d\mu = \tfrac{1}{3}u(\mathrm{Q}). \tag{119}$$

Thus, our treatment shows that the classical relation recalled at the beginning of the present Subsection, though rigorously true for an *isotropic* stationary field, represents in a more general case of a *non-isotropic* (anisotropic) stationary radiation field, only *a first approximation*.

Indeed, to derive in the general case a relation similar to Equation (119), we must take into account the '*smoothing effect*' of the factor μ^2 in the integration with respect to μ, and replace in Equation (117) $u(\mathrm{Q}, \boldsymbol{\omega})$ by its *average* value, i.e. by the right-hand side of Equation (118). This yields the *approximate* relation:

$$\text{(General field)} \quad P_{\text{rad}} \approx \tfrac{1}{3}u(\mathrm{Q}). \tag{120}$$

In stellar interiors, in addition to the approximation leading to Equation (120), we can use the approximation expressed by Equation (101). This yields the usual approximate expression for the radiation pressure at the level where the local temperature is equal to T:

$$\boxed{P_{\text{rad}} \approx \tfrac{1}{3}aT^4 \quad (a = 7.56 \times 10^{-15}\,\text{c.g.s.})} \quad . \tag{121}$$

Note that this result is quite rigorous when it is applied to the radiation field inside a black body at the temperature T.

4.5. The Opacity as a Function of Density, Temperature and Composition of a Stellar Mixture

Let us drop the subscript R in the definition (105) of the opacity and call it simply $\bar{k}$ for shortness sake. We can then write:

$$\frac{1}{\bar{k}} \int_0^\infty \frac{\partial B_\nu}{\partial T} d\nu = \int_0^\infty \frac{1}{\Sigma_\nu} \frac{\partial B_\nu}{\partial T} d\nu. \tag{122}$$

It was shown (Section 2) that usually the coefficient of removal Σ_ν depends both on the *composition* and on the *physical state* of the material medium.

On the other hand, it was pointed out (Subsection 2.4.1.1) that continuous removals can be associated with *two* types of transitions: the bound–free and the free–free transitions.

In stellar interiors bound–free transitions occur when elements of atomic weight A greater than about 16 are still incompletely ionized (in the outer layers of the star), whereas free–free transitions occur when the ionization is complete.

In this last case the continuous removals correspond mainly to modification of non-quantized energy states – hyperbolic orbits in the old quantum theory – of the free electrons moving in the electric field of protons and α-particles left out by the ionization of H and He atoms.

The very complex, and somewhat uncertain, theoretical determination of Σ_ν is rather similar for both types of transitions.

However, the detailed calculations will be given below only for the free–free transitions, because in this case the *physical meaning* of the corresponding equations appears more clearly. The reader who feels frustrated by this selection is referred to more advanced textbooks (see for instance Aller, 1953, Ch. 7).

A classical result obtained, in 1924, by H. A. Kramers with help of the 'correspondence principle' of Niels Bohr, gives for free–free transitions an *elementary* coefficient of continuous removal a_ν^{ff} defined as a statistical mean for a single electron in the field of a proton:

$$a_\nu^{\mathrm{ff}} = C_{\mathrm{ff}} v^{-1} \nu^{-3} \quad \text{(for protons; charge 1 e)}, \tag{123}$$

or in the field of an α-particle:

$$a_\nu^{\mathrm{ff}} = 4C_{\mathrm{ff}} v^{-1} \nu^{-3} \quad \text{(for } \alpha\text{-particles; charge 2 e)}; \tag{124}$$

present in a cm^3 of the mixture.

In Equations (123) and (124) C_{ff} is a constant coefficient, which depends on (e, c, h and m_e), and v is the speed of the electron removing the photon $h\nu$.

The formal analogy between the greater efficiency of *low* energy photons and slow *electrons* in this type of interaction and the well known greater efficiency of *slow* neutrons in interactions leading to nuclear fission is remarkable.

Let us now consider a stellar mixture of a given overall mass density ρ containing X grams of H and Y grams of He *per gram* of mixture.

If N_A denotes the Avogadro's number, each cm^3 of the mixture – after complete ionization – will contain $X\rho N_A$ protons and $\frac{1}{4}Y\rho N_A$ α-particles, (see Chapter 2, Section 5, Table 2.1).

Thus the contribution to Σ_ν from *each single* free electron (of rest mass m_e) and from *all* protons and α-particles, present in each cm^3, will take the form:

$$C'\rho v^{-1} \nu^{-3}, \tag{125}$$

where C' is a constant equal to:

$$C' = (X + Y) N_A C_{\mathrm{ff}}. \tag{126}$$

Equation (125) shows that the contribution to Σ_ν from each of the free electrons present in the considered cm^3 depends, for given frequency ν and mass density ρ, only on the reciprocal $1/v$ of its individual speed v. Thus the total contribution to Σ_ν corresponding to the n_e electrons present in the cm^3 will be proportional to the integral of $(1/v)$ weighted by the number (dn_e) of electrons present in each cm^3 having their speeds between v and $(v + dv)$; i.e. to $\int_0^\infty (1/v)\, dn_e$.

In the state of thermodynamic equilibrium dn_e is given by Maxwell's distribution:

$$dn_e = C''(kT)^{-3/2} n_e\, e^{-W/kT} v^2\, dv. \qquad (127)$$

In Equation (127) T is the equilibrium temperature of the system, $W = \frac{1}{2} m_e v^2$ is the kinetic energy of each electron of speed (v, dv) and C'' is a constant given by

$$C'' = 4\pi m_e^{3/2} (2\pi)^{-3/2}. \qquad (128)$$

Thus we find that the total contribution to Σ_ν of all free electrons, of different speed, present in each cm^3 will be proportional to

$$\int_0^\infty \frac{1}{v}\, dn_e = C'' n_e (kT)^{-3/2} \int_0^\infty e^{-W/kT}\, v\, dv. \qquad (129)$$

Let us introduce the variable x defined by

$$x = \frac{W}{kT}. \qquad (130)$$

Then $v\, dv$ will be given by

$$v\, dv = \frac{1}{m_e} kT\, dx. \qquad (131)$$

The integral at the right-hand side of Equation (129) is thus given by

$$\frac{1}{m_e} kT \int_0^\infty e^{-x}\, dx = \frac{1}{m_e} kT. \qquad (132)$$

Finally:

$$\int_0^\infty \frac{1}{v}\, dn_e = C'' n_e \frac{1}{m_e} (kT)^{-1/2}. \qquad (133)$$

Combining this result with the coefficient of $1/v$ in Equation (125) we find that the coefficient of continuous removal for the free–free transitions Σ_ν^{ff} is given by:

$$\Sigma_\nu^{\mathrm{ff}} = C' C'' \rho n_e m_e^{-1} \nu^{-3} (kT)^{-1/2} \text{ per cm}^3. \qquad (134)$$

On introducing a new constant C''' defined by

$$C' C'' = C''' m_e, \qquad (135)$$

and a new variable u defined by

$$\mathrm{u} = h\nu / kT, \qquad (135')$$

we finally obtain:

$$\boxed{\frac{1}{\Sigma_\nu^{\mathrm{ff}}} = \frac{\mathrm{u}^3 (kT)^{3.5}}{C''' n_e \rho h^3}}\,. \qquad (136)$$

Let us now return to Equation (122). The variable u introduced by Equation (135′) must be considered as representing physically the *frequency* ν for a given constant value of T. It is therefore necessary to calculate $\partial B_\nu(T)/\partial T$ which appears in Equation (122) *before* replacing ν by its expression as a function of u.

Planck's function $B_\nu(T)$ is given by Equation (43), viz.

$$B_\nu(T) = \frac{2h\nu^3}{c^2}\frac{1}{(e^{h\nu/kT}-1)}, \tag{137}$$

which yields:

$$\frac{\partial B_\nu}{\partial T} = -\frac{2h\nu^3}{c^2}(e^{h\nu/kT}-1)^{-2}\, e^{h\nu/kT}(-h\nu k^{-1}T^{-2}). \tag{138}$$

Henceforth we can reintroduce the variable u and rewrite Equation (138) in a more condensed form:

$$\frac{\partial B_\nu}{\partial T}d\nu = \varphi(T)\mathrm{u}^3(e^{\mathrm{u}}-1)^{-2}\mathrm{u}e^{\mathrm{u}}\,d\mathrm{u} = \varphi(T)f(\mathrm{u})\,d\mathrm{u}. \tag{139}$$

The function $\varphi(T)$ depends – besides the universal constants h, c and k – only on the temperature T, whereas the function $f(\mathrm{u})$ depends only on u.

Let us now substitute into the definition (122) of $\bar{k}$ the expressions for $(1/\Sigma_\nu^{\mathrm{ff}})$ and $\partial B_\nu(T)/\partial T$ given by Equations (136) and (139). The function $\varphi(T)$ goes outside the integral *on both sides* of Equation (122) and cancels out.

Finally we are left with equation

$$\frac{1}{\bar{k}_{\mathrm{ff}}}\int_0^\infty f(\mathrm{u})\,d\mathrm{u} = \frac{(kT)^{3.5}}{C'''n_e\rho h^3}\int_0^\infty \mathrm{u}^3 f(\mathrm{u})\,d\mathrm{u}, \tag{140}$$

where $\bar{k}_{\mathrm{ff}}$ denotes the part of $\bar{k}$ corresponding to free–free transitions.

Clearly each of the two integrals in Equation (140) represents a numerical constant. Thus, introducing a new constant C_0, we can write:

$$\frac{1}{\bar{k}_{\mathrm{ff}}} = \frac{(kT)^{3.5}}{C_0 n_e\rho}. \tag{141}$$

Finally $\bar{k}_{\mathrm{ff}}$ is given by

$$\bar{k}_{\mathrm{ff}} = C_0 n_e\rho(kT)^{-3.5}. \tag{142}$$

It will be shown in Chapter 2 (see Section 5, Table 2.1) that in a completely ionized stellar mixture characterized by the composition defined above by the values of X and Y, the number density of free electrons n_e is equal to

$$n_e = \tfrac{1}{2}(1+X)\rho N_A. \tag{143}$$

Thus, introducing the constant $\bar{k}_0$ defined by

$$\bar{k}_0 = \tfrac{1}{2}C_0 N_A k^{-3.5}, \tag{144}$$

we obtain the following expression for $\bar{k}_{\mathrm{ff}}$ as a function of ρ, T and X:

$$\bar{k}_{\mathrm{ff}} = \bar{k}_0\rho^2(1+X)T^{-3.5}. \tag{145}$$

Although this expression for $\bar{k}_{\text{ff}}$ is often sufficient in a first approximation, a more thorough treatment (taking into account the fact that ionization is not total near the surface of the star) yields an *empirical* expression (see Schwarzschild, 1958):

$$\bar{k}_{\text{ff}} = (1.08 \times 10^{23})\rho^{1.75}(1 + X)^{0.75}T^{-3.5} \text{ per cm}^3, \tag{146}$$

with a little modified powers of ρ and of $(1 + X)$.

Equation (146) corresponds to the temperature range proper to the interior of the Sun.

When in the case of the Sun the bound–free transitions are taken into account the complete $\bar{k}$ keeps the same form as in Equation (146), but the numerical coefficient of course becomes greater:

$$\bar{k} = (2.40 \times 10^{23})\rho^{1.75}(1 + X)^{0.75}T^{-3.5}. \tag{147}$$

4.6. THE FINAL EXPRESSION FOR L_r

Let us return to the expression for L_r given by Equation (106). On taking into account the obvious relation (where r' represents the ratio r/R of r to the radius R of the star):

$$\left|\frac{d(\log_{10} T)}{dr'}\right| = -\frac{(0.434)}{T}\frac{dT}{dr'}, \tag{148}$$

and on replacing $\bar{k}_R = \bar{k}$ by its expression given by Equation (147), we obtain for the Sun (with $R_\odot = 6.95 \times 10^{10}$ cm):

$$L_r = L_\odot L_r' = L_\odot(6.75 \times 10^{-49})r'^2 T^{7.5}\rho^{-1.75}(1 + X)^{-0.75}\left|\frac{d \log_{10} T}{dr'}\right|. \tag{149}$$

In Equation (147) we have introduced the dimensionless ratio $L_r' = L_r/L_\odot$, between L_r and the luminosity $L_\odot$ of the Sun, equal to 3.78×10^{33} erg s^{-1}. On solving the equation (149) for $d \log_{10}(T/10^6)/dr'$ we find:

$$\boxed{-\frac{d \log_{10}(T/10^6)}{dr'} = (1.48 \times 10^3)r'^{-2}(T \times 10^{-6})^{-7.5}\rho^{7/4}(1 + X)^{3/4}L_r'}\,. \tag{150}$$

Sufficiently far from the center of the Sun, X can be replaced by its value deduced from the observations of the superficial layers, viz. $X = 0.744$. This yields:

$$\boxed{L_r' = \frac{L_r}{L_\odot} = (4.46 \times 10^{-49})r'^2 T^{7.5}\rho^{-1.75}\left(-\frac{d \log_{10} T}{dr'}\right)}\,. \tag{151}$$

Appendix: The Relations between Einstein's Coefficients

Let us consider an enclosure in thermodynamic equilibrium at the temperature T and neglect the width of the *discrete* energy levels.

For each pair of levels, j and k, there will be in this case a complete compensation (microreversibility) between upward and downward transitions.

With the notation of Section 2 we have:

$$N^{jk} = N^{kj} + N^{kj}_{\text{sp}}. \tag{A1}$$

According to Equations (23), (29) and (48) this is equivalent to

$$B_{jk}N^j u_r(P,t) = B_{kj}N^k u_r(P,t) + A_{kj}N^k. \tag{A2}$$

Let us introduce the notations:

$$\frac{B_{jk}g_j}{B_{kj}g_k} = x; \qquad \frac{A_{kj}g_k}{B_{kj}g_k} = y; \qquad \frac{8\pi h\nu^3}{c^3} = \alpha_\nu; \qquad \frac{k}{h\nu} = \beta_\nu. \tag{A3}$$

If $h\nu/kT$ is sufficiently small we can expand the exponential in Equation (44) and keep only the first two terms. We obtain in this way the well known *Rayleigh–Jeans formula*, (which *historically* antedates Equation (44) and can be established 'classically', i.e. without quantum statistics):

$$\tilde{u}^*_\nu(T) = \alpha_\nu\beta_\nu T = \frac{8\pi k}{c^3}\nu^2 T. \tag{A4}$$

On the other hand, for small values of $h\nu/kT$ Boltzmann law given by Equation (32), viz.

$$\frac{N^j}{N^k} = \frac{g_j}{g_k}e^{h\nu/kT}, \tag{A5}$$

takes the form

$$\frac{N^j/g_j}{N^k/g_k} = 1 + \frac{1}{\beta_\nu T} + \cdots. \tag{A6}$$

On dividing both sides of Equation (A2) by $B_{kj}g_k$ and on replacing u_r by $\tilde{u}^*_\nu$ defined by Equation (A4) we obtain:

$$x(N^j/g_j)\tilde{u}^*_\nu(T) = (N^k/g_k)\tilde{u}^*_\nu(T) + y(N^k/g_k). \tag{A7}$$

On dividing both sides of Equation (A7) by (N^k/g_k) and on taking into account Equations (A4) and (A6) we find:

$$x\left(1 + \frac{1}{\beta_\nu T}\right)\alpha_\nu\beta_\nu T = \alpha_\nu\beta_\nu T + y. \tag{A8}$$

Equation (A8) can be written as

$$(x-1)\alpha_\nu\beta_\nu T = y - \alpha_\nu x. \tag{A9}$$

The left-hand side of Equation (A9) is proportional to T; the right-hand side is independent of T. This relation must be therefore of the form $0 \cdot T = 0$. Hence

$$x = 1; \quad y = \alpha_\nu x \qquad \text{or} \qquad x = 1; \quad y = \alpha_\nu; \tag{A10}$$

and

$$\boxed{B_{jk} g_j = B_{kj} g_k; \qquad A_{kj} = B_{kj} \frac{8\pi h\nu^3}{c^3}} \tag{A11}$$

Equation (A11) gives Equation (34).

References

1. WORKS CITED IN THE TEXT
(Works that can be used for further study are marked by an asterisk.)
Aller, L. H.: 1953, *The Atmospheres of the Sun and Stars*, Ronald, New York.
Kittel, C.: 1969, *Thermal Physics*, Wiley, New York.
Kourganoff, V.: 1969, *Introduction to the General Theory of Particle Transfer*, Gordon and Breach, New York.
Feynman, R.: 1964, *Lectures on Physics*, (Vol. I. Sect. 40), Addison-Wesley, Reading.
Schwarzschild, M.: 1958, *Structure and Evolution of the Stars*, Princeton University Press. Princeton.

2. REFERENCES FOR FURTHER STUDY
(See first the work cited in 1. (marked by an asterisk.)
Chandrasekhar, S.: 1950, *Radiative Transfer*, Dover, New York.
Kourganoff, V.: 1964, *Basic Methods In Transfer Problems*, Dover, New York.
Sampson, D. H.: 1965, *Radiative Contributions to Energy and Momentum Transport in a Gas*, Wiley, New York.
Sobolev, V. V.: 1963, *A Treatise on Radiative Transfer*, Van Nostrand, Toronto.
Weinberg, A. M. & Wigner, E. P.: 1958, *The Physical Theory of Neutron Chain Reactors*, Univ. of Chicago Press.

CHAPTER 2

ELEMENTARY INTRODUCTION TO THE PHYSICS OF STELLAR INTERIORS

1. General Conditions of Mechanical Equilibrium

Let us consider a spherical star of radius R, in equilibrium, stratified in homogeneous concentric layers, whose center is O. We introduce an outward-directed unit vector $\mathbf{r}$ originating at O. (See Figure 2.1).

Since the pressure, the temperature and the density increase towards the center (a statement which will be justified later), the gradients of all these quantities are inward-directed, and their projections on $\mathbf{r}$ are negative.

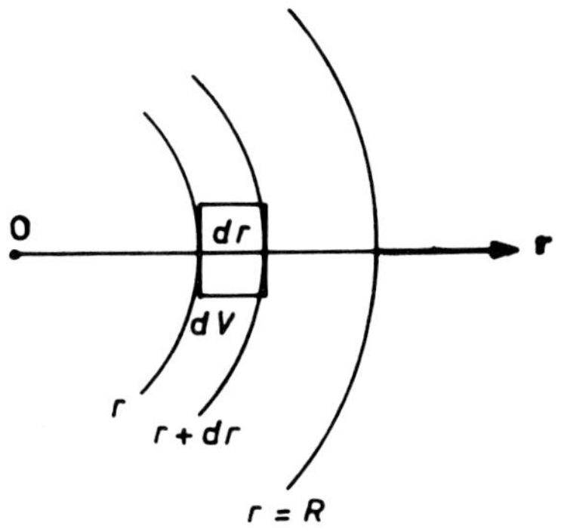

Fig. 2.1. The definition of the cylindrical element (dV).

Consider a *strictly cylindrical* volume element (dV) of cross-section dA and length dr, whose axis is $\mathbf{r}$ and whose generatrix is parallel to $\mathbf{r}$. This element (dV) is subject to a gravitational interaction *at a distance* with the rest of the star and mechanical equilibrium is guaranteed by a *local* interaction with the neighbouring elements which exert a *pressure* on (dV). The pressures on the lateral parts of the cylinder balance each other by symmetry; but the bases, located at different levels (r) and $(r + dr)$, experience different pressures because the densities and the temperatures at these levels are different. This gives a net outward-directed force which can balance the resultant of the gravitational forces attracting (dV) towards the center of the star.

In this elementary study, we naturally assume that there are no internal motions, not even stationary ones (*static* equilibrium).

Let us try to specify the conditions of this equilibrium. Note that in postulating the existence of an equilibrium state, we make it possible to determine the pressure gradient from the parameters governing the gravitational field and the density, *without calling on* a detailed analysis of the *physical processes* responsible for the pressure, i.e. the bombardment of the bases of (dV) by photons and by particles coming from the neighbouring elements.

2. The Equilibrium between the Gradient of the Total Pressure and the Gravitational Force per Unit Volume

The gravitational interaction between the star as a whole and the element (dV) located at a distance r from the center O of the star obeys the law of universal gravitation.

Now, in potential theory (a simplified version of which is given below) one proves 'Newton's theorem', according to which the *attraction exerted on an element* (dV) *inside the star is equivalent to that of the entire mass* M_r *of the part located inside the sphere* (Σ_r) *of radius* r *centered on* O; *the effect of the layers* $(r; R)$ *between* r *and* R *is zero*. We shall begin with a proof of this fundamental theorem.

2.1. Newton's Theorem. The Gravitational Force per Unit Volume

In its modern version, the proof of Newton's theorem rests on a transposition to gravitational forces of *Gauss's theorem for electrostatics*. (The historical order is obviously just the opposite.)

According to the latter theorem, given the field E created by a positive charge e_c located at C, and a closed surface (Σ_r) made up of elements (dA_r) oriented by means of outward-directed vectors $\mathbf{r}$ normal to (Σ_r), the *emergent flux* of E across (Σ_r) is given by:

$$\int_{\Sigma_r} (dA_r)(\mathbf{r} \cdot E) = \begin{bmatrix} 0 & \text{if C is outside } (\Sigma_r) \\ \dfrac{4\pi}{k} e_c & \text{if C is inside } (\Sigma_r) \end{bmatrix} \tag{1}$$

In this expression, k is the dielectric constant in Coulomb's law of repulsion

$$E = \frac{e_c}{kd^2} \mathbf{r}, \tag{2}$$

which corresponds to the law of attraction by universal gravitation

$$g = -\frac{Gm_c}{d^2} \mathbf{r}. \tag{3}$$

Here G is the constant of gravitation (6.67×10^{-8} c.g.s.); m_c the mass in grams; d the distance in cm; and g the gravitational field, i.e. the force exerted by the point-mass m_c on unit mass, in dynes per gram.

When we transpose Gauss's theorem (1), keeping in mind the change from repulsion to attraction, i.e. the fact that g is always directed towards the mass responsible for the gravitational field, we find that the emergent flux of the field g is given by

$$\int_{\Sigma_r} (dA_r)(\mathbf{r} \cdot g) = \begin{bmatrix} 0 & \text{if } m_c \text{ is outside } (\Sigma_r) \\ -4\pi G m_e & \text{if } m_e \text{ is inside } (\Sigma_r) \end{bmatrix} \tag{4}$$

Here is a *direct proof* of Equation (4) for those of our readers who may not be familiar with Equation (1). (Advanced readers may jump to p. 49).

We say that the element $(dA_r)(\mathbf{r} \cdot g)$ of the emergent flux is given in absolute value by $Gm_c(d\Omega)$, where $(d\Omega)$ is the solid angle subtended by (dA_r) at the point C.

For, let $\boldsymbol{\Omega}$ be the unit vector in the direction from (dA_r) to C. Then if C is outside (Σ_r), at C_1, we have (Figure 2,2):

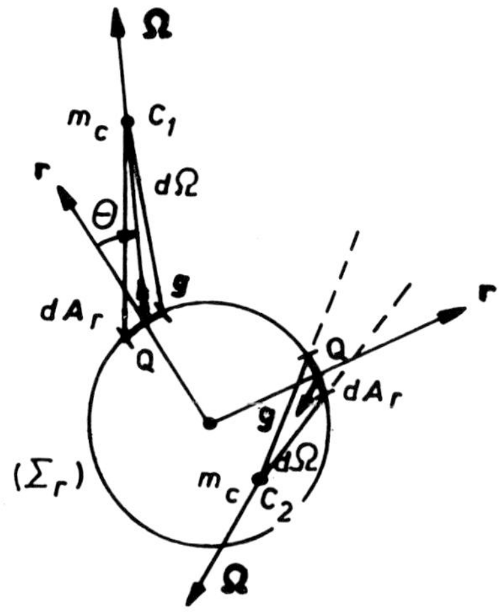

Fig. 2.2. The proof of Equation (6).

$$g = g\boldsymbol{\Omega} = + Gm_c d^{-2}\boldsymbol{\Omega}, \tag{5}$$

and, denoting the projection of (dA_r) on the plane normal to $\boldsymbol{\Omega}$ by (dA_Ω):

$$(dA_r)(\mathbf{r} \cdot g) = Gm_c(dA_r)d^{-2} \cos\theta = Gm_c d^{-2}(dA_\Omega) = Gm_c(d\Omega). \tag{6}$$

It is easily verified that this formula remains valid in absolute value when C is at C_2, inside (Σ_r).

Let us apply Equation (6) to the case in which m_c is located outside (Σ_r), dividing (Σ_r) into two parts (Σ') and (Σ'') by the curve of intersection with the cone tangent to (Σ_r) whose vertex is C_1 and whose aperture is the solid angle Ω.

Then (Figure 2.3) we have for *the emergent flux of the field g*:

$$\int_{\Sigma'} Gm_c\, d\Omega - \int_{\Sigma''} Gm_c\, d\Omega = Gm_c\Omega - Gm_c\Omega = 0, \tag{7}$$

for as we see in Figure 2.3 – all the elements of *emergent* flux are negative over (Σ'') and positive over (Σ').

On the other hand, for a point C_2 inside (Σ_r), all the angles between **r** and *g* are obtuse, and all the elements of emergent flux are negative. The sum of all the $(d\Omega)$ is equal to 4π, and we have:

$$\int_{\Sigma_r} (dA_r)(\mathbf{r} \cdot g) = -Gm_c \int_{\Sigma_r} d\Omega = -4\pi Gm_c. \tag{7'}$$

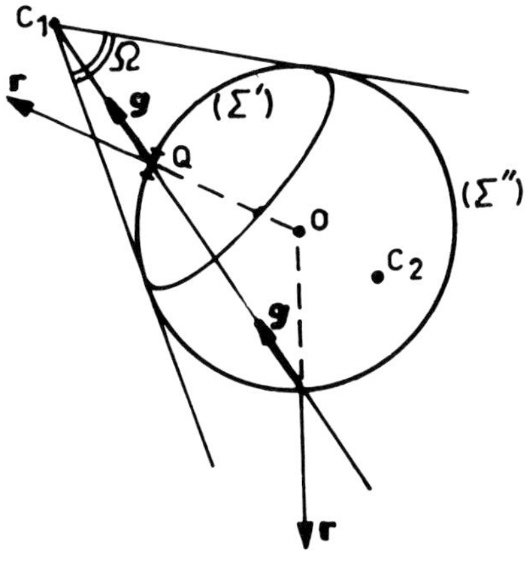

Fig. 2.3. The proof of Equations (7) and (7').

After this brief review of Gauss's theorem, let us return to Newton's theorem, which we intend to prove.

By symmetry, under the hypothesis of *homogeneous* spherical layers the field $\mathbf{g}^i$ produced at a given point at a distance r from the center O by all the mass points *inside* (Σ_r) must be *radial* and its absolute value g^i must be the same at all points of (Σ_r) at the distance r from the center O (Figure 2.4).

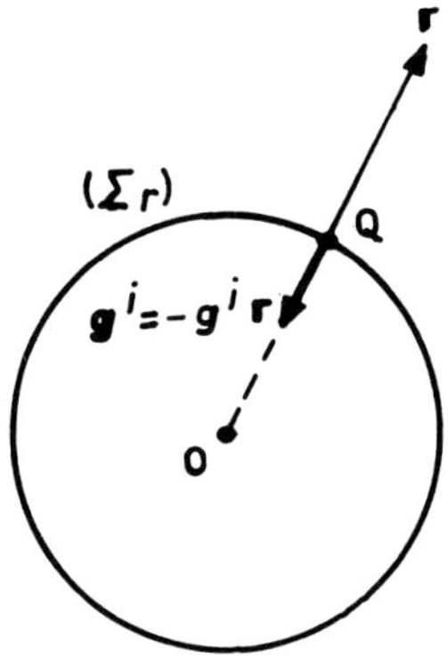

Fig. 2.4. The proof of Equation (9).

Thus we have:

$$\mathbf{g}^i = -g^i\mathbf{r}; \qquad g^i = |\mathbf{g}^i|. \tag{8}$$

And for the (negative) emergent flux of the field $\mathbf{g}^i$ over the sphere (Σ_r), we have:

$$\int_{\Sigma_r} (dA_r)(\mathbf{r}\cdot\mathbf{g}^i) = -g^i \int_{\Sigma_r} (dA_r) = -g^i 4\pi r^2. \tag{9}$$

But the sum of the quantities $(-4\pi G m_c)$ referring to the mass points enclosed in (Σ_r) is equal to $(-4\pi G M_r)$. Thus we have, by Equations (4) and (9),

$$-g^i 4\pi r^2 = -4\pi G M_r, \tag{10}$$

that is,

$$g^i = G M_r r^{-2}. \tag{11}$$

Let us now consider the field $\mathbf{g}^e$ produced by the set of mass points *outside* the sphere (Σ_r) of radius r.

As a result of the hypothesis of stratification in homogeneous spherical layers, the field $\mathbf{g}^e$ (whatever it may be) for these external mass points must also have the same absolute value g^e at all points on the sphere (Σ_r) of radius r. Moreover, this field must by symmetry be radially directed (to see this, one has only to associate two by two the mass points which are symmetric with respect to $\mathbf{r}$). Since we do not know *a priori* in which sense $\mathbf{g}^e$ is directed, we shall set

$$\mathbf{g}^e = \epsilon g^e \mathbf{r} \quad \text{with } \epsilon = \pm 1. \tag{12}$$

A calculation of the total flux, applying the definition of the emergent flux, gives as in Equation (9): $+\epsilon g^e 4\pi r^2$. It follows, according to Equation (4), that $g^e = 0$.

Qualitatively, this result can be physically interpreted as a balance (Figure 2.5) between the component $\mathbf{g}^e_+$ of the total field $\mathbf{g}^e$ which is created, at Q, by the diagonally hatched area (directed along $\mathbf{r}$ in the sense of $\mathbf{r}$), and the component $\mathbf{g}^e_-$ of $\mathbf{g}^e$ created by the vertically hatched area (directed along $-\mathbf{r}$). The component $\mathbf{g}^e_+$ is produced by a mass composed of a *small number* of elements *close* to Q, while the component $\mathbf{g}^e_-$ is produced by a mass composed of a *large number* of elements *far* from Q.

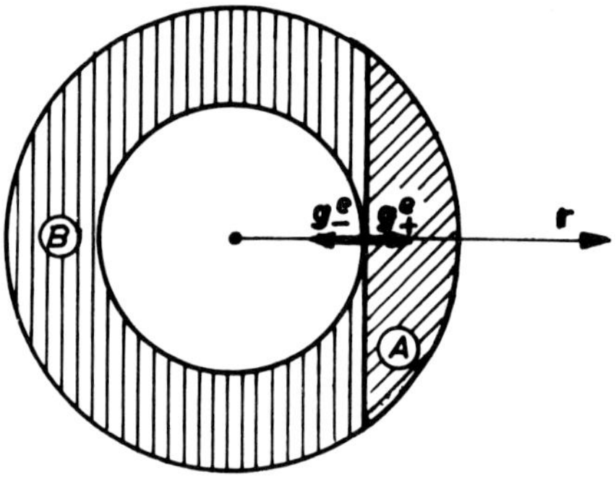

Fig. 2.5. The balance between $\mathbf{g}^e_+$ produced by (A) and $\mathbf{g}^e_-$ produced by (B).

But the rigorous nature of this balance is one of those extraordinary simplifications with which Nature sometimes gratifies the scientist, and whose astonishing and almost miraculous quality was emphasized by Newton and Einstein.

We also note that one can, of course, prove Newton's theorem directly, without recourse to Gauss's theorem; but the method we have followed has the advantage of providing an immediate proof for the many readers who are well acquainted with classical electrostatics.

If we apply Newton's theorem at a point Q of the stellar interior, located at a distance r from the center, we have:

$$g = g^i = \frac{GM_r}{r^2}; \qquad \mathbf{g} = -g\mathbf{r}. \tag{13}$$

Then, denoting the density by $\rho = \rho(r)$ (a function of r alone, under the hypothesis of homogeneous layers) and the gravitational force acting on the mass of (dV) by $d\mathbf{F}_{\text{grav}}$, we have

$$d\mathbf{F}_{\text{grav}} = (\rho\, dV)\mathbf{g}. \tag{14}$$

Hence, the gravitational force per unit volume is:

$$d\mathbf{F}_{\text{grav}}/dV = -\frac{G\rho M_r}{r^2}\mathbf{r}. \tag{15}$$

This force per unit volume is directed towards the interior (more exactly, towards the center O), and its absolute value is:

$$|d\mathbf{F}_{\text{grav}}/dV| = \frac{G\rho M_r}{r^2}. \tag{15'}$$

2.2. The Force per Unit Volume Produced by the Pressure Gradient

The exchange of material particles and of photons between the element (dV) and the *adjacent* elements of the hot gas forming the star produces dynamical effects, i.e. forces which – when reduced to unit area – appear as *pressures*.

In order to clarify this idea of dynamical pressure, it is convenient to imagine the element (dV) as separated from the adjacent elements by *a break* (a 'no man's land'), formed by an empty space across which the exchange of particles (Figure 2.6) in the broad sense of the word (including photons) takes place.

The break at the level r makes it possible to distinguish the surface (r^+) of (dV), facing the center O, from the surface (r^-) of the medium adjacent to (r^+) but outside (dV). The same situation naturally occurs at the level $(r + dr)$, where the surface of (dV) facing the 'outside' should be called $(r + dr)^-$, in contrast to the surface $(r + dr)^+$ of the adjacent medium outside (dV).

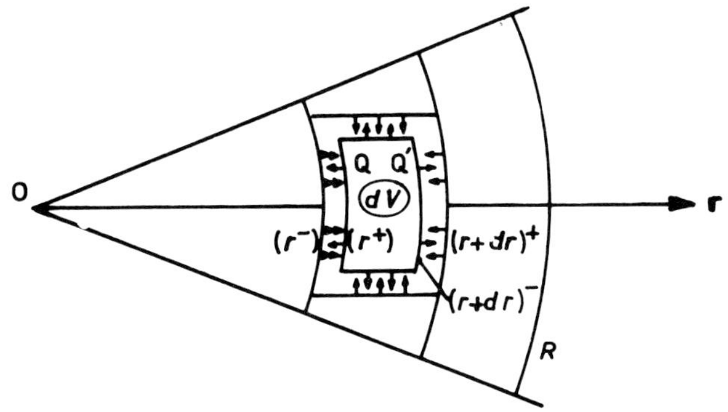

Fig. 2.6. The active and the reactive components of pressure acting on different faces of (dV).

Consequently, we shall denote by $P^+(r)$ the pressure produced on the surface (r^+) of (dV) by the exchange of particles between (r^-) and (r^+). Likewise, we shall denote by $P^-(r+dr)$ the pressure produced on the surface $(r+dr)^-$ of (dV) by the exchange of particles between $(r+dr)^-$ and $(r+dr)^+$.

As a specific example, let us begin by comparing $P^+(r)$ and $P^-(r)$, where the latter expression denotes the pressure produced on the surface (r^-) of the adjacent medium by the exchange with (dV). Physically, $P^+(r)$ results from the addition of *two* components:

(1) an *'active'* component produced by the impact of the particles (double arrows in the Figure 2.6) coming from (r^-);

(2) a *'reactive'* component produced by the ejection of praticles (single arrows) from (dV) towards the center O – an ejection which tends to push (dV) towards the outside by reaction (as with the ejection of gas from the nozzle of a rocket or a jet plane).

But the same particles appear, with reversed roles, in the sum of which $P^-(r)$ is composed: in this case, the single arrows correspond to the active component and the double arrows to the reactive component. The result is that $P^+(r) = P^-(r)$, and their common value may be denoted by $P(r)$.

This makes it possible to express the absolute value of the total pressure applied to the base of (dV) facing outwards at the level $(r+dr)$ as $P(r+dr)$ (see Figure 2.7).

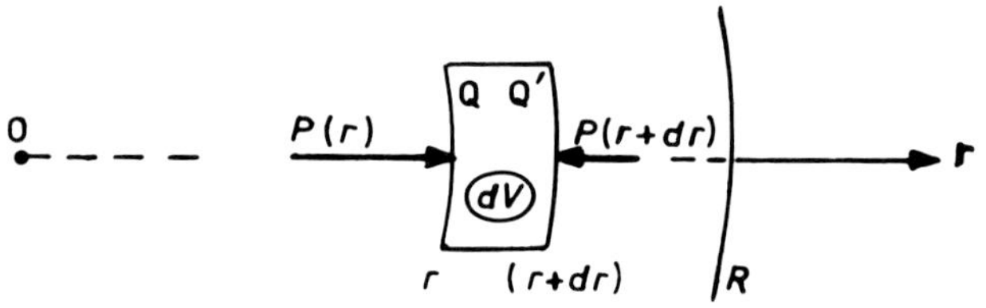

Fig. 2.7. The difference between $P(r)$ and $P(r+dr)$.

Thus the force dF_{press} exerted on the whole of the cylindrical element (dV) is reduced to:

$$dF_{\text{press}} = P(r)\,dA\mathbf{r} - P(r+dr)\,dA\mathbf{r} = -[P(r+dr) - P(r)]\,dA\mathbf{r}, \tag{16}$$

where dA is the common area of the two bases.

Since

$$[P(r+dr) - P(r)] = \frac{dP}{dr}\,dr, \tag{16'}$$

we immediately deduce that

$$\frac{dF_{\text{press}}}{dV} = -\frac{dP}{dr}\,dr\,dA\mathbf{r}/dV = -\frac{dP}{dr}\mathbf{r} = -\textit{grad}\ P. \tag{17}$$

As P depends only on r, the vector ***grad P*** reduces to $(dP/dr)\mathbf{r}$.

2.3. THE EQUATION EXPRESSING THE MECHANICAL EQUILIBRIUM OF THE STAR

Since dP/dr is negative, the absolute value of the force of *pressure* per unit volume is $(-dP/dr)$. It is outward-directed. Now, the force of *gravity* per unit volume is inward-directed, and according to Equation (15′) its absolute value is $GM_r\rho/r^2$. For equilibrium, it is thus required that:

$$-\frac{dP}{dr} = \frac{G\rho M_r}{r^2} \tag{18}$$

This equation is equivalent to the vector equation

$$\textbf{grad}\, P = \rho \boldsymbol{g}, \tag{18′}$$

where the vector $\boldsymbol{g}$ given by Equation (13) is *inward*-directed, as is the vector ***grad*** *P* (but the equilibrium is between the outward-directed − ***grad*** *P* and the inward-directed $\boldsymbol{g}\rho$!).

We have reasoned as if we knew that dP/dr is negative, but in reality it is precisely from Equation (18) that we learn this fact – just as we learn, more generally, the value of dP/dr (that is, $[-GM_r r^{-2}\rho]$) without examining in detail the physical mechanism responsible for the existence of a pressure $P(r)$.

3. The Relation between M_r and the Density ρ at a Distance r from the Center

Let us express M_r as a function of ρ, the density at a distance r from the center. The mass of a layer $(\xi, d\xi)$ inside (Σ_r), the sphere of radius r, is obtained by multiplying the volume $4\pi\xi^2 d\xi$ of that layer by ρ.

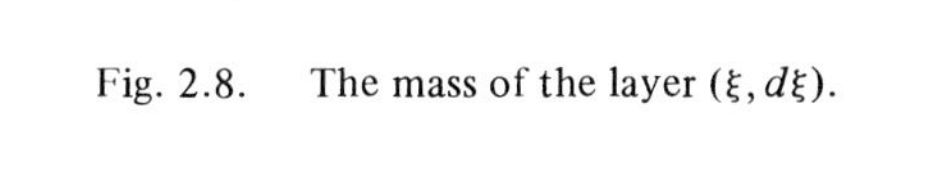

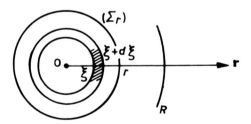

Fig. 2.8. The mass of the layer $(\xi, d\xi)$.

Thus we have (Figure 2.8):

$$M_r = \int_0^r 4\pi\rho\xi^2\, d\xi. \tag{19}$$

Similarly, the increase (dM_r) in M_r when r increases by (dr) is

$$dM_r = 4\pi\rho r^2 dr, \quad \text{hence} \quad \boxed{dM_r/dr = 4\pi\rho r^2} \tag{20}$$

On combining Equations (20) and (13), we immediately obtain the following relation:

$$d(r^2 g_r)/dr = d(GM_r)/dr = GdM_r/dr = 4\pi G\rho r^2, \tag{21}$$

in which we have placed a subscript r on g to remind ourselves that we are dealing with the absolute value of the field g at the point Q at a distance r from the center.

4. The Expression for div g as a Function of the Local Density ρ. Poisson's Equation

Let us apply the definition of the divergence of a vector field at a point Q of the field. According to the physical definition of the divergence, div $\boldsymbol{g}$ is merely the flux *density* of $\boldsymbol{g}$ (flux per unit volume) emerging from a small volume element containing Q. Let us take as the volume element in question the spherica layer (r, dr) (Figure 2.9).

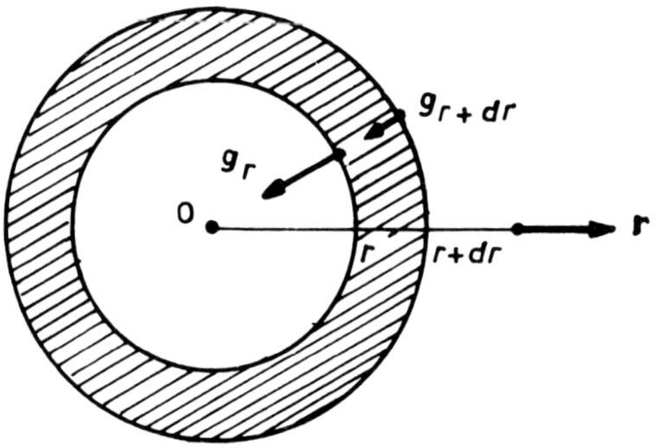

Fig. 2.9. The flux of $\boldsymbol{g}$ emerging from (r, dr).

Dividing *the flux of* $\boldsymbol{g}$ *emerging from* (r, dr) by the volume of (r, dr) and letting dr tend to zero, we have

$$\text{div}\, g = \lim \frac{4\pi r^2 g_r - 4\pi(r + dr)^2 g_{r+dr}}{4\pi r^2\, dr} = -\frac{1}{r^2}\frac{d(r^2 g_r)}{dr}. \tag{22}$$

According to Equation (21), this expression reduces to the formula

$$\text{div}\, g = -4\pi\rho G, \tag{23}$$

where div $\boldsymbol{g}$ and ρ refer to the same point Q at a distance r from the center.

This equation corresponds to the first of Maxwell's equations.

$$\text{div}\, E = 4\pi\rho\frac{1}{k}, \tag{23'}$$

where the electric field $\boldsymbol{E}$ has been replaced by the gravitational field $\boldsymbol{g}$; the dielectric constant k has also been replaced by $(-G^{-1})$, and the charge density ρ by the mass density.

On the other hand, since the field $\boldsymbol{g}$ is purely radial, its *circulation* about any small curve traced on (Σ_r) is zero; thus, according to Stokes's theorem, the flux of ***rot g*** across the area enclosed by this curve will also be zero. This means that ***rot g*** is identically zero, and we can consider $\boldsymbol{g}$ as the gradient of some scalar field, writing:

$$\boldsymbol{g} = -\boldsymbol{grad}\ \varphi_g. \tag{24}$$

The function $\varphi_g(\mathrm{Q})$, which in the present case reduces to $\varphi_g(r)$, is called the '*gravitational potential*' at the point Q.

Note that some authors set

$$\boldsymbol{g} = +\boldsymbol{grad}\ \varphi_g, \tag{24'}$$

but only the formula with the minus sign preserves the analogy with the formulae of electrostatics, when k is changed to $(-1/G)$.

When $\boldsymbol{g}$ is expressed in this way as a function of the gravitational potential φ_g, the fundamental Equation (23) takes the form of *Poisson's equation*:

$$\boxed{\operatorname{div}(\boldsymbol{grad}\ \varphi_g) = \Delta\varphi_g = +4\pi G\rho}\ , \tag{25}$$

where $\Delta\varphi_g$ is the Laplacian of φ_g. This equation is analogous to Poisson's equation for electrostatics:

$$\varphi_e = -4\pi\frac{1}{k}\rho, \tag{25'}$$

where φ_e is the electrical potential and ρ is the charge density.

5. The Calculation of the Gas Pressure P_{gas}. The Concept of the Mean Mass μ of a Particle of the Mixture, in Units of m_B (where m_B is the Mass in Grams of a Baryon)

According to the perfect gas law, in the especially simple form suggested by the kinetic theory of gases, we have

$$P_{gas} = n_{tot}kT. \tag{26}$$

Here n_{tot} is the *number* density (i.e. the total number of particles per cm^3 of the mixture), T is the absolute temperature, and k is a universal constant (the Boltzmann constant, of course different from the dielectric constant considered above) whose numerical value will be given below.

According to this formula, P_{gas} does not depend *fundamentally* on the *nature* of the particles, nor, in particular, on the *mass* in grams m_g of each of them.

However, the nature of the particles comes in (through their mass m_g) as soon as we seek the pressure P_{gas} corresponding to a given value of the density ρ of the mixture – or, to put it another way, as soon as we calculate the ratio P_{gas}/ρ.

In fact, the mean mass in grams $\bar{m}_g$ of each particle of the mixture is obtained by dividing the mass ρ of one cm^3 of the gas by the number n_{tot} of particles contained in this cm^3:

$$\bar{m}_g = \frac{1}{n_{tot}}\rho. \tag{27}$$

Then let $m_B = 1.66 \times 10^{-24}$ grams be the approximate mass of a baryon, proton or neutron, (actually, *the mass of unit atomic weight*; but in the present approximation, setting this mass equal to that of a baryon, or of a real H atom, is acceptable and makes possible a more concrete description – see Chapter 4, Section 7 and Chapter 6, Section 1). To evaluate $\bar{m}_g$ in units of m_B, we have only to set

$$\boxed{\mu = \frac{1}{m_B}\bar{m}_g} \ . \tag{28}$$

Then we have

$$n_{\text{tot}} = \frac{1}{\mu m_B}\rho, \tag{29}$$

and Equation (26) takes the form

$$P_{\text{gas}} = \frac{\rho}{\mu}RT, \tag{30}$$

where R is given by

$$R = \frac{k}{m_B}. \tag{31}$$

In order to find the value of the constant k, we have only to apply Equation (26) to one gram of atomic H under 'standard' conditions of temperature:

$$(T^0 = 0°\text{C} = 273\,\text{K}) \tag{31'}$$

and pressure

$$(P^0_{\text{gas}} = 1 \text{ atmosphere} = 1.013 \times 10^6 \,\text{dyne cm}^{-2}). \tag{31''}$$

We know that the corresponding volume V^0 is 22.4 L (22 400 cm^3), and that the total number of H atoms contained in this volume is equal to Avogadro's number, $N_A = 1/m_B \approx 6.02 \times 10^{23}$. Under these conditions, the number density n is equal to $n_0 = N_A/V^0 = 2.69 \times 10^{19}\,\text{cm}^{-3}$, and we find

$$k = \frac{P^0_{\text{gas}}}{n_0 T^0} = 1.38 \times 10^{-16} \text{ c.g.s.} \tag{32}$$

Equation (31) then gives, in c.g.s. units,

$$R = kN_A = 8.32 \times 10^7 = \text{'gas constant'}. \tag{33}$$

If, instead of considering Equation (26) or Equation (30), which are valid for any mass of gas, we consider the more particular case of a mass equal to N_A particles of the mixture, this mass will be $(\mu m_B)N_A = \mu$ grams.

If we suppose that it occupies a volume V, we will have $n_{\text{tot}} = N_A/V$, so that the expression (26) for P_{gas} will take on the classical form of the perfect gas law for *one mole*:

$$P_{\text{gas}}V = RT. \tag{34}$$

In stellar interiors the atoms are almost completely *ionized* and we can assume, in the first approximation, that we are dealing with a state of *complete* ionization.

Let us examine the effect of such an ionization on the number density of free electrons and on the value of μ. Following the usual notation, we shall let X be the mass in grams of H contained (before ionization) in one gram of the mixture, and Y be the mass in grams of He contained (before ionization) in one gram of the mixture. Finally, let $Z = 1 - X - Y$ be the mass of all the elements *other than* H *and* He contained in one gram of the mixture. We shall say, somewhat loosely, that the sum of these 'other elements' represents the 'heavy elements'. We can then draw up the following table, in which lines (1), (2), (3), etc., give respectively:

(1) the mass in grams of H, He, and 'heavy elements' in one gram of the mixture;
(2) the mass in grams of *one* nucleus of a certain 'species';
(3) the number of nuclei of this species in one gram of the mixture;
(4) the number of electrons liberated by each completely ionized atom;
(5) the number of electrons liberated by the atoms of each species present in one gram of the mixture;
(6) the total number of electrons and nuclei for each species;
(7) the total number of particles of all kinds per gram of the mixture;
(8) the total number of free electrons per gram of the completely ionized mixture.

According to the result given in the next to last line of the table, the mean mass (in grams)

TABLE 2.1
The details of the computation of μ and of N_e

	Hydrogen	Helium	'Heavy elements'	Notes
(1)	X	Y	Z	
(2)	$m_H = 1/N_A$	$4m_H$	$\mu_Z m_H$	a
(3)	$X/m_H = XN_A$	$\frac{1}{4}YN_A$	$(1/\mu_Z)ZN_A$	
(4)	1	2	$\frac{1}{2}\mu_Z$	b
(5)	XN_A	$\frac{1}{2}YN_A$	$\frac{1}{2}ZN_A$	
(6)	$2XN_A$	$\frac{3}{4}YN_A$	$\frac{1}{2}ZN_A$	c
(7)		approximately $(2X + \frac{3}{4}Y)N_A$		d
(8)		$\frac{1}{2}(2X + Y + Z)N_A = \frac{1}{2}(1 + X)N_A$		e

[a] Here we denote by μ_Z the mean mass of the 'heavy elements' expressed in units of m_H, and we neglect the mass of the electrons belonging to an atom in comparison with the mass of the nucleus.
[b] The nucleus of a neutral atom contains as many protons as there are orbital electrons. But in the nucleus there are *approximately* as many neutrons as protons. If the mean atomic mass in units of m_H is μ_Z, this means that the sum of the number of protons and the number of neutrons is equal to μ_Z. Therefore there are approximately $\mu_Z/2$ protons in the nucleus and $\mu_Z/2$ orbital electrons.
[c] When μ_Z is greater than or equal to 16, the term ZN_A/μ_Z will be negligible in comparison with the remaining term, equal to $\frac{1}{2}ZN_A$.
[d] Since X is of the order of 0.74, Y of the order of 0.24, and Z of the order of 0.02, we can neglect $Z/2$ in comparison with $(2X + \frac{3}{4}Y)$.
[e] Here we use the fact that $(X + Y + Z) = 1$ by definition.

$\bar{m}_g$ of a particle of the mixture is given by:

$$\bar{m}_g = \frac{\text{One gram}}{\text{Total number of particles}} = \frac{1}{(2X + 0.75Y)N_A} . \quad (35)$$

Consequently, the definition of μ given in Equation (28) yields (taking into account the fact that $m_B N_A = 1$):

$$\mu = \frac{1}{2X + \frac{3}{4}Y} . \quad (36)$$

Considering Z to be negligible in comparison with Y, we can replace Y by $(1 - X)$, giving as a final result

$$\boxed{\mu = \frac{1}{\frac{3}{4} + \frac{5}{4}X}} . \quad (37)$$

Moreover, since each cm^3 of the mixture contains by definition ρ grams, the number n_e of free electrons per cm^3 is obtained by multiplying the last line of the table by ρ, hence

$$\boxed{n_e = \tfrac{1}{2}(1 + X)\rho N_A} . \quad (38)$$

5.1. Remarks

5.1.1. The perfect gas law is applied to stellar interiors because the high temperature of the densest regions causes them to be very far from the *critical conditions* for liquefaction of the gas, despite the enormous pressures that prevail. (We will discuss this problem, in Chapter 4, Section 9).

5.1.2. The gas pressure P_{gas} is only one part (although generally strongly predominant in normal stars) of the total pressure P, which results from the combined effect of gas pressure (involving material particles) and of radiation pressure (involving photons) P_{rad}. The calculation of P_{rad}, taking into account the properties of the *radiation field* in the interior of a star (see Subsection 4.4 of Chapter 1), shows that to a very good approximation we have (where a is a constant equal to 7.56×10^{-15} c.g.s.):

$$\boxed{P_{rad} = \tfrac{1}{3}aT^4} . \quad (39)$$

When we have calculated P_{gas} and T, Equation (39) will show that P_{rad} is very small in comparison with P_{gas} in the interior of a star like the Sun. In fact, we shall find that although the ratio P_{rad}/P_{gas} increases towards the center of the Sun, it never exceeds 6×10^{-4}.

6. A Model of the Sun at Constant Density $\rho = \bar{\rho}$

To get an idea of the order of magnitude of the principal parameters, let us try to construct a model of the Sun in the very first approximation (the only approximation that can be guessed) by assuming that the density of the Sun is independent of the distance r from the center, and therefore equal to the mean density $\bar{\rho}$. This $\bar{\rho}$ is obtained by diving the total mass of the Sun ($M_\odot = 2 \times 10^{33}$ g) by its volume ($V_\odot = \frac{4}{3}\pi R_\odot^3$), where $R_\odot = 7 \times 10^{10}$ cm is the solar radius. With these figures, the density $\bar{\rho}$ is equal to 1.4 g cm^{-3}. In this section we have $\rho = \bar{\rho} = \text{const.}$

Let us set $r' = r/R_\odot$. At the center $r' = 0$. At the surface $r' = 1$.

In this notation we have

$$M_r = \tfrac{4}{3}\pi r^3 \rho = \tfrac{4}{3}\pi (R_\odot^3 r'^3)\rho = M_\odot r'^3. \tag{40}$$

Substituting this expression for M_r into the equation of mechanical equilibrium (18) and replacing r by $r' R_\odot$, we find

$$-\frac{dP}{dr'} = (GM_\odot/R_\odot)\rho r'. \tag{41}$$

Let us denote by A the constant coefficient of r' on the right-hand side of Equation (41). Limiting ourselves to a single significant figure, we have

$$A \approx (7 \times 10^{-8}) \times (2 \times 10^{33}) \times (1.4) \times (7 \times 10^{10})^{-1} \approx 3 \times 10^{15} \text{ c.g.s.} \tag{42}$$

Let us integrate the equation

$$-\frac{dP}{dr'} = Ar', \tag{43}$$

making use of the physically obvious condition $P(1) = 0$ (zero pressure at the surface $r' = 1$ of the Sun).

We find at once

$$P(r') = \tfrac{1}{2}A(1 - r'^2). \tag{44}$$

We immediately deduce that the pressure P_c at the center ($r' = 0$) of the Sun must be of the order of

$$P_c = P(0) = \tfrac{1}{2}A \approx 1.5 \times 10^{15} \text{ dynes cm}^{-2} \approx 1.5 \times 10^9 \text{ atm.} \tag{45}$$

Thus, the pressure at the center appears to be of the order of 1.5 billion atmospheres. Now, a more exact calculation – which we shall make later on – gives 220 *billion atmospheres*. The error obviously results from the fact that $\rho(r')$ actually varies with r', and increases *slowly* towards the center. In reality, we shall later find that $\bar{\rho} = \rho(0.45)$. Nevertheless, we note that the relative error in (45) is fairly small.

It is useful to put Equation (44) in the form

$$P(r') = P_c(1 - r'^2), \tag{46}$$

which shows explicitly that the pressure *decreases* from P_c to zero when one goes from the center O to the surface.

Despite the rough nature of the approximation $\rho = \bar{\rho}$, this hypothesis not only provides a reasonable order of magnitude for P_c, but also makes it possible (with the addition of a few plausible hypotheses concerning the composition of the stellar mixture and its physical state) to determine the order of magnitude of the central temperature T_c and the temperature distribution $T(r')$.

To begin with, let us make the additional hypothesis that the Sun is *composed of pure hydrogen* entirely dissociated into protons and electrons (*complete ionization*).

In this case, the mean mass μ of a particle of the mixture in units of m_B equals, according to Equation (36) (or by direct reasoning),

$$\mu = \tfrac{1}{2} \quad (\text{since } X = 1 \text{ and } Y = 0); \tag{47}$$

and according to Equation (29) the number density n_{tot} in particles per cm^3 equals

$$n_{\text{tot}} = \frac{1}{\mu m_B}\rho \approx \frac{1.4}{(0.5) \times (6 \times 10^{23})^{-1}} = 2 \times 10^{24}\,\text{particles cm}^{-3}. \tag{48}$$

Neglecting radiation pressure and using relation (26) between P and T, we immediately find from the approximation formula (46)

$$T(r') = \frac{1}{kn_{\text{tot}}} = T_c(1 - r'^2), \tag{49}$$

with

$$T_c = T(0) = \frac{P_c}{kn_{\text{tot}}} \approx 5 \times 10^6\,\text{K}. \tag{50}$$

This value of T_c is not very different from the value given by the most exact calculation (14.6×10^6 K). The order of magnitude is entirely correct. This in itself justifies the use of the perfect gas law (26) ('agitation' balancing compression), and to a certain extent it justifies and explains the validity of the hypothesis $\rho = \bar{\rho}$ (the center is more compressed but hotter). However, this mention of 'compression', very frequent in books devoted to internal structure, is worth a brief commentary.

Indeed, the mechanical equilibrium of a star is often presented as the result of a balance between the '*compression*' of each layer (dV) by the layers located above (dV) and the *gradient* of the gas pressure (when one does not make the additional error of speaking of the *pressure* without qualification).

Now, the gradient of P does have a 'supporting' effect, but what it balances is not the supposed 'compression' (for according to Newton's theorem the effect of the outer layers is simply *zero*), but the attraction exerted on (dV) by M_r, which *pulls* (dV) towards the center (the gravitational force on dV).

As we see, the model constructed by assuming that the density is independent of the distance r' from the center provides an entirely correct idea of what takes place in the interior of the Sun, revealing the enormous magnitude of the central temperature. But this model does not explain the origin of this enormous temperature. Nor does it explain the origin of the energy which the Sun emits each second. The missing explanation will be given in Chapter 3.

References

REFERENCES FOR FURTHER STUDY

Chandrasekhar, S.: 1967, *An Introduction to the Study of Stellar Structure*, Dover, New York.
Kourganoff, V.: 1973, *Introduction to the Physics of Stellar Interiors*, Reidel, Dordrecht, Holland.
Schwarzschild, M.: 1965, *Structure and Evolution of the Stars*, Dover, New York.

CHAPTER 3

THE PHYSICS OF THE INTERIORS OF THE MAIN SEQUENCE STARS

1. Introduction

From now on we shall consider the problem of the *internal structure* of 'normal' stars (those which belong to the main sequence of the HR diagram – about 99% of all stars). However, let us start with *a summary* of the properties of mechanical equilibrium, valid for all static stars, given by the following Subsection and Equations of Chapter 2.

(1) Newton's theorem, (Subsection 2.1).

(2) The expression for the force per unit volume produced by the pressure gradient, given by Equation (17).

(3) The expression for the gravitational force per unit volume, given by Equation (15′):

$$\left|\frac{dF_{\text{grav}}}{dV}\right| = GM_r\rho r^{-2}. \tag{1}$$

(4) Equation (18) expressing the mechanical equilibrium of an element of the star:

$$\boxed{-\frac{dP}{dr} = G\rho M_r r^{-2}}\,, \tag{2}$$

where G is the constant of gravitation (6.67×10^{-8} c.g.s.), r the distance from the center, M_r the mass enclosed in the sphere of radius r, $\rho = \rho(r)$ the density of the stellar mixture at a distance r from the center, and $P = P(r)$ is the total pressure (produced by material particles and by photons).

(5) The relation between M_r, r and ρ given by Equation (20):

$$\boxed{dM_r/dr = 4\pi\rho r^2}\,. \tag{3}$$

(6) The expression for the mean mass μ of a particle of the mixture in units m_B, given by Equations (28) and (27), where

$$m_B \approx 1.66 \times 10^{-24}\,\text{g}. \tag{4}$$

The respective number densities of free electrons, protons, and α-particles (after a complete ionization) being n_e, n_p, n_α; n_{tot} is given by

$$n_{\text{tot}} = n_e + n_p + n_\alpha. \tag{5}$$

Thus μ is given by Equation (37):

$$\mu = (\tfrac{3}{4} + \tfrac{5}{4}X)^{-1}, \qquad (6)$$

where $X = X(r)$ is the mass *in grams* of H contained before ionization in one gram of the mixture (supposed in a first approximation to be composed solely of H and He).

(7) The expression for the gas pressure P_{gas} (the perfect gas law) given by Equation (26) (with $n = n_{tot}$), and by Equation (30) (with R replaced by $k/m_B = kN_A$ in order to avoid confusion with the radius of the star):

$$P_{gas} = n_{tot}kT = (\rho/\mu)(k/m_B)T = \frac{1}{\mu}\rho N_A kT, \qquad (7)$$

where

$$k \approx 1.4 \times 10^{-16}\,\text{c.g.s.}; \qquad N_A = \frac{1}{m_B} \approx 6.02 \times 10^{23}\,\text{c.g.s.}. \qquad (8)$$

(8) Finally, the radiation pressure P_{rad} is given by Equation (39):

$$P_{rad} = \tfrac{1}{3}aT^4, \qquad (9)$$

where $a = 7.56 \times 10^{-15}$ c.g.s..

2. The Equation of Energy Equilibrium

Let $L_r = L_r(r)$ be the function representing the *net* flow of integrated radiation across a sphere of radius r, already considered in Subsection 4.3 of Chapter 1.

L_r obviously remains *constant* (and equal to its value L at the surface) as long as we remain outside the central region where nuclear reactions take place. In the present state of evolution of the Sun, this central region corresponds to about $r' \leqslant 0.3$.

Let us now consider a layer (r, dr) located at some distance r from the center. The energy equilibrium *in* this layer (r, dr) requires a *net outflow of energy per second* crossing the boundaries of (r, dr) equal to any *energy generation* (by nuclear reactions) in (r, dr). More explicitly, let $\epsilon = \epsilon(r)$ be the average energy produced per second by nuclear reactions in each *gram* of the stellar mixture making up (r, dr). The energy produced per second in each cm^3 of (r, dr) will be given by $\epsilon(r)\rho(r)$, where ρ is the mass density at a distance r from the center (in g cm^{-3}).

The energy produced per second throughout (r, dr) will be

$$\epsilon(r)\,dM_r = \epsilon(r)4\pi r^2\rho(r)\,dr. \qquad (10)$$

The balance between gains and losses per second then requires that we have (using obvious notation, corresponding to that introduced in Subsection 3.3 of Chapter 1):

$$\epsilon(r)\,dM_r + L_r^+ + L_{(r+dr)}^- = L_r^- + L_{(r+dr)}^+, \qquad (11)$$

or, introducing the *net* flow L_r equal by definition to $(L_r^+ - L_r^-)$:

$$\epsilon(r)\,dM_r = L_{(r+dr)} - L_r = (dL_r/dr)\,dr. \qquad (12)$$

Dividing by dr and letting dr approach zero, we obtain the *equation of energy equilibrium*:

$$\boxed{\frac{dL_r}{dr} = \epsilon(r)\frac{dM_r}{dr}}\ . \tag{13}$$

It can also be put, according to Equation (3), into the form:

$$\frac{dL_r}{dr} = 4\pi r^2 \rho(r)\epsilon(r). \tag{14}$$

In the regions (r' greater than about 0.3) where $\epsilon(r) = 0$, energy equilibrium implies that $dL_r/dr = 0$, and we recover $L_r = \text{Const}$.

3. The Expression for $\epsilon(r)$ in the Case when Energy is Produced by the *p–p* Chain or the C–N Cycle

It can be shown (see for instance Kourganoff, 1973, IV, 3) that the energy generation per gram per second $\epsilon(r)$ depends essentially on the composition, the temperature T, and the mass density ρ of the stellar mixture at a distance r from the center.

It also can be shown (Kourganoff, 1973, IV, 3) that for temperatures of the order of 10^7 K (more precisely between 4×10^6 and 50×10^6 K), the only long-term reactions possible under stellar conditions are:

(1) *the p–p chain* (between 4×10^6 and 25×10^6 K);
(2) *the C–N cycle* (between about 12×10^6 and 50×10^6 K).

The *p–p* chain (or proton–proton chain) and the C–N cycle (or carbon–nitrogen cycle) are two different modes of *fusion* of protons into α-particles. The details of these reactions are given in all popular books on astrophysics (also given in Kourganoff, 1973, IV, 2).

In both cases, one finds (considering *exact* masses of reacting particles) that the fusion of four protons into an α-particle produces an energy of $Q_{pp} = 26.204$ MeV by the *p–p* chain and an energy of $Q_{\text{CN}} = 25.02$ MeV by the C–N cycle.

These results do not take into account the energy released in the form of *neutrinos*, which escape from the star without contributing to the energy balance considered above. (The problem of the discrepancy between the present theory and the measured neutrino flux from the Sun is too complex to be considered here.)

The determination of the energy $\epsilon(r)$ produced per second by each gram of the stellar material consists of the following steps:

(1) The calculation of the average number of reactions per cm^3 per second for each part of a particular chain or cycle.
(2) The theoretical or experimental determination of certain constants appearing in this calculation.
(3) The theoretical or experimental determination of the quantity of energy released by each reaction in the chain or cycle.
(4) The addition of energies released by all the reactions of *one* chain or *one* cycle, taking into account certain losses (neutrino losses).
(5) Analysis of the adjustment of the various reactions of a given cycle to a stationary

state, and calculation of the *equilibrium abundances* of the various constituents of of the mixture. (All these calculations can be found in Kourganoff, 1973, pp. 45–73).

Here we shall limit ourselves to a summary of the results thus obtained, distinguishing the energy ϵ_{pp} released per gram per second by the p–p chain, and the energy ϵ_{CN} released per gram per second by the C–N cycle. ϵ_{pp} can be put into the form:

$$\epsilon_{pp} = (\epsilon_{pp})_1 \rho(r) X^2(r) T_6^{\nu}(r), \tag{15}$$

where $(\epsilon_{pp})_1$ is simply a numerical constant given (as a function of the temperature) by Table 3.1. $X(r)$ is the mass in grams of H contained (before ionization) in one gram of the mixture at a distance r from the center. As to T_6 it represents the temperature in billions of degrees: $T_6 = T/10^6$. Finally, ν represents here a constant depending, as $(\epsilon_{pp})_1$, on the domain of temperature, (see Table 3.1).

TABLE 3.1

The values of ν and of $\log_{10}(\epsilon_{pp})_1$ for the p–p chain

T_6	$\log_{10}(\epsilon_{pp})_1$	ν
4–6	−6.84	6
6–10	−6.04	5
9–13	−5.56	4.5
11–17	−5.02	4
16–24	−4.40	3.5

Similarly, ϵ_{CN} is given by

$$\epsilon_{CN} = (\epsilon_{CN})_1 \rho(r) X(r) X_{CN} T_6^{\nu}(r), \tag{16}$$

where ρ, as usual, is the mass density of the mixture (in c.g.s units), X is defined as above, and $X_{CN} = X_C + X_N$, with X_C being the number of grams of C^{12} and X_N being the number of grams of N^{14} per gram of the mixture. The constants $(\epsilon_{CN})_1$ and ν, to be used in Equation (16), according to the domain of temperature are given in Table 3.2.

TABLE 3.2

The values of ν and of $\log_{10}(\epsilon_{CN})_1$ for the C–N cycle

T_6	$\log_{10}(\epsilon_{CN})_1$	ν
12–16	−22.2	20
16–24	−19.8	18
21–31	−17.1	16
24–36	−15.6	15
36–50	−12.5	13

4. The System of Differential Equations and of Boundary Conditions which Determine the Internal Structure of a Normal Star

There are at least four main structural parameters of a normal main sequence star:

(1) The *mass* M_r enclosed in the sphere of radius r: $M_r(r)$;
(2) The *total pressure* P at a distance r from the center: $P(r)$;
(3) The *temperature* T at a distance r from the center: $T(r)$; and
(4) The *net outflow of energy per second* L_r through the sphere of radius r: $L_r(r)$.

The study of the internal structure of a star such as the Sun, consists in the search of the variations of quantities such as the total pressure P, the temperature T, the density ρ, the H abundance X, etc., as functions of the distance r to the center of the star.

Mathematically, this means that we have to find the unknown functions $P(r)$, $T(r)$, $\rho(r)$, $X(r)$, etc., satisfying all physical relations depending on these quantities. Most of these relations are *differential equations*, such as, for instance, Equation (2) of mechanical equilibrium.

When trying to solve the system of these equations we have to satisfy some boundary conditions, some of these are 'theoretical' and evident such as (R being the radius of the star):

$$P(R) = 0. \tag{17}$$

Others are less obvious, but permitted with a good approximation (taking into account the enormous temperature at the center: 6000 K at the surface is negligible compared with 15 000 000 K at the center):

$$T(R) = 0. \tag{18}$$

By analysing what happens near the center it can also be easily proved that:

$$M_r(0) = 0. \tag{19}$$

Indeed, in the vicinity of the center (ρ_c being the finite value of the mass density of the mixture near the center):

$$\lim_{r\to 0} M_r(r) = \lim_{r\to 0} \tfrac{4}{3}\pi r^3 \rho(r) = \lim_{r\to 0} \rho_c \tfrac{4}{3}\pi r^3 = 0. \tag{19'}$$

Similarly, making use of Equation (106) of Chapter 1 we see that in the vicinity of the center L_r is proportional to r^2 with a finite value of the constant of proportionality. Hence

$$L_r(0) = 0. \tag{20}$$

Other boundary conditions are given by a theoretical interpretation of *observations*. Thus, for instance, for the Sun, we know the value of $M_r(R)$ which is simply the total mass $M_\odot$ of the Sun ($M_\odot \approx 2 \times 10^{33}$ g). If more generally we call M the total mass of the star, then

$$M_r(R) = M. \tag{21}$$

On the other hand, observation of the brightness of a star of known distance gives its luminosity L, which, for an isolated star, is equal to $L_r(R)$; (for the Sun $L_r(R) = L_\odot \approx 4 \times 10^{33}$ c.g.s.)

$$L_r(R) = L. \tag{22}$$

Naturally, for the Sun, we also know the value of the radius: $R_\odot = 6.96 \times 10^{10}$ cm.

It is again the observation which gives the boundary conditions for abundances $X(r)$, $Y(r)$, etc., of H, He, etc. Quantities such as $X(R)$, $Y(R)$, etc. are given by the quantitative analysis of spectra of stellar atmospheres.

Let us indicate immediately these boundary values for the Sun:

$$X(R)_\odot = 0.744; \qquad Y(R)_\odot = 0.236. \tag{23}$$

Thus we can neglect $Z(R)_\odot$ which is in most cases of the order of 0.020.

We shall show in Section 5 how the study of the *evolution* of the stellar structure due to nuclear reactions allows one to overcome the absence of an immediate knowledge of $X(0)$, $Y(0)$, etc., near the center.

Finally, we have to take into account some '*constitutive relations*', such as:

$$P(r) = P_{\text{gas}}(r) + P_{\text{rad}}(r); \tag{24}$$

$$P_{\text{gas}}(r) = n_{\text{tot}}(r)kT(r) = \frac{1}{\mu}\rho N_A kT(r); \tag{25}$$

$$\frac{1}{\mu(r)} = \frac{3}{4} + \frac{5}{4}X(r); \tag{26}$$

$$P_{\text{rad}}(r) = \frac{a}{3}T^4(r). \tag{27}$$

These are simply relations (7), (6) and (9) reproduced here for the convenience of the reader.

Other constitutive relations concern the *opacity* $\bar{k}$ and can be written – according to Subsection 4.5 of Chapter 1 – as

$$\bar{k}(r) = \tilde{k}_0\rho^2(r)[1 + X(r)]T^{-3.5}(r), \tag{28}$$

where $\tilde{k}_0$ is a constant.

Finally, a last constitutive relation is given by Equation (15) *or* (16) for $\epsilon(r)$, which can be summarized by

$$\epsilon(r) = \tilde{\epsilon}_0\rho(r)X^2(r)T^\nu(r). \tag{29}$$

The constant $\tilde{\epsilon}_0$ depends generally on the ratio X_{CN}/X which is chosen by considerations too 'advanced' to be given here.

The integration of the set of differential equations is not very difficult when modern electronic computers are used. However, finding a large *excess of physical parameters* capable of describing the same state of a given star at a distance r from the center, we are faced with a preliminary problem of the choice of some '*fundamental*' unknown functions – knowing which we can entirely determine the distribution of all the other parameters.

Thus, for instance, it is not reasonable to keep both as unknown functions $\rho(r)$ and $M_r(r)$, since they are related by the very simple relation (3).

Oddly enough, although *physically* the function $\rho(r)$ is the keystone of the internal structure of a star (this is illustrated, for instance by Section 6 of Chapter 2), it is advantageous from the *mathematical* point of view to eliminate it as soon as possible. Indeed, it is the only one of the important unknown functions which does *not* appear as a derivative in any of the fundamental relations (2), (3), (14) to which we must add the relation between L_r and dT/dr established in Subsection 4.3 of Chapter 1 (Equation (106)).

Thus, before elimination of ρ we have the following system of *four* differential equations, between the 'fundamental functions' $M_r(r), P(r), T(r)$ and $L_r(r)$:

$$\frac{dM_r}{dr} = 4\pi r^2 \rho(r); \tag{30}$$

$$\frac{dP(r)}{dr} = -G\rho(r) r^{-2} M_r(r); \tag{31}$$

$$\frac{dT}{dr} = -\frac{3\bar{k}(r) r^{-2} T^{-3}(r)}{16\pi ac} L_r(r); \tag{32}$$

$$\frac{dL_r(r)}{dr} = 4\pi r^2 \rho(r) \epsilon(r). \tag{33}$$

Solving the purely algebraic system of four equations (24), (25), (26), (27) for $\rho(r)$, we find $\rho(r)$ as a *known function* $F(P, T, X)$ of the three variables P, T and X, given by

$$\rho = \left(P - \frac{a}{3} T^4\right) \Big/ \left[\left(\frac{3}{4} + \frac{5}{4} X\right) N_A k T\right] = F(P, T, X). \tag{34}$$

We can therefore eliminate ρ, by means of Equation (34), from Equations (30), (31), (32), (33) and use Equations (28) and (29) to obtain the system of differential equations:

$$\frac{dM_r}{dr} = 4\pi r^2 F(P, T, X); \tag{35}$$

$$\frac{dP}{dr} = -G r^{-2} M_r F(P, T, X); \tag{36}$$

$$\frac{dT}{dr} = -\frac{3(1 + X) r^{-2} T^{-6.5} F^2(P, T, X)}{16\pi ac} \tilde{k}_0 L_r; \tag{37}$$

$$\frac{dL_r}{dr} = 4\pi \tilde{\epsilon}_0 r^2 X^2 T^\nu F^2(P, T, X). \tag{38}$$

But this system contains only *four* equations, which seems insufficient for the determination of *five* unknown functions: M_r, P, T, L_r and X.

The solution of this difficulty is given in the next Section.

5. Evolutionary Models and Solution of the Problem Concerning the Function $X(r)$

The search for the unknown function $X(r)$ is based upon an important physical concept which has been introduced in astrophysics by Martin Schwarzschild (Princeton University, U.S.A.).

One must not isolate the different *evolutionary stages* of a star in an individual (instantaneous) manner, but rather one must follow *step by step*, beginning with a *'pre-stellar'* object – of a composition $X = X_0$, $Y - Y_0$, etc., *independent of r* – the evolution of a gaseous sphere of *constant* mass M_0.

In the initial state this mass has just finished its gravitational contraction from the interstellar material and is just beginning to 'burn' its main 'nuclear fuel': H.

The determination of the *four* functions $M_r(r)$, $P(r)$, $T(r)$, $L_r(r)$ by the system of differential equations ((35), (36), (37), (38)) for these 'homogeneous models' – homogeneous with respect to the *space* distribution, but not homogeneous chemically – obviously does not require a knowledge of the *functions* $X(r)$, $Y(r)$, . . . since in the initial state these functions are purely *numerical constants* X_0, Y_0, . . .

Nevertheless, it seem *a priori* that it is impossible to know the initial composition of a star, such as it was far in the past.

Fortunately, when we want to know the initial composition of a real star – which is *actually observable* in an advanced evolutionary stage – we can take advantage of the following circumstances.

Thermonuclear reactions are possible in a star only at sufficiently high temperature. But the gradient of temperature is related to the gradient of pressure in such a way that only a limited region near the center of the star – called the 'core' of the star – can be the place of nuclear reactions and consequently of *transformation of relative abundances*. For instance, as already stated above, in the Sun the core is limited presently to about one third of the total radius of the Sun.

We can, therefore, be certain that the *outer* layers of the star *today* have the same chemical composition (X_0, Y_0, Z_0), and even $(X_{CN})_0$, as (X_0, Y_0, Z_0) of the initial homogeneous star, provided that no *'general mixing'* affects the star (by convection connecting the core with outer layers) during its evolution from initial to the present state.

Now, it happens that application of fluid mechanics to stellar interiors shows that *no* such *'meridian circulation'*, connecting the core to outer layers, is possible in normal stars. However, this does not exclude the possibility that the core of the stars more massive than $1.7M_\odot$ – *or* an outer thin *envelope* of the stars less massive than $1.7M_\odot$ – can be stirred by *convective* motions due to a temperature gradient leading to a mechanical instability, (see Section 10).

This *absence* of general mixing means that the 'external' composition (that of an eventual outer envelope) of the *present* observable evolved phase – which can be determined by quantitative spectral analysis – is none other than the *'fossilized'* (prehistoric) composition of the *initial phase*.

In other words, (X_0, Y_0, Z_0) and $(X_{CN})_0$, as now observed at the surface, are the same as (X_0, Y_0, Z_0) of the *initial homogeneous star*, with which we must *begin* the study of 'evolutionary sequences'. For the Sun these values have been indicated in Equation (23).

Of course, space homogeneity – i.e. the fact that X_0 and Y_0 do not depend on r, does not mean *chemical purity*: already in the initial state, before the start of nuclear reactions *in the star* some He can be present besides H. This is due to the fact that the 'beginning' of a star formation, from the diffuse gas, occurs later than the initial phase of the expansion of the Universe. During this very hot and very dense phase, a certain amount of He could have been formed. This explains why one finds He at the *surface* of very old stars – in the absence of 'general mixing'.

However, it should be noted that this question is presently the object of numerous theoretical and observational investigations, and the problem cannot be considered to be entirely solved.

Note, also, that space homogeneity of X_0 and Y_0 does not prevent variation of M_r, P, T and L_r with distance r from the center. We have already encountered this phenomenon in the model of the Sun with ρ independent of r considered in Section 6 of Chapter 2.

Evolution causes the radius R and the luminosity L of the star to change very slowly with time, whereas the mass M can be considered constant, in spite of the mass lost in the form of radiation or through *the escape* or *ejection* of particles. In any case, there is no factor that can influence either the mass M or the conformity of the surface composition with the initial composition in the *initial phases of evolution* (approximately 10^8 yr) of normal main sequence stars. We shall limit ourselves to such stars. (A more advanced and complete treatment can be found in Schwarzschild's (1958) remarkable treatise.)

After the initial model whose '*homogeneous*' composition X_0, Y_0, Z_0 is *known*, and whose structure will be found by solving the system of differential equations ((35), (36), (37), (38)), Schwarzschild considers a *discrete* series of stellar models $\{E\}_t$. These models correspond to a time step Δt small enough so that the composition $X(r)_{t+\Delta t}$, $Y(r)_{t+\Delta t}$, etc., at each 'point' r of the model $\{E\}_{t+\Delta t}$ can be determined from that of the previous model $\{E\}_t$ by the differential relationship

$$X(r)_{t+\Delta t} - X(r)_t = \left[\frac{\partial X(r)}{\partial t}\right]_t \Delta t, \tag{39}$$

and by analogous relations for $Y(r)$ and $Z(r)$.

However, we must take into account the fact that (in addition to the variation of composition in each element of the star where nuclear reactions take place) evolution is generally accompanied by an *expansion* or *contraction*. Thus, one cannot use the variable r or even r' in equations such as (39). This difficulty is surmounted by a physical–mathematical artifice which consists of taking M_r instead of r as the variable that determines the position of an element. This is equivalent to defining a sphere (Σ_r) by the mass of the matter it encloses, rather than by its radius r. Equations (39) are then replaced by equations such as:

$$X(M_r)_{t+\Delta t} - X(M_r)_t = [\partial X(M_r)/\partial t]_t \Delta t. \tag{40}$$

The advantage of this change of variables is obvious. Indeed, the transmutations transform protons into α-particles, or produce other nuclear transformations with conservation of mass (except for the binding energies). Thus the mass M_r of *a given set of particles* remains constant in time, even when r varies because of contraction or expansion or when the particles themselves change.

When the functions $r(M_r)$, $P(M_r)$, $T(M_r)$ and $L_r(M_r)$ have been found, a purely mathematical 'inversion' makes it possible to return to the variable r.

In other words, we shall use the variable M_r to connect two successive evolutionary stages. But for the discussion of each stage we can use the variable r, which is physically clearer.

We note, in particular, that the functions $X(M_r)$, $Y(M_r)$ and $Z(M_r)$ in a given evolutionary stage are determined not only by the usual (mechanical, energy, etc.) equilibrium equations, but also by the results of the nuclear transmutations that took place during the *preceding evolutionary phases*. This makes it necessary to follow *all* these phases, from a 'homogeneous' mixture of given composition up to the phase under consideration.

Let us return to Equation (40), and to the analogous equations for $Y(M_r)$ and $Z(M_r)$. The coefficient of Δt on the right-hand side is *completely known* (if one has used the prescribed methods, i.e. examined all the intermediate phases between $t = 0$ and t). Actually, the *variation* $|\Delta X(M_r)|$ of $X(M_r)$ during the time Δt inside a layer (M_r, dM_r) located between spheres of mass M_r and $(M_r + dM_r)$ is proportional to the energy $[\epsilon(M_r)]_t$ produced per gram per second near the time t in the corresponding layer; the constant of proportionality is easily obtained by the following argument.

In the p–p chain, for example, the energy Q liberated during the formation of each He *atom* by the transmutation of four H *atoms*, i.e. by using up a mass of $4m_B$ grams of H, is equal to $Q_{pp} = 26.2\,\text{MeV} = 42.1 \times 10^{-6}\,\text{erg}$.

Thus, if we follow Schwarzschild's notation and let ϵ^* be the energy in ergs supplied by each gram of H consumed, we will have

$$\epsilon^* = \frac{Q\text{ ergs}}{4(m_B\text{ grams})} = \tfrac{1}{4}QN_A = (1.5 \times 10^{23})Q, \tag{41}$$

where ϵ^* is expressed in ergs per gram of consumed H.

Thus ϵ^* for the p–p chain is equal to $\epsilon^*_{pp} = 6.34 \times 10^{18}\,(\text{erg g}^{-1})$ of consumed H. Similarly Equation (41) gives, with $Q_{\text{CN}} = 25.02\,\text{MeV}$, a value of ϵ^*_{CN} equal to $\epsilon^*_{\text{CN}} = 6.0 \times 10^{18}\,(\text{erg g}^{-1})$ of consumed H.

Now we know ϵ (without the asterisk), i.e. the energy supplied *per second* by each gram of the stellar mixture. But in each gram of this mixture in the layer (M_r, dM_r) there are (in a first approximation) at the time t, $X(M_r)$ grams of H and $Y(M_r)$ grams of He; and the (negative) variation $\Delta X(M_r)$ of $X(M_r)$ during Δt seconds is connected to ϵ, Δt, and ϵ^* by the relation

$$\epsilon\Delta t = -\epsilon^*\Delta X(M_r). \tag{42}$$

Thus Equation (40) can be replaced by

$$X(M_r)_{t+\Delta t} = X(M_r)_t - (\epsilon/\epsilon^*)_t\Delta t, \tag{43}$$

and the corresponding Equation for $Y(M_r)$ can be written as

$$Y(M_r)_{t+\Delta t} = Y(M_r)_t + (\epsilon/\epsilon^*)_t\Delta t, \tag{44}$$

where the change in sign is obviously due to the fact that we have a *conservation law*

$$X(M_r)_{t+\Delta t} + Y(M_r)_{t+\Delta t} = X(M_r)_t + Y(M_r)_t = 1, \tag{45}$$

(if we neglect Z).

Naturally, if fusions of type p–p and C–N were operating *at the same time*, which is the case when T is near 16×10^6 degrees, it would be necessary to replace in Equations (43) and (44) the coefficient of Δt by $[(\epsilon_{pp}/\epsilon^*_{pp}) + (\epsilon_{\text{CN}}/\epsilon^*_{\text{CN}})]$. This procedure can be easily extended to other types of nuclear reactions, such as, for instance, the conversion of three alpha particles into a C^{12} nucleus, which can take place near temperatures greater than about 10^8 K.

Relations (43) and (44) provide the necessary connecting link between two successive evolutionary stages at times t and $(t + \Delta t)$.

6. Utilization of Boundary Conditions in the Study of Initial Models

One assumes as known the mass M of the model. This mass is that of the star under investigation: given by observation. As stated previously, M is considered as constant during a star's evolution.

Relative abundance (chemical composition) X_0, Y_0, Z_0 of the initial model are also given by observation (the 'fossilized composition').

One of the difficulties of the problem lies in the fact that for the initial model we do not know the *radius* R and the *luminosity* L of the model. Accordingly we start with the following boundary conditions for our four unknown functions $M_r(r)$, $P(r)$, $T(r)$ and $L_r(r)$:

$$M_r(0) = 0; \qquad P(R) = 0; \qquad T(R) = 0; \qquad L_r(0) = 0. \tag{46}$$

Thus two of the four boundary conditions are relative to the *center* $(r = 0)$ and the other two are relative to the *surface* $(r = R)$.

To face this difficulty we can proceed, in principle (the practical procedure is slightly different) as follows.

We start our numerical integrations from the *surface*. However, for being able to do so we choose *arbitrarily* the value R_0 of the radius and the value L_0 of the luminosity of our initial model. Thus we begin with two *arbitrary parameters* R_0 and L_0, and we can construct, (see Figure 3.1 overleaf), the points S_M, S_P, S_L and S_T of the curves representing the variations of $M_r(r)$, $P(r)$, $L_r(r)$ and $T(r)$ at the point $r = R_0$ – that is at the surface S of the star; for we will have the common abscissa $r = R_0$ of these points, as well as the corresponding ordinates $M, 0, L_0, 0$.

Then, using Equations ((35), (36), (37), (38)) which give the derivatives dM_r/dr, dP/dr, dL_r/dr and dT/dr in terms of $M_r(R_0) = M$, $P(R_0) = 0$, $L_r(R_0) = L_0$ and $T(R_0) = 0$ we can compute, in principle, the values of the derivatives in question at the points S_M, S_P, S_L and S_T. Thus we will have the *slopes of the tangents to the curves* representing our four unknown functions, at the surface S.

This is indicated in Figure 3.1 by the arrows originating at the points S_M, S_P, S_L and S_T; in this figure we have deliberately ignored in the interest of clarity, the fact that $M_r(r)$ remains very close to M and that $L_r(r)$ remains very close to L_0 when r is close to R_0.

We have also neglected, for the moment, an eventual existence of a narrow *convective layer* near the surface, which is considered in the Section 10.

Continuing in this way, step by step, that is applying the classical methods for the numerical integration of systems of differential equations (transposed to electronic computers), we can follow the graph of our four curves towards $r = 0$, all the way to the center. But in general – if R_0 and L_0 are truly arbitrary – the final situation will be as indicated by the dashed curves in Figure 3.1. Thus for $r = 0$, the curve $M_r(r)$ will pass through a point m_0 whose ordinate $M_r(0)$ is different from zero, and the curve $L_r(r)$ will pass through a point l_0 whose ordinate $L_r(0)$ is different from zero.

On the other hand, if we seek solutions for which $M_r(0) = 0$ and $L_r(0) = 0$, we have *no preconceived* constraints concerning the central value of the initial pressure $P(0)$, and the central value of the initial temperature $T(0)$. Thus the position of the points p_0 and t_0 on the curves representing $P(r)$ and $T(r)$ is *indifferent* to us, for the time being.

At this point, it is obvious that a *methodical* search for the solution could be reduced

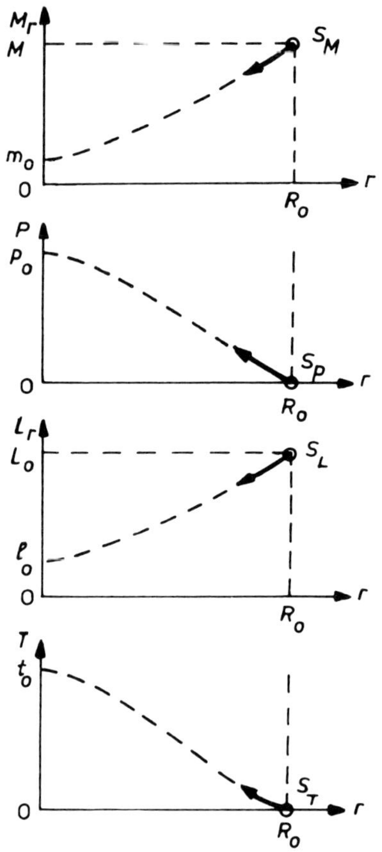

Fig. 3.1. The curves representing the variations of $M_r(r)$, $P(r)$, $L_r(r)$, $T(r)$ corresponding to an arbitrary choice of R_0 and L_0.

to the construction of a grid of curves like those in the figure, for a *rectangular array* of values of R_0 and L_0 which are chosen arbitrarily but in sufficient number, and with a reasonable order of magnitude. An interpolation between the values of R_0 and L_0 which give m_0 and l_0 *near* the origin then gives the pair $[(R_0)_s, (L_0)_s]$ corresponding to the 'correct solution' – the one which makes the two curves, the one representing M_r and the one representing L_r, pass through the origin of the coordinate system.

The curves representing $P(r)$ and $T(r)$ which correspond to $(R_0)_s$ and $(L_0)_s$ will give the exact values, hitherto unknown, of the central pressure $P(0)$, and of the central temperature $T(0)$, corresponding respectively to the ordinates of p_0 and t_0 on the 'right curves'.

In general, the method we have just described 'converges' poorly when it is applied beginning at $r = R_0$, where $P = 0$ and $T = 0$. This is especially true for the function $T(r)$, for T appears with a very large exponent in the denominator of the expression (37) of dT/dr. Schwarzschild (1958) surmounts this difficulty by looking for an *analytical* expression for the limiting value of the right-hand side of Equation (37) in the neighborhood of $r = R_0$.

Moreover, Equation (31) shows that in the neighborhood of the center, where the mass density ρ remains *finite* and takes on the value $\rho_c = \rho(0)$, we can consider (replacing M_r by $\frac{4}{3}\pi r^3 \rho_c$) that dP/dr varies as $r^3/r^2 = r$, and consequently tends to *zero* when $r \to 0$.

However, numerical computation shows that dP/dr passes through a very sharp maximum before going to zero for $r = 0$. It is therefore difficult to determine the central pressure $P(0)$ with precision by the method which consists of starting from $r = R_0$ and approaching the 'difficult' region near the center with a certain accumulation of numerical errors, making the large value of $|dP/dr|$ especially uncertain.

This is why, instead of taking R_0 and L_0 as the only free parameters, it is preferable to take $P(0)$ and $T(0)$ as *additional* provisional parameters; then, starting from $r = 0$, one constructs (Figure 3.2) curves (C_{center}) for each of the four unknown functions, for which all the ordinates are now known from the outset.

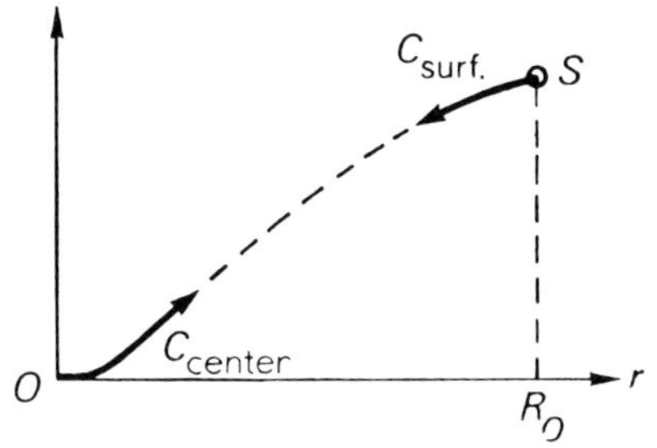

Fig. 3.2. The curves (C_{center}) and (C_{surface}) corresponding to an arbitrary choice of R_0, L_0, $P(0)$ and $T(0)$.

These new curves (C_{center}) do not necessarily join the curves (C_{surface}) which have been previously discussed, and which can be drawn starting from $r = R_0$ once R_0 and L_0 have been assumed.

Since the curves (C_{center}) originating at the center already satisfy the conditions $M_r(0) = 0$ and $L_r(0) = 0$, the test for the right values of R_0, L_0, $P(0)$ and $T(0)$ will now be able to make the four curves (C_{center}) join the corresponding curves (C_{surface}) at some point of abscissa $r = r_j$.

7. From Initial Models to Models Corresponding to the Present State. Determination of the Age of a Star

For more advanced stages of evolution we can apply the methods just described, but interchange the roles played by r and M_r, and use Equations such as (43) and (44).

Now let t be the *sum* of the (Δt) corresponding to the (ΔX), (ΔY), and (ΔZ) which give a solution R_t and L_t, where R_t and L_t represent the values observed *at the present time* for certain stars (for very distant stars, one must of course correct for the propagation time of the light). Then $t = \Sigma(\Delta t)$ represents the *'age'* of the star – that is, the interval separating its present stage from its initial stage. In this way, we find, for instance, that the present radius and luminosity of the Sun are reached – starting with $X_0 = 0.744$, $Y_0 = 0.236$, $Z_0 = 0.020$ (and $X_{\text{CN}}/Z = \frac{1}{5} =$ const. independent of t) – in $t \approx 5 \times 10^9$ yr, with $X_t(\text{center}) \approx 0.50$.

Details of the results concerning the Sun are given (following Schwarzschild, 1958) in the Tables and Figures explained below.

8. The Present Internal Structure of the Sun

8.1. THE DENSITY DISTRIBUTION IN THE SUN AS A FUNCTION OF THE DISTANCE FROM THE CENTER

TABLE 3.3

The density distribution in the Sun as a function of the distance from the center

$r' = r/R$	0.0	0.1	0.2	0.3	0.4	0.5	0.6	0.7	0.8
ρ (g/cm^3)	135	86	36	13	4.1	1.3	0.40	0.12	0.036
$r' = r/R$	0.90	0.92	0.94	0.96	0.98	1.00			
$10^3 \rho$ g/cm^3	9.45	6.55	4.14	2.16	0.752	0.00			

Watch out for the factor 1000 in the second part of the table.

As one can read from Table 3.3, the mass density of the Sun decreases monotonically, from the center, where it is equal to 135 g cm^{-3}, to the surface, where it is equal to zero.

(A better insight into the trend of these variations of ρ with r, will be provided below by the logarithmic representation of Figure 3.4.)

8.2. THE DISTRIBUTION OF THE MASS CONTAINED IN A SPHERE OF RADIUS R

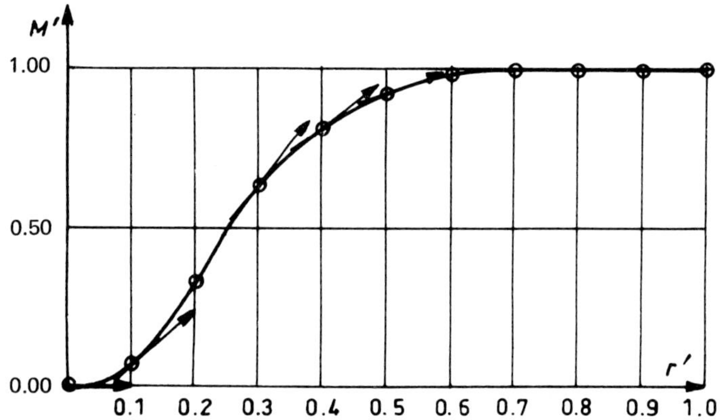

Fig. 3.3. The distribution of the mass M_r contained in a sphere of radius r.

Figure 3.3 indicates the variations of $M' = M_r/M_\odot$ as a function of $r' = r/R_\odot$, where $M_\odot$ and $R_\odot$ are the present total mass and the radius of the Sun.

We see in Figure 3.3 that *most of the total mass of the Sun is concentrated near the 'central half'*, since M' remains very close to 1 from the surface to about $r' = 0.60$. This is due, of course, to the very low density (see Table 3.3) in the 'outer half' of the Sun.

(We shall take advantage of these particularities in Section 10, when studying the outer 'convective zone' of the Sun.)

8.3. The Distribution of Temperatures in the Sun

TABLE 3.4
The distribution of temperatures T in the Sun

r'	1.00	0.98	0.96	0.94	0.92	0.90	0.80	0.60	0.40	0.20	0.00
T_6	–	0.11	0.23	0.35	0.47	0.61	1.27	2.50	4.74	9.35	14.6

We see from Table 3.4 that the temperature $T_6 = T/10^6$ K presents *a very abrupt increase near the surface* (where its value is of the order of 6000 K, according to observations).

But as soon as we leave the immediate neighbourhood of the surface, the variation of T becomes *very slow*. One can even consider, in the very first rough approximation, that the temperature in the interior of the Sun *is constant* and of the order of 10^6 K.

(A more precise description will be given below, in comments on Figure 3.4.)

8.4. The Distribution of Pressures in the Sun

TABLE 3.5
The distribution of pressures P in the Sun

r'	1.00	0.98	0.96	0.94	0.92	0.90	0.80	0.60	0.40	0.20	0.00
$\log_{10} P$	–	10.1	10.8	11.3	11.6	11.9	12.8	14.1	15.4	16.7	17.35

We see from Table 3.5 that the pressure P (given in c.g.s. units) *varies very rapidly* with the distance r' to the center. The pressure at the center (or, as we say, *the central pressure*) P_c is enormous, for it reaches 22×10^{16} c.g.s. – that is, 220 *billion atmospheres* (1 atm $= 1.013 \times 10^6$ c.g.s.).

The computation of P_{gas} and of P_{rad} shows that everywhere in the Sun $P_{\text{rad}}/P_{\text{gas}}$ is less than 6×10^{-4}. One can therefore assimilate P with P_{gas}.

8.5. The Empirical Representation of the Functions $g(r')$, $\rho(r')$, $T(r')$ and $P(r')$

A graphical representation, (Figure 3.4 overleaf), of the gravitational field strength g, of the density ρ, of the temperature T and of the pressure P, as *logarithmic* functions of the distance r' to the center, shows that the logarithms of these functions *vary linearly* with r' everywhere except in the immediate neighbourhood of the center and of the surface.

We see that the variations of g and T are the slowest: the lines representing them have slopes equal respectively to (-3.10) and (-3.27), (on the $\log_e$ scale). The variation of $\log_e \rho$ is more rapid, with a slope of (-11.4). It is $\log_e P$ that varies most quickly, with a slope of (-14.7). Measuring the 'intercepts', we obtain the following c.g.s. formulae:

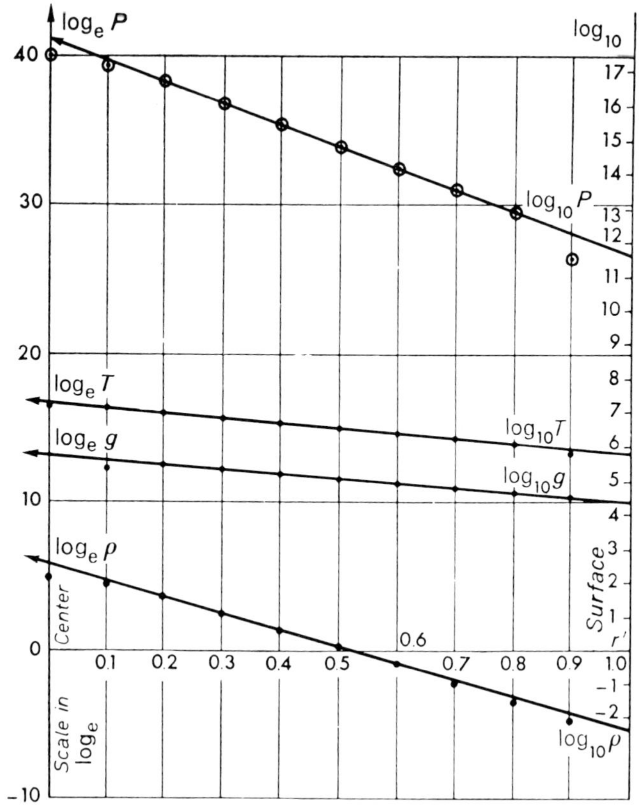

Fig. 3.4. The empirical representation of the functions $g(r')$, $\rho(r')$, $T(r')$, $P(r')$.

$$P(r') = (891 \times 10^{15}) e^{-(14.7)r'} \text{ c.g.s.}; \tag{47}$$

$$T(r') = (20 \times 10^{6}) e^{-(3.27)r'} \text{ K}; \tag{48}$$

$$\rho(r') = (331) e^{-(11.4)r'} \text{ c.g.s.}; \tag{49}$$

$$g(r') = (4.36 \times 10^{5}) e^{-(3.10)r'} \text{ c.g.s.}. \tag{50}$$

We must take care not to confuse the coefficients of the exponentials with the exact values for $r' = 0$, since these empirical representations are not valid for r' less than 0.15, or greater than 0.85.

Real, exact, central values are: $P_c = 2.24 \times 10^{17}$; $T_c = 14.6 \times 10^{6}$; $\rho_c = 135$.

8.6. THE DISTRIBUTION OF ENERGY PRODUCTION PER GRAM IN THE SUN

Table 3.6 shows that the energy production per gram of the mixture decreases very rapidly when r' increases (the contribution of ϵ_{CN} becomes negligible as soon as r' attains 0.2; this explains why the uncertainty concerning X_{CN} is *not* important). The total energy

TABLE 3.6

The distribution of energy production per gram, ϵ, in the Sun

r'	0.0	0.1	0.2	0.3	0.4
ϵ_{pp}	14.4	7.9	1.2	0.1	0.0
ϵ_{CN}	2.1	0.2	0.0	0.0	0.0

production ϵ per gram of the stellar mixture is already negligible for $r' = 0.4$. This gives the extension of the 'core' where nuclear energy is produced.

8.7. The Distribution of the Net Flow of Energy through the Sphere of Radius r in the Sun

TABLE 3.7

The distribution of the net flow of energy, $L_r(r)$, through the sphere of radius r, in the Sun

r'	ρ	ϵ_r	dL'/dr'	L'
0.0	135	16.5	0.00	0.000
0.1	86	8.1	7.75	0.396
0.2	36.4	1.2	1.9	0.909
0.3	13.0	0.1	0.1	0.994
0.4	4.1	0.0	0.0	1.000

Equations (20) and (14) of the energy equilibrium (and the values of ρ and ϵ in Table 3.7) show that $L' = L_r/L_\odot$ equals zero at the center, and the same is true for dL'/dr'. But according to the same equation dL'/dr' also becomes negligible as soon as the energy production ϵ becomes negligible, that is as soon as r' becomes greater than 0.4.

Table 3.7 also shows that dL'/dr' has a very sharp maximum in the vicinity of $r' = 0.1$, which means that the slope of the curve A, representing the variations of $L'(r')$ on Figure 3.5 (overleaf), is maximum for r' approximately equal to 0.1. For $r' > 0.4$ the derivative dL'/dr' is zero (no energy production) and consequently L' is constant (and equal by definition to 1) up to the surface ($r' = 0$).

Curve B, on Figure 3.5 (overleaf), illustrates the distribution of the temperatures T which are given here as logarithms of $T_6 = T/10^6$ K. The scale of the ordinates is the same for $\log_{10} T_6$ and for L'.

8.8. The Distribution of Hydrogen Abundances in the Sun

Table 3.8 shows the distribution of hydrogen (more specifically *proton*) abundances X inside the core of the Sun (in the present stage of its evolution). We see that X becomes very near to its observed value (0.744) at the surface, as soon as r' becomes greater than 0.2.

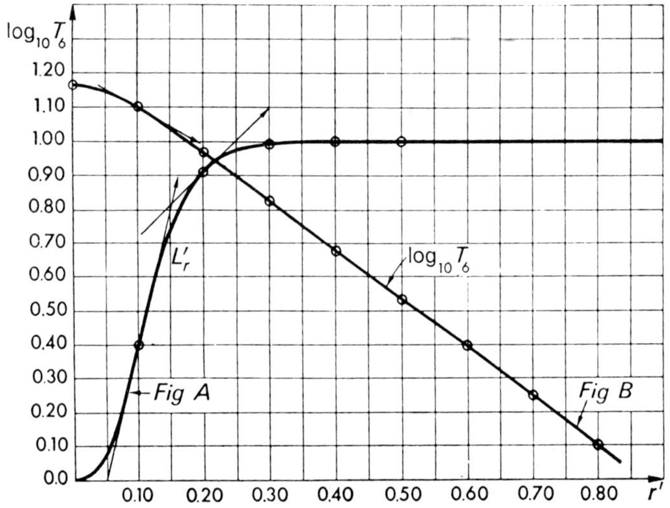

Fig. 3.5. The variations of $L'(r')$ (curve A) and the variation of the temperatures (curve B).

Consequently, the mean mass μ of one particle of the stellar mixture – in units m_B – as given by Equation (6), and indicated in Table 3.8, varies very slowly from the value 0.734 at the center to the value 0.595 at the surface, remaining constant throughout the whole region between $r' = 0.3$ and the surface.

TABLE 3.8

The distribution of hydrogen abundances, X, in the Sun

r'	0.0	0.1	0.2	0.3	0.4
X	0.494	0.611	0.723	0.744	0.744
μ	0.734	0.660	0.606	0.595	0.595

9. Comparison between the Present Structure of the Sun and its Structure at Age Zero

Table 3.9 compares the initial stage (zero age) of the Sun and the present (evolved) stage.

It is self explanatory, except for r'_z which represents the lower boundary of the 'convective zone', which extends from r'_z up to the surface.

TABLE 3.9

A comparison between the initial state and the present state of the Sun

Initial state	*Present state*
$X_0 = 0.744 \quad Z_0 = 0.02$	$X_0 = 0.744 \quad Z_0 = 0.02$
$M_\odot = 1$	$M_\odot = 1$
$r'_z = 0.88$	$r'_z = 0.86$
$M'(r'_z) = 0.999$	$M'(r'_z) = 0.998$
$T_6(r'_z) = 0.72$	$T_6(r'_z) = 0.88$
$T_6(\text{center}) = 12.39$	$T_6(\text{center}) = 14.6$
$R_0/R_\odot = 1.021$	$R_\odot$
$\bar{\rho} = 1.32\ \text{g cm}^{-3}$	$\bar{\rho} = 1.41\ \text{g cm}^{-3}$
$\rho_c = 77\ \text{g cm}^{-3}$	$\rho_c = 135\ \text{g cm}^{-3}$
$L_0/L_\odot = 0.578$	$L_\odot$
$X_c = X_0 = 0.744$	$X_c = 0.50$

(The theory of convection in stars – given in Section 10 – is slightly more advanced than the general level of our textbook, and can be omitted in a first reading.)

Table 3.9 shows that the principal changes produced by an evolution which decreases the value of X at the center from its zero-age value of 0.744 to its present value of about 0.50, are: a slight increase in the central temperature, which goes from T_6(center) = 12.39 to T_6(center) = 14.6; a slight decrease (hardly noticeable) in the radius R from $R_0 = 1.021 R_\odot$ to $R_\odot$; a slight increase in the mean density, which goes from $\bar{\rho} = 1.32$ g cm^{-3} to $\bar{\rho} = 1.41$ g cm^{-3}; finally, a relatively large increase in central density ρ_c which goes from 77 g cm^{-3} to 135 g cm^{-3} (a change essentially due to a decrease in X_c, which increases μ and the central pressure P_c, whose common logarithm goes from 17.13 to 17.35 in c.g.s. units). But the most spectacular change (in contrdiction to an oversimplified '*mass-luminosity relation*') is in the luminosity L, which goes from $L_0 = 0.578 L_\odot$ to $L_\odot$

As for the relative position of the convective zone defined by r'_z, it remains practically unaffected by the evolution.

10. The Superficial Convective Zone of the Sun

10.1. Preliminary Remarks

It is well known that an adiabatic transformation can be described by setting $dQ = 0$ in the 'thermodynamic identity' (see, e.g. Kittel, 1969, p. 111)

$$dU = dQ - PdV \tag{51}$$

describing reversible transformations with a constant number N of particles of a thermodynamic system.

For a monoatomic *ideal* (perfect) gas the average kinetic energy (of translation) of *one* particle is equal to $\frac{3}{2}kT$, so that we have

$$dU = N\tfrac{3}{2}kdT. \tag{52}$$

On the other hand, again for a *perfect gas* occupying a volume V, the number density n is equal to N/V, and the pressure P is given by:

$$P = \frac{N}{V}kT. \tag{53}$$

Let us substitute Equations (52) and (53) into Equation (51) with $dQ = 0$: we obtain:

$$\frac{3}{2}\frac{dT}{T} = -\frac{dV}{V}. \tag{54}$$

Conservation of the mass of the system, ρ being the mass density, can be expressed by:

$$\rho V = \text{Const.} \tag{55}$$

or by

$$\frac{d\rho}{\rho} = -\frac{dV}{V}. \tag{56}$$

Therefore, we can write Equation (54) as:

$$\frac{(d\rho/\rho)}{(dT/T)} = \frac{3}{2}. \tag{57}$$

Now, taking into account Equation (56) we get for the logarithmic derivative of P:

$$\frac{dP}{P} = \frac{dT}{T} - \frac{dV}{V} = \frac{dT}{T} + \frac{d\rho}{\rho}; \tag{58}$$

and, eliminating (dT/T) between Equations (57) and (58), we obtain:

$$\frac{dP}{P} = \left(1 + \frac{1}{3/2}\right)\frac{d\rho}{\rho}. \tag{59}$$

This relation is a particular case, for $n' = \frac{3}{2} = 1.5$, of a *general* relation defining, according to Emden, the '*polytropic index*' n', namely:

$$\frac{dP}{P} = \left(1 + \frac{1}{n'}\right)\frac{d\rho}{\rho}, \tag{60}$$

which is equivalent, *when n' is constant*, to

$$P = \text{Const.}\, \rho^{(1+1/n')}. \tag{61}$$

Thus, taking into account the definition (61), we can summarize the relation (59), by saying that adiabatic transformations of a (non-relativistic) perfect monoatomic gas are described by a polytropic index n' equal to

$$n'_{\text{ad}} = 1.5. \tag{62}$$

(N.B. – The usual notation for the polytropic index is n, but our notation avoids the danger of confusion with the number density.)

10.1.1. Remark. It is, of course, possible to obtain those of the relations written above concerning adiabatic transformation, by starting from the well-known formula

$$PV^{\gamma_{\text{ad}}} = \text{Const.} \tag{63}$$

with γ_{ad} equal to $\frac{5}{3}$ for a perfect (non-relativistic) monoatomic gas.

Some authors use the symbol γ in a general sense, as representing $(1 + 1/n')$ in the general formula (60), (we come back to this point in the remark of Subsection 10.3). With that definition one can write

$$n' = \frac{1}{\gamma - 1} \quad \text{and} \quad n'_{\text{ad}} = \frac{1}{\gamma_{\text{ad}} - 1} = 1.5. \tag{64}$$

10.2. The Direct Forms of the Condition of Convective Instability

Let us consider a macroscopic '*bubble*' of volume V initially situated at the level r, (r being the distance from the center of the star), where absolute value of the gravitational field is equal to g, (Figure 3.6).

This bubble is simply an element of the gas constituting the star, considered (in a

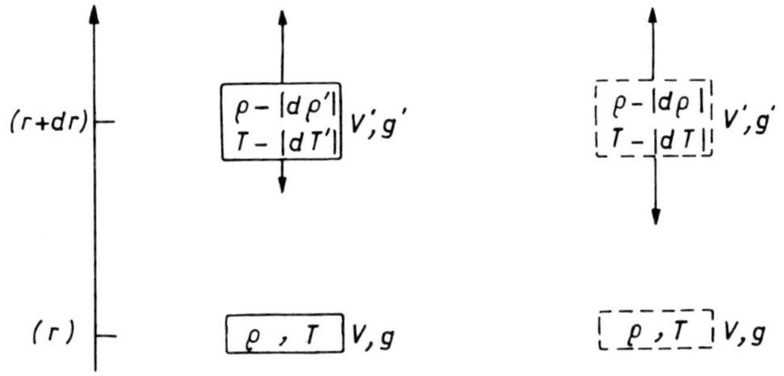

Fig. 3.6. The schematic representation of the data concerning the bubble (left) and of the data concerning the 'reference elements' of the external medium (right).

purely *conceptual* way) as separated (by a *geometrical* 'envelope', deprived of any stress or mass but capable of extension) from the remaining material (called 'external medium'), in order to give some individuality to this element, thus allowing one to consider its *mass* and the corresponding *number* N of particles of the bubble as constant.

Now, suppose that under the influence of some 'perturbation' the bubble suffers a change of level (on Figure 3.6 the bubble is supposed to *rise*, but the same type of discussion would be valid for the opposite case). Let $\rho = \rho(r)$ be the 'initial' mass density of the bubble, equal to that of the external medium, at the level r, since initially the bubble is simply a 'piece' of this medium. On the other hand, let $(\rho + d\rho) = \rho - |d\rho|$ (since $d\rho$ is generally *negative*) be the density, and $(P + dP) = P - |dP|$ (since dP is generally *negative*) be the pressure of the 'external medium' at the level $(r + dr)$. The corresponding temperature is $(T + dT) = T - |dT|$.

The left part of Figure 3.6 concerns the bubble, the right part concerns the external medium. The dotted box at the right lower part of the figure concerns an element of the external medium *identical to the 'initial' bubble* (the same volume V, mass, density ρ, temperature T, and the same intensity of the gravitational field g).

Arriving at the level $(r + dr)$, the *pressure inside* the bubble adjusts itself to the *external pressure* $(P - |dP|)$, and since we supposed, to fix our ideas, that the bubble is *rising*, this external pressure on the bubble is *smaller* than in the initial position.

If we suppose that the rise of the bubble is sufficiently rapid to be *adiabatic*, the *decrease* of the pressure P – interpreted in Equation (63) as internal pressure inside the bubble – will correspond to an *increase* of its volume (an *expansion*), which instead of V takes a *greater value* V', at the level $(r + dr)$. Consequently, since its mass remains constant, the mass density of the bubble will *decrease* and become $\rho + d\rho'$ with $d\rho'$ *negative*, so that the new density of the bubble can be written $(\rho - |d\rho'|)$. (Note the prime over ρ'.) The corresponding temperature inside the bubble (as usual, in an adiabatic expansion) will become *smaller*: $(T + dT')$, with dT' *negative*, so that the new temperature can be written $(T - |dT'|)$. All these results are indicated in or near the box representing the bubble at the *upper left* side of Figure 3.6.

Now let us fix our attention on an element of the *external* medium, (represented by the *dotted* box at the right upper side of the figure), at the *same upper level* $(r + dr)$ as the upper level of the bubble and of *same volume* V' as the volume occupied by the bubble at its upper position. This 'reference element' of external medium, in a static normal star, is in a state of mechanical equilibrium between the *gravitational force* (per unit volume) pulling down, and the *gradient of the pressure* pushing up.

In the absence of a 'contamination' of the external medium by the bubbles, the pressure, the density, and the temperature of the reference element, as that of the external medium, would be those of a normal star at the level $(r + dr)$, *determined by the methods and equations described in* Chapter 2 and the previous Sections of the present Chapter. They would be, respectively $(P - |dP|)$, $(\rho - |d\rho|)$ and $(T - |dT|)$, since in a normal star the gradients of P, ρ and T are all *negative*. Taking $P(r)$, $\rho(r)$, and $T(r)$, to be the functions describing respectively the variations of P, ρ and T, with r, in the *corresponding model of the normal star*, we obviously have: $|dP| = (-dP/dr)\,dr$; $|d\rho| = (-d\rho/dr)\,dr$ and $|dT| = (-dT/dr)\,dr$.

Now, at the upper level $(r + dr)$, the bubble is *pushed up* with the same force (due to the pressure gradient) *per unit volume* as the reference element of the external medium, and – since it has the same volume V' – the same total force as the reference element. The gravitational field, of course, has the same modulus g' at the level $(r + dr)$, for the bubble and for the reference element.

But the total gravitational force acting on the bubble, and pulling it down, equal to $g'(\rho - |d\rho'|)\,V'$, will now be *different* from the gravitational force acting on the reference element, which is equal to $g'(\rho - |d\rho|)\,V'$ [if $|d\rho'|$ is different from $|d\rho|$, which is highly probable, because these two variations are determined by quite different physical circumstances].

Since the forces pushing up are equal, and since the reference element is motionless, the bubble would *go back* to its initial position if the gravitational force pulling it down were greater than that acting on the reference element. The whole situation would be *stable*.

On the contrary, the bubble would continue to rise, if the gravitational force pulling it down were smaller than that acting on the reference element – which would mean the onset of a *'convective motion'* – *the condition for such a motion being*, according to the explanations given above:

$$\boxed{|d\rho'| > |d\rho| \quad \text{or} \quad d\rho' < d\rho} \tag{65}$$

Indeed, it is easily found that the same condition (65) ensures the continuation of a descent for the descending bubbles, the combination of ascending and descending motions producing a circulation of the matter in a certain *'convective layer'* or *'convective zone'*.

Now, if we use the general Equation (60) to define a *polytropic index of the external medium*, n'_{ext}, that is if we write

$$\frac{dP}{P} = \left(1 + \frac{1}{n'_{\text{ext}}}\right)\frac{d\rho}{\rho} \tag{66}$$

where P, dP, ρ, $d\rho$ are *those of the external medium*, (contrary to $P, dP, \rho, d\rho'$ proper to the bubble), and if we apply the same general Equation (60) to the adiabatic transformation in the bubble, that is if we write

$$\frac{dP}{P} = \left(1 + \frac{1}{n'_{\text{ad}}}\right)\frac{d\rho'}{\rho}, \tag{67}$$

we find that the condition of convective circulation (65) can be put (after division by the *negative* quantity $\rho(dP/P)$) first in the form:

$$\left(1 + \frac{1}{n'_{\text{ad}}}\right)^{-1} > \left(1 + \frac{1}{n'_{\text{ext}}}\right)^{-1} \tag{68}$$

and finally in the elegant form:

$$\boxed{n'_{\text{ext}} < n'_{\text{ad}}} \quad \text{or} \quad \boxed{n'_{\text{ext}} < 1.5}\ . \tag{69}$$

Eliminating dP/P between Equations (58) and (60), we get a new *general* definition of the polytropic index, n', (for a non-relativistic perfect gas):

$$n' = \frac{(d\rho/\rho)}{(dT/T)}. \tag{70}$$

Applied respectively to the external medium (characterized by $\rho, d\rho, T, dT$) and to the bubble (characterized by $\rho, d\rho', T, dT'$) this gives:

$$n'_{\text{ext}} = \frac{(d\rho/\rho)}{(dT/T)} \tag{71}$$

and

$$n'_{\text{ad}} = \frac{(d\rho'/\rho)}{(dT'/T)} = 1.5. \tag{72}$$

The last equation shows (since $d\rho'$ is negative) that dT' is *negative* (and we already know that dT is negative). Therefore, the identity of dP/P between the external medium and the bubble can be expressed, through Equation (58), by

$$(-|dT|/T) + (-|d\rho|/\rho) = (-|dT'|/T) + (-|d\rho'|/\rho). \tag{73}$$

It is easily found, after writing all terms relative to the temperatures on the left-hand side, and all terms relative to the densities on the right-hand side, that the condition (65) is equivalent to

$$\boxed{|dT'/dr| < |dT/dr|} \quad \text{or} \quad \boxed{(T + dT') > (T + dT)}\ . \tag{74}$$

This means that convection will take place if, and only if, the bubble is *hotter*, at its *higher level*, than the external medium, which is almost obvious physically, since in that case it will be *lighter* (of a smaller density) than the reference element of the external medium, and will behave like a cork immersed in water.

10.3. Indirect Forms of the Condition of Convective Instability

In most treatises on astrophysics, the direct relations established above, namely the conditions (65), or (69), or (74), are presented in different, indirect, forms more suitable for numerical determination of stellar structure, but less directly connected with description of the physical situation. Conserving the notation introduced above we shall simply write successive relations leading to these indirect conditions of convective instability.

Eliminating $(d\rho/\rho)$ between Equations (70) and (60) we get a new *general* definition of the polytropic index n' (for a perfect non-relativistic gas):

$$n' + 1 = \frac{(dP/P)}{(dT/T)}, \tag{75}$$

which is equivalent to

$$\frac{dT}{dr} = \frac{1}{(n' + 1)} \frac{T}{P} \frac{dP}{dr}. \tag{76}$$

Applied to the bubble, this general relation takes the form:

$$\frac{dT'}{dr} = \frac{1}{(n'_{\text{ad}} + 1)} \frac{T}{P} \frac{dP}{dr}, \quad \text{or} \quad \left|\frac{dT'}{dr}\right| = \frac{1}{(n'_{\text{ad}} + 1)} \frac{T}{P} \left|\frac{dP}{dr}\right|, \tag{77}$$

since both dT' and dP are negative.

Thus condition (74) takes the form:

$$\frac{1}{(n'_{\text{ad}} + 1)} \frac{T}{P} \left|\frac{dP}{dr}\right| < \left|\frac{dT}{dr}\right|, \tag{78}$$

or, using the second relation (64):

$$\boxed{\left(1 - \frac{1}{\gamma_{\text{ad}}}\right) \frac{T}{P} \left(-\frac{dP}{dr}\right) < \left(-\frac{dT}{dr}\right)}\,. \tag{79}$$

Both conditions (78) or (79) of *convective instability* are useful because they contain only quantities relative to the *external medium*, namely: T, P, dT/dr, and dP/dr, whereas n'_{ad} and γ_{ad} are simply numerical constants respectively equal to 1.5 and $\frac{5}{3}$.

On the other hand, writing the condition (65) as

$$\frac{d\rho'}{\rho} < \frac{d\rho}{\rho}, \tag{80}$$

use of the second relation (64), and elimination of $d\rho'/\rho$ by means of (67), yield a new indirect form of the condition of convective instability:

$$\frac{1}{\gamma_{ad}} \frac{1}{P} \frac{dP}{dr} < \frac{1}{\rho} \frac{d\rho}{dr} . \tag{81}$$

Equivalently, taking into account the negativity of dP and of $d\rho$:

$$\boxed{\frac{1}{\gamma_{ad}} \frac{1}{P} \left(-\frac{dP}{dr}\right) > \frac{1}{\rho} \left(\frac{-d\rho}{dr}\right)} . \tag{82}$$

10.3.1. Remark.

The use of the subscript 'ad' in γ_{ad}, which id omitted by some authors, (for instance Weinberg, 1972) is very important. Indeed, Weinberg denotes by γ not the adiabatic exponent of the relation (63), as does Schwarzschild, but more generally the quantity defined by the *first relation* (64), which is equivalent to writing the general relation (61) defining the polytropic index (when n' is constant) as

$$P = \text{Const.}\ \rho^{\gamma} . \tag{83}$$

This definition lets γ to play the role of a (simple) function of the *general* polytropic index n', and not the particular role of γ_{ad}.

Since the theory of convective instability presented above is essentially based on the distinction between n', n'_{ad} and n'_{ext}, it is obvious that one must avoid confusion between the general γ and γ_{ad}.

10.4. THE PASSAGE FROM n'_{ext} TO THE 'EXTRAPOLATED POLYTROPIC INDEX'

It is quite obvious that the concept of a complete separation between the 'bubble' and the 'external medium' can represent only a first approximation.

Indeed, at least for two reasons the external medium, considered above, will be 'contaminated' by the moving bubbles. First of all, we have seen (in our comments on Equation (74)) that the bubble arriving at the level $(r + dr)$ is *hotter* than the external medium at the same level. We have assumed, certainly, that the expansion of the bubble associated to its rise from the level r to the level $(r + dr)$ is *adiabatic*, that is without any exchange of heat with the external medium. However, since the bubble is not separated from its surroundings by any thermostatic wall and since it is hotter all along its path than the surrounding medium, it will have a tendency to loose some heat to the external medium. Thus the medium will acquire, at the level $(r + dr)$ a temperature $(T + dT')$ *higher* than the temperature $(T + dT)$ it would have in the absence of a convective circulation.

Next, in the absence of a thermostatic wall the bubble can suffer an approximately adiabatic expansion (i.e. only with a *negligible* heat exchange with its surroundings), only if these surroundings are *almost* at the same temperature as the bubble, all along its path.

Thus, everything conspires for a (thermal) contamination of the external medium, and for a participation of this medium in the convective circulation, this medium acquiring a *thermal gradient* (dT/dr) *infinitely near* the *adiabatic gradient* (dT'/dr), but still *very slightly different*, and obeying the condition (74) of convective instability, in order to ensure the validity of the whole theory.

Consequently, the meaning of n'_{ext}, defined previously is modified: it represents no longer the polytropic index of a really 'external' medium, crossed by convective bubbles. Instead it will represent (Figure 3.7) the value of the polytropic index n', (defined in a general way by Equation (60), with actual value of P, dP, ρ, $d\rho$, *outside* the convective layer – i.e. *within* the region where n' is *greater* than *its adiabatic value* 1.5) *extrapolated* to the region (r' greater than 0.86 for the Sun) where it *would* become smaller than 1.5 if temperature gradients were *not* equalized by convection.

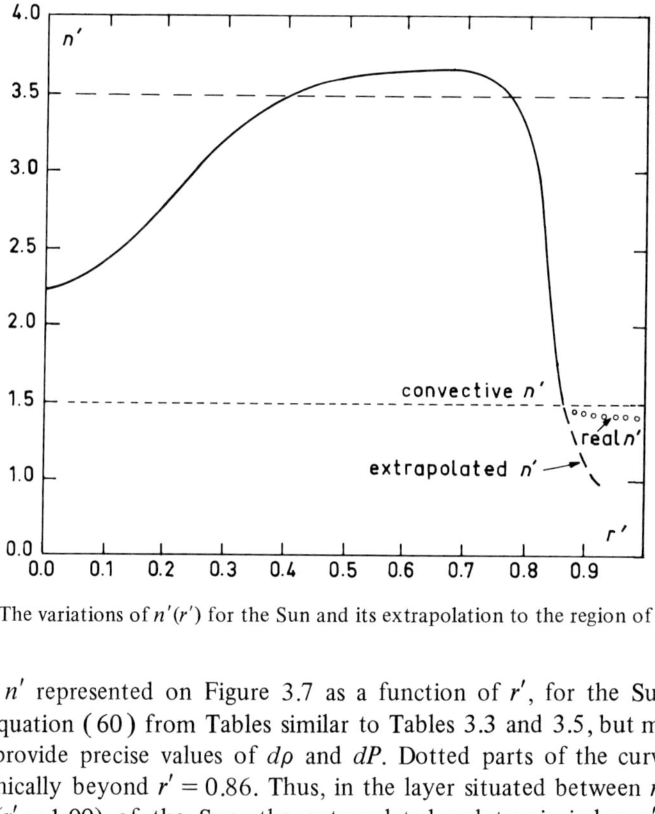

Fig. 3.7. The variations of $n'(r')$ for the Sun and its extrapolation to the region of convection.

Values of n' represented on Figure 3.7 as a function of r', for the Sun, have been derived by Equation (60) from Tables similar to Tables 3.3 and 3.5, but more detailed, in order to provide precise values of $d\rho$ and dP. Dotted parts of the curves are extrapolated graphically beyond $r' = 0.86$. Thus, in the layer situated between $r' = 0.86$ and the surface ($r' = 1.00$) of the Sun, the extrapolated polytropic index n'_{ext} would be smaller than 1.5 and a 'convective zone' must take place, with a *real* value of n' infinitely near the adiabatic value $n' = n'_{\text{ad}} = 1.5$, but *very slightly smaller than* 1.5 (schematically indicated by *open circles* on the figure).

Neglecting this very small difference between dT' and dT, and using the variable r' instead of the variable r, Equation (77) can be applied in the convective zone and (with $n'_{\text{ad}} = 1.5$) we get:

$$\frac{dT}{dr'} = \frac{2}{5}\frac{T}{P}\frac{dP}{dr'} \quad (\text{for } 0.86 \leqslant r' \leqslant 1.00). \tag{84}$$

Now, according to the explanations of Figure 3.3, $M' = M_r/M_\odot$ in the convective zone is very near 1.00, so that in this region Equation (2) of mechanical equilibrium can be written in the form (in c.g.s. units)

$$\frac{dP}{dr} = -GM_{\odot}\rho/r^2 \tag{85}$$

or, numerically,

$$\frac{dP}{dr'} = -1.91 \times 10^{15}\rho/r'^2. \tag{85'}$$

On the other hand, we still can apply here Equation (25) which, taking into account the value of μ given by Table 3.8 ($\mu = 0.595$), becomes numerically:

$$\frac{T}{P} = (7.2 \times 10^{-9})\rho^{-1} \quad \text{(in c.g.s.).} \tag{86}$$

Substitution of Equations (85) and (86) into (84) gives:

$$\frac{dT}{dr'} = (-5.50 \times 10^6)\frac{1}{r'^2}. \tag{87}$$

Integrating this equation, with $T(1.00) = 0$ as boundary condition, we find:

$$T = T(r') = (5.50 \times 10^6)\left(\frac{1}{r'} - 1\right) \quad (T \text{ in K}). \tag{88}$$

This last relation gives the distribution of temperatures in the convective zone ($0.86 \leqslant r' \leqslant 1.00$) of the Sun, and consequently, through Equations (84) and (86), the distribution of other parameters such as P or ρ.

10.4.1. Remark.

In the convective zone of the Sun, H is only *partly* ionized. Thus an important part of the work of the bubble on the external medium, instead of adiabatically cooling the bubble, is used in recombination of H (and He) ions, so that γ_{ad} takes a more complicated value than (5/3), and Equations (63) is replaced by a more complicated law.

References

1. WORKS CITED IN THE TEXT

(Works that can be used for further study are marked by an asterisk.)

Kittel, C.: 1969, *Thermal Physics*, Wiley, New York.

*Kourganoff, V.: 1973, *Introduction to the Physics of Stellar Interiors*, Reidel, Dordrecht, Holland/Boston, U.S.A.

*Schwarzschild, M.: 1958, *Structure and Evolution of the Stars*, Princeton University Press, (Reprinted (1965) by Dover, New York).

Weinberg, S.: 1972, *Gravitation and Cosmology*, Wiley, New York, p. 309.

2. REFERENCES FOR FURTHER STUDY

(See first the works cited in 1 marked by an asterisk)

Chandrasekhar, S.: 1967, *An Introduction to the Study of Stellar Structure*, Dover, New York.

Kuiper, G. P.: 1965, (ed.) *Stars and Stellar Systems*, Vol. VIII, Univ. of Chicago Press.

Meadows, A. J.: 1967, *Stellar Evolution*, Pergamon, Oxford.

PART II

WHITE DWARFS, NEUTRON STARS AND PULSARS

CHAPTER 4

ELEMENTARY PROPERTIES OF A DEGENERATE FERMI GAS

1. Different 'Energy Parameters' of an Isolated Particle. Energy Groups (NR), (UR) and (RR)

1.1. Introduction

We deal in this section with relatively elementary and well known results of the theory of special relativity. However, most of the corresponding *notations* used either in astrophysics or in statistical physics are rather unusual. Therefore a few words about these notations can be useful even for some advanced readers.

Let us consider *one* isolated particle (electron, proton, neutron, or a more complex nucleus) not necessarily at rest but whose mass *at rest* is m. Instead of speaking about the 'rest mass', we shall call this mass at rest simply the '*mass*' m. Indeed, in the problems considered below the rest mass of the particle is the only one appearing in different equations, and our terminology and notation greatly simplify the language and most formulae. It allows us, in particular, to drop the subscript (o) often used to specify the rest mass. Of course, when we say that a particle is at rest it means simply that the modulus p of its momentum is zero in some specific reference system.

A particle in motion with respect to some reference system possesses a momentum $\mathbf{p}$, whose modulus p is *non-zero*, and an energy ϵ. This '*total*' energy must not be confused with the corresponding kinetic energy ϵ_k, which will be defined below.

Since we will be interested only in the *modulus* p, we shall simplify our terminology by saying that the particle has the momentum p.

1.2. The (Total) Energy as a Function of the Momentum

The '(*total*) *energy*' ϵ of a particle of mass m and of momentum p is given, according to the theory of special relativity, by the fundamental relation (see for instance Kittel *et al.*, 1962, chap. 12):

$$\epsilon^2 = (mc^2)^2 + (pc)^2. \tag{1}$$

This relation shows that at rest the energy ϵ_0 (the 'rest energy') of a particle of mass m is equal to

$$\epsilon_0 = mc^2. \tag{1'}$$

Thus the (total) energy ϵ of a particle in motion can be expressed as a function $\epsilon(p)$ of its momentum p by

$$\epsilon = \epsilon(p) = \epsilon_0[1 + (p/mc)^2]^{1/2}. \tag{2}$$

1.3. THE DIMENSIONLESS PARAMETER x CORRESPONDING TO MOMENTUM

On introducing the *dimensionless parameter* x, defined by

$$x = \frac{p}{mc}, \tag{3}$$

(x measures *the momentum* p in units mc), we obtain the (total) energy ϵ as a function $\epsilon(x)$ of x, (depending on m through ϵ_0):

$$\boxed{\epsilon = \epsilon(x) = \epsilon_0(1+x^2)^{1/2}.} \tag{4}$$

On introducing next the parameter Θ defined by

$$x = sh\Theta, \tag{5}$$

we obtain the (total) energy ϵ as a very simple function $\epsilon(\Theta)$ of Θ:

$$\epsilon = \epsilon(\Theta) = \epsilon_0 ch\Theta. \tag{6}$$

1.4. THE KINETIC ENERGY

The '*kinetic energy*' ϵ_k of a particle, defined in a *general* way, is the difference between the total energy ϵ and the rest energy ϵ_0 of the particle by:

$$\epsilon_k = \epsilon - \epsilon_0. \tag{7}$$

Thus, the kinetic energy ϵ_k of the particle is given by the 'classical' relation $\frac{1}{2}mV^2$, where V is the modulus of the velocity of the particle, only for very small, 'non-relativistic', values of V or p.

The kinetic energy ϵ_k is a function $\epsilon_k(x)$ of x given by

$$\epsilon_k = \epsilon_k(x) = \epsilon_0[(1+x^2)^{1/2} - 1] = \epsilon(x) - \epsilon_0, \tag{8}$$

which can be expressed as a function $\epsilon_k(\Theta)$ of Θ by

$$\epsilon_k = \epsilon_k(\Theta) = \epsilon_0(-1 + ch\Theta). \tag{9}$$

1.5. THE ENERGY PARAMETERS

Thus, we see that the energy state of a given moving particle (characterized at rest by mc^2 or by ϵ_0) can be described by *any* of the following '*energy parameters*': $\epsilon, p, x, \epsilon_k$ or Θ. The last one Θ, contrary to the others, has no simple physical meaning, but it can be useful (as will be shown later) in some computations.

The knowledge of *one* of these energy parameters allows us to obtain any of the others by the general formulae written in the form of a chain of equalities:

$$\boxed{\epsilon = \epsilon_0 + \epsilon_k = \epsilon_0(1+x^2)^{1/2} = \epsilon_0[1+(p/mc)^2]^{1/2} = \epsilon_0 ch\Theta.} \tag{10}$$

1.6. THE CRITICAL VALUE $x = 1$. APPROXIMATIONS (NR) AND (UR)

Each of the above energy parameters is particularly well adapted to one of the specific problems considered below.

This already can be seen in the following example. Let us suppose that the motion is sufficiently slow for x to be *very much smaller than 1*. In this case we can expand $(1 + x^2)^{1/2}$ in powers of x^2, and limit our expansion to the term proportional to x^2. This approximation is called '*non-relativistic*' and is indicated by the symbol (NR). Thus on using the energy parameter x, we obtain very simple expressions for $\epsilon_k(\text{NR})$ and $\epsilon(\text{NR})$, viz.

$$\epsilon_k(\text{NR}) = \tfrac{1}{2}\epsilon_0 x^2 \quad (x \ll 1), \tag{11}$$

and

$$\epsilon(\text{NR}) = \epsilon_0 + \tfrac{1}{2}\epsilon_0 x^2. \tag{11'}$$

Besides, Equation (11) is identical with the 'classical' relation which gives the kinetic energy ϵ_k of the particle as a function of its momentum p and mass m:

$$\epsilon_k(\text{NR}) = \frac{1}{2}\frac{p^2}{m}. \tag{12}$$

On the other hand, for x *much greater than 1* we can write in (8) the term $(1 + x^2)^{1/2}$ as $x(1 + x^{-2})^{1/2}$ and limit the expansion of $(1 + x^{-2})^{1/2}$ to its first term, i.e. to *1*. This approximation, called '*ultra relativistic*' (or 'extreme relativistic' by some authors) is indicated by the symbol (UR). It gives (neglecting 1 in comparison with x):

$$\epsilon_k(\text{UR}) = \epsilon_0 x \quad (x \gg 1), \tag{13}$$

and

$$\epsilon(\text{UR}) = \epsilon_k(\text{UR}) = \epsilon_0 x = (mc^2)(p/mc) = cp. \tag{14}$$

Thus we see that a description of the energy state of a particle by means of the energy parameter x (m being represented through ϵ_0) is particularly convenient if one wishes to distinguish between the non-relativistic case (NR) when $x \ll 1$ and the ultra-relativistic case (UR) when $x \gg 1$.

1.7. THE RIGOROUS RELATIVISTIC (RR) RELATIONS

The distinction between (NR) and (UR) is much more a *mathematical* distinction (between two types of simple *formulae*) than a physical one. Physically, the passage between the two extreme cases is quite continuous, all physical situations being represented, quite generally, by the '*rigorous relativistic*' formulae indicated by the symbol (RR). These formulae are rewritten in Equation (16) with explicit mention of this symbol (RR).

Near the '*critical*' value $x = 1$ one cannot use the relations marked (NR) or (UR) but must use the general relations marked (RR). On the other hand, relations marked (RR) can be applied for any value of x.

Thus we can summarize all possible relations and definitions given above by the following chains of equations:

$$\epsilon_k(\text{NR}) = \epsilon(\text{NR}) - \epsilon_0 = \frac{p^2}{2m} = \frac{1}{2}\,\epsilon_0 x^2 \quad (x \ll 1); \tag{15}$$

$$\epsilon(\text{RR}) = \epsilon_0 + \epsilon_k(\text{RR}) = \epsilon_0 ch\Theta = \epsilon_0(1 + x^2)^{1/2}$$
$$= \epsilon_0[1 + (p/mc)^2]^{1/2}; \tag{16}$$

$$\epsilon(\text{UR}) = \epsilon_k(\text{UR}) = cp = \epsilon_0 x \quad (x \gg 1); \tag{17}$$

$$x = \frac{p}{mc}; \qquad x = sh\Theta; \qquad \epsilon_k = \epsilon - \epsilon_0. \tag{18}$$

Note first that the expression for ϵ(RR) is simpler than that for ϵ_k(RR). For this reason the chain (16) starts with ϵ(RR), whereas the chain (15) starts with ϵ_k(NR). Note next that we have introduced the parameter Θ only in the general formulae (16) because all relations (NR) and (UR) are simple enough not to necessitate any change of variables. Note finally that we have not introduced p and x in the same order in the chains (15), (16) and (17), because p is used very seldom in the case (RR).

2. The Number of Independent Identical Particles (Confined in a Macroscopic Unit of Volume) whose Momentum Lies between p and $(p + dp)$

2.1. Introduction

Like the preceding section, this section deals with relatively elementary and well known results of statistical physics and quantum mechanics.

However, the *terminology* used in statistical physics is not always the same as that used in quantum mechanics. Moreover, the differences in *notation* used by different authors are so considerable that it can be useful to define, in a precise way, the notation used below.

We consider a collection of *independent* (i.e. with no other mutual interaction than elastic collisions) *identical* particles, of individual mass m, confined in a macroscopic volume V supposed for more simplicity to be equal to 1 cm^3. This last choice allows us to identify the total number of particles of our system with the *number density* n of the considered particles.

Each *energy* state of *one* particle of the system can be expressed by *any* one of the energy parameters (p, ϵ_k, ϵ, x or Θ), defined in Section 1, and can be visualized by a *single* horizontal line on a diagram of energy levels (with as many *scales* as different energy parameters).

2.2. The Occupancy of the Accessible Quantum States. The 'Orbitals'

The fact that the considered system of particles is quantized means that each particle of the system can only be in a *quantum* state perfectly determined and characterized by a discrete set of quantum numbers depending (for the independent particles considered here) only on the boundary conditions imposed by the mode of confinement (volume V) of the system (no continuous spectra are considered below).

However, we must distinguish between the virtual quantum states *allowed* by quantification, the only *possible* or 'accessible' states, and the *effective*, real, occupancy of an accessible state by a particle. This concept of virtuality of accessible states is now usually described in terms of '*orbitals*' (see, for instance, Kittel, 1969, p. 137).

The term '*orbital*' (a very misleading and unhappy one!) means *one* of the virtual quantum states, i.e. a solution of the wave equation (without any distinction between eigenvalues and eigenfunctions) for one *single* particle (a model particle) confined in the volume V.

Since the system is supposed to be formed by independent identical particles, the virtual quantum states of *each* particle must be identical and can, indeed, be determined by the study of the behaviour of just one particle considered as the model of all particles.

The difference between a *virtual* quantum state and the real occupancy corresponds, in our case, to the fact that there are usually an *infinite* number of orbitals (virtual states) available for occupancy, whereas the number n of particles of the system is always practically *finite*; n particles will occupy only n virtual states, they cannot occupy all of the infinite number of orbitals (see, for instance, Kittel 1969, p. 137).

2.3. THE NUMBER $D(p, dp)$ OF VIRTUAL QUANTUM STATES BETWEEN p AND $(p + dp)$

Let us recall that all particles can be divided in two large groups: *fermions* and *bosons*, according to the value of their *spin* s. The most important fermions considered below (electrons, protons, neutrons) have in units $h/2\pi$ a spin s equal to $\frac{1}{2}$. The most important bosons considered below (photons, α-particles) have a spin s equal to zero.

Now, the virtual *quantum* states are, in principle, characterized, in our case, *both* by some discrete value of the momentum p, or any other equivalent energy parameter $\epsilon(p)$, $\epsilon_k(p)$, $x(p)$, . . . , *and* by a quantum number expressing the orientation of the spin (when s is not zero) with respect to a virtual magnetic field. So, for fermions of spin $s = \frac{1}{2}$, such as those considered below, this spin can be *vectorially* parallel or antiparallel to the virtual magnetic field, hence (for a given p) *two* possible quantum states according to the state of spin. Thus, for our fermions each virtual value of an *energy* parameter, such as p, corresponds to *two* different virtual *quantum* states. (Certainly a difference in 'spin state' when a real magnetic field is present also corresponds to a difference in potential *energy*, but this aspect of the problem is always deliberately neglected in the problems of statistical physics considered below.)

For the moment let us simply note that the horizontal lines indicating the virtual values of energy of our model particle on a level diagram always present (in virtue of quantization) some spacing.

Now, because of the macroscopic character of the volume V, enclosing our model particle, the spacing of these levels is *so narrow* that one is led to describe the situation *statistically* by the *number* $D'(p, dp)$ of virtual *energy* levels whose momentum lies between p and $(p + dp)$. One can also consider the 'density of virtual energy levels' $D'(p)$ defined by $D'(p, dp)/dp$.

Let us take the well known expression of $D'(p, dp)$ as a function of p and dp (see for instance Reif, 1964), viz.

$$D'(p, dp) = (4\pi p^2\, dp)/h^3, \tag{19}$$

where h is Planck's constant.

Thus, if we call g_s the number of possible spin states (equal to *1* for photons and α-particles; and equal to *2* for electrons, protons, and neutrons) the number $D(p, dp)$ of (complete) *quantum* states accessible to our model particle, when its momentum is lying between p and $(p + dp)$, will be given by

$$D(p, dp) = g_s(1/h^3)\, 4\pi p^2\, dp. \tag{19'}$$

Of course, in quantum mechanics one could say that in absence of a real magnetic field each of the energy levels $\epsilon(p)$ is g_s-fold *degenerate*, but this would introduce a dangerous confusion with the concept of '*degeneracy*' as it is used in statistical physics.

2.4. THE OCCUPANCY RATE

Let us now return to the problem of effective occupancy of a given virtual (complete) quantum state by some of the particles of our system or, more precisely, to the problem of the real distribution of the n particles contained in 1 cm^3 between the different virtual quantum states (this distribution necessarily includes, entirely, unoccupied virtual states).

We must first recall (restricting for the moment the problem to *fermions* alone) that the problem of an eventual occupancy of virtual quantum states is governed by the Pauli *exclusion principle*, according to which two or more fermions never can occupy the same virtual *quantum* state.

Consequently, if some of the virtual *quantum* states (*infinite* in number) are occupied (by *1* fermion at maximum) others will remain necessarily *empty* and the number $n(p, dp)$ of particles in each cm^3 whose momentum lies between p and $(p + dp)$ will be generally less than, or at maximum equal to, the number $D(p, dp)$ of virtual *quantum* states accessible in this interval of momentum. In other words, if we write

$$f_p = \frac{n(p, dp)}{D(p, dp)} = \frac{n(p) \cdot dp}{D(p) \cdot dp} = \frac{n(p)}{D(p)}, \tag{20}$$

the number f_p will obey (for fermions) the condition $f_p \leqslant 1$. This quantity f_p (f being here the first letter of the word *factor* and not of the word *function*) is a number depending not only on p but, as will be shown later, on very many other parameters. It can also be shown – but this property will not be used below – that for *bosons* f_p can become greater than unity.

2.5. A PROBLEM OF TERMINOLOGY

Conceptual difficulties and difficulties in terminology, usually encountered in astrophysical applications of statistical physics, are generated by the fact that one usually *speaks* only about *energy* states (the value of p, or ϵ, or ϵ_k, or x, etc) and not about 'complete' *quantum* states including the spin (although these are naturally taken into account).

We have just defined f_p as the *occupancy rate* of each (complete) virtual *quantum* state in the vicinity of an *energy* level corresponding to p. But in statistical physics and in advanced thermodynamics the following general procedure is adopted, usually without a clear explanation: one expresses the exclusion principle in a way voluntarily incorrect from the point of view of quantum mechanics by saying that 'two or more fermions can never occupy the same *energy* state' (or, equivalently, the same *energy level*), but immediately compensating this error by another deliberate error which consists in saying that $D(p, dp)$ given by Equation (19′) is the number of (virtual) *energy* levels whose momentum lies between p and $(p + dp)$.

Adopting this procedure, Equations (19′) and (20) still hold, but one ceases to speak about *quantum* states, restricting the terminology only to *energy* levels (these latter being understood as related *exclusively* to p without any reference to the spin). Thus, henceforth f_p will represent the occupancy rate of each *energy* level in the vicinity of an energy corresponding to the momentum p. (Those who prefer to skip over all these difficulties simply speak about the '*occupancy rate*'.)

In any case, whatever terminology one adopts (either that of quantum mechanics or that of statistical physics) $n(p, dp)$ will be given by the fundamental relation

$$\boxed{n(p,dp) = f_p D(p,dp) = f_p g_s (1/h^3)\, 4\pi p^2\, dp.} \tag{20'}$$

2.6. THE FERMI–DIRAC DISTRIBUTION FUNCTION

It is shown in sufficiently advanced textbooks on thermodynamics (see for instance Kittel, 1969, Chapter 9) that in a state of thermodynamic equilibrium, f_p is given for *fermions* by the '*Fermi–Dirac Distribution Function*'

$$\boxed{f_p = \frac{1}{1 + e^{[\epsilon_k(p) - \mu_k]/kT}},} \tag{21}$$

where T is the temperature of the system, $\epsilon_k(p)$ is the kinetic energy of each particle of momentum p, k (when not used as a subscript) is the Boltzmann constant ($k \approx 1.38 \times 10^{-16}$ c.g.s.), and μ_k a parameter *independent of* p, whose relation to other parameters of the problem will be given below by Equation (22′)

This parameter μ_k is called the '*chemical potential*' because of its role in the theory of chemical reactions; but this term is anachronistic and somewhat misleading since it does not indicate how important μ_k is in the theory of nuclear reactions.

On integrating $n(p, dp)$, given by (20′), for all virtually possible values of p, i.e. from $p = 0$ to $p \to \infty$, we obtain, on taking into account Equation (21) and the fact that the total number n of particles in 1 cm^3 is a given constant quantity:

$$n = \int_0^\infty f_p g_s \frac{4\pi p^2 dp}{h^3} = \varphi(m, g_s, T, \mu_k). \tag{22}$$

The function φ at the right-hand side of Equation (22) depends on m because of the dependence of $\epsilon_k(p)$ expressed by Equation (16). Considering (22) as an equation to be solved for μ_k, it is immediately seen that μ_k is generally a function of n, m, g_s and T:

$$\mu_k = \mu_k(n, m, g_s, T). \tag{22'}$$

2.6.1. Remark

The chemical potential μ_k introduced above by Equation (21) obviously corresponds to the *kinetic* energy ϵ_k of fermions (whence the subscript k). However, it is sometimes more convenient to consider the chemical potential $\tilde{\mu}$ corresponding to the *total* energy ϵ. If ϵ_0 denotes the rest energy of the fermion, Equation (21), holds when we replace μ_k and ϵ_k by $\tilde{\mu} = \mu_k + \epsilon_0$ and $\epsilon = \epsilon_k + \epsilon_0$. (We write $\tilde{\mu}$ in order to avoid confusion with μ defined in Chapter 2).

3. The General Trend of the Fermi–Dirac Distribution Function. Definition of the Complete Degeneracy. The Fermi Level

3.1. THE 'TOBOGGAN'

Equation (21) represents the analytical expression for the Fermi–Dirac distribution function, i.e. the occupancy rate f_p as a function of p (for fermions).

This function can also be written in an other way, more convenient for several reasons.

Let us introduce an additional 'energy parameter', particularly well fitted to our present problem, the parameter X defined by

$$X = \frac{\epsilon_k(p)}{\mu_k}, \tag{23}$$

where μ_k is the *chemical potential* of the system of particles (see Subsection 2.6). According to Equation (16) X becomes zero for $p = 0$.

On the other hand, let Δ be another new parameter defined by

$$\Delta = \mu_k/kT, \tag{24}$$

where k is again the Boltzmann constant and T the temperature of the system.

In terms of X and Δ, Equation (21) takes the form

$$f_p = F_\Delta(X) = \frac{1}{1 + e^{(X-1)\Delta}}. \tag{25}$$

The function $F_\Delta(X)$, considered as a function of X for a given value of the parameter Δ, is schematically plotted in Figure 4.1 for a value of Δ equal to 50.

We see from Equation (25) that the curve representing the variations of $F_\Delta(X)$ can be separated more or less arbitrarily into three sections. A first section AB corresponding to $1.00 \leqslant F \leqslant 0.99$, and consequently to $0 \leqslant X \leqslant X_B$, where $X_B = 1 - \eta_B$, is defined by equation

$$F_\Delta(1 - \eta_B) = 0.99. \tag{26}$$

A second section CD corresponding to $0.01 \leqslant F \leqslant 0.00$, and consequently to $X_C \leqslant X \leqslant \infty$, where $X_C = 1 + \eta_C$, is defined by equation

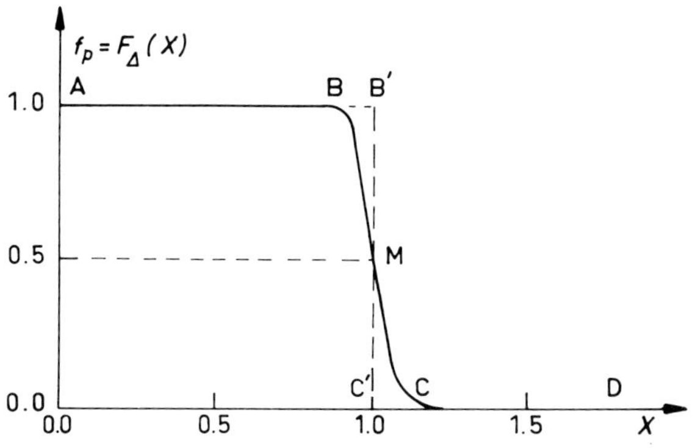

Fig. 4.1. The general trend of the Fermi–Dirac distribution function.

$$F_\Delta(1+\eta_C) = 0.01. \tag{26'}$$

And finally a third section BC, in a form of a 'toboggan', centred on $X = 1$.

On numerically solving Equations (26) and (26′) for η_B and η_C respectively, one finds:

$$\eta_B\Delta = \log_e 100 = 4.605, \tag{27}$$

and

$$\eta_C\Delta = \log_e 99 = 4.595. \tag{27'}$$

In spite of a slight inequality between η_B and η_C, these quantities can be assimilated to a common value η given approximately by

$$\eta \approx \eta_B \approx \eta_C \approx (4.6)/\Delta. \tag{28}$$

Thus, for a given value of Δ, the toboggan BC will be limited to a domain of X equal to 2η:

$$2\eta \approx (X_C - X_B) \approx (9.2)/\Delta. \tag{29}$$

(In Figure 4.1, with a choice of Δ equal to 50, the toboggan BC will correspond approximately to $X_B \approx 0.9$ and $X_C \approx 1.1$, with $2\eta \approx 0.2$).

Note that according to the general relation (28) the domain of the toboggan becomes *more narrow* when Δ increases, i.e. when, for a given value of μ_k, *T decreases*.

3.2. The State of Complete Degeneracy

When Δ becomes greater than 500, the toboggan reduces itself, with an approximation better than 1%, to a *'vertical wall'* (see Figure 4.1) B′C′ (at $X = 1$) between the point B′($X = 1; f_p = 1$) and the point C′($X = 1; f_p = 0$). The whole of the diagram takes then the form of a *step* (on a stair).

According to the definition (23) of X, the vertical wall B′C′ corresponds to a value of $\epsilon_k(p)$ equal to μ_k. In such a situation (when Δ is very much greater than one) the system is very near a state called a *'state of complete degeneracy'* corresponding to an *infinite* value of Δ. The rest of this Chapter (with the exception of Section 9) will be devoted to the study of the properties of the state of complete degeneracy.

3.3. The Fermi Level

Let us consider the limiting case of complete degeneracy defined in Subsection 3.2. Then for $\epsilon_k \leqslant \mu_k$ the occupancy rate f_p is equal to 1 and for $\epsilon_k > \mu_k$ the occupancy rate is zero: all virtual energy levels, from the level of zero kinetic energy up to the level (called the '*Fermi level*') of kinetic energy $\epsilon_{k,F} = \mu_k$, are occupied (according to the terminology defined in 2.5) at a rate of *one* particle per energy level, and all virtual energy levels of kinetic energy greater than $\epsilon_{k,F}$ remain unoccupied.

In the limiting case of the complete degeneracy, the totality of the general relations (10), identical with the RR relations (16), can be applied, with *a subscript F in all energy parameters*, in order to indicate that in Equation (30) we deal with a *particular* energy corresponding to the Fermi level:

$$\begin{aligned}\tilde{\mu} &= \epsilon_F = \epsilon_0 + \epsilon_{k,F} = \epsilon_0[1 + (p_F/mc)^2]^{1/2} \\ &= \epsilon_0(1 + x_F^2)^{1/2} = \epsilon_0 ch\Theta_F.\end{aligned} \tag{30}$$

More particularly we can write:

$$\mu_k = \epsilon_{k,F}(\mathrm{RR}) = \epsilon_0[(1 + x_F^2)^{1/2} - 1] = \epsilon_0[-1 + ch\Theta_F]. \tag{30'}$$

Note again that in terms of the parameter X the Fermi level corresponds to $X = 1$. So in the limiting case of a complete degeneracy we can replace the definition (24) of Δ by

$$\Delta = \epsilon_{k,F}/(kT). \tag{31}$$

3.4. The Degeneracy Parameter

Thus we can express the condition of an almost complete degeneracy ($\Delta \ggg 1$) by

$$kT \lll \epsilon_{k,F}, \tag{32}$$

or by

$$\frac{\epsilon_{k,F}}{kT} \ggg 1. \tag{32'}$$

The left-hand side of (32′) is sometimes called *the 'degeneracy parameter'*. If this parameter is sufficiently *large* the situation becomes *very near to* a state of a complete degeneracy. Obviously the condition (32) is satisfied for an absolute temperature T equal to *zero*; but it is very important from the astrophysical point of view to note that a situation of a practically complete degeneracy can take place for temperatures of the order of several billion degrees, if at the same time the number density n is sufficiently large (since $\epsilon_{k,F}$ is an increasing function of n, as we show it in the next Section).

4. Relations between the Number Density of a System of Fermions in a State of Complete Degeneracy and Energy Parameters of the Fermi Level

4.1. THE NUMBER DENSITY AS A FUNCTION OF THE FERMI MOMENTUM

Let us take into account the fact, explained in Subsection 3.3, that in a state of complete degeneracy the occupation rate f_p is equal to 1 for $0 < p \leqslant p_F$ and to *zero* for $p > p_F$. (Of course, p_F is the value of p corresponding to the Fermi level of energy). On restricting our analysis to those fermions (electrons, protons, neutrons) for which $s = \frac{1}{2}$ (and consequently $g_s = 2s + 1 = 2$) let us suppose that a state of a complete degeneracy takes place in a system of number density n. If we compute n by integration of Equation (20′) for all virtually possible values of p, i.e. from $p = 0$ to p infinite, we find

$$n = \int_0^{p_F} n(p, dp) = \int_0^{p_F} \frac{8\pi p^2\, dp}{h^3} = \frac{8\pi}{h^3}\left(\tfrac{1}{3} p_F^3\right). \tag{33}$$

According to the definition (3) applied to the Fermi level, the value of p_F and the corresponding value x_F of the energy parameter x are connected by

$$p_F = mcx_F, \tag{34}$$

and Equation (33) can be written as

$$n = \frac{8\pi m^3 c^3}{3h^3} x_F^3. \tag{35}$$

4.2. THE CRITICAL NUMBER DENSITY

As already mentioned in Subsection 1.6, from the point of view of the division of particles in the three groups (NR), (UR) and (RR), the value $x = 1$ of the parameter x plays a critical role. For this reason we may call the value of n corresponding to $x_F = 1$ the '*critical number density*', and denote this critical density by n_{cr}.

Thus according to Equation (35):

$$n_{\mathrm{cr}} = 8\pi m^3 c^3/(3h^3) = C^{-3} m^3 c^3, \tag{36}$$

where

$$C^3 = \frac{3h^3}{8\pi}. \tag{36′}$$

A substitution of numerical values $c = 3 \times 10^{10}$ cm s^{-1}; $h = 6.63 \times 10^{-27}$ c.g.s.; m_e = the mass of an electron $= 9.1 \times 10^{-28}$ g; m_n = the mass of a neutron $= 1.66 \times 10^{-24}$ g; yields, after some rather tedious computations, for electrons and neutrons:

$$\boxed{n_{\mathrm{cr},e} = 5.9 \times 10^{29}\ \mathrm{cm}^{-3},} \tag{36, e}$$

and

$$n_{\mathrm{cr},n} = 3.9 \times 10^{39}\,\mathrm{cm}^{-3}. \tag{36, n}$$

Note that Equation (36, n) can also be applied to protons.

Equation (36) allows us to replace Equation (35) by the fundamental very simple relation:

$$n = n_{\mathrm{cr}} x_F^3. \tag{37}$$

This relation is quite general. It can be applied for any value of x_F and is perfectly satisfactory from the mathematical point of view. However, such as it stands in Equation (37) it is *not* perfectly satisfactory from the point of view of a physicist. It might suggest (by itself and by the way it was obtained) that n is conditioned by x_F, whereas actually, for each real system, n is a *given quantity* and x_F is physically determined by the value of n. Moreover, as often happens with 'elegant' relations, Equation (37) hides the fundamental fact (shown by Equation 36) that n_{cr} depends on the (rest) mass m of each particle of the system. This property can be put into evidence by writing n_{cr} as $n_{\mathrm{cr}}(m)$. For all these reasons, a more complicated, but mathematically equivalent, relation (38) is much more satisfactory:

$$x_F = x_F(n, m) = n_{\mathrm{cr}}(m)^{-1/3} n^{1/3}. \tag{38}$$

We have written x_F as $x_F(n, m)$ in order to stress the fundamental fact of dependence of x_F on *two physical variables n* and *m*.

In a more general way, for particles of g_s different from 2, Equation (38) remains valid, but in the definition (36) of n_{cr} the factor 8 must be replaced by $4g_s$.

4.3. Fermi Momentum and Fermi Energy as a Function of the Number Density

On replacing $n_{\mathrm{cr}}^{-1/3}$ by its numerical value and on denoting the respective number densities of electrons or neutrons by n_e or n_n, Equation (38) takes one of its numerical forms:

$$x_{F,e} = 1.2 \times 10^{-10} n_e^{1/3}, \tag{38, e}$$

and

$$x_{F,n} = 6.5 \times 10^{-14} n_n^{1/3}. \tag{38, n}$$

Let us now return to relations in which the particular nature (electrons or neutrons) of particles is not specified, and let us express ϵ_F, p_F, and $\epsilon_{k,F}$ as an explicit function of n (and an implicit function of m). On substituting Equation (38) into Equation (30), and on pointing out the rigorously relativistic (RR) character of the result, we obtain:

$$\epsilon_F(\mathrm{RR}) = \epsilon_0(1 + x_F^2)^{1/2} = \epsilon_0[1 + (n/n_{\mathrm{cr}})^{2/3}]^{1/2}. \tag{39}$$

This emphasizes that $\epsilon_F(\text{RR})$ is generally (for a given value of g_s) a function $\epsilon_F(n, m)$ of n and m, since ϵ_0 and n_{cr} depend on m. This shows also, according to Subsection 3.3, that in the state of complete degeneracy, when $\mu_k = \epsilon_{k,F}$, the chemical potential μ_k depends only on n and m (and eventually on g_s), whereas in other states μ_k can also depend, as shown by Equation (22′), on T.

Conversely, whereas ϵ_F depends *both* on n and m, it results from Equation (33) and that, for $g_s = 2, p_F$ depends *only* on n. Indeed, it is given by

$$\boxed{p_F = p_F(n) = Cn^{1/3},} \tag{40}$$

where C is a universal constant defined by Equation (36′) and equal to

$$\boxed{C = \frac{h}{2}\left(\frac{3}{\pi}\right)^{1/3} \approx (3.25 \times 10^{-27}) \text{ c.g.s..}} \tag{40′}$$

This shows again how some energy parameters (here p), rather than others, can describe a given physical situation by particularly simple and general relations. Compare, for instance, Equation (40) with the expression (41) for $\epsilon_{k,F}(\text{RR})$ given by the substitution of Equation (38) into Equation (30′):

$$\epsilon_{k,F}(\text{RR}) = \epsilon_0\{[1 + (n/n_{\text{cr}})^{2/3}]^{1/2} - 1\} = \epsilon_{k,F}(n, m). \tag{41}$$

4.4. The (NR) and (UR) Approximations

Fortunately, expressions (39) and (41) for ϵ_F and $\epsilon_{k,F}$, relatively complicated in the general case (RR), take a more simple form when the mathematical approximations (NR) or (UR) are permitted by the physical situation.

Consider a system where the actual number density n of particles of some particular mass m is very much inferior to the corresponding critical density $n_{\text{cr}}(m)$. Then according to Equation (38) x_F will be much *less* than 1, and the non-relativistic approximation (NR) will be acceptable. In this case all of the usual energy parameters corresponding to the Fermi level will be given with a good approximation by Equation (15) applied to the Fermi level, i.e., through Equation (40), by:

$$\epsilon_{k,F}(\text{NR}) = \frac{1}{2m}p_F^2 = \frac{1}{2}\epsilon_0 x_F^2 = \frac{1}{2m}C^2 n^{2/3} = \epsilon_F(\text{NR}) - \epsilon_0. \tag{42}$$

Conversely, suppose that the number density n of a system of particles of individual mass m is very much *larger* than the corresponding critical density $n_{\text{cr}}(m)$. Then according to Equation (38) x_F will be much larger than 1, and the ultra-relativistic approximation (UR) will be acceptable. In this case, all of the usual energy parameters corresponding to the Fermi level will be given by Equation (17) applied to the Fermi level, i.e., through Equation (40), by

$$\epsilon_F(\text{UR}) = \epsilon_{k,F}(\text{UR}) = cp_F = \epsilon_0 x_F = cCn^{1/3}. \tag{43}$$

C being independent of m, the ultra-relativistic chain of relations (43) shows that in this particular situation, contrary to the general case (RR), ϵ_F and $\epsilon_{k,F}$ depend *only* on the number density n and are independent from m. Therefore, in this case of very large density n, ϵ_F and $\epsilon_{k,F}$ take, for a given value of n, the same value for electrons, neutrons or protons. For this reason the case (UR) can be considered as the simplest one.

4.5. The Limiting Temperature

Let us now define a '*limiting temperature*' (for degeneracy or a 'limit of temperature') $T_{\text{lim}}^{\text{deg}}$ by the relation

$$kT_{\text{lim}}^{\text{deg}} = \epsilon_{k,F}. \tag{44}$$

Then, the condition (32) of an (almost) complete degeneracy will take the form

$$T \lll T_{\text{lim}}^{\text{deg}}. \tag{45}$$

Taking advantage of the simplicity of relations (42) or (43) which give $\epsilon_{k,F}$ for the situations when approximations (NR) or (UR) are allowed, it is easy to be explicit about the dependence of $T_{\text{lim}}^{\text{deg}}$ on n, and eventually on m, since $T_{\text{lim}}^{\text{deg}}(\text{NR})$ and $T_{\text{lim}}^{\text{deg}}(\text{UR})$ are then given respectively by

$$\boxed{T_{\text{lim}}^{\text{deg}}(\text{NR}) = \frac{1}{2mk} C^2 n^{2/3},} \tag{46}$$

and

$$\boxed{T_{\text{lim}}^{\text{deg}}(\text{UR}) = \frac{cC}{k} n^{1/3}.} \tag{46'}$$

These limiting temperatures always increase more or less rapidly, for a given value of m, with the number density n. For a *given value* of n sufficiently small to allow the approximation (NR), the limiting temperature $T_{\text{lim}}^{\text{deg}}(\text{NR})$ is *inversely proportional to the individual mass* m of the particles of the system.

Thus, for instance, the (almost) complete degeneracy for equal *number* densities [smaller than the corresponding *critical* density $n_{\text{cr}}(m)$] can take place for temperatures T much *larger* for a system of electrons than for a system of neutrons (or protons).

More precisely Equation (40′) and the Table on p. 467 yield, after some rather tedious calculations:

$$\frac{1}{2km_e} C^2 = 3.4 \times 10^{-11}; \tag{47}$$

$$\frac{1}{2km_n} C^2 = 2.1 \times 10^{-14}; \tag{48}$$

$$\frac{cC}{k} = 0.70. \tag{49}$$

Thus finally for electrons and neutrons respectively, n_e and n_n being the corresponding number densities, the conditions of an (almost) complete degeneracy are:

$$T \lll (4.2 \times 10^{-11})\, n_e^{2/3} \qquad \text{if } n_e \ll (5.9 \times 10^{29})\,\text{cm}^{-3}; \tag{50}$$

or

$$T \lll (0.70)\, n_e^{1/3} \qquad \text{if } n_e \gg (5.9 \times 10^{29})\,\text{cm}^{-3}; \tag{51}$$

for electrons; and

$$T \lll (2.3 \times 10^{-14})\, n_n^{2/3} \qquad \text{if } n_n \ll (3.7 \times 10^{39})\,\text{cm}^{-3}; \tag{52}$$

or

$$T \lll (0.70)\, n_n^{1/3} \qquad \text{if } n_n \gg (3.7 \times 10^{39})\,\text{cm}^{-3} \tag{53}$$

for neutrons.

Let us illustrate these conditions by the following examples.

Consider a number density n of the order of $6 \times 10^{23}\,\text{cm}^{-3}$. In the case of *electrons*, a density n_e equal to $6 \times 10^{23}\,\text{cm}^{-3}$ corresponds to the non-relativistic situation described by the conditions (50). Since $(6 \times 10^{23})^{2/3} \approx 71 \times 10^{14}$ one finds that for electrons the temperature T can reach a value of the order of 300 000 K when n_e is as *high* as 6×10^{23} cm^{-3} (a situation easily encountered in some cosmic conditions), without preventing an (almost) complete degeneracy. At a temperature as high as 300 000 K this electron gas behaves as if it were at a temperature of *zero* absolute.

Consider now the same density of 6×10^{23} particles per cm^3, in the case of neutrons. A density n_n of $6 \times 10^{23}\,\text{cm}^{-3}$ corresponds to the non-relativistic situation described by the conditions (52). Now, the limiting temperature would be only of $(2.3 \times 10^{-14}) \times (71 \times 10^{14}) \approx 160\,\text{K}$. Thus, for the same number density (lower than the corresponding critical density), a neutron gas demands, as a condition for an almost complete degeneracy, a much *lower* temperature (a temperature nearer to zero) than an electron gas!

Finally, for a very high number density of the order of 6×10^{35} particles per cm^3 (that is for $n_e = n_n = 6 \times 10^{35}\,\text{cm}^{-3}$), the limiting temperature for electrons would be given by the conditions (51) and would be equal to $5.9 \times 10^{11}\,\text{K}$ whereas the limiting temperature for neutrons would *not* be given by the conditions (53) but by the conditions (52), and equal only to $1.6 \times 10^{10}\,\text{K}$, again (slightly) lower than $5.9 \times 10^{11}\,\text{K}$.

Attention must be paid to the fact that the limitations concerning the number densities n_e and n_n in conditions (50), (51), (52) and (53) are only of a *mathematical* nature: they only discriminate different *mathematical* formulae, but a limiting temperature $T_{\text{lim}}^{\text{deg}}$ physically exists for any value of n, and can be calculated by its definition (44), through the use of the general expression (41) for $\epsilon_{k,F}$ as a function of n and m.

On returning to the numerical examples considered above, we can summarize (in obvious notation) our results by

$$T_{\text{lim},e}^{\text{deg}}(n_e = 6 \times 10^{23}) \approx 230\,000\,\text{K}; \tag{54}$$

$$T_{\text{lim},e}^{\text{deg}}(n_e = 6 \times 10^{35}) \approx 5.9 \times 10^{11}\,\text{K}; \tag{55}$$

$$T_{\text{lim},n}^{\text{deg}}(n_n = 6 \times 10^{23}) \approx 160\,\text{K}; \tag{56}$$

$$T_{\text{lim},n}^{\text{deg}}(n_n = 6 \times 10^{35}) \approx 1.6 \times 10^{10}\,\text{K}. \tag{57}$$

5. Energy Density (for Total and for Kinetic Energy) of a Completely Degenerate System of Independent Identical Fermions

5.1. The General (RR) Case

The relation (7) between the (total) energy ϵ of an individual particle and the kinetic energy ϵ_k of the same particle, can be written, when both energies are considered as functions of the parameter $x = p/mc$:

$$\epsilon(x) = \epsilon_0 + \epsilon_k(x). \tag{58}$$

With the terminology defined in Subsection 2.5 we can say that in a system of fermions *each* particle of the system occupies a different *energy* level between the fundamental level and the Fermi level. Each particle of the system is characterized by a *different* energy parameter x, p, ϵ, ϵ_k or Θ.

Equation (20′) gives us the number $n(p, dp)$ of particles per cm^3 corresponding to the interval between p and $(p + dp)$, as a function of p, dp, f_p and g_s. In a state of complete degeneracy, the occupation rate f_p is equal to 1 in the domain (F) defined by $0 < p \leqslant p_F$ (or by $0 < x \leqslant x_F$), and equal to *zero* for $p > p_F$ (or $x > x_F$). In the rest of the present Section we suppose $s = \frac{1}{2}$ (thus $g_s = 2$) but the theory is easily extended to any value of g_s.

On taking into account the definition (36′) of the constant C, $n(p, dp)$ takes in the domain (F) the form:

$$n(p, dp) = (1/h^3)\, 8\pi p^2\, dp = (3/C^3)\, p^2\, dp. \tag{59}$$

If we express p in terms of the parameter $x = p/mc$, the number $n[x, dx]$ of particles corresponding to the interval (x, dx) will be given, in the domain (F), by

$$n[x, dx] = (1/h^3)\, 8\pi m^3 c^3 x^2\, dx = 3 n_{\mathrm{cr}} x^2\, dx, \tag{59'}$$

where we have introduced the critical density n_{cr} defined by Equation (36).

The '*energy density*' u is by definition the sum of *different* (total) energies of *all* particles contained in each cm^3 of the system. This sum is obviously given by addition of all (total) energies $\epsilon(p) n(p, dp) = \epsilon(x) n[x, dx]$ of particles whose p lies between p and $(p + dp)$ [or, whose x lies between x and $(x + dx)$], from the fundamental level $p = x = 0$, up to the Fermi level characterized by p_F or by x_F. Thus u will be given, depending on the energy parameter used in integration, by one of the two integrals of (60):

$$u = \int_0^{p_F} \epsilon(p) n(p, dp) = \int_0^{x_F} \epsilon(x) n[x, dx]. \tag{60}$$

Similarly the '*kinetic energy density*' u_k, representing by definition the sum of all *different* kinetic energies of *all* particles contained in each cm^3 of the system, will be given by one of the two integrals of (60′):

$$u_k = \int_0^{p_F} \epsilon_k(p) n(p, dp) = \int_0^{x_F} \epsilon_k(x) n[x, dx]. \tag{60'}$$

It is obvious, taking into account the relation (58) between the total and the kinetic energy of an individual particle, that the (total) energy density u and the kinetic energy density u_k are related by

$$u = u_k + \epsilon_0 n, \tag{61}$$

which can also be written, taking into account the relation (37) between n and x_F:

$$u_k = u - \epsilon_0 n_{\text{cr}} x_F^3. \tag{61'}$$

A substitution of the explicit expressions for $n(p, dp)$ or $n[x, dx]$ given respectively by Equations (59) or (59′) into the integrals (60) and (60′) yields the following rigorous (RR) expressions of u(RR) and u_k(RR):

$$u(\text{RR}) = (3/C^3)\int_0^{p_F} \epsilon(p)p^2\,dp = 3n_{\text{cr}}\int_0^{x_F} \epsilon(x)x^2\,dx, \tag{62}$$

$$u_k(\text{RR}) = (3/C^3)\int_0^{p_F} \epsilon_k(p)p^2\,dp = 3n_{\text{cr}}\int_0^{p_F} \epsilon_k(x)x^2\,dx. \tag{62'}$$

Obviously, the rigorous expressions for ϵ(RR) and ϵ_k(RR), given by Equation (16), must be used for $\epsilon(p)$, or $\epsilon(x)$, and for $\epsilon_k(p)$, or $\epsilon_k(x)$, under the integral sign. The rather long and tedious, though quite elementary, calculations given in the Appendix to this Chapter show that u(RR) is given by

$$u(\text{RR}) = \tfrac{3}{8} n_{\text{cr}} \epsilon_0 G(x_F), \tag{63}$$

where $G(x)$ is defined by Equation (A12) of the Appendix, i.e. by

$$G(x) = x(2x^2+1)(1+x^2)^{1/2} - \log_e[x + (1+x^2)^{1/2}]. \tag{63'}$$

On introducing with Chandrasekhar, 1939, the function $g(x)$ defined by (64′) and (63′), Equation (61′) gives immediately:

$$u_k(\text{RR}) = \tfrac{1}{8} n_{\text{cr}} \epsilon_0 g(x_F), \tag{64}$$

where

$$g(x) = 3x(2x^2+1)(1+x^2)^{1/2} - 8x^3 - 3\log_e[x + (1+x^2)^{1/2}]$$
$$= 3G(x) - 8x^3. \tag{64'}$$

Since n_{cr} and ϵ_0 depend just as x_F on n and on m, in the general case (RR) both u and u_k depend on n and m. This can be made more explicit by writing $u(n, m)$ and $u_k(n, m)$.

5.2. The (NR) and (UR) Approximations

If the physical situation allows the use of the non-relativistic approximation (NR), or of the ultra-relativistic approximation (UR), the corresponding expressions of u or u_k can be obtained through limited expansion of $G(x)$ near $x = 0$ or through an asymptotic expansion of $G(x)$ for $x \to \infty$ given in the Appendix. However it is never a good procedure to derive particular results from complicated general formulae. Here these simple particular results can be obtained *directly*:

Indeed, when $x_F \ll 1$ (approximation NR corresponding to $n \ll n_{\text{cr}}$) we can replace $\epsilon_k(p)$ in Equation (60′) by its approximate (NR) value $p^2/2m$ or $\frac{1}{2}\epsilon_0 x^2$ given by Equation (15).

On taking into account Equations (59) or (59′), this gives:

$$u_k(\mathrm{NR}) = (3/C^3)\frac{1}{2m}\int_0^{p_F} p^4\,dp = \frac{3C^{-3}}{10m}p_F^5 = \frac{3}{10}n_{\mathrm{cr}}\epsilon_0 x_F^5. \tag{65}$$

On replacing p_F by its value $Cn^{1/3}$, given by Equation (40), $u_k(\mathrm{NR})$ takes the form

$$u_k(\mathrm{NR}) = \frac{3C^2}{10m}n^{5/3} = (3.15\times 10^{-54})\frac{1}{m}n^{5/3}, \tag{66}$$

where the numerical factor is easily obtained through the use of Equation (40′). Then, by Equation (61) applied to the case (NR), and on replacing in Equation (65) $n_{\mathrm{cr}}x_F^3$ by n, according to Equation (37), we obtain:

$$u(\mathrm{NR}) = \epsilon_0 n\,(1 + \tfrac{3}{10}x_F^2). \tag{66′}$$

Thus when $n \ll n_{\mathrm{cr}}$, the principal term of $u(\mathrm{NR})$ reduces simply to $\epsilon_0 n$.

When $x_F \gg 1$, (approximation UR corresponding to $n \gg n_{\mathrm{cr}}$) we can replace $\epsilon_k(p)$ in Equation (60′) by its approximate (UR) value cp or $\epsilon_0 x$:

$$\epsilon_k(\mathrm{UR}) = cp = \epsilon_0 x. \tag{67}$$

It seems at first sight that the procedure just indicated for the case (UR) is incorrect, since the integral (60′) (with x as variable) extends from $x = 0$ to $x = x_F$, thus enclosing the values of x for which the approximation (UR) cannot be applied. However, if we divide this integral in two parts, a first one where x varies between 0 and 1 (where the approximation NR can be applied), and a second one where x varies between 1 and x_F (where the approximation UR can be applied), it is easily found that for large values of x_F the contribution of the first part to the whole is negligible.

Thus, in the case (UR) we have the right to use Equation (67) and we immediately obtain the following result (confirmed by the use of an asymptotic expansion given in the Appendix):

$$u_k(\mathrm{UR}) = (3/C^3)c\int_0^{p_F} p^3\,dp = (3/C^3)(c/4)p_F^4 = \tfrac{3}{4}n_{\mathrm{cr}}\epsilon_0 x_F^4, \tag{68}$$

or, on using once more (40) and (40′),

$$u_k(\mathrm{UR}) = \tfrac{3}{4}Ccn^{4/3} = (7.2\times 10^{-17})n^{4/3}. \tag{69}$$

In this exceptional situation (when $n \gg n_{\mathrm{cr}}$) u_k depends only on n and is independent of m.

More specifically, for a gas of electrons and for a gas of neutrons the corresponding values of $u_{k,e}$ and $u_{k,n}$, obtained by the use of the numerical value of m_e in Equation (66), are given by

$$u_{k,e}(\mathrm{NR}) = (3.45\times 10^{-27})n_e^{5/3}; \tag{70}$$

$$u_{k,e}(\mathrm{UR}) = (7.2\times 10^{-17})n_e^{4/3}; \tag{71}$$

$$u_{k,n}(\mathrm{NR}) = (1.9\times 10^{-30})n_n^{5/3}; \tag{72}$$

$$u_{k,n}(\mathrm{UR}) = (7.2\times 10^{-17})n_n^{4/3}. \tag{73}$$

5.2.1. Remark

For a system of atomic nuclei of atomic mass A (in units $m_B = 1.66 \times 10^{-24}$ g) and of spin $s = \frac{1}{2}$, (n_{nuc} being the number density of the system) Equation (66) immediately yields the value of the corresponding kinetic energy density $u_{k,\text{nuc}}$ in a state of complete degeneracy:

$$u_{k,\text{nuc}}(\text{NR}) = (1.9 \times 10^{-30}) \frac{1}{\text{A}} n_{\text{nuc}}^{5/3}. \tag{74}$$

Similarly, in a mixture of such nuclei and electrons, resulting from a complete ionization of neutral atoms of one single chemical and isotopic nature (of a given atomic mass A and a given atomic number Z), we have $n_e = \text{Z} n_{\text{nuc}}$. Therefore, in a state of complete degeneracy of both electrons and nuclei, Equation (66) gives:

$$\frac{u_{k,e}(\text{NR})}{u_{k,\text{nuc}}(\text{NR})} = \frac{\text{A} m_B}{m_e} (n_e/n_{\text{nuc}})^{5/3} = \text{A}(1840)\,\text{Z}^{5/3}, \tag{75}$$

where m_B is, as above, the approximate mass of a *baryon* such as a neutron or a proton.

6. The Mean (Total) Energy and the Mean Kinetic Energy of One Particle of a Completely Degenerate System of Independent Identical Fermions. Relations with the Fermi Level

One can introduce the concept of the '*mean (total) energy of one particle*' $\bar{\epsilon}(n, m)$ of a completely degenerate system of independent identical fermions, (of individual masses m and number density n), by Equation

$$\bar{\epsilon}(n, m) = \frac{u(n, m)}{n}, \tag{76}$$

i.e. by dividing the (total) energy $u(n, m)$ contained in each cm^3 by the number n of particles contained in the same cm^3.

Similarly, one can introduce the concept of the '*mean kinetic energy of one particle*' $\bar{\epsilon}_k(n, m)$ of the same system, by Equation

$$\bar{\epsilon}_k(n, m) = \frac{u_k(n, m)}{n}. \tag{77}$$

According to Equation (61) we have:

$$\bar{\epsilon}(n, m) = \epsilon_0 + \bar{\epsilon}_k(n, m), \tag{78}$$

where $\epsilon_0 = mc^2$ is the rest energy of each particle.

A general expression $\bar{\epsilon}(\text{RR})$ for $\bar{\epsilon}(n, m)$ can be derived from Equation (63), after elimination of n_{cr} by means of Equation (37). This yields:

$$\bar{\epsilon}(\text{RR}) = [(3\epsilon_0)/(8x_F^3)]\,G(x_F), \tag{79}$$

where $G(x)$ is given either by Equation (63′) or by Equation (A12) of the Appendix.

When the physical situation allows the use of the approximations (NR) *or* (UR), an application of approximate expressions (66) for $u_k(\text{NR})$, or (69) for $u_k(\text{UR})$, to the definition (77) of $\bar{\epsilon}_k(n, m)$ immediately gives

$$\bar{\epsilon}_k(\text{NR}) = \frac{3C^2}{10m} n^{2/3} = (3.15 \times 10^{-54}) \frac{1}{m} n^{2/3}, \tag{80}$$

or

$$\bar{\epsilon}_k(\text{UR}) = \frac{3Cc}{4} n^{1/3} = (7.2 \times 10^{-17}) n^{1/3}. \tag{81}$$

Let us compare these relations with the expression for $\epsilon_{k,F}(\text{NR})$ given by Equation (42), or with the expression for $\epsilon_{k,F}(\text{UR})$ given by Equation (43), viz.

$$\epsilon_{k,F}(\text{NR}) = \frac{C^2}{2m} n^{2/3}; \qquad \epsilon_{k,F}(\text{UR}) = cCn^{1/3}. \tag{82}$$

One finds the following fundamental and very simple relation between the *mean* kinetic energy of *one* particle and the kinetic energy of *the particle occupying the Fermi level* of the system, for the case $n \ll n_{\text{cr}}$ (NR approximation):

$$\boxed{\bar{\epsilon}_k(\text{NR}) = \tfrac{3}{5}\epsilon_{k,F}(\text{NR}),} \tag{83}$$

and a similar relation for the case $n \gg n_{\text{cr}}$ (UR approximation):

$$\boxed{\bar{\epsilon}_k(\text{UR}) = \tfrac{3}{4}\epsilon_{k,F}(\text{UR}).} \tag{84}$$

One may be tempted to look for relations similar to (83) or (84) in the general case (RR). Consider then first the expression for $\bar{\epsilon}_k(\text{RR})$

$$\bar{\epsilon}_k(\text{RR}) = \frac{1}{n} u_k(\text{RR}) = [\epsilon_0/(8x_F^3)]g(x_F), \tag{85}$$

given by an application of Equations (64) and (37) to the definition (77) of $\bar{\epsilon}_k$, where $g(x)$ is given by Equation (64′). Consider next the expression for $\epsilon_{k,F}(\text{RR})$

$$\epsilon_{k,F}(\text{RR}) = \epsilon_0[(1 + x_F^2)^{1/2} - 1] \tag{85′}$$

given by Equation (30′). It is obvious that no relation as simple as (83) or (84) can exist between $\bar{\epsilon}_k(\text{RR})$ and $\epsilon_{k,F}(\text{RR})$.

However, if the result (85) of the *division* of $u_k(\text{RR})$ by n is very complicated, on the contrary the *partial derivation* of $u_k(\text{RR}) = u_k(n, m)$ with respect to n gives a function $\partial n_k(n, m)/\partial n$ which is *extremely simple*, and can be obtained through a very elementary physical argument.

The kinetic energy $u_k(\text{RR})$ contained in each cm^3 of a completely degenerate system of identical fermions can increase by an amount ∂u_k only by the addition of a 'layer' of ∂n new fermions to the 'stack' of the n fermions already contained in this cm^3 and occupying all levels up to the Fermi level of kinetic energy $\epsilon_{k,F}$. The kinetic energy of each of these additional ∂n particles is necessarily individually very near $\epsilon_{k,F}$. Each cm^3 of the system thus receives an additional kinetic energy ∂u_k equal to $\epsilon_{k,F}\partial n$. Hence

$$\boxed{\frac{\partial u_k(n, m)}{\partial n} = \epsilon_{k, F}(n, m).} \tag{86}$$

This Equation (86) is quite general: it is valid for any value of n, even if for more simplicity we do not mention (RR).

It is easy to verify that the approximate expressions for u_k and $\epsilon_{k, F}$, given by Equations (66) and (42), for the case (NR), satisfy the general relation (86). A similar statement is true for the approximate expressions for u_k and $\epsilon_{k, F}$, given by Equations (69) and (43), for the case (UR).

The same general argument can be applied to the partial derivative of the density u of the (total) energy with respect to the number density n. Hence, a relation similar to the relation (86):

$$\boxed{\frac{\partial u(n, m)}{\partial n} = \epsilon_F(n, m).} \tag{86$'$}$$

which is again a (RR) relation valid for any value of n.

7. Expressions for Partial Number Densities of Different Components of a Stellar Mixture as a Function of the Mass Density of the Mixture. Parameters μ, μ_e and μ_e'

Only systems of *identical* particles have been considered in the preceding sections. However, the next two chapters deal first with *white dwarfs* composed mainly of a mixture of free electrons and α-particles, and next with *neutron stars* composed of free electrons and, at a certain phase of their evolution, of free neutrons and of nuclei more complex than α-particles.

Therefore, after a brief summary of the properties of mixtures proper to normal stars already considered in Section 5 of Chapter 2, we study first in Subsection 7.1 the mixtures deprived of free neutrons (normal stars and white dwarfs), and in Subsection 7.2 the mixtures corresponding to neutron stars.

In the whole of this Section we consider as *mass density* ρ the sum of the *rest* masses of free particles and of particles composing the nuclei of the mixture contained in each cm^3. In other words, we deliberately neglect the mass u_k/c^2 equivalent to the density of kinetic energy u_k of the particles of the mixture. Certainly in a situation of complete degeneracy, each u_k component of the mixture increases rather rapidly with the corresponding numerical density n, as was shown in preceding sections. However, because of the very large divisor c^2 the corresponding contribution to ρ is generally negligible, except in the case of very dense neutron stars, a case which should be considered separately.

7.1. Partial Number Densities in Mixtures Composed of Protons, Electrons and α-Particles, as a Function of the Mass Density of the Mixture. Parameters μ and μ_e

Let us first consider a stellar mixture of a normal star, already studied in Chapter 2, Section 5, so that we can simply recall the main definitions and the main results of that section.

In the case of a normal star of solar type, the stellar mixture is mainly composed of H nuclei (protons). He nuclei (α-particles) and free electrons, produced by the total ionization of neutral H and He.

Neglecting the mass m_e of an electron in comparison with the 1840 times larger mass m_p of a proton (and in comparison with the even larger mass m_α of an α-particle) we can characterize our mixture, as in Chapter 2, by the number X of *grams* of protons and the number Y of *grams* of α-particles contained in each *gram* of the mixture.

X and Y are not independent since

$$Y = 1 - X. \tag{87}$$

Thus *one* single parameter, for instance X, can describe the proportion of protons and α-particles in the mixture.

Let N_A, approximately equal to $6.0 \times 10^{23}\ \text{cm}^{-3}$, denote the Avogadro number, and let m_B, approximately equal to $1/N_A$, denote the approximate mass of a baryon such as a proton or a neutron. The difference between the mass m_p of a proton and the mass m_n of a neutron, essential in the theory of nuclear reactions, is negligible in the theory to be developed in this section. All these approximations can be summarized by the following relations:

$$m_e = m_B/1840; \qquad m_p = m_B; \qquad m_n = m_B; \qquad m_\alpha = 4m_B. \tag{88}$$

Let n_p, n_α and n_e denote respectively the partial number densities of free protons, α-particles and electrons, i.e. the respective number of these particles in 1 cm^3 of the mixture. In Table 2.1 the following expressions for n_p, n_α and n_e as a function of ρ and X have been found:

$$n_p = \rho X/m_p = \rho X/m_B = \rho X N_A; \tag{89}$$

$$n_\alpha = \rho Y/m_\alpha = \rho Y/(4m_B) = \tfrac{1}{4}\rho Y N_A = \rho(1-X)/(4m_B); \tag{89'}$$

$$n_e = 1n_p + 2n_\alpha = \rho N_A\left(X + \frac{Y}{2}\right) = \tfrac{1}{2}(1+X)\,\rho N_A; \tag{90}$$

where the last one is simply Equation (38) of Chapter 2.

Let n_{tot} denote, as in Chapter 2, the total number of particles per cm^3 of the mixture (free protons, free electrons and α-particles). For the case considered in this subsection n_{tot} is equal to

$$n_{\text{tot}} = n_p + n_\alpha + n_e = (\tfrac{3}{4} + \tfrac{5}{4}X)\,\rho N_A. \tag{91}$$

On introducing the *mean mass in grams* $\bar{m}_g$ of *one* particle of the mixture and the value μ of $\bar{m}_g$ in units equal to m_B, we obtain Equations (27) and (37) of Chapter 2:

$$\bar{m}_g = \rho/n_{\rm tot} = \frac{1}{(\frac{3}{4}+\frac{5}{4}X)N_A} = m_B/(\tfrac{3}{4}+\tfrac{5}{4}X), \tag{92}$$

and

$$\mu = \bar{m}_g/m_B = \frac{\rho}{m_B n_{\rm tot}} = \frac{1}{(\frac{3}{4}+\frac{5}{4}X)}. \tag{93}$$

Some authors call the parameter μ the '*mean atomic mass*' or even the '*mean molecular mass*' of the mixture, but both terms (especially the last one) are rather improper and misleading.

Another useful concept, not always clearly defined, is that of the *total number density* n_B of baryons, where free baryons (here free protons) are totalized with the protons and neutrons contained in the complex nuclei (here α-particles). Thus, for the present case n_B is given by

$$n_B = n_p + 4n_\alpha. \tag{94}$$

More generally we have the physically obvious relation

$$n_B = \rho/m_B = \rho N_A. \tag{94'}$$

The parameter μ_e defined by

$$\mu_e = \frac{n_B}{n_e}, \tag{95}$$

is also often used in astrophysics.

The difference between the parameter μ, defined by Equation (93), and the parameter μ_e becomes very clear if we take into account Equation (94′), which allows us to write μ as

$$\mu = \frac{n_B}{n_{\rm tot}}. \tag{96}$$

The definition (95) of μ_e is quite general and, since Equation (94′) is quite general too, the parameter μ_e can also be defined, in a general way, by

$$\mu_e = \rho N_A \frac{1}{n_e}. \tag{97}$$

On expressing the parameter μ_e as a function of X, by means of Equations (95), (90) and (94′), we obtain:

$$\mu_e = \frac{2}{1+X}. \tag{98}$$

A comparison with the corresponding expression (93) of the parameter μ shows again (in a particular case) the difference between μ and μ_e.

Sometimes the following expression for n_α as a function of μ_e, given by Equations (89′) and (98), can be useful:

$$n_\alpha = \rho N_A \tfrac{1}{2}\left(1 - \frac{1}{\mu_e}\right). \tag{99}$$

From the point of view of its composition a white dwarf is characterized by an absence of free protons (i.e. $X = 0$), all protons having been transformed by thermonuclear reactions into α-particles or into more complex nuclei.

If the α-particles are the only nuclei present in the mixture the preceding relations take (for $X = 0$) the following particular form, which can also be established directly by elementary physical argument:

$$\begin{array}{lllll} X = 0; & Y = 1; & \mu = \frac{4}{3}; & \mu_e = 2; & n_B = \rho N_A; \\ n_p = 0; & n_\alpha = \frac{1}{4}\rho N_A; & n_e = \frac{1}{2}\rho N_A; & n_{tot} = \frac{3}{4}\rho N_A. & \end{array} \tag{100}$$

For white dwarfs with nuclei heavier than α-particles see below, Subsection 7.2.

7.2. PARTIAL NUMBER DENSITIES IN MIXTURES COMPOSED OF FREE ELECTRONS, FREE NEUTRONS AND COMPLEX NUCLEI, AS A FUNCTION OF THE MASS DENSITY OF THE MIXTURE. PARAMETERS μ_e AND μ_e'

In the study of neutron stars we shall consider stellar mixtures of free electrons, free neutrons, and nuclei (A, Z) of a single 'chemical' and isotopic species (formed of N neutrons and Z protons, with N + Z = A).

These stellar mixtures result from a complete ionization of neutral atoms, followed by a succession of nuclear reactions leading to a nuclear equilibrium (by processes described in Chapter 6). They are characterized by partial number densities n_e of free electrons, n_n of free neutrons, and n_{nuc} of nuclei (A, Z).

Mixtures present in white dwarfs can be considered as a particular case of these more general mixtures with $n_n = 0$.

Thus, the stellar mixture considered in the present subsection is characterized, at a given moment, by the values of n_n, n_e, and n_{nuc} (which, as will be shown below, are not independent) and not, as for the case considered in Subsection 7.1, by the value of X, μ or μ_e.

Before ionization the initial atoms of a stellar mixture are electrically neutral, and nuclear reactions do not modify this neutrality. Therefore, the positive charge $Z n_{\text{nuc}}$ (in units equal to the absolute value of the charge of an electron) of the n_{nuc} nuclei (A, Z) contained in each cm^3 of the mixture, must be equal to the absolute value n_e of the charge $(-1)n_e$ of the n_e free electrons of the same cm^3. So we must have

$$n_e = Z n_{\text{nuc}}, \tag{101}$$

a relation which already reduces the number of independent parameters from three to two, and which can also be written as

$$n_{\text{nuc}} = \frac{1}{Z} n_e. \tag{101'}$$

On the other hand, if n_B denotes, as in Subsection 7.1, the *total* number density of *baryons*, [both free, as the *free neutrons* of the mixture, and those, *protons and neutrons*, *contained in* complex *nuclei* (A, Z) of the mixture], this definition will be expressed by

$$n_B = n_n + A n_{\text{nuc}}. \tag{102}$$

The general expression of n_B as a function of the mass density ρ of the mixture, already given by Equation (94'), viz.

$$n_B = \frac{1}{m_B} \rho = \rho N_A \tag{102'}$$

still holds.

On applying the general definition of the parameter μ_e given by Equation (95), we immediately obtain from Equations (102) and (101'):

$$\mu_e = n_B/n_e = \frac{A}{Z} + \frac{n_n}{n_e}. \tag{103}$$

More generally we have also a relation identical to Equation (97):

$$\mu_e = \rho N_A/n_e. \tag{103'}$$

A method of determination of the parameter μ_e will be described in Chapter 6. Here we consider μ_e as a known quantity, and n_e is given as a function of the mass density ρ by

$$\boxed{n_e = \rho N_A/\mu_e.} \tag{104}$$

Let us now introduce a new parameter μ_e' defined in a general way by the ratio:

$$\mu_e' = \frac{n_B - n_n}{n_e}. \tag{105}$$

Equations (101) and (102) show that μ_e' is equal, for the case considered here, to A/Z:

$$\mu_e' = \frac{A}{Z}, \tag{105'}$$

and Equation (103) shows that

$$\mu_e' = \mu_e - \frac{n_n}{n_e}. \tag{105''}$$

For stars with a composition similar to that of white dwarfs, where there are no free neutrons ($n_n = 0$) $\mu_e' = \mu_e = A/Z$. The term 'composition' used above deliberately does not take into account the difference in physical *state*, to be discussed later, between white dwarfs and some neutron stars.

Since for the general case of the present section μ_e and ρ are considered as known quantities, the partial number density n_{nuc} of nuclei (A, Z) will be given, on applying Equations (101′) and (104), by

$$n_{\text{nuc}} = \frac{1}{Z}\rho N_A \frac{1}{\mu_e}. \tag{106}$$

Finally, for the same case, the partial number density of free neutrons n_n will be given, on applying Equations (102), (102′) and (106), by

$$n_n = \rho N_A \left(1 - \frac{A}{Z}\frac{1}{\mu_e}\right), \tag{107}$$

or in a more elegant form, taking into account Equation (105′), by

$$n_n = \rho N_A \left(1 - \frac{\mu_e'}{\mu_e}\right). \tag{107'}$$

If beside the *total* mass density ρ of the mixture we consider the *partial* (collective) *mass density* of nuclei ρ_{nuc} and the partial mass density ρ_n of free neutrons, these partial densities will be given, on using Equations (106) and (105′), by

$$\rho_{\text{nuc}} A m_B n_{\text{nuc}} = \frac{A}{Z}\rho\frac{1}{\mu_e} = \rho\frac{\mu_e'}{\mu_e}, \tag{108}$$

and

$$\rho_n = \rho - \rho_{\text{nuc}} = \rho\left(1 - \frac{\mu_e'}{\mu_e}\right). \tag{109}$$

7.2.1. *Remark*

In white dwarfs and in neutron stars some of the components of the mixture are in a state of complete degeneracy (the electron gas in white dwarfs; the neutron gas and the electron gas in some neutron stars). For each case it is important to know whether the non-relativistic approximation or the ultra-relativistic one can be used for a given value of the total mass density ρ of the mixture.

If $\rho_{\text{cr},e}$ denotes the critical value of ρ corresponding to the critical value $n_{\text{cr},e}$ of the number density of the electron component given by Equation (36, e), we find, on applying Equation (104):

$$\rho_{\text{cr},e} = (9.7 \times 10^5)\mu_e \text{c.g.s.}. \tag{110}$$

Similarly, if $\rho_{\text{cr},n}$ denotes the critical value of ρ corresponding to the critical value $n_{\text{cr},n}$ of the number density of the neutron component given by Equation (36, n), we find, on applying Equation (107′):

$$\rho_{\text{cr},n} = (6.1 \times 10^{15})\left(1 - \frac{\mu_e'}{\mu_e}\right)^{-1} \text{c.g.s.}. \tag{111}$$

When one of the components of the mixture is in a state of an (almost) complete degeneracy, it can be useful to have an expression for the values of $x_{F,e}$ and $x_{F,n}$ of the parameter x corresponding to the Fermi level of each component, not as a function of n_e or n_n, as given by Equations (38, e) or (38, n), but as a function of the total mass density ρ.

Equation (104) gives

$$\boxed{x_{F,e} \approx (10^{-2})(\rho/\mu_e)^{1/3},} \tag{112}$$

and Equations (107′) and (109) give

$$x_{F,n} = (5.4 \times 10^{-6})\rho^{1/3}\left(1 - \frac{\mu_e'}{\mu_e}\right)^{1/3} = (5.4 \times 10^{-6})\rho_n'^{1/3}. \tag{113}$$

8. The Pressure Produced by a System of Independent Identical Fermions in a State of Complete Degeneracy

8.1. THE FUNDAMENTAL GENERAL RELATION

Let us return to systems of *identical* particles, instead of the mixtures considered in the preceding section, and let N be the total constant number of particles of the system, occupying an eventually variable volume of $V\,\text{cm}^3$. The system is supposed to consist of fermions of spin $s = \frac{1}{2}$ in a state of complete degeneracy.

Since N is supposed to be invariable, the number density n of the system can be considered as a function $n(V)$ of V:

$$n = n(V) = \frac{N}{V}, \tag{114}$$

hence

$$\frac{dn(V)}{dV} = -N/V^2, \tag{115}$$

and

$$-N\frac{dn}{dV} = N^2/V^2 = n^2. \tag{115'}$$

It was shown in Section 6 that the mean (total) energy $\bar{\epsilon}$ of *one* particle of the system is generally a function $\bar{\epsilon}(n, m)$ of the number density n and of the individual mass m of each particle.

Thus, U being the internal (total) energy of the whole system when its volume is V, and u being the corresponding (total) energy density, we have

$$\bar{\epsilon} = \bar{\epsilon}(n, m) = \frac{u}{n} = \frac{uV}{nV} = \frac{U}{N}. \tag{116}$$

It is known (see for instance Kittel, 1969, p. 111) that generally the internal energy U of a system of an invariable number of particles depends on *two* independent variables, for instance the volume V and the entropy S. The temperature of the system being T, and the pressure being P, any infinitesimal variations dS of S and dT of T correspond to a variation dU of U, given by the thermodynamic identity:

$$dU = dQ - P\,dV = T\,dS - P\,dV. \tag{117}$$

According to Equation (116)

$$U = \bar{\epsilon}(n, m)N, \tag{118}$$

and, since m and N are constant, U depends only (through n) on a *single variable* V. Thus the relation (117) reduces to

$$dU = -P\,dV. \tag{119}$$

This allows us to express P simply by $(-dU/dV)$. Physically U depends not only on V but also on the value of m, so that mathematically $U = U(n, m)$ appears as a function of two variables n and m. Therefore P will be given by the *partial* derivative:

$$P = -\frac{\partial U(n, m)}{\partial V} = -\frac{\partial U[n(V), m]}{\partial V} \tag{120}$$

Let us substitute in Equation (120) the expression (118) for U. We obtain:

$$P = -N\frac{\partial \bar{\epsilon}[n(V), m]}{\partial V} = -N\frac{\partial \bar{\epsilon}(n, m)}{\partial n}\frac{dn(V)}{dV}, \tag{121}$$

which, by Equation (115′), can be written

$$P = n^2 \frac{\partial \bar{\epsilon}(n, m)}{\partial n}. \tag{122}$$

According to Equation (116) we have

$$u(n, m) = n\bar{\epsilon}(n, m), \tag{123}$$

and taking the derivative of this relation with respect to n:

$$\frac{\partial u(n, m)}{\partial n} = \bar{\epsilon}(n, m) + n\frac{\partial \bar{\epsilon}(n, m)}{\partial n}, \tag{124}$$

which is equivalent to

$$n^2 \frac{\partial \bar{\epsilon}(n, m)}{\partial n} = n\left(\frac{\partial u}{\partial n} - \bar{\epsilon}\right). \tag{125}$$

If we take into account Equation (86′), Equation (125) becomes

$$n^2 \frac{\partial \bar{\epsilon}(n, m)}{\partial n} = n(\epsilon_F - \bar{\epsilon}), \tag{126}$$

and the expression (122) for P takes the form:

$$\boxed{P = P(\mathrm{RR}) = n(\epsilon_F - \bar{\epsilon}).} \tag{127}$$

It is relatively easy to transform this fundamental relation into a relation expression the pressure P is terms of *kinetic* energies. Let us first rewrite Equations (30) and (78): viz.

$$\epsilon_F = \epsilon_0 + \epsilon_{k,F}, \tag{128}$$

and

$$\bar{\epsilon} = \epsilon_0 + \bar{\epsilon}_k. \tag{129}$$

The substitution of these expressions for ϵ_F and $\bar{\epsilon}$ into Equation (127) yields the most important relation of the present Chapter, viz.

$$P = P(\mathrm{RR}) = n(\epsilon_{k,F} - \bar{\epsilon}_k) = n[\epsilon_{k,F}(n, m) - \bar{\epsilon}_k(n, m)]. \tag{130}$$

This relation shows that generally the pressure P is a function $P(n, m)$ of n and m.

On replacing ϵ_F in Equation (127) by its general expression given by Equation (39), and $\bar{\epsilon}$ by its general expression given by Equations (79) and (63′), we find by Equation (37), after some elementary calculations:

$$P(\mathrm{RR}) = \tfrac{1}{8}\epsilon_0 n_{\mathrm{cr}} f(x_F), \tag{131}$$

where $f(x)$ is the function defined by

$$f(x) = x(2x^2 - 3)(1 + x^2)^{1/2} + 3 \log_e [x + (1 + x^2)^{1/2}] \tag{132}$$

and tabulated by Chandrasekhar, 1939, (p. 361).

8.2. THE PRESSURE AS A FUNCTION OF THE ENERGY DENSITY

We might find the approximate expressions $P(\mathrm{NR})$ and $P(\mathrm{UR})$ for P by application of the limited expansions of $f(x)$ for $x \to 0$ and for $x \to \infty$ given in the Appendix to this Chapter. However, it is more reasonable, and very easy, to determine $P(\mathrm{NR})$ and $P(\mathrm{UR})$ *directly* from Equation (130) on replacing $\epsilon_{k,F}$ and $\bar{\epsilon}_k$ by their respective (NR) and (UR) approximate expressions.

Thus, taking into account the definition (77) of $\bar{\epsilon}_k$ and the relation (83) between $\epsilon_{k,F}(\mathrm{NR})$ and $\bar{\epsilon}_k(\mathrm{NR})$, we immediately find the very simple and important relation

$$P(\mathrm{NR}) = \tfrac{2}{3}u_k(\mathrm{NR}). \tag{133}$$

In the particular case of a system of electrons, this relation with obvious notation takes the form:

$$P_e(\mathrm{NR}) = \tfrac{2}{3}u_{k,e}(\mathrm{NR}). \tag{133, e}$$

Similarly, taking into account the definition (77) of $\bar{\epsilon}_k$ and the relation (84) between $\epsilon_{k,F}(\mathrm{UR})$ and $\bar{\epsilon}_k(\mathrm{UR})$, we find

$$P(\mathrm{UR}) = \tfrac{1}{3}u_k(\mathrm{UR}), \tag{134}$$

and

$$P_e(\mathrm{UR}) = \tfrac{1}{3}u_{k,e}(\mathrm{UR}). \tag{134, e}$$

Note that Equation (133) is very similar to the 'classical' relation between the pressure produced by a perfect gas and the kinetic energy density of the gas, whereas Equation (134) is very similar to the relation between the radiation pressure of a photon gas and the (kinetic) energy density of this gas (i.e. of the radiation field, necessarily ultra-relativistic since the velocity of photons is c). But these analogies have mainly a mnemonic value, since the physical situation (that of a degenerate gas of fermions) described by Equations (133) and (134), is quite different, either from that of a perfect gas or from that of a gas of bosons such as photons.

8.3. THE PRESSURE AS A FUNCTION OF THE NUMBER DENSITY

The relations (133) and (134) are important, but it is even more useful for most applications to express $P(\mathrm{NR})$ and $P(\mathrm{UR})$ as explicit functions of n (and eventually of m). This is achieved by the use of the expressions (66) and (69) for $u_k(\mathrm{NR})$ and $u_k(\mathrm{UR})$.

The result for $P(\mathrm{NR})$ is:

$$P(\mathrm{NR}) = \frac{C^2}{5}\frac{1}{m}n^{5/3} = (2.1 \times 10^{-54})\frac{1}{m}n^{5/3}. \tag{135}$$

If we introduce the numerical values of m for electrons and neutrons ($m_e = 9.1 \times 10^{-28}$ g and $m_n \approx 1.66 \times 10^{-24}$ g) Equation (135) yields the corresponding pressures $P_e(\mathrm{NR})$ and $P_n(\mathrm{NR})$ for electrons and neutrons:

$$\boxed{P_e(\mathrm{NR}) = (2.3 \times 10^{-27})n_e^{5/3},} \tag{135, e}$$

and

$$\boxed{P_n(\mathrm{NR}) = (1.3 \times 10^{-30})n_n^{5/3}.} \tag{135, n}$$

$P(\mathrm{UR})$, independent of m, is given both for electrons and for neutrons by

$$P(\mathrm{UR}) = \frac{cC}{4}n^{4/3} = (2.4 \times 10^{-17})n^{4/3}; \tag{136}$$

but it can be more convenient to write separately

$$\boxed{P_e(\mathrm{UR}) = (2.4 \times 10^{-17})n_e^{4/3},} \tag{136, e}$$

and

$$\boxed{P_n(\mathrm{UR}) = (2.4 \times 10^{-17})n_n^{4/3}.} \tag{136, n}$$

Similarly $P(\mathrm{RR})$, given by the general Equations (131) and (132) as a function of x, can be transformed into a function of n by Equation (38), viz.

$$x_F = (n/n_{\mathrm{cr}})^{1/3}. \tag{137}$$

Some authors consider that in this general case (RR) P is given by a *parametric representation* as a function of n:

$$\begin{cases} P(\mathrm{RR}) = \tfrac{1}{8}\epsilon_0 n_{\mathrm{cr}} f(x_F); \\ n = n_{\mathrm{cr}} x_F^3, \end{cases} \tag{138}$$

where x_F plays the role of parameter. This can be convenient, but (as shown above) is not absolutely necessary.

8.4. Preparation of Applications to White Dwarfs and Neutron Stars

For some applications it can be useful to write separately the relations (138) for electrons and for neutrons, with the explicit values of the numerical coefficients:

$$P_e(\text{RR}) = (6.01 \times 10^{22}) f(x_{F,e}); \qquad n_e = (5.9 \times 10^{29}) x_{F,e}^3, \qquad (138, e)$$

and

$$P_n(\text{RR}) = (6.9 \times 10^{35}) f(x_{F,n}); \qquad n_n = (3.7 \times 10^{39}) x_{F,n}^3. \qquad (138, n)$$

Finally, we can express the approximate values for P_e and P_n, given respectively by Equations (135, e), (135, n), (136, e) and (136, n), as a function of the total mass density ρ of the mixture.

Thus one finds, after some numerical computations where use is made of Equation (104), that the pressure produced in a white dwarf by the electron component in a state of complete degeneracy admitting the NR approximation, i.e. $P_e(\text{NR})$ is given as a function of ρ and μ_e by

$$\boxed{P_e(\text{NR}) = (9.9 \times 10^{12})(\rho/\mu_e)^{5/3}.} \qquad (139)$$

When the electron gas is supposed to be in a state of complete degeneracy admitting the UR approximation, the pressure $P_e(\text{UR})$ is given by

$$\boxed{P_e(\text{UR}) = (1.2 \times 10^{15})(\rho/\mu_e)^{4/3}.} \qquad (140)$$

For the particular case of white dwarfs in which the nuclei are simply α-particles (A = 4, Z = 2) the parameter μ_e is equal to 2, and Equations (139), (140) take respectively an even more simple form:

$$P_e(\text{NR}) \approx (3.1 \times 10^{12})\, \rho^{5/3}, \qquad (139')$$

and

$$P_e(\text{UR}) \approx (4.8 \times 10^{14})\, \rho^{4/3}. \qquad (140')$$

For neutrons stars in the particular state when the mixture contains *only free neutrons* and electrons (no complex nuclei any more), $\rho_n = \rho$ and $n_n = \rho N_A$. Then the pressure produced by the free neutron component in a state of complete degeneracy admitting the NR approximation, i.e. $P_n(\text{NR})$, is given by

$$\boxed{P_n(\text{NR}) = (5.6 \times 10^{9})\, \rho^{5/3}.} \qquad (141)$$

8.5. The Function u_k/P

Relations (133) and (134) can be put respectively in a form particularly convenient for the study of white dwarfs by introducing the function n'' (which is *not* any kind of number density!) defined by

$$n'' = u_k/P, \tag{142}$$

thus allowing us to write (133) as

$$P(\mathrm{NR}) = n''(\mathrm{NR})^{-1}u_k(\mathrm{NR}), \tag{143}$$

where

$$n''(\mathrm{NR}) = \tfrac{3}{2} = 1.5 = \text{const.}; \tag{144}$$

and to write (134) as

$$P(\mathrm{UR}) = n''(\mathrm{UR})^{-1}u_k(\mathrm{UR}), \tag{145}$$

where

$$n''(\mathrm{UR}) = 3 = \text{const.}. \tag{146}$$

In the general case (RR), Equations (64) and (131), applied to the definition (142) of n'', show that in this case n''(RR) is no longer a constant, but a complicated function of x_F (tabulated by Chandrasekhar, 1939, p. 361), given by

$$n''(\mathrm{RR}) = g(x_F)/f(x_F). \tag{147}$$

Thus n''(NR) and n''(UR) are simply the values of n''(RR) when $x \to 0$ and $x \to \infty$.

9. The Domain of Separation between the State of a Perfect Gas and the State of Complete Degeneracy. Application to the Sun

9.1. General Definition and General Properties of a Perfect Gas

We have already given the condition (32), viz.

$$kT \lll \epsilon_{k,F}, \tag{148}$$

to be satisfied to insure an (almost) complete *degeneracy*. And we have shown that this condition can also be written as Equation (45), viz.

$$T \lll T_{\lim}^{\deg}. \tag{148'}$$

$T_{\lim}^{\deg}$(NR) and $T_{\lim}^{\deg}$(UR) are given by Equations (46) and (46′), viz.

$$T_{\lim}^{\deg}(\mathrm{NR}) = \frac{1}{2mk} C^2 n^{2/3}, \tag{149}$$

and

$$T_{\lim}^{\deg}(\mathrm{UR}) = \frac{cC}{k} n^{1/3}. \tag{150}$$

Equations (149) and (150) show that, for a given temperature T, an (almost) complete degeneracy demands in the case (UR) *a number density n as large as possible*, and in the case (NR) still demands a *large number density n* (smaller than n_{cr}), but is *favoured by a small value of the individual mass m of fermions*.

However, it is very useful to analyse a somewhat inverse case of a *perfect* (or 'ideal') *non-relativistic gas*. One can define a *perfect gas* (before introducing any distinction between the non-relativistic and the ultra-relativistic case) as a system of independent identical particles in which the occupancy rate of *quantum* states (see our discussion in Section 2) f_p is *very much smaller than one*: ($f_p \ll 1$), (see e.g. Kittel, 1969, Ch. 11).

We have already shown, in Subsection 3.1 that for fermions the general expression of f_p can be written as Equation (25), viz.

$$f_p = F_\Delta(X) = [e^{\Delta(X-1)} + 1]^{-1}, \tag{151}$$

where

$$\Delta = \mu_k/kT \quad \text{and} \quad X = \epsilon_k(p)/\mu_k. \tag{151'}$$

Thus the statement $f_p \ll 1$ is equivalent to the statement that $e^{\Delta(X-1)} \gg 1$, so that we can neglect (+ 1) in the square bracket of Equation (151) and write (indicating that we deal with a perfect gas by a superscript PG):

$$f_p^{\mathrm{PG}} = e^{-\Delta(X-1)} = e^{\Delta}\, e^{-X\Delta} = e^{\mu_k/kT} e^{-\epsilon_k/kT}. \tag{152}$$

Let us introduce a new parameter λ_k defined by

$$\lambda_k = e^{\mu_k/kT}, \tag{153}$$

and note that λ_k is independent of p and ϵ_k. This allows us to write f_p^{PG} as

$$f_p^{\mathrm{PG}} = \lambda_k\, e^{-\epsilon_k/kT}. \tag{154}$$

Advanced readers will recognize the *Boltzmann distribution* characteristic of a thermodynamic equilibrium in classical thermodynamics (in which the difference between fermions and bosons is ignored).

Let us now make use of the general expression (valid for fermions and for bosons) for the number $D(p, dp)$ of (complete) *quantum* states accessible for a particle whose momentum lies between p and $(p + dp)$, given by Equation (19′), viz.

$$D(p, dp) = g_s(1/h^3)\, 4\pi p^2\, dp, \tag{155}$$

where g_s is as previously the number of possible spin states.

Then, for a perfect gas the number of particles $n^{\mathrm{PG}}(p, dp)$ whose momentum lies between p and $(p + dp)$ will be given by

$$n^{\mathrm{PG}}(p, dp) = n^{\mathrm{PG}}(p)\,dp = f_p^{\mathrm{PG}} D(p, dp) = \lambda_k g_s\, e^{-\epsilon_k/kT}\left(\frac{4\pi p^2\, dp}{h^3}\right). \tag{156}$$

Let us compute the corresponding total number density n^{PG} of the system, which will be given by

$$n^{\mathrm{PG}} = \lambda_k(4\pi g_s/h^3) \int_0^\infty e^{-\epsilon_k(p)/kT} p^2\, dp. \tag{157}$$

Similarly, the corresponding kinetic energy density u_k^{PG} of the system, will be given by

$$u_k^{\mathrm{PG}} = \int_0^\infty \epsilon_k(p)\, n^{\mathrm{PG}}(p)\,dp = \lambda_k(4\pi g_s/h^3) \int_0^\infty \epsilon_k\, e^{-\epsilon_k/kT} p^2\, dp. \tag{158}$$

9.2. THE MAIN PROPERTIES OF A NON-RELATIVISTIC PERFECT GAS

Hitherto the distinction between different possible approximations (NR) or (UR) has been ignored. Considering, henceforth, a more particular case of a physical situation allowing the non-relativistic approximation (NR), we can make use of the 'classical' expression for $\epsilon_k(p)$ given by Equation (12), viz.

$$\epsilon_k(\mathrm{NR}) = \epsilon_k(p) = \frac{p^2}{2m}. \tag{159}$$

On introducing, for more brevity, the quantities defined by

$$A = (4\pi g_s/h^3)\lambda_k, \tag{160}$$

and

$$p = (2mkT)^{1/2}y, \tag{160'}$$

we obtain:

$$n^{\mathrm{PG}}(\mathrm{NR}) = A(2mkT)^{3/2}I, \tag{161}$$

and

$$u_k^{\mathrm{PG}}(\mathrm{NR}) = A(2mkT)^{3/2}(kT)J, \tag{162}$$

where I and J represent rather well known integrals (see e.g. Kittel, 1969, p. 23):

$$I = \int_0^\infty e^{-y^2}y^2\,dy = \tfrac{1}{4}\pi^{1/2}, \tag{163}$$

and

$$J = \int_0^\infty e^{-y^2}y^4\,dy = \tfrac{3}{8}\pi^{1/2}. \tag{164}$$

Hence

$$u_k^{\mathrm{PG}}(\mathrm{NR}) = A(2mkT)^{3/2}(kT)\tfrac{3}{8}\pi^{1/2} = \tfrac{3}{2}kT\lambda_k g_s(2\pi mkT/h^2)^{3/2}, \tag{165}$$

and

$$n^{\mathrm{PG}}(\mathrm{NR}) = A(2mkT)^{3/2}\tfrac{1}{4}\pi^{1/2} = \lambda_k g_s(2\pi mkT/h^2)^{3/2}. \tag{166}$$

Dividing $u_k^{\mathrm{PG}}(\mathrm{NR})$ by $n^{\mathrm{PG}}(\mathrm{NR})$ we find the mean kinetic energy $\bar{\epsilon}^{\mathrm{PG}}(\mathrm{NR})$ of one particle of a perfect non-relativistic gas:

$$\bar{\epsilon}_k^{\mathrm{PG}}(\mathrm{NR}) = (u_k/n)^{\mathrm{PG}}(\mathrm{NR}) = \tfrac{3}{2}kT. \tag{167}$$

This result is identical with the 'classical' expression for an ideal gas.

Equation (166) gives us the value of λ_k as a function of n, T, m and g_s:

$$\lambda_k^{\mathrm{PG}}(\mathrm{NR}) = \frac{n}{g_s}(2\pi mkT/h^2)^{-3/2}, \tag{168}$$

from which we deduce, by application of the definition (153), the corresponding expression for μ_k as a function of n, T, m and g_s:

$$\boxed{\mu_k^{\mathrm{PG}}(\mathrm{NR}) = (kT)\log_e\left[\frac{n}{g_s}h^3(2\pi mkT)^{-3/2}\right].} \tag{169}$$

This result is physically very important since it expresses the theoretical parameter μ_k as a function of the physically significant quantities n, m, g_s and T, which can be considered as given for each concrete case.

Note further that in Equation (154) the 'Boltzmann factor' $e^{-\epsilon_k/kT}$ can vary only between *zero* and 1. Thus a *sufficient* condition for a state of perfect gas is generally:

$$\lambda_k \ll 1. \tag{170}$$

On taking into account Equation (168), this result is equivalent (for a state of a perfect *non-relativistic* gas) to

$$n \ll g_s h^{-3}(2\pi mk)^{3/2} T^{3/2}. \tag{171}$$

9.3. Comparison between the Limiting Temperatures for Degeneracy and for the State of Perfect Gas (at the NR Approximation)

According to Equation (171) a state of a perfect non-relativistic gas (for given values of m, g_s and n) demands a sufficiently *high temperature T*, higher than the limiting temperature $T_{\text{lim}}^{\text{PG}}(\text{NR})$ defined by

$$T_{\text{lim}}^{\text{PG}}(\text{NR}) = (2\pi mk g_s^{2/3} h^{-2})^{-1} n^{2/3}. \tag{172}$$

The same inequality (171) also shows that for given values of m, g_s and T a state of a perfect non-relativistic gas demands a sufficiently *low number density n*, lower than the right-hand side of (171).

However, it is the proper *combination* of values of n and T, expressed by the inequality (171), which represents the general condition for a state of a perfect non-relativistic gas.

Let us now reintroduce, in order to facilitate a comparison with the results concerning the complete degeneracy, the constant C defined by Equation (40′), viz.

$$C = (h/2)(3/\pi)^{1/3} \approx 3.25 \times 10^{-27} \text{ c.g.s.} \tag{173}$$

and take into account the approximate value 0.83 of $(16/9\pi)^{1/3}$. Then $T_{\text{lim}}^{\text{PG}}(\text{NR})$ defined by Equation (172) can be written as

$$T_{\text{lim}}^{\text{PG}}(\text{NR}) = (0.83)\frac{1}{2mk} C^2 (2/g_s)^{2/3} n^{2/3}. \tag{174}$$

Let us compare this result with the expression (149) for the *limiting temperature for degeneracy* $T_{\text{lim}}^{\text{deg}}(\text{NR})$. We see that for fermions such as electrons, protons or neutrons, for which $g_s = 2$, the only difference between $T_{\text{lim}}^{\text{deg}}(\text{NR})$ and $T_{\text{lim}}^{\text{PG}}(\text{NR})$ is the presence of the factor 0.83 instead of 1.00. Thus, for given values of m and n, the condition for a state of complete degeneracy of a gas of fermions (for a number density n sufficiently *low* to permit the NR approximation), viz. $T \lll T_{\text{lim}}^{\text{deg}}(\text{NR})$, and the condition for a state of a perfect non-relativistic gas, viz. $T \ggg T_{\text{lim}}^{\text{PG}}(\text{NR})$, involve (for a fermion gas) *a limiting temperature of the same order of magnitude* with reversed inequalities.

9.3.1. *Remarks*

9.3.1.1. Note incidentally that, according to the expression of $T_{\text{lim}}^{\text{PG}}(\text{NR})$ as a function of m, the state of a perfect non-relativistic gas is realized for a lower temperature or (and) higher numerical density with *heavy* fermions, such as protons or neutrons, than with *light* ones such as electrons.

Similarly, according to the expression of $T_{\text{lim}}^{\text{deg}}(\text{NR})$ as a function of m, the almost complete degeneracy is realized for a higher temperature or (and) lower number density with *light* fermions (electrons) than with *heavy* ones.

Both results contradict the false current *intuition* of most students.

9.3.1.2. Note finally that the condition (171) obviously holds, with the proper value of g_s, for *bosons* such as α-particles, if considered as a sufficient condition for a state of perfect non-relativistic gas, but *cannot* be used, with a reversed sign of inequality, as a condition for complete degeneracy (a state proper only to fermions).

9.4. Application to the Sun

Let us apply the condition (171) to the present state of the stellar mixture at the center of the Sun. The mass density of the mixture is there of the order of 135 g cm^{-3}; proton and α-particle abundances X and Y are both of the order of 0.50 grams per gram of the mixture. One finds, after some tedious numerical computations, that the respective limiting temperatures $T_{\text{lim},e}^{\text{PG}}$, $T_{\text{lim},p}^{\text{PG}}$ and $T_{\text{lim},\alpha}^{\text{PG}}$ above which the gas of electrons, protons and alpha particles, can be considered as a perfect gas are: $T_{\text{lim},e}^{\text{PG}} \approx 4 \times 10^6$ K; $T_{\text{lim},p}^{\text{PG}} \approx 2 \times 10^6$ K; $T_{\text{lim},\alpha}^{\text{PG}} \approx 320$ K. All these temperatures are below the central temperature of about 15×10^6 K found with the assumption that all components of the solar mixture are in a state of perfect gas. This insures the consistency of the standard theory of interiors of normal stars.

Appendix: Establishment of the Rigorous (RR) Expressions for $u(n, m)$ and $u_k(n, m)$

According to Equation (62), viz.

$$u(\text{RR}) = 3n_{\text{cr}} \int_0^{x_F} \epsilon(x) x^2 \, dx, \tag{A1}$$

we must take for $\epsilon(x)$ in the general case (RR) the expression $\epsilon_0(1 + x^2)^{1/2}$ given by Equation (16).

On introducing the parameter Θ defined as in Subsection 1.3 by

$$x = sh\Theta, \tag{A2}$$

the following expression for the energy density $u(\text{RR})$ of a completely degenerate system of fermions of spin $s = \frac{1}{2}$ is obtained:

$$u(\text{RR}) = \tfrac{3}{8} n_{\text{cr}} \epsilon_0 G(x_F), \tag{A3}$$

where the function $G(x_F)$ is defined by

$$G(x_F) = \int_0^{\Theta_F} 8 \sinh^2 \Theta \cosh^2 \Theta \, d\Theta. \tag{A4}$$

We have the well known identities for hyperbolic functions:

$$\cosh^2 \Theta = 1 + \sinh^2 \Theta; \tag{A5}$$

$$\sinh 2\Theta = 2 \sinh \Theta \cosh \Theta; \tag{A6}$$

$$\cosh 2\Theta = \sinh^2 \Theta + \cosh^2 \Theta = 2\cosh^2 \Theta - 1 = 1 + 2\sinh^2 \Theta; \tag{A7}$$

from which one can deduce the new identity:

$$8 \sinh^2 \Theta \cosh^2 \Theta = -1 + \cosh 4\Theta. \tag{A8}$$

On performing the integration in Equation (A4) we find:

$$G(x_F) = -\Theta_F + \tfrac{1}{4} \sinh 4\Theta_F, \tag{A9}$$

where, according to the definition (A2) of Θ:

$$\Theta_F = \sinh^{-1} x_F = \log_e [x_F + (1 + x_F^2)^{1/2}]. \tag{A10}$$

An application of identities (A6) and (A7) gives $\sinh \Theta_F$ as a function of x_F:

$$\tfrac{1}{4} \sinh 4\Theta_F = x_F(1 + x_F^2)^{1/2}(1 + 2x_F^2). \tag{A11}$$

Thus the final expression of $G(x)$ is:

$$G(x) = x(2x^2 + 1)(1 + x^2)^{1/2} - \log_e [x + (1 + x^2)^{1/2}], \tag{A12}$$

which is equivalent to Equation (63′).

The corresponding Equation (63) is equivalent to Equation (A3).

Let us introduce, as in Equation (64) the expression for u_k(RR) as a function of $g(x_F)$, viz.

$$u_k(\mathrm{RR}) = \tfrac{1}{8} n_{\mathrm{cr}} \epsilon_0 g(x_F); \tag{A13}$$

where

$$g(x) = 3G(x) - 8x^3, \tag{A14}$$

and, as in Equation (131), the expression for P(RR) as a function of $f(x)$, viz.

$$P(\mathrm{RR}) = \tfrac{1}{8} n_{\mathrm{cr}} \epsilon_0 f(x_F); \tag{A15}$$

where

$$f(x) = 8x^3[(1 + x^2)^{1/2} - 1] - g(x)$$
$$= x(2x^2 - 3)(1 - x^2)^{1/2} + 3 \log_e [x + (1 + x^2)^{1/2}]. \tag{A16}$$

One finds, after relatively elementary but rather long and tedious calculations based on well known expansions of usual functions, the following expansions in the vicinity of $x = 0$:

$$g(x) = \tfrac{12}{5} x^5 (1 - \tfrac{5}{28} x^2 + \ldots); \tag{A17}$$

$$f(x) = \tfrac{8}{5} x^5 (1 - \tfrac{5}{14} x^2 + \ldots). \tag{A17′}$$

Expansions of the same functions in the vicinity of very large values of x (asymptotic expansions) are:

$$g(x) = 6x^4 \left(1 - \frac{4}{3}\frac{1}{x} + \ldots\right); \tag{A18}$$

$$f(x) = 2x^4(1 - \tfrac{3}{2} x^{-2} + \ldots). \tag{A18′}$$

Expansions of $G(x)$ are easily deduced from those given above.

References

WORKS CITED IN THE TEXT

(Works that can be used for further study are marked by an asterisk.)

* Chandrasekhar, S.: 1939, *An Introduction to the Study of Stellar Structure*, Univ. of Chicago Press. (Reprinted 1967), Dover, New York).
* Kittel, C.: 1969, *Thermal Physics*, Wiley, New York.
* Kittel, C., Knight, W. D., Ruderman, M. A.: 1962, *Mechanics* (Berkeley Physics Course, Vol. I), McGraw-Hill, New York.
* Reif, F.: 1964, *Statistical Physics* (Berkeley Physics Course, Vol. V), (Section 3.5 (p. 121)), McGraw-Hill, New York.

CHAPTER 5

WHITE DWARFS

1. A Few Historical Remarks

The history of astronomy can be characterized by intellectual achievements of different kinds: discovery of astronomical objects by means of theoretical interpretation of apparently anomalous observations, or the extrapolation of known physical principles to anticipate 'new' and more or less strange objects whose existence is or may be confirmed observationally later.

The discovery of the 'invisible companion' of Sirius, by the German astronomer Bessel, like the discovery of Neptune by the French astronomer Le Verrier, belongs to the first category. By application of Newtonian theory of binary stars to a theoretical interpretation of irregularities in the proper motion of Sirius, Bessel concluded (1844) that these irregularities correspond to a motion of Sirius around the center of gravity of a binary system composed by Sirius itself and an invisible companion (today called 'Sirius B').

According to observations of Sirius and their interpretation by Bessel, this companion was a relatively massive star. It remained unobserved for about 20 years until the progress in astronomical instrumentation permitted its observational discovery (1862) by the American astronomer Clark.

We now know that Sirius B is a star of about the same mass as the Sun, but with a very small radius (in comparison with a normal star). Indeed, the radius of Sirius B is of the order of 17 500 km. Thus, its *mass density is of several tons per* cm^3. A possibility of such high densities was accepted only after the application (in 1926) by Fowler of the theory of *degeneracy* of the electron gas (see Chapter 4) to stellar interiors.

Today, we know a relatively great number of stars similar to Sirius B. They are called '*white dwarfs*' ('dwarf' by their radius, and 'white' because of the colour of the stars of this type observed originally – though now several *yellow* or *red* 'white dwarfs' are known).

Nowadays, the great density of white dwarfs has lost its strangeness, for we know today '*neutron stars*' (see Chapters 6 and 9), observed as '*pulsars*' and we suspect the existence of '*black holes*' (see Chapter 10), observed as some of the '*X-ray Sources*', whose mass densities considerably exceed those of white dwarfs.

The theory of white dwarfs, to be presented in this Chapter, already interesting by itself, represents a useful introduction to the study of neutron stars. Both theories make use, in a first approximation, of the '*virial theorem*' and of the concept of '*polytrope*' discussed in the next Section.

2. Polytropes and the Virial Theorem. Application to an Elementary Theory of White Dwarfs

2.1. Introduction

The theory of white dwarfs, to be developed below, shows that the mass density ρ of these stars increases monotonically (as that of normal stars) towards the center. On the other hand, one finds that the '*central density*' ρ_c, (apart from very exceptional cases discussed separately) usually lies between values of the order of $\rho_c^{(1)} = 2.4 \times 10^5 \text{ g cm}^{-3}$ and $\rho_c^{(2)} = 2.0 \times 10^9 \text{ g cm}^{-3}$.

In a first approximation, white dwarfs can be considered as a mixture of *α-particles* and *free electrons* (the α-particles being replaced by more complex nuclei at a further stage of evolution) of respective number densities n_α and n_e, with the obvious relation (imposed by electric neutrality):

$$n_e = 2n_\alpha. \tag{1}$$

Neglecting, as in Chapter 4, the mass m_e of each electron in comparison with the mass m_α of each α-particle ($m_\alpha = 7360 m_e$) (and neglecting, in a first approximation, masses corresponding to kinetic energies of particles) we get immediately as limits in $n_{\alpha,c}$, corresponding to the limits in ρ_c, the number densities:

$$n_{\alpha,c}^{(1)} = 3.6 \times 10^{28} \text{ cm}^{-3}, \tag{2}$$

and

$$n_{\alpha,c}^{(2)} = 3.0 \times 10^{32} \text{ cm}^{-3}. \tag{3}$$

Hence, according to Equation (1), the corresponding limits in number densities ($n_{e,c}$) of the electron component of the mixture, at the centre:

$$n_{e,c}^{(1)} = 7.2 \times 10^{28} \text{ cm}^{-3}, \tag{4}$$

and

$$n_{e,c}^{(2)} = 6.0 \times 10^{32} \text{ cm}^{-3}. \tag{5}$$

Now, according to Equation (36, e) of Ch. 4, the critical number density for an electron gas, $n_{\text{cr},e}$, is of the order of $6 \times 10^{29} \text{cm}^{-3}$. Thus, an eventual complete degeneracy of the electron component of a white dwarf will be non-relativistic (NR) near the limit (1) and ultra-relativistic (UR) near the limit (2). On the other hand, according to Equations (50) and (51) of Ch. 4, the (almost) *complete degeneracy* of the electron gas demands, near the limit (1), a central temperature T_c lower than the limiting temperature of $(4.2 \times 10^{-11})[n_{e,c}^{(1)}]^{2/3} \approx 7 \times 10^8$ K and, near the limit (2), a central temperature lower than the limiting temperature of $(0.70)[n_{e,c}^{(2)}]^{1/3} \approx 6 \times 10^{10}$ K.

If near the limit (1) T_c is of the order of 10^6 K, the electron component of the mixture will be in a state of an (almost) complete degeneracy. But, conversely, such a temperature will be sufficiently *high* to insure to the α-particles component a state of a perfect non-relativistic (ideal) gas. Indeed, according to Equation (174) of Ch. 4, applied to α-particles ($s = 0, g_s = 1$), the limiting temperature which must be *exceeded* in this case is of the order of

$$T_{\text{lim},\alpha}^{\text{PG}} = (0.83)\frac{1}{2m_\alpha k} C^2 n_\alpha^{2/3} 2^{2/3} \approx (7 \times 10^{-15}) n_\alpha^{2/3}. \tag{6}$$

With a value of $n_{\alpha,c}^{(1)}$ of the order of $3.6 \times 10^{28}\,\text{cm}^{-3}$ given by Equation (2), this indicates a limiting temperature of the order of 8×10^4 K. Thus, a central temperature of the order of 10^6 K is sufficiently high. It is relatively easy to verify that at a temperature of about 10^6 K the gas of α-particles is non-relativistic, so that Equation (174) of Ch. 4 can be used, since at this temperature the mean kinetic energy $\bar{e}_k = \frac{3}{2}kT$ of any particle of an ideal gas is very much lower than the rest energy $m_\alpha c^2$ of an alpha particle (which is of the order of 6×10^{-3} erg).

If near the limit (2) the central temperature T_c is of the order of 10^8 K, the electron gas still will be in an (almost) completely degenerate state. But such a temperature will be again quite sufficient to insure to the α-particle component a state of a perfect non-relativistic gas. Indeed, with a value of $n_{\alpha,c}^{(2)}$ of the order of $3.0 \times 10^{32}\,\text{cm}^{-3}$ given by Equation (3), the limiting temperature given by Equation (6) will be of the order of 3×10^7K. And again, it is relatively easy to verify that the non-relativistic relation (6) holds, by the same argument as above.

Thus, in all cases the partial pressure of α-particles, in the vicinity of the center, will be given by the classical relation for a perfect gas:

$$P_\alpha = n_\alpha kT = (2.1 \times 10^7)\,\rho T \text{ c.g.s.,} \tag{7}$$

since in a mixture where the only heavy particles are α-particles

$$n_\alpha = \rho/m_\alpha = (1/4)\,\rho N_A = (1.5 \times 10^{23})\,\rho. \tag{7'}$$

As to the partial pressure of electrons P_e, in the vicinity of the center, one must separate case (1) from case (2). Near the limit (1) P_e will be given by Equation (139′) of Ch. 4, corresponding to the non-relativistic complete degeneracy of electrons, with a value of the parameter μ_e equal to 2 (see Chapter 4, Section 7), i.e. by Equation

$$P_e(\text{NR}) = P_e^{(1)} \approx (3.1 \times 10^{12})\rho^{5/3}. \tag{8}$$

Near the limit (2), P_e will be given by Equation (140′) of Ch. 4, corresponding to the ultra-relativistic complete degeneracy of electrons, with $\mu_e = 2$, i.e. by Equation

$$P_e(\text{UR}) = P_e^{(2)} \approx (4.8 \times 10^{14})\rho^{4/3}. \tag{9}$$

One thus finds, after some elementary calculations, that near the limit (1) the ratio P_e/P_α near the center is of the order of $(600 \times 10^6)/T_c \approx 600$ and that near the limit (2) the ratio P_e/P_α is of the order of $(300 \times 10^8)/T_c \approx 300$. Therefore, near the center (with reasonable assumptions made above), P_α is *always negligible in comparison with* P_e.

Obviously the preceding discussion cannot be directly extended far from the center, into a region where the density can become too low for the corresponding temperature for the maintenance of the degeneracy of the electron gas, and similarly the temperature can become too low for the corresponding density for the maintenance of the perfect gas state of α-particles. However, since *both* temperature and density *decrease* with increasing distance from the center, the situation which exists at the center also can be expected to hold further out. This is confirmed by a more advanced analysis, for instance that of Chandrasekhar (1939) showing, in addition, that the temperatures quoted above are quite reasonable.

Henceforth, directing our attention to the degeneracy of the electron gas, we note that white dwarfs near the limit (1) correspond (as shown later) to stars of a mass of about

$0.22M_\odot$, (where $M_\odot$ is the mass of the Sun). In this domain, the approximation (NR) *at the center*, conditioned by $n_{e,c} < n_{\mathrm{cr},e}$ [or by $\rho < 2 \times 10^6$ according to Eqs. (110) and (100) of Ch. 4], is permitted. The corresponding condition of a *non-relativistic* degeneracy of the electron gas, that is $n_e < n_{\mathrm{cr},e}$ (equivalent to $\rho < 2 \times 10^6$) will certainly be satisfied *everywhere* in the star if it is satisfied at the center, since ρ is *a decreasing* function of the distance from the center.

However, near the limit (2), which corresponds (as shown later) to white dwarfs of a mass about $1.4M_\odot$, the *ultra-relativistic* approximation (UR) which can be used in the study of the complete degeneracy of the electron gas *near the center*, is certainly forbidden near the surface, where the decreasing mass density tends to zero, and where n_e falls below the critical value $n_{\mathrm{cr},e}$. Therefore, 'massive' white dwarfs (of a mass larger than about $\frac{1}{2}M_\odot$) must be studied by means of the rigorously relativistic (RR) expressions of the physical state. This has been done by Chandrasekhar in his fundamental investigations, and we quote his results at the end of this chapter. Fortunately, a comparison of his rigorous results with those obtained by an application of the (UR) approximation *up to the surface* of massive white dwarfs shows that the contribution of the non-relativistic superficial layers to the equilibrium of the star is small enough to be negligible in a rough analysis aiming at a first approximation. (A similar phenomenon was already encountered in Section 5.2 of Chapter 4 in connection with a direct calculation of the ultra-relativistic expression of the kinetic energy density u_k).

2.2. Polytropes and the Emden–Lane Equation

In a physical situation in which the pressure P produced by a system of particles depends *on the mass density* ρ *alone* by a relation of the form

$$P = K\rho^{(1+1/n')}, \tag{10}$$

where *both* K and n' are *numerical constants*, the system is called a '*polytrope*' (of order n') and n' is called the '*polytropic index*' of the system.

In our stellar mixture of free electrons and α-particles (or heavier nuclei) the total pressure P is equal to the partial pressure P_e of the electron component. Thus, we can apply, when respectively the (NR) or the (UR) approximations are allowed, Equations (133, e) or (134, e) of Chapter 4. This gives:

$$P(\mathrm{NR}) = \tfrac{2}{3}u_{k,e}(\mathrm{NR}), \tag{11}$$

and

$$P(\mathrm{UR}) = \tfrac{1}{3}u_{k,e}(\mathrm{UR}). \tag{11'}$$

Thus, for these extreme approximations (NR) and (UR), (but *not* in the rigorous treatment using the RR relations) we can combine Equations (11) and (11′) in a single more general relation, by writing

$$P = \frac{1}{n'}u_{k,e}, \tag{12}$$

if, at the same time, we specify that

$$n'(\mathrm{NR}) = 1.5, \tag{13}$$

and

$$n'(\mathrm{UR}) = 3.0. \tag{13'}$$

Equation (12) is consistent with Equation (10), and the symbol n' has the same meaning in both Equations. Indeed, Equations (139′) and (140′) of Ch. 4, applied with $P = P_e$. can be written respectively (for a mixture of α-particles and electrons):

$$P(\mathrm{NR}) = K_1 \rho^{5/3} = K(\mathrm{NR})\, \rho^{(1+1/(3/2))} \tag{14}$$

with

$$K_1 = K(\mathrm{NR}) \approx (3.1 \times 10^{12}) \text{ c.g.s.}; \tag{14'}$$

and

$$P(\mathrm{UR}) = K_2 \rho^{4/3} = K(\mathrm{UR})\, \rho^{(1+1/3)} \tag{15}$$

with

$$K_2 = K(\mathrm{UR}) \approx (4.8 \times 10^{14}) \text{ c.g.s.}, \tag{15'}$$

which shows that $P(\mathrm{NR})$ and $P(\mathrm{UR})$ are given by polytropic relations of the form (10), with $n'(\mathrm{NR})$ and $n'(\mathrm{UR})$ equal to $\frac{3}{2}$ and 3 respectively, in conformity with (13) and (13′).

Eqs. (139) and (140) of Ch. 4 applied, with $P = P_e$, to a mixture of electrons and nuclei (A, Z) heavier than α-particles, with $\mu_e = \mu_e' = A/Z = \text{const.}$, (as explained in Chapter 4, Section 7) are still of the polytropic form (10), with $n'(\mathrm{NR})$ and $n'(\mathrm{UR})$ given respectively by Equations (13) and (13′), the respective values of the *constant* K being given by more general formulae

$$K(\mathrm{NR}) = (9.9 \times 10^{12})\, \mu_e^{-5/3} \text{ c.g.s.}, \tag{14''}$$

and

$$K(\mathrm{UR}) = (1.2 \times 10^{15})\, \mu_e^{-4/3} \text{ c.g.s.}. \tag{15''}$$

Referring to Chapter 4, Subsection 8.5, we discover that the constant values of the function n'' in the cases (NR) and (UR) are respectively identical to the polytropic index $n'(\mathrm{NR})$ and $n'(\mathrm{UR})$ given by Equations (13) and (13′).

Let us now consider a white dwarf in mechanical equilibrium. As for any normal star, we can express this equilibrium by Equation (18) of Chapter 2

$$-\frac{dP}{dr} = GM_r \rho r^{-2}, \tag{16}$$

which can be written:

$$-\frac{r^2}{\rho}\frac{dP}{dr} = GM_r. \tag{16'}$$

If we take the derivative of Equation (16′) with respect to r, and replace (dM_r/dr) by its expression given by Equation (20) of Chapter 2, we obtain, after division by r^2:

$$(r^{-2})\frac{d}{dr}\left(\frac{r^2}{\rho}\frac{dP}{dr}\right) = -4\pi G\rho. \tag{17}$$

Eliminating P by means of the polytropic relation (10) we obtain the equation of mechanical equilibrium of a polytropic white dwarf, which can be simplified by an introduction of two dimensionless variables Θ and ξ, defined respectively by

$$\Theta^{n'} = \rho(r)/\rho(0) = \rho/\rho_c, \tag{18}$$

and

$$\xi = r/\alpha, \tag{18'}$$

where α is a constant parameter with dimension of a length, defined by Equation

$$\alpha = \left[\frac{K}{4\pi G}(n'+1)\rho_c^{(1-n')/n'}\right]^{1/2}. \tag{18''}$$

Then Equation (17) takes the form of the '*Emden–Lane equation*':

$$\boxed{(\xi^{-2})\frac{d}{d\xi}\left(\xi^2\frac{d\Theta}{d\xi}\right) = -\Theta^{n'}.} \tag{19}$$

This equation describes the mechanical equilibrium of a polytropic white dwarf. One finds, after some tedious calculations, that in order to transform Equation (17) into Equation (19), after the substitutions defined by (18) and (18'), α must satisfy the relation (18").

The Emden–Lane Equation (19) will be discussed in Section 3, but for the moment, let us present another method of analysis of the properties of white dwarfs, the method based on the use of the '*virial theorem*'.

2.3. A CONCEPTUAL PROBLEM AND ITS SOLUTION BY THE VIRIAL THEOREM

Before starting the application of this method, let us discuss an interesting general conceptual problem.

We have shown in Section 6 of Chapter 2 that a model of the Sun at *constant density* $\rho(r) = \bar{\rho} =$ const. yields a very satisfactory order of magnitude of the principal parameters of the internal structure of the Sun, such as the central temperature T_c and the central pressure P_c, especially when some reasonable assumptions are made concerning the chemical composition and the physical state of the Sun.

One could be tempted to apply a similar method of first approximation to a rough analysis of a polytropic white dwarf, that is, to start with a study of a model of a white dwarf of constant mass density.

Unfortunately, from a *physical* point of view, for a polytropic white dwarf it is not possible to make such an assumption. Indeed, according to Equation (10), with constant values of K and n', to assume that ρ is independent of r *means that* we assume that *the pressure P is independent* of r too. Then, the discussion presented in Chapter 2, Section 2 shows that no mechanical equilibrium would be possible, since *without a pressure gradient* the gravitational force per unit volume could not be compensated, producing a collapse of the star. Thus, a model of a polytropic white dwarf based on a *physical assumption* of a constant density is physically absurd.

Note, that such is *not* the case for the Sun, in which the stellar mixture is in a state of a *perfect gas*, so that the pressure P is dependent [see for instance Eqn. (30) of Ch. 2] *both* on the density ρ *and the temperature T*. Assuming that the density is independent of r, the variation of T with distance to the center provides a pressure *gradient* which prevents a gravitational collapse.

However, one can reach a simplification as efficient as in the model of the Sun at constant density, even in the case of a polytropic white dwarf, without any encounter with absurdity, by the following procedure.

Let us start *without any physical assumption* about the function $\rho(r)$ and let us express the condition of mechanical equilibrium, in a quite general way, once more, by Equation (16). Then, multiplying both sides of this equation by the volume $V_r = \frac{4}{3}\pi r^3$ of a sphere of radius r, we obtain an equation which, taking into account Equation (20) of Chapter 2 becomes:

$$- V_r \frac{dP}{dr} = \frac{1}{3}\frac{G}{r} M_r 4\pi r^2 \rho = \frac{1}{3}\frac{G}{r} M_r (dM_r/dr). \tag{20}$$

Integrating by parts, with respect to r, between the center and the surface, i.e. between $r = 0$ and $r = R$, and taking into account that $V_r = 0$ for $r = 0$ and that $P = 0$ for $r = R$, we obtain a quite general integral relation

$$\boxed{\int_{r=0}^{R} P\,dV_r = \frac{1}{3}\int_{r=0}^{R} \frac{G}{r} M_r (dM_r/dr)\,dr,} \tag{21}$$

which holds for any star in a state of mechanical equilibrium.

It is easy to show that this equation is equivalent to the well known 'virial theorem' of Newtonian mechanics (see, for instance, Kittel *et al.*, 1962, Ch. 9 or Kittel, 1969, p. 388). Indeed, as we have explained in Chapter 4, Section 8, both in the case of a non-relativistic perfect gas *and* in the case of a (non-relativistic) completely degenerate fermion gas, the pressure P can be expressed by Equation (133) of Chapter 4 as a very simple function of the kinetic energy density u_k, viz.

$$P = \tfrac{2}{3} u_k. \tag{22}$$

Let us now represent by T the *total kinetic energy of all particles* of the star (no confusion with the temperature is possible, since the temperature never appears explicitly in the mechanical equilibrium of a white dwarf). Then we see that Equation (22) is equivalent to

$$T = \int_{r=0}^{R} u_k\,dV_r = \frac{3}{2}\int_{r=0}^{R} P\,dV_r. \tag{22'}$$

On the other hand, it can be shown (see the Appendix to this Chapter) that the integral at the right-hand side of Equation (21) represents the *gravitational binding energy* B_g^+ of the star, assumed to be stratified in homogeneous concentric layers:

$$B_g^+ = \int_{r=0}^{R} \frac{G}{r} M_r (dM_r/dr)\,dr. \tag{23}$$

Moreover, as shown in the Appendix, B_g^+ represents the absolute value $(-\Omega)$ of the (negative) potential energy of self-gravitation Ω of the star (supposed to vanish for an infinite dispersion of its matter). Therefore, Equation (21) can be written

$$\boxed{2T = -\Omega,} \tag{24}$$

which is precisely the usual form of the virial theorem, or, in a form more convenient in astrophysical applications:

$$2T = B_g^+. \tag{24'}$$

We are now able to solve the conceptual problem set by the impossibility of a physical model of a polytropic white dwarf of constant mass density. Indeed, we can now, without any particular assumption concerning the function $\rho(r)$, take advantage of the fact that the function $P(r)$, instead of appearing by its derivative as in Equation (16), now *appears under the integral sign* in Equation (21) and in the expression (22′) for T. This allows a *mathematical approximation*, which consists of a computation of these integrals with a suitable *mean* value $\bar{P}$ of the function $P(r)$, instead of making a *physical assumption*.

2.4. Application of the Virial Theorem to an Elementary Model of a White Dwarf

According to Equations (14) and (15), in a polytropic white dwarf the pressure P is proportional to $\rho^{5/3}$ or to $\rho^{4/3}$. Therefore, we can take as a suitable mean value $\bar{P}$ of P, that one which by Equation (14) or (15) corresponds to the mass density $\bar{\rho}$ of a homogeneous sphere of the same mass M and the same radius R as that of the white dwarf, i.e. $\bar{\rho}$ defined by

$$M = \int_0^R 4\pi r^2 \rho(r)\,dr = \bar{\rho}\int_0^R 4\pi r^2\,dr = \bar{\rho}\,\frac{4\pi R^3}{3} = \bar{\rho}V. \tag{25}$$

Applying this procedure to the integral expression (22′) of T, we get an approximate value $T(\mathrm{NR})_{\mathrm{app}}$ of T in the case (NR):

$$T(\mathrm{NR})_{\mathrm{app}} = \tfrac{3}{2}\int_0^R P(r)\,dV_r = \tfrac{3}{2}\bar{P}V = \tfrac{3}{2}K_1(\bar{\rho})^{5/3}V. \tag{26}$$

Applying a similar procedure to the integral expression (23) of B_g^+, we first take as an approximate expression $(M_r)_{\mathrm{app}}$ of M_r:

$$(M_r)_{\mathrm{app}} = \frac{4\pi}{3}\,r^3\bar{\rho}, \tag{27}$$

(without any distinction between the cases NR and UR), and we get an approximate value $(B_g^+)_{\mathrm{app}}$ of B_g^+:

$$(B_g^+)_{\mathrm{app}} = \int_{r=0}^R \frac{G}{r}\,(M_r)_{\mathrm{app}}(dM_r)_{\mathrm{app}} = \frac{(4\pi)^2}{3}\,G(\bar{\rho})^2\,\frac{R^5}{5} = \frac{3}{5}\frac{G}{R}M^2, \tag{28}$$

which evidently represents physically the gravitational binding energy of a *homogeneous* sphere of mass M and radius R.

Expressing $(\bar{\rho})^{5/3}V$ as a function of M and R, by means of Equation (25) we obtain

$$(\bar{\rho})^{5/3}V = \left(\frac{3}{4\pi}\right)^{2/3} M^{5/3}R^{-2} \approx (0.384)M^{5/3}R^{-2}, \tag{29}$$

which allows us to transform the expression (26) of $T(\mathrm{NR})_{\mathrm{app}}$ into

$$T(\mathrm{NR})_{\mathrm{app}} \approx (0.576)K_1M^{5/3}R^{-2}. \tag{30}$$

Introducing the approximate values of T and B_g^+, given by Equations (30) and (28), into the expression (24′) of the virial theorem, we find the following approximate relation between the mass M and the radius R of a white dwarf in which the electron gas is in a state of *non-relativistic* degeneracy:

$$R = QM^{-1/3}. \tag{31}$$

Q is a constant, approximately equal to

$$Q_{\mathrm{app}} = (1.92)(K_1/G). \tag{32}$$

G is the constant of gravitation. K_1 is the constant given by Equation (14′) for a white dwarf made of a mixture of electrons and α-particles, or given by Equation (14″) for a mixture of electrons and nuclei heavier than α-particles. In the extreme case of such nuclei, that of *iron* nuclei ($A = 56; Z = 26$), $\mu_e = A/Z = 2.15$, which is very near the value of $\mu_e = 2$ corresponding to α-particles. Therefore K_1 remains very near the value given by Equation (14′) in all stages of evolution of a non-relativistic white dwarf.

Replacing K_1 by this value, and expressing M in solar masses $M_\odot$, and R in solar radii $R_\odot$, we find after some numerical calculations:

$$\boxed{\frac{R}{R_\odot} \approx \left(\frac{M}{M_\odot}\right)^{-1/3} (0.010).} \tag{33}$$

This relation discloses two very interesting (and unexpected) properties of white dwarfs considered here:

(1) Their radius R *decreases* when their mass M becomes larger (whereas the converse takes place for normal stars of the main sequence). It will be shown below that this property holds for *all* white dwarfs.

(2) According to relation (33), a white dwarf with a mass equal to $(0.6)^3 = 0.216 \approx 0.22$ solar masses would have a radius of the order of $(0.6)^{-1} \approx 0.017$ solar radii, that is, of the order of 12 000 km (or about *twice the radius of the earth*). (A more precise treatment mentioned below gives, for a white dwarf of 0.22 solar masses, a radius of 14 000 km).

2.4.1. *Remarks*

2.4.1.1. Let us try to evaluate the upper limit of M beyond which the relation (33) cannot be applied. Since, according to this relation, R decreases when M becomes larger, an increase of M produces (within the limits of validity of the relation (33)) an *increase* of the mean density $\bar{\rho} = M/V$. On the other hand, the gravitational force acting on each element of a star being always directed towards the center, mechanical equilibrium demands a pressure force directed towards the surface, i.e. a *negative* value of dP/dr. But in a polytropic white dwarf, in which the pressure P is related to the mass density ρ by the relation (10), dP/dr has *the same sign* as $d\rho/dr$. Therefore, no equilibrium is possible unless $d\rho/dr$ is *negative*; i.e., unless the mass density increases monotonically towards the center. Consequently, the central mass density ρ_c will always be greater than the mean density ρ, and will increase with increasing mass M.

Now, relation (33) was based on the assumption of a *non-relativistic* degeneracy of the electron component of the stellar mixture *everywhere* inside the polytropic white dwarf, which means that the parameter $x_{F,e}$ of the theory of degeneracy exposed in Chapter 4, was implicitly expected to satisfy everywhere the condition $x_{F,e} \ll 1$. Since, according to Eqn. (112) of Ch. 4, $x_{F,e} = (1/126)\rho^{1/3}$ (for $\mu_e = 2$), and since ρ is maximum at the center, the degeneracy will be non-relativistic everywhere if the central value $(x_{F,e})_c$ of $x_{F,e}$ satisfies the condition $(x_{F,e})_c \ll 1$.

Integrating Equation (16) between $r = 0$ and $r = R$, and taking into account the fact that $P(R) = 0$, the approximate value $(P_c)_{\text{app}}$ of the central pressure $P_c = P(0)$ will be given, [if we replace M_r by its approximate value $(M_r)_{\text{app}}$ given by Equation (27)], by

$$(P_c)_{\text{app}} = \int_0^R G\,(M_r)_{\text{app}}\bar{\rho}r^{-2}dr = \frac{3G}{8\pi}(M^2/R^4). \tag{34}$$

Evaluating the approximate value Q_{app} of the coefficient Q defined by Equation (32) for $\mu_e = 2$, we obtain

$$Q_{\text{app}} \approx 0.89 \times 10^{20} \text{ c.g.s.}, \tag{35}$$

and replacing in Equation (34) the radius R by its expression (31), with Q given by Equation (35), we find:

$$(P_c)_{\text{app}} \approx \frac{3G}{8\pi} Q_{\text{app}}^{-4} M^{10/3} \approx (0.127 \times 10^{-87}) M^{10/3}. \tag{36}$$

Now, utilization of $\bar{\rho}$ as an approximation to $\rho(r)$ in the computation of $(P_c)_{\text{app}}$ does not prevent us from taking this approximate value of the central pressure P_c in a model of white dwarf where physically ρ varies with r, and where the central value ρ_c of ρ takes the value given by Equation (14) applied at the center:

$$P_c(\text{NR}) = K_1 \rho_c^{5/3}, \tag{37}$$

with K_1 given by Equation (14′) for $\mu_e = 2$ or by Equation (14″) for a more general value of μ_e.

However, according to the relation $x_{F,e} = (1/126)\rho^{1/3}$, (for $\mu_e = 2$), applied at the center, $(x_{F,e})_c$ can be expressed, by utilization of Equation (37), as a function of (P_c), by

$$(x_{F,e})_c \approx (10^{-23})^{1/5} P_c^{1/5}. \tag{38}$$

On replacing (P_c) by its approximate expression (38) we find, after some numerical computations:

$$(x_{F,e})_c \approx (0.66 \times 10^{-22}) M^{2/3} \approx (1.05)\left(\frac{M}{M_\odot}\right)^{2/3}. \tag{39}$$

If we impose the condition that $(x_{F,e})_c$ not exceed 0.6, in order to satisfy the inequality $(x_{F,e})_c \ll 1$. Equation (39) demands a value of M *lower* than about $(0.5)M_\odot$. For values of M *much larger* than $(0.5)M_\odot$ the degeneracy of the electron gas would cease to be non-relativistic and one could no longer apply the relation (33).

2.4.1.2. Let us now consider the opposite case of a polytropic white dwarf sufficiently dense near its center to have $(x_{F,e})_c \gg 1$, that is the case of *ultra-relativistic* degeneracy of the electron gas. Then, following the same procedure as before, but inserting in Equation (22′) the value of P(UR) given by Equation (15), we obtain, instead of Equations (26) and (30):

$$T(\mathrm{UR})_{\mathrm{app}} = \tfrac{3}{2}K_2(\bar{\rho})^{4/3}V \approx (0.93)K_2M^{4/3}R^{-1}, \tag{40}$$

and, by application of the virial theorem (24′), instead of Equations (31) and (32):

$$M = (5.5)(K_2/G)^{3/2}, \tag{41}$$

or, replacing K_2 by its value for $\mu_e = 2$, given by Equation (15′):

$$M \approx (1.7)M_\odot, \tag{41′}$$

or more generally, replacing K_2 by its value for any μ_e, given by Equation (15″):

$$M \approx (6.8)\mu_e^{-2}M_\odot. \tag{41″}$$

Thus we find, that in the ultra-relativistic case, the radius R disappears from our equations, after application of the virial theorem. Instead of a relation between the radius and the mass, similar to Equation (31) for the non-relativistic case, we get simply a particular value of the mass given by Equations (41), (41′) or (41″).

A more rigorous treatment, presented in the next Section, will explain this unexpected result. Indeed, we will find that no white dwarf can exceed a '*limiting mass*' equal to that corresponding to the ultra-relativistic degeneracy at the center.

This limiting mass, usually called M_3 in order to recall that (by Equation 13′) the ultra-relativistic case corresponds to the value 3 of the polytropic index n', will be shown to be equal (for a general value of μ_e) to $(5.75)\mu_e^{-2}M_\odot$, and to $(1.44)M_\odot$ for $\mu_e = 2$. Thus, we immediately see, by comparison with Equations (41′) and (41″), that our first rough treatment yields a quite satisfactory order of magnitude for the limiting mass corresponding to the ultra-relativistic case.

2.4.1.3. Returning to the non-relativistic case, that is to white dwarfs with a mass lower than about $(0.5)M_\odot$, we can also mention that the more rigorous treatment of the next section shows that, for these white dwarfs of small mass, the functional relation (31) between the radius and the mass holds without modification. But the value of the constant Q is replaced by a more exact value

$$Q_{\mathrm{exact}} = (2.36)(K_1/G) \tag{42}$$

which differs from its approximate value, given by Equation (32), by less than about 20%.

If the exactness of the relation (31) is not fortuitous, the relative correctness of Q_{app} given by Equation (32) is due to an easily explained compensation of errors. Indeed, rigorous expressions for B_g^+ and T, not too difficult to obtain, are given by Chandrasekhar, (1939, pp. 423, 424). They are

$$(B_g^+)_{\text{exact}} = \frac{6}{7}\frac{G}{R}M^2, \tag{43}$$

and

$$T_{\text{exact}} = \left(\frac{6}{7}\right)\left(\frac{1}{2}\right)\frac{G}{R}M^2. \tag{44}$$

Taking into account Equations (31), (44) and (42) we get

$$T_{\text{exact}} = \tfrac{3}{7}QGM^{5/3}R^{-2} \approx (1.01)K_1M^{5/3}R^{-2}. \tag{44'}$$

A comparison of the exact expressions (43) and (44′) of B_g^+ and T, with the corresponding approximate expressions (28) and (30), (while confirming the functional dependence with respect to R and M), shows that the numerical coefficients of the approximate expressions are wrong by about 30% for B_g^+ and by about 43% for T. But a compensation of these errors occurs for the numerical coefficient of (K_1/G) in the approximate expression (32) of Q, because this coefficient depends on the *quotient* of the division of T by B_g^+, both errors affecting the approximate values of T and B_g^+ in the same direction (both approximate values being *too small*).

2.5. Comparison with Observations

It can be instructive to compare the approximate theoretical results derived above with the observations compiled by Allen, (1973, p. 226):

TABLE 5.1

The values of the masses and of radii of some white dwarfs deduced from observations

White dwarf	$M/M_\odot$	$R^{\text{km}}/1000$
40 Eri B	0.36	11.9
Wolf 1346	0.40	11.4
Procyon B	0.43	8.8
Luyten 770-3	0.48	10.1
He 3=Ci_{20}398	0.50	10.4
Luyten 532-81	0.63	8.0
Van Maanen 2	0.63	8.6
Ross 627	0.66	7.0
Luyten 870-2	0.69	9.0
Luyten 1512-34B	0.81	8.8
Sirius B	0.98	17.5

The only observational result which displays a strong discrepancy with the theory concerns Sirius B. It may be due to the difficulty of the corresponding observations. (The comparison of the observational data of Table 5.1 with the more precise theoretical results of Section 3 is given in Figure 5.2).

3. Polytropic White Dwarfs Studied by Means of the Emden–Lane Equation

As a second approximation to the theory of white dwarfs, one can limit oneself, once more, to the two limiting cases of non-relativistic and ultra-relativistic degeneracy (masses respectively near *zero* and near M_3) corresponding to the validity of the *polytropic* relation (10), but replace the rough mathematical treatment of Section 2 by a more rigorous mathematical treatment based on a *numerical* solution of the Emden–Lane equation (19).

Indeed, the Emden–Lane equation is a 'classical' equation from the point of view of applied mathematics. Numerical tables of the function $\Theta(\xi)$ for different values of n' (with physically suitable boundary conditions), and especially for $n' = 1.5$ and $n' = 3.0$, are very numerous: (see for instance, Wrubel, 1958, and the references given by Chandrasekhar, 1939, p. 95). The mathematical properties of the Emden–Lane equation, and of the functions $\Theta(\xi)$ are discussed very extensively by Chandrasekhar, (1939, pp. 84–182).

Here we shall use only the following results. Let us, first of all, recall the definitions introduced by Equations (18), (18′), and (18″) of Subsection 2.2 viz.

$$\Theta = [\rho(r)/\rho(0)]^{1/n'}, \tag{45}$$

$$\xi = \frac{r}{\alpha}, \tag{46}$$

$$\alpha = \left[\frac{K}{4\pi G}(n'+1)\,\rho_c^{(1-n')/n'}\right]^{1/2}, \tag{47}$$

where K and n' are the constants of the polytropic relation (10), viz.

$$P = K\rho^{(1+1/n')}. \tag{48}$$

The function Θ *physically* corresponds, since $\rho(r)/\rho_c = \Theta^{n'}$, to the ratio of the mass density $\rho(r)$ to the central density ρ_c. The variable ξ is *physically* similar to the dimensionless variable r' used in Chapter 2, but measures the *distance r to the center* in units α different from R.

According to the limiting case chosen for investigation, the case (NR) or the case (UR), n' will have the constant value of 1.5 or 3.0, given by Equations (13) and (13′). For $\mu_e = 2$ (an 'α-particle' white dwarf) the values of K_1 and K_2 of K will be given by Equations (14′) and (15′). More generally, K_1 and K_2 will be given by Equations (14″) and (15″) for white dwarfs with nuclei (A, Z) heavier than α-particles.

Thus, physically, to look for a solution of the Emden–Lane equation (for given values of the constants n' and K):

$$\boxed{(\xi^{-2})\frac{d}{d\xi}\left(\xi^2\frac{d\Theta}{d\xi}\right) = -\Theta^{n'}.} \tag{49}$$

is equivalent to looking for the way the ratio $\rho(r)/\rho_c$ varies with the distance r to the center, for a given value of $\mu_e = A/Z$, for a given choice between the proximity to the limits $M = 0$ or $M = M_3$, and for physically suitable boundary conditions.

Equation (45) shows that at the boundary $r = 0$, that is at the center, we must have

$$\Theta(0) = 1. \tag{50}$$

However, because the Emden–Lane equation is an equation of second order, we need a second boundary condition. Let us show that this second condition is

$$\left[\frac{d\Theta}{d\xi}\right]_{\xi=0} = 0. \tag{51}$$

Indeed, on applying Equation (48) at the center, we obtain

$$P_c = K\rho_c^{(1+1/n')}, \tag{52}$$

then, taking into account Equation (48) itself, and the definition (45) of Θ, we find (dividing P by P_c):

$$P = P_c\Theta^{(n'+1)}. \tag{53}$$

Let us now differentiate $P(r)$, given by Equation (53), with respect to r, and divide the result by ρ. Eliminating $\Theta^{n'}$ by means of Equation (45), we obtain:

$$\frac{1}{\rho}\frac{dP}{dr} = P_c(n'+1)\frac{1}{\rho_c}\frac{d\Theta}{dr}. \tag{54}$$

If we write the equation of mechanical equilibrium (16) in the form

$$M_r = -\frac{1}{G}r^2\frac{1}{\rho}\frac{dP}{dr}, \tag{55}$$

and replace $(1/\rho)(dP/dr)$ by the right-hand side of Equation (54), we obtain

$$M_r = -\frac{1}{G}r^2P_c(n'+1)\frac{1}{\rho_c}\frac{d\Theta}{dr}, \tag{56}$$

and, expressing M_r as a function M_ξ of ξ defined by Equation (46), we finally find

$$M_\xi = -\frac{1}{G}\alpha\xi^2P_c(n'+1)\frac{1}{\rho_c}\frac{d\Theta}{d\xi}, \tag{57}$$

or, with A defined by

$$A = \frac{1}{G}\alpha P_c(n'+1)\frac{1}{\rho_c}, \tag{58}$$

a very simple expression of M_ξ:

$$M_\xi = A\left(-\xi^2\frac{d\Theta}{d\xi}\right). \tag{59}$$

Now, it is obvious that M_r can be represented near the center (for $r \to 0$) by

$$M_r = \frac{4\pi}{3} r^3 \rho_c, \tag{60}$$

which means that M_ξ can be represented for $\xi \to 0$ by

$$M_\xi = \frac{4\pi}{3} \alpha^3 \rho_c \xi^3. \tag{61}$$

On equating the values of M_ξ given by Equations (59) and (61), we obtain after division by ξ^2 (for $\xi \to 0$) the announced boundary condition (51).

Table 5.2 gives (for $n' = 1.5$) a rough idea of the trend of the functions $\Theta(\xi)$ and $(-1)\xi^2(d\Theta/d\xi)$, solutions of the Emden–Lane equation with the boundary conditions (50) and (51). The *trend* is the same for $n' = 1.5$ and $n' = 3.0$. (We recommend that the reader draw the graph corresponding to Table 5.2).

TABLE 5.2
The trend of the functions $\Theta(\xi)$ and $(-\xi^2)(d\Theta/d\xi)$ for $n' = 1.5$

ξ	Θ	$-\xi^2 \frac{d\Theta}{d\xi}$
0.0	1.00	0.00
0.5	0.96	0.04
1.0	0.85	0.29
1.5	0.68	0.81
2.0	0.50	1.49
2.5	0.32	2.13
3.0	0.16	2.56
3.5	0.03	2.710
ξ_1	0.00	β_1

ξ_1 (for $n' = 1.5$) = 3.653 75
β_1 (for $n' = 1.5$) = 2.714 06

This example shows that for a certain value ξ_1 of ξ (depending on n') the function $\Theta(\xi)$ *reduces to zero*. But, since Θ reduces to zero at the same time as $\rho(r)$, which is zero at the surface of the star, this means that the value $\xi_1(n')$ of ξ corresponds, in terms of r, to the radius R of the white dwarf. Thus

$$R = \alpha \xi_1(n'). \tag{62}$$

According to Equation (59), the total mass M of the star, equal to M_ξ for $\xi = \xi_1(n')$, will be given by

$$M = A\beta_1(n'), \tag{63}$$

where $\beta_1(n')$ represents the numerical value of $(-1)\xi^2(d\Theta/d\xi)$ for $\xi = \xi_1(n')$. The numerical values of both ξ_1 and β_1, for $n' = 1.5$, are indicated in Table 5.2.

In Equations (62) and (63), the value ρ_c of the central mass density is implicitly involved (beside n'), since α depends, by Equation (47), on ρ_c and A depends, by

Equation (58), directly on ρ_c, and indirectly on ρ_c through α and P_c (see Equation (52)).

Replacing α and P_c in the definition (58) of A by their expressions (47) and (52) (and avoiding deliberately unnecessary algebra) we first obtain:

$$A = \frac{1}{G}\left[\frac{K}{4\pi G}(n'+1)\,\rho_c^{(1-n')/n'}\right]^{1/2} K\rho_c^{(1+1/n')}(n'+1)\frac{1}{\rho_c}, \tag{64}$$

then, by substitution of α by its expression (47) into Equation (62), we find:

$$R = \xi_1\left[\frac{K}{4\pi G}(n'+1)\,\rho_c^{(1-n')/n'}\right]^{1/2}. \tag{65}$$

Thus, we see that R is a function of ρ_c, depending on constants or parameters (ξ_1, K, n', G):

$$R = f(\rho_c; \xi_1, K, n', G) \propto \rho_c^{(1-n')/2n'}, \tag{66}$$

whereas, in a similar way, M given by Equations (63) and (64), is another function of ρ_c depending on constants or parameters (β_1, K, n', G):

$$M = \psi(\rho_c; \beta_1, K, n', G) \propto \rho_c^{(3-n')/2n'}. \tag{67}$$

Eliminating ρ_c between Equations (66) and (67), a very long and tedious, but quite elementary, calculation yields:

$$M = \varphi(\xi_1, \beta_1, K, n', G)\,R^{(3-n')/(1-n')}, \tag{68}$$

where φ is a certain function of the quantities indicated between parentheses. But it is relatively easy to verify the proportionality of M to $R^{(3-n')/(1-n')}$, which results from the last parts of Equations (66) and (67).

Chandrasekhar, 1939, (p. 98) writes the relation (68) in the equivalent form:

$$K = N_{n'}GM^{(n'-1)/n'}R^{(3-n')/n'}, \tag{69}$$

where $N_{n'}$ is a purely numerical coefficient, depending on n', equal to 0.424 22 for $n' = 1.5$.

Let us now consider separately the case (1) of white dwarfs of very small mass, corresponding to $n' = 1.5$ (non-relativistic degeneracy), and the case (2) of white dwarfs with a mass near the limiting mass M_3, corresponding to $n' = 3.0$ (ultra-relativistic degeneracy).

Case (1)

In this case $n' = 1.5$ and $N_{n'} = 0.424\,22$, so that Equation (69) gives, with $K = K_1$:

$$K_1 \approx (0.424)\,GM^{1/3}R, \tag{70}$$

which is equivalent to

$$R = QM^{-1/3}, \tag{71}$$

with

$$Q = Q_{\text{exact}} = (2.36)(K_1/G)\,\text{c.g.s.}, \tag{72}$$

a result already mentioned in Equation (42).

For $\mu_e = 2$ Equations (72) and (14') yield:

$$Q_{\text{exact}} \approx (1.1 \times 10^{20})\,\text{c.g.s.}\ . \tag{72'}$$

Thus, for instance, for a mass M equal to $(0.22)M_\odot$ Equations (71) and (72') indicate a radius of the order of 14 000 km. And, as already found in Subsection 2.4, the radius R decreases when the mass M increases.

Case (2)

In the case $n' = 3.0$ the numerical tables of the function $\Theta(\xi)$ indicate that $\xi_1(3) = 6.90$ and $\beta_1(3) = 2.02$. On the other hand, for $n' = 3$, ρ_c disappears from Equation (67), and M given by Equations (63) and (64) takes the form (called M_3 because of the value of n'):

$$M_3 = \frac{4\beta_1(3)}{\pi^{1/2}}(K_2/G)^{3/2}, \tag{73}$$

where K was replaced by K_2, the value of K for the case (UR).

Since ρ_c does not appear any more in Equation (73), we can no longer eliminating ρ_c among Equations (66) and (67) obtain a relation of the form (68). However, on replacing K by K_2 and n' by 3 in (69), we see that it becomes

$$K_2 = N_3 G M^{2/3}, \tag{74}$$

which is equivalent to

$$M_3 = (N_3)^{-3/2}(K_2/G)^{3/2}. \tag{74'}$$

Thus, Equation (69) becomes identical with Equation (73) if, for $n' = 3$, we take for $N_{n'}$ the value defined by

$$(N_3)^{3/2} = \frac{\pi^{1/2}}{4\beta_1(3)}, \tag{75}$$

that is

$$N_3 \approx 0.364. \tag{75'}$$

This means that, though Equation (69) cannot be established for $n' = 3$ by the method leading to Equation (68), it holds even for $n' = 3$, provided that we take for N_3 the value given by (75'). On replacing K_2 by its value given for $\mu_e = 2$ by Equation (15'), and expressing M_3 in units equal to $M_\odot$ we find

$$M_3 = (1.44)M_\odot. \tag{76}$$

On replacing K_2 by its more general value, given by Equation (15''), we obtain a more general relation:

$$M_3 = (5.75)\mu_e^{-2}M_\odot. \tag{76'}$$

For $n' = 3$ Equation (66) shows that R is proportional to $(1/\rho_c)^{1/3}$, but the method followed hitherto does not disclose the limiting value of R or of ρ_c for $n' = 3$. These limiting values of R and ρ_c can be found by the following procedure due to Chandrasekhar.

4. Chandrasekhar's 'Rigorous' Theory of White Dwarfs

We have seen in Sections 2 and 3 that the possibility of reduction of the theory of the structure of white dwarf to polytropes was based on the fact that, for the limiting cases of non-relativistic or ultra-relativistic complete degeneracy of the electron component of a stellar mixture, the relation between electron pressure P_e and mass density ρ of the mixture is of the polytropic form (10).

If one adopts a more general treatment based on the rigorous relativistic (RR) relation between electron pressure P_e and mass density ρ, resulting from the theory of degeneracy presented in Chapter 4, no simple relation between $P = P_e$ and ρ, of the form (10), exists any more and the polytropic approximation cannot be used.

Indeed, according to Eq. (138, e) of Ch. 4, the pressure $P = P_e(\mathrm{RR})$ of the free electrons of the mixture in the general case (RR) is related to the number density n_e by the pair of Equations (77) and (77′):

$$P_e(\mathrm{RR}) = (6.01 \times 10^{22}) f(x_{F,e}); \tag{77}$$

$$n_e = (5.9 \times 10^{29}) x_{F,e}^3, \tag{77'}$$

where $f(x)$ is the Chandrasekhar's function

$$f(x) = x(2x^2 - 3)(1 + x^2)^{1/2} + 3 \log_e [x + (1 + x^2)^{1/2}], \tag{78}$$

defined by Equation (132) of Ch. 4.

However, according to the physically obvious Eq. (104) of Ch. 4, in a mixture of electrons and nuclei (A, Z), n_e and $x_{F,e}$ are related to the mass density ρ of the mixture and to the parameter $\mu_e = \mathrm{A/Z}$ by

$$n_e = \rho N_A / \mu_e, \tag{79}$$

hence

$$\rho = \rho(x_{F,e}) = (9.82 \times 10^5) \mu_e x_{F,e}^3. \tag{79'}$$

Thus, for a given value of μ_e, the pair of relations (77) and (79′), can be considered as a parametric representation of P as a function of ρ, $x_{F,e}$ playing the role of the parameter. From this point of view, the physical meaning of $x_{F,e}$ is irrelevant and we can drop for a while the subscripts (F, e) and, write the pair of Equations (80) and (80′):

$$P = (6.01 \times 10^{22}) f(x); \tag{80}$$

and

$$\rho = (9.82 \times 10^5) \mu_e x^3, \tag{80'}$$

with $f(x)$ given by Equation (78).

After this reminder of statistical physics let us come back to the astrophysical problem, and consider the general Equation (17), (derived from the condition of mechanical equilibrium), viz.

$$(r^{-2}) \frac{d}{dr}\left(\frac{r^2}{\rho}\frac{dP}{dr}\right) = -4\pi G\rho. \tag{81}$$

Taking the derivative of $f(x)$ with respect to x and taking into account the first of Equations (80), we find:

$$\frac{dP}{dx} = (4.8 \times 10^{23}) x^4 (1 + x^2)^{-1/2} = (4.8 \times 10^{23}) x^3 \frac{dy}{dx}, \tag{82}$$

where y is defined by

$$y = (1+x^2)^{1/2}. \tag{83}$$

Multiplying both sides of (82) by dx/dr and taking into account the second of Equations (80) we obtain

$$\frac{1}{\rho}\frac{dP}{dr} = (4.9 \times 10^{17})\frac{1}{\mu_e}\frac{dy}{dr}. \tag{84}$$

Let y_c be the value of y corresponding, by Equations (83) and (80′), to the central density ρ_c and let α' be a quantity defined by

$$\alpha' = \frac{(7.71 \times 10^8)}{\mu_e y_c}. \tag{85}$$

This choice of α' simplifies the numerical coefficients of the equation obtained by substitution of expressions (84) for $(1/\rho)(dP/dr)$ and (80′) for ρ into Equation (81):

$$(r^{-2})\frac{d}{dr}\left(r^2\frac{dy}{dr}\right) = -\frac{x^3}{(\alpha')^2y_c^2} = -\frac{(y^2-1)^{3/2}}{(\alpha')^2y_c^2}. \tag{86}$$

If we introduce, with Chandrasekhar, (1939, p. 416), the dimensionless function defined by

$$\Phi = y/y_c, \tag{87}$$

and the variable η, analogous to the variable ξ, defined by

$$\eta = r/\alpha', \tag{88}$$

Equation (86) takes the form:

$$(\eta^{-2})\frac{d}{d\eta}\left(\eta^2\frac{d\Phi}{d\eta}\right) = -\left(\Phi^2-\frac{1}{y_c^2}\right)^{3/2}, \tag{89}$$

rather similar to the Emden–Lane equation (49).

Comparing the two equations we see that the Emden–Lane equation corresponding to $n' = 3.0$ mathematically represents a particular case of the general Equation (89) when y_c tends to infinity, so that for $y_c \to \infty$, $\Phi(\eta)$ becomes the same function as $\Theta(\xi)$ for $n' = 3.0$.

Now, let us recall that x in the calculations starting with Equations (80) and (80′) physically represents $x_{F,e}$, so that the limit $y_c \to \infty$ – corresponding by definition (83) to $x_c \to \infty$ – physically corresponds to $(x_{F,e})_c \to \infty$, i.e. to the ultra-relativistic degeneracy at the center.

Thus, at the ultra-relativistic limit, (our case 2), the rigorous (RR) treatment of the problem leads (as could be expected physically) to a vanishing (boundary) value of $\Phi(\eta)$ at the surface $(r = R)$ for a value of η_1 of η numerically equal to $\xi_1(3)$, i.e. to 6.90; hence:

$$R = (6.90)\alpha'. \tag{90}$$

However, according to the definition (85) of α' this parameter *tends to zero* when y_c tends to infinity. Thus we find, with Chandrasekhar, that according to Equations (83), (79′) and (90), at the ultra-relativisitic limit:

(1) The *central density* ρ_c becomes *infinite*.
(2) The *radius* R tends to *zero*.

Both properties are consistent with Equation (66).

Naturally, the rigorous theory based on Equation (89) also leads to the existence of a 'limiting mass' M_3, at the ultra-relativistic limit (M_3 is called *'Chandrasekhar's limit'*) but the polytropic approximation becoming identical with the rigorous treatment at the ultra-relativistic limit, we do not need to resume the calculation since Equations (76) and (76′) give the same values of M_3 as the rigorous discussion.

In his fundamental treatise of 1939, Chandrasekhar gives the numerical tables of the function $\Phi(\eta)$ for different values of the parameter y_c, i.e. physically for different values of the central mass density ρ_c. In Table 5.3 a part of his results is presented (for $\mu_e = 2$, i.e. for white dwarfs composed of a mixture of free electrons and α-particles), in a way aiming to put into evidence the main *physical* characteristics of these white dwarfs.

TABLE 5.3
The main physical characteristics of white dwarfs (for $\mu_e = 2$) as a function of their mass or as a function of the central density ρ_c

$M/M_\odot$	R km	ρ_c g cm^{-3}	$\rho_c/\bar{\rho}$	$\bar{\rho}$ g cm^{-3}	y_c^{-2}	$(x_{F,e})_c$	n''	M/M_3
0.00	–	–	6.0	–	1.0	0.0	1.50	0.00
0.22	14 000	2.4×10^5	6.4	3.8×10^4	0.8	0.5	1.56	0.15
0.40	11 000	1.1×10^6	7.0	1.5×10^5	0.6	0.8	1.64	0.28
0.50	9 700	2.0×10^6	7.4	2.7×10^5	0.5	1.0	1.69	0.35
0.61	8 500	3.6×10^6	7.9	4.6×10^5	0.4	1.2	1.75	0.42
0.88	6 500	15.8×10^6	9.9	1.6×10^6	0.2	2.0	1.96	0.61
1.08	5 000	5.3×10^7	12.6	4.2×10^6	0.1	3.0	2.16	0.75
1.33	3 200	6.8×10^8	21.5	3.1×10^7	0.02	7.0	2.54	0.92
1.38	2 000	2.0×10^9	26.2	7.4×10^7	0.01	10.0	2.66	0.95
1.44	0	∞	54.2	∞	0.00	∞	3.00	1.00

Table 5.3 is self-explanatory, except, perhaps, the column relative to the function n'', defined by Eq. (142) of Ch. 4 as the ratio of the kinetic energy density u_k to the pressure P of the electron component of the mixture. This function of $x_{F,e}$ tends at the non-relativistic limit (NR), for $x_{F,e} \to 0$, to the polytropic index $n' = 1.5$, and tends at the ultra-relativistic-limit (UR), for $x_{F,e}$ infinite, to the polytropic index $n' = 3.0$.

Note, the general smallness of the radius R (for a star); the very high value of the central density ρ_c and of the mean density $\bar{\rho}$ (both of the order of several tons per cm^3); and the relatively small range of masses M, of the same order, on the average, as the mass $M_\odot$ of the Sun.

Note equally, a good confirmation of approximate results yielded by the polytropic models, except for masses M between about $(0.5)M_\odot$ and $(1.0)M_\odot$.

Figure 5.1 represents the variations of $(x_{F,e})_c$ as a function of $M/M_\odot$, (for $\mu_e = 2$). The dotted line corresponds to the polytropic model $n' = 1.5$ and shows how this model, rather satisfactory for masses M below $(0.5)M_\odot$, becomes quite wrong for masses larger than $(0.5)M_\odot$.

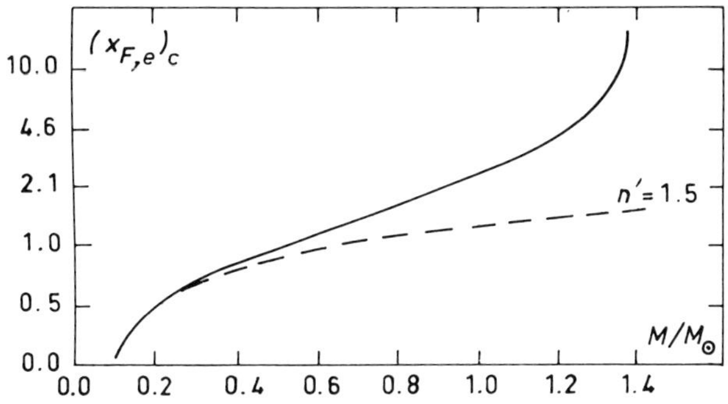

Fig. 1. The variations of $(x_{F,e})_c$ as a function of $M/M_\odot$ for a white dwarf.

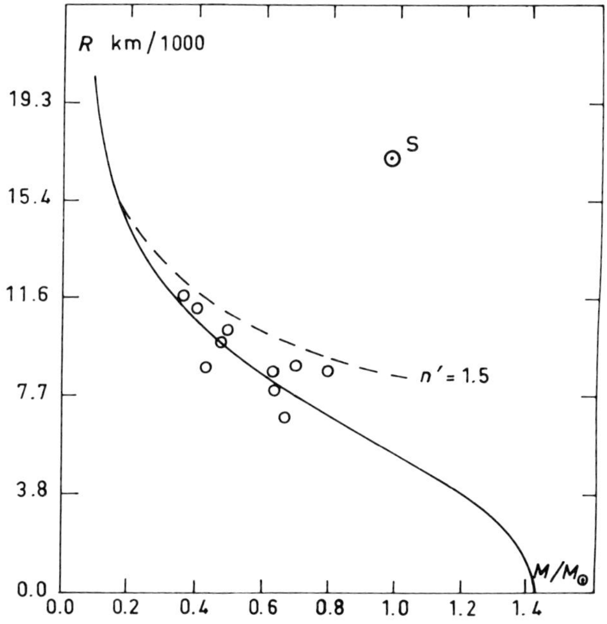

Fig. 5.2. The variations of the radius R of a white dwarf as a function of its mass M.

Figure 5.2 represents the variations of R as a function of $M/M_\odot$. Once more, the dotted line corresponds to the polytropic model $n' = 1.5$. The circles around different points correspond to the *observed* values given in Table 5.1; we find again that the rigorous theory represents very well the observations, except for Sirius-B marked S on the Figure.

Figure 5.3 (overleaf) represents, for some typical values of $M/M_\odot$, and consequently of $(x_{F,e})_c$, the variations of ρ/ρ_c as a function of $r' = r/R$. The limiting cases $n' = 1.5$ and $n' = 3.0$ are indicated by dotted lines.

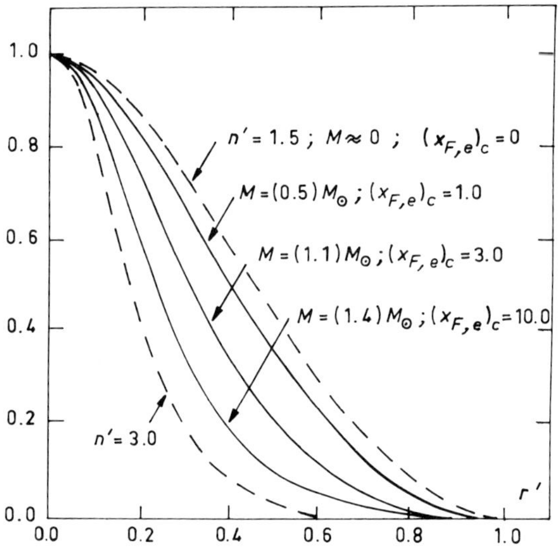

Fig. 5.3. The variations of the mass density ρ/ρ_c as a function of $r' = r/R$, for some typical values of the mass M of a white dwarf.

4.1. Remark

Actually, recent advanced theories of white dwarfs are more complicated than the 'rigorous' theory of Chandrasekhar presented above. As noted, in particular by Weinberg, (1972, p. 317), and as will be shown in Chapter 6 dealing with 'neutron stars', for a sufficiently large mass (below M_3) of a white dwarf, the number density n_e of the free electrons becomes so large that the nuclei of the mixture (α-particles, or heavier nuclei) begin to *capture* electrons which transform more and more protons of the nuclei *into neutrons*. Since this process does not change the number density n_B of baryons (defined in Chapter 4, Section 7) these captures of free electrons decrease the value of n_e, thus (slowly) increasing the value of the parameter μ_e equal to n_B/n_e (according to Eq. (95) of Ch. 4). Hence, *a decrease of M for a given central density* ρ_c. This is also easily seen from Equations (63), (64), (14″) and (15″), which show that M is (in the polytropic model), for a given value of ρ_c, proportional to $K^{3/2}$, which is always proportional to a *negative* power of μ_e.

One of the consequences is that considering, in the first and the third columns of Table 5.3, ρ_c as the independent variable and $M/M_\odot$ as a function of ρ_c, the monotonic increase of M with ρ_c in reality stops *before* Chandrasekhar's limit $M_3 = (1.44)M_\odot$ is reached and ends by a decrease.

A thorough analysis, by Harrison *et al.* (1965), shows that the true limit is equal to about $(1.2)M_\odot$, actually not very far from Chandrasekhar's value. Anyhow, since electron captures producing a 'neutronization' of nuclei, combined with thermonuclear reactions, result in a mixture of electrons and iron (A = 56; Z = 26) nuclei, the value of μ_e increases from its value 2 corresponding to α-particles, up to 56/26, changing already the value of M_3 given by Equation (76′), from $(1.44)M_\odot$ to $(1.26)M_\odot$. The more advanced theory discloses that the corresponding radius is not zero but of the order of 4000 km. Thus for masses larger than about $(1.2)M_\odot$ or for radii smaller than 4000 km no white dwarfs exist, and after an eventual 'domain of instability' (discussed later) we enter the domain of the neutron stars considered in Chapter 6.

N.B. – Recent investigations on the *magnetic* properties of some white dwarfs can be found in Angel (1978).

Appendix: The Gravitational Binding Energy of a Star

Let us consider a material sphere (a star) of radius R stratified in homogeneous concentric layers. The gravitational forces tend to hold together the particles of the star: they represent 'cohesive forces'. The degree of cohesion of these particles can be measured by the *absolute value* B_g^+ of the *negative* work W to be done by the gravitational forces in a process of dispersion of the material particles of the star. A dispersion from its condensed state of a sphere of radius R to a final state of a gas infinitely diluted (with an infinite distance between all particles). W is negative since gravitational forces would exert a *braking* action during the process of dispersion.

$B_g^+ = -W = |W|$ represents the *'gravitational binding energy'*. The superscript (+) is useful in order to recall that this energy is *essentially positive*, and to avoid confusion with similar notations used in nuclear physics. The subscript g can also be useful in order to stress the difference between the *gravitational* binding energy and the *nuclear* binding energy considered in Chapter 6.

Let M_r represent (as in Chapter 2) the mass enclosed in the sphere (Σ_r) of a radius r smaller than or equal to the radius R of the star. The mass enclosed between the sphere (Σ_r) and the sphere (Σ_{r+dr}) will be denoted by dM_r.

The main 'trick' in the computation of B_g^+ consists in a conception of the dispersion of the star as a kind of a 'stellar striptease', layer by layer: each new 'exterior' layer being dispersed to infinity *without touching any of the remaining 'interior' layers*.

Let us then apply Newton's Theorem established in Chapter 2, Subsection 2.1. According to this theorem the gravitational field created, at a distance s from the center O, by the mass M_r, stratified in homogeneous concentric layers, is equivalent to that of the entire mass M_r placed at O. Thus the modulus g_s of the gravitational force exerted on unit mass at a distance s from O is simply

$$g_s = GM_r/s^2. \tag{A1}$$

Consider now the layer (s, ds) of mass dM_r, *primitively* situated between (Σ_r) and (Σ_{r+dr}), which *in the course of its dispersion* occupies a position between (Σ_s) and (Σ_{s+ds}) (when all layers primitively situated farther from the center than (Σ_r) have already been dispersed).

The modulus $|dF|$ of the gravitational force exerted on *one* of the mass elements (dm) of the layer (s, ds) is equal to

$$|dF| = (dm)g_s. \tag{A2}$$

The element $dW_{dm,ds}$ of the (negative) work, corresponding to a displacement of (dm) along a radial distance (ds), (conditioned, for the distance s, by the conservation of the mass dM_r), away from O, is given by

$$dW_{dm,\,ds} = -|dF|\,ds = -(dm)g_s\,ds. \tag{A3}$$

If we totalize the contributions of all the elements (dm) which constitute the mass (dM_r), the corresponding value $dW_{dM_r,\,ds}$ of dW will be given by

$$dW_{dM_r,\,ds} = -(dM_r)g_s\,ds = -(dM_r)\frac{GM_r}{s^2}\,ds. \tag{A4}$$

Consider the element δW of W corresponding to the displacement of the layer (s, ds), primitively situated between r and $(r + dr)$, from a distance r to infinity. δW is given by an integration of $dW_{dM_r,\,ds}$, with respect to s, from $s = r$ to $s = \infty$; an integration in which (dM_r) and M_r must be considered *as constant*:

$$\delta W = \int_{s=r}^{\infty} dW_{dM_r,\,ds} = -GM_r dM_r \int_r^{\infty} \frac{ds}{s^2} = -G\frac{1}{r}M_r dM_r. \tag{A5}$$

In order to get W, we apply this procedure to a complete 'striptease' of all layers (r, dr) of the primitive star from its surface $r = R$ up to its center $r = 0$. Then we must integrate δW, with respect to r, from $r = 0$ to $r = R$ (if, instead of $-dr$ we use $+dr$) hence (replacing dM_r by $(dM_r/dr)\,dr$ in order to put into evidence the element of integration dr):

$$W = \int_{r=0}^{R} -G\frac{1}{r}M_r(dM_r/dr)\,dr. \tag{A6}$$

Taking into account the fact that $B_g^+ = -W$, the final general result (for homogeneous layers) is:

$$\boxed{B_g^+ = \int_0^R \frac{G}{r}M_r(dM_r/dr)\,dr.} \tag{A7}$$

In the case of a *homogeneous* material *sphere* of mass M and of radius R, the local mass density ρ is a constant, equal to $M/((4\pi/3)R^3)$, independent of the distance r to the center. Then one easily finds that

$$M_r = (r/R)^3 M. \tag{A8}$$

A substitution of this value of M_r into Equation (A7) shows that, in this particular case of a homogeneous sphere, the gravitational binding energy B_g^+ is given by

$$\boxed{B_g^+\text{ (homogeneous sphere)} = \frac{3}{5}\frac{GM^2}{R}.} \tag{A9}$$

References

WORKS CITED IN THE TEXT

(Works that can be used for further study are marked by an asterisk.)

Allen, C. W.: 1973, *Astrophysical Quantities* (3rd Edn.), Athlone Press, London.
* Angel, J. R. P.: 1978, Magnetic White Dwarfs, *Ann. Rev. Astron. Astrophys.* **16**, 487.
* Chandrasekhar, S.: 1939, *An Introduction to the Study of Stellar Structure*, Univ. of Chicago Press, Chicago (Reprinted in 1967 by Dover Publ., New York).
* Harrison, B. H., Thorne, K. S., Wakano, M., and Wheeler, J. A.: 1965, *Gravitational Theory and Gravitational Collapse*, (Ch. 9 and 10), Univ. of Chicago Press, Chicago.
Kittel, C.: 1969, *Thermal Physics*, Wiley, New York.
Kittel, C., Knight, W. D., and Ruderman, M. A.: 1962, *Mechanics* (Berkeley Physics Course, Vol. I), McGraw-Hill, New York.
Weinberg, S.: 1972, *Gravitation and Cosmology*, Wiley, New York.
* Wrubel, M. H.: 1958, In *Encyclopedia of Physics*, Vol. 51, Springer, Berlin.

CHAPTER 6

NEUTRON STARS

1. Introduction: (R_Z) Reactions (Electron Captures and β^- Disintegrations)

The theory of '*neutron stars*' is one of the most difficult astrophysical theories. It involves simultaneously, beside more or less classical thermodynamics and statistical physics, some basic laws of nuclear physics (which will be summarized in this introduction) and different properties of the solid state related to general relativity (these last properties will not be considered here: they appear only in the references mentioned at the end of this Chapter).

It is rather widely known that the masses of elementary particles, such as electrons, protons, and neutrons or the masses of complex nuclei are commonly expressed in nuclear physics in two different systems of units: the 'physical system' (or the system $O^{16} = 16.000\,000$) and the system $C^{12} = 12.000\,000$. In the system $O^{16} = 16$ (the physical system) it is not the mass of the *nucleus* O^{16} but that of the *atom* O^{16} (with its eight electrons) that equals 16.000 000. Similarly, in the system $C^{12} = 12$ it is the mass of the *atom* C^{12} with its six electrons that equals 12.000 000.

We shall use below only the system $O^{16} = 16$. In this system, the unit of mass, which will be called mA_{16}, equals $1.659\,81 \times 10^{-24}$ g. According to Einstein's equation $\epsilon_0 = mc^2$, this represents an energy of $1.491\,77 \times 10^{-3}$ erg. One also uses as unit of atomic mass *the thousandth* part of mA_{16}, which will be called mmA_{16}. Expressing the energy corresponding to mA_{16} in MeV one finds that $mA_{16} = 931.145$ MeV.

Then we have for the mass of the electron (β^+ or β^-): $m_e = A_e = 0.548\,761\ mmA_{16} = 0.510\,976$ MeV.

Expressing the masses in mA_{16} we can introduce the purely *mathematical* notion of '*mass excess*' (analogous to the fractional part of a logarithm). A mass excess Δm is defined by *the difference* between the mass m of a nucleus and $A \times (mA_{16})$, where $A = (N + Z)$ is the '*atomic mass*' (an integer!) of the nucleus composed of N neutrons and Z protons. One can express the respective masses, m_p, m_n, m_d and m_α of the proton, neutron, deuteron and α-particle (in mmA_{16} or MeV units) by the following table of 'mass excesses':

TABLE 6.1

The 'mass excesses' of protons, neutrons, deuterons and alpha-particles

Δm_p	=	$(m_p - 1)(mA_{16})$	=	7.59687 mmA_{16}	=	7.07379 MeV
Δm_n	=	$(m_n - 1)(mA_{16})$	=	8.976 mmA_{16}	=	8.358 MeV
Δm_d	=	$(m_d - 2)(mA_{16})$	=	14.19375 mmA_{16}	=	13.21645 MeV
Δm_α	=	$(m_\alpha - 4)(mA_{16})$	=	2.7786 mmA_{16}	=	2.5873 MeV.

The mass excesses given above should not be confused with the *physical* concept of '*mass defect*' defined as the difference between the mass m of a nucleus and the sum of the masses of its constituent parts, which represents the energy lost during nuclear 'fusion', and which is equivalent to the *opposite* of the (nuclear) '*binding energy*' B^+ of the nucleus.

Let us be more precise, and consider a nucleus [N, Z] of mass $m_{[N,Z]}$, formed of N neutrons and Z protons; this nucleus also can be denoted by (A, Z). If m_n represents the mass of one neutron, and if m_p represents the mass of one proton, then the mass defect of the nucleus [N, Z] = (A, Z) will be defined by

$$m_{[N,Z]} - (Nm_n + Zm_p) = -B^+[N, Z] = -B^+(A, Z), \tag{1}$$

and the binding energy of the nucleus [N, Z] = (A, Z) will be defined by

$$B^+[N, Z] = B^+(A, Z) = c^2\{(Nm_n + Zm_p) - m_{[N,Z]}\} \tag{2}$$

when m_n, m_p and $m_{[N,Z]}$ are expressed in units of mass, and (more usually) by

$$B^+[N, Z] = Nm_n + Zm_p - m_{[N,Z]} \tag{3}$$

or

$$B^+(A, Z) = Am_n + Z(m_p - m_n) - m_{(A,Z)}, \tag{3'}$$

when m_n, m_p, $m_{[N,Z]}$, $m_{(A,Z)}$ are expressed in units of energy. Hence

$$\boxed{m_{(A,Z)} = Am_n + Z(m_p - m_n) - B^+(A, Z).} \tag{3''}$$

Some authors call the left-hand side of Equation (1) (expressed in units of energy) 'binding energy'; that is the reason why, to avoid confusion, we add a superscript (+) to the symbol B defining the right-hand side of Equations (2) and (3).

As an example, one readily verifies that the binding energy $B^+[2, 2] = B^+(4, 2)$ of an α-particle, computed by means of Equation (3) and Table 6.1, is equal to 28.28 MeV, which means that an energy of 28.28 MeV would have to be *supplied* to an α-particle to 'disperse' its four constituent particles at an infinite distance from each other.

Obviously, in nuclear physics, contrary to the procedure adopted in Chapter 4, Section 7, one must take into account the *precise masses* of particles and, in particular, distinguish between the mass of a neutron and the mass of a proton (see Table 6.1).

Since the nature of nuclear forces is not yet known, the binding energy $B^+[N, Z] = B^+(A, Z)$ of a nucleus [N, Z] = (A, Z) as a function of N and Z (or of A and Z) cannot be established (as was done, in Chapter 5, for the gravitational binding energy B_g^+ of a star of mass M and radius R with a known distribution of mass densities) in a *purely* theoretical way. But, guided by different nuclear models, one can use the *observed* values of nuclear mass defects (thus, of nuclear binding energies) of nuclei of different composition [N, Z] = (A, Z) to establish *semi-empirical* relations which best represent *analytically* $B^+[N, Z] = B^+(A, Z)$ as a function of N and Z (or of A and Z).

One of the most usual, though not the most precise, of such semi-empirical relations is (see, for instance, Blatt and Weisskopf, (1952, pp. 227)) the '*Bethe–Weizsäcker formula*', which (with notations of Bethe and Morrison, (1956), see our Table 6.2) has the following form when A and Z are used as variables:

$$\boxed{B^{+}(\mathrm{A},\mathrm{Z}) = +a_v\mathrm{A} - a_s\mathrm{A}^{2/3} - a_c\mathrm{Z}^2\mathrm{A}^{-1/3} - a_T(\mathrm{A}-2\mathrm{Z})^2\mathrm{A}^{-1},} \qquad (4)$$

where a_v, a_s, a_c and a_T are numerical constants. According to Equation (3″) this is equivalent to:

$$m_{(\mathrm{A},\mathrm{Z})} = (m_n - a_v)\mathrm{A} + (m_p - m_n)\mathrm{Z} + a_s\mathrm{A}^{2/3} + a_c\mathrm{Z}^2\mathrm{A}^{-1/3} + + a_T(\mathrm{A}-2\mathrm{Z})^2\mathrm{A}^{-1}. \qquad (4')$$

Table 6.2 gives the *notations* used by different authors, or the *numerical values* of the corresponding coefficients (the latter, of course, 'evolve' with the progress of precision in observations). All references can be found in Tondeur (1971). Note especially, that Langer, Salpeter and Tondeur call B our $(-B^{+})$, whereas Bethe–Morrison and Blatt call B our $(+B^{+})$; this difference justifying a superscript (+) in order to avoid confusion. (All numerical values are in MeV.)

TABLE 6.2

The notations and the numerical values of the coeffients in 'Bethe–Weizsäcker formula' for the binding energy $B^{+}(\mathrm{A}, \mathrm{Z})$ of a nucleus (A, Z)

Bethe–Morrison	Green (1954)	Ayres (1962)	Langer (1969)	Bethe (1970)	Tondeur (1971)
a_v	15.756	14.7	$-\alpha$	$C_1 = 16$	a_1
a_c	0.710	0.63	$+\delta$	$C_2 = 0.72$	a_3
a_T	23.694	15.1	$+\beta$	$C_3 = 24$	a_4
a_s	17.794	14.8	$+\gamma$	$C_4 = 18$	a_2

The physical origin of the form of Equation (4) and the physical meaning of different terms are explained in Blatt and Weisskopf, (1952, Ch. VI), and a discussion of the rather poor precision of relation (4), with the numerical coefficients used by different authors, is given (with proposals for a more precise representation) by Tondeur, (1971).

Let us now agree to represent the negative electron, β^- or e^-, by $\boxminus$, the proton by $\oplus$, the neutron by $\bigcirc$, and the more complex nuclei by a rectangle 'filled' with the appropriate components, (the part of these components which does not partake in a nuclear reaction being represented by :::).

One of the most important nuclear reactions in the theory of neutron stars is that of '*electron capture*' (or '*neutronization*'), in which *one of the protons* of a complex nucleus is transformed by reaction with a free electron *into a neutron*.

Usually (unless some particular physical situation, discussed later, hinders this process) an *inverse* reaction is possible, that of 'β^- *disintegration*' (or 'β^- decay') in which a complex nucleus spontaneously *emits a negative electron* (β^-), this emission being accompanied by the *transformation of one of the neutrons of the nucleus into a proton*.

Both reactions are schematically represented in Figure 6.1; electron captures corresponding to the arrow directed to the right, β^- disintegrations corresponding to the arrow directed to the left.

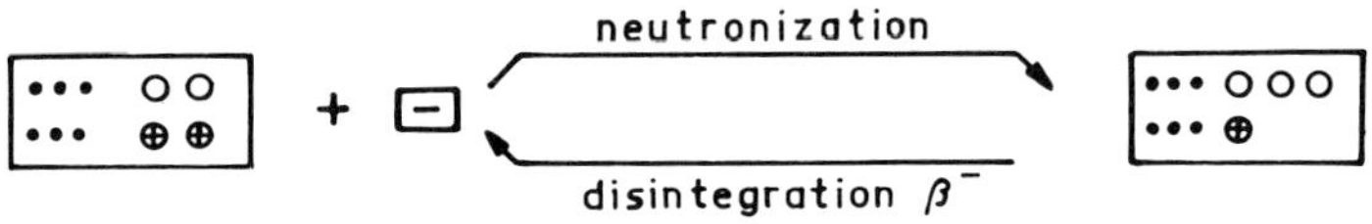

Fig. 6.1. A schematic representation of electron captures and β^- disintegrations.

(Some authors use an unnecessarily complicated terminology speaking, instead of electron capture, of 'inverse β^- disintegration', or 'inverse β^- decay'.)

Actually, an electron capture is usually associated with an emission of *neutrinos* ν_e, so that a symbolic representation of a reaction of neutronization is given (the negative electron being symbolized by e^-) by

$$[\mathrm{N}, \mathrm{Z}] + e^- \rightarrow [\mathrm{N}+1, \mathrm{Z}-1] + \nu_e, \tag{5}$$

or by

$$(\mathrm{A}, \mathrm{Z}) + e^- \rightarrow (\mathrm{A}, \mathrm{Z}-1) + \nu_e, \tag{5'}$$

according to the parameters, N and Z, or A and Z, used to specify the nucleus involved in the reaction. (In this symbolic representation the complete energy balance, discussed later, is not yet considered.)

Conversely, a β^- disintegration is usually associated to an emission of *antineutrinos* $\bar{\nu}_e$ so that a symbolic representation of a β^- disintegration is given (without taking into account the complete energy balance) by

$$[\mathrm{N}+1, \mathrm{Z}-1] \rightarrow [\mathrm{N}, \mathrm{Z}] + e^- + \bar{\nu}_e \tag{6}$$

or by

$$(\mathrm{A}, \mathrm{Z}-1) \rightarrow (\mathrm{A}, \mathrm{Z}) + e^- + \bar{\nu}_e. \tag{6'}$$

Equations (6) and (6′) are written in such a way that the correspondence between the reactions described by (5) or (5′), and the inverse reaction described by (6) or (6′) is best put into evidence.

Sometimes the reaction of β^- disintegration is represented by Equations (equivalent from a general point of view):

$$[\mathrm{N}, \mathrm{Z}] \rightarrow [\mathrm{N}-1, \mathrm{Z}+1] + e^- + \bar{\nu}_e, \tag{7}$$

or

$$(\mathrm{A}, \mathrm{Z}) \rightarrow (\mathrm{A}, \mathrm{Z}+1) + e^- + \bar{\nu}_e. \tag{7'}$$

Though the neutrinos and antineutrinos produced by nuclear reactions give rise to several important theoretical and observational problems, they are usually neglected in a first approximation.

Note that both reactions of neutronization and the inverse reactions of β^- disintegration *do not modify* the total number A of baryons of the nuclei involved in these reactions. For this reason, all these reactions are called *'isobaric'* and the corresponding nuclei (A, Z − 1), (A, Z) or (A, Z + 1) are called *'isobars'*.

The term 'neutronization' applied to electron captures comes obviously from the fact that in reactions described by Equations (5) or (5′) the number of neutrons of the nucleus *increases* by *one* unit.

Very often, reactions of neutronization and inverse reactions of β^- disintegration, called '(R_Z) *reactions*', take place in the same stellar mixture, and the result is a certain state of nuclear *equilibrium* characterized by some respective abundances of free electrons, nuclei (A, Z) and nuclei (A, Z − 1).

On the other hand, some particular circumstances (discussed in the next section), can produce a displacement of this equilibrium, increasing the abundance of 'neutronized' nuclei at the expense of a normal equilibrium. This possibility is based on the following general properties of nuclear reactions of electron capture and β^- disintegration.

Let us start with reactions of β^- disintegration. The negative electron emitted in a reaction described for instance by Equation (6), generally possesses some kinetic energy $\epsilon_{k,e}$, but this kinetic energy cannot exceed a certain easily determined maximum value, $(\epsilon_{k,e})_{\max}$. Assuming indeed that we consider only nuclei in the fundamental (not the excited state) and neglecting the kinetic energy of nuclei with respect to that of electrons, the principle of conservation of energy can be expressed (when energy units are used for all masses) by Equation:

$$m_{[N+1,Z-1]} = m_{[N,Z]} + m_e + \epsilon_{k,e} + \epsilon_{\bar{\nu}_e}. \tag{8}$$

Since a part of the available energy is taken away by the antineutrino, $\epsilon_{k,e}$ cannot exceed a maximum value which corresponds to a *zero* value of $\epsilon_{\bar{\nu}_e}$. Thus

$$(\epsilon_{k,e})_{\max} = m_{[N+1,Z-1]} - \{m_{[N,Z]} + m_e\} = M_{(A,Z-1)} - M_{(A,Z)} \tag{9}$$

where, again, all masses are expressed in energy units (for instance in MeV), and $M_{(A,Z)}$ is the mass of the (A, Z) *atom* (including the Z electrons).

But, according to Equation (3), the masses of the nuclei [N + 1, Z − 1] and [N, Z] can be expressed in terms of the corresponding binding energies $B^+[N+1, Z-1]$ and $B^+[N, Z]$, so that $(\epsilon_{k,e})_{\max}$ can also be given by

$$(\epsilon_{k,e})_{\max} = B^+[N,Z] - B^+[N+1,Z-1] + (m_n - m_p - m_e). \tag{10}$$

The numerical value (in MeV) of $(m_n - m_p)$, given by Table 6.1 is

$$m_n - m_p = \Delta m_n - \Delta m_p = 1.284\,\text{MeV}, \tag{11}$$

so that finally $(\epsilon_{k,e})_{\max}$ is given in MeV by

$$(\epsilon_{k,e})_{\max} = B^+[N,Z] - B^+[N+1,Z-1] + 0.773. \tag{12}$$

Some authors express this relation in terms of the (total) energy ϵ_e, equal to $(m_e c^2 + \epsilon_{k,e})$, of the electron. Then Equation (12) is replaced (in MeV) by

$$(\epsilon_e)_{\max} = B^+[N,Z] - B^+[N+1,Z-1] + 1.284. \tag{13}$$

Conversely, in a reaction of neutronization, the kinetic energy of the neutronizing electron must exceed some specific *threshold*: otherwise the electron capture cannot take place. This is due to the fact that the sum on the left-hand side of Equation (5) generally represents an energy *smaller* than the rest energy $m_{[N+1,Z-1]}$ of the nucleus [N + 1, Z − 1] to be formed. Thus, in the most favourable case in which the energy ϵ_{ν_e} of the corresponding neutrino would be zero, *an additional contribution* in the form of *kinetic energy of the incident electron* is necessary, in order to satisfy the principle of conservation of energy. And this additional energy must be *at least equal* to the difference between $m_{[N+1,Z-1]}$ and the sum $\{m_{[N,Z]} + m_e\}$.

Thus, if we call $(\epsilon_{k,e})_{\text{thr}}$ this indispensable minimal threshold kinetic energy of the incident electron, this energy will be defined by

$$(\epsilon_{k,e})_{\text{thr}} = m_{[\text{N}+1,\text{Z}-1]} - \{m_{[\text{N},\text{Z}]} + m_e\}. \tag{14}$$

Now, comparison between Equations (9) and (14) shows that

$$(\epsilon_{k,e})_{\max} = (\epsilon_{k,e})_{\text{thr}}. \tag{15}$$

Introducing a new notation, in order to stress the fact that both sides of Equation (15) represent a link between (A, Z) and (A, Z − 1), in all (R_Z) reactions, we can write (in MeV):

$$\begin{aligned}\epsilon_{k,e}(\text{A}, \text{Z}, -1) &= (\epsilon_{k,e})_{\max} = (\epsilon_{k,e})_{\text{thr}} \\ &= B^{+}(\text{A}, \text{Z}) - B^{+}(\text{A}, \text{Z}-1) + 0.773\end{aligned} \tag{16}$$

where, of course, we have used Equation (12) with $B^{+}(\text{A}, \text{Z}) = B^{+}[\text{N}, \text{Z}]$ and $B^{+}(\text{A}, \text{Z}-1) = B^{+}[\text{N}+1, \text{Z}-1]$.

Similarly, expressing the same relation in terms of the (total) energy ϵ_e, and using Equation (13) we can write for all (R_Z) reactions (in MeV):

$$\epsilon_e(\text{A}, \text{Z}, -1) = B^{+}(\text{A}, \text{Z}) - B^{+}(\text{A}, \text{Z}-1) + 1.284, \tag{17}$$

or using Equation (9), and remembering that (in MeV) $\epsilon_e = m_e + \epsilon_{k,e}$:

$$\epsilon_e(\text{A}, \text{Z}, -1) = m_{(\text{A},\text{Z}-1)} - m_{(\text{A},\text{Z})}. \tag{17$'$}$$

2. Neutronization by a Degenerate Gas of Free Electrons

In order to show clearly how the presence of a degenerate gas of free electrons can in (R_Z) reactions inhibit an equilibrium between electron captures and β^- disintegrations, and favour a formation of nuclei with an excess of neutrons, let us consider this process applied to iron nuclei Fe^{56}, which, in our notation, are represented by [30, 26].

A very convenient way of discussing possible electron captures or β^- disintegrations is offered by the scheme of Table 6.3:

TABLE 6.3

A 'matrix representation' of the chemical symbols, of the excess (in MeV) of the rest energy over the energy of 55 mA_{16}, and of the binding energy $B^{+}(56, \text{Z})$, of some nuclei produced by a neutronization of Fe^{56}

	N = 30	N = 31	N = 32	N = 30	N = 31	N = 32	N = 30	N = 31	N = 32
Z = 24			Cr^{56}			880.66			487.709
Z = 25		Mn^{56}			878.55			488.536	
Z = 26	Fe^{56}			873.84			491.964		

In Table 6.3, the first 'matrix' indicates the names (chemical symbols) of the corresponding nuclei, the second matrix indicates the *excess* (in MeV) of the rest energy of the same nuclei over the energy of 55 units (mA_{16}), in order to avoid negative mass excesses which would be introduced if the usual mass excesses (over 56 units mA_{16}), defined in

Section 1, were used. Thus, for instance,

$$m_{[30,26]} - 55(mA_{16}) = 931.14 - 57.30 = 873.84\,\text{MeV} \qquad (18)$$

whereas

$$m_{[30,26]} - 56(mA_{16}) = -57.30\,\text{MeV}. \qquad (19)$$

The third matrix indicates the binding energy B^{+} (in MeV) of the corresponding nuclei.

A similar matrix (Table 6.4) gives the values (in MeV) of the corresponding kinetic energies $\epsilon_{k,e}(A, Z, -1)$, defined by Equation (16), involved in electron captures (represented by a displacement along a diagonal rising towards the right) and in β^- disintegrations (represented by a displacement along a diagonal going down towards the left). These value of $\epsilon_{k,e}(A, Z, -1)$ obviously result from Table 6.3 either by Equation (14), or, more easily, by Equation (16).

TABLE 6.4

A 'matrix' representation' of the kinetic energies $\epsilon_{k,e}(A, Z, -1)$, defined by Equation (16), involved in electron captures and in β^- disintegrations of the nuclei Fe^{56}, Mn^{56} and Cr^{56}

	N = 30		N = 31		N = 32
Z = 24					Cr^{56}
				(1.600)	
Z = 25			Mn^{56}		
		(4.200)			
Z = 26	Fe^{56}				

It is instructive to compare the figures of Table 6.4 to the corresponding figures relative to nuclei with the values of A equal respectively to 24, 28 and 32, given in Table 6.5:

TABLE 6.5

A 'matrix representation' of the kinetic energies $\epsilon_{k,e}(24, Z, -1)$, $\epsilon_{k,e}(28, Z, -1)$ and $\epsilon_{k,e}(32, Z, -1)$ involved in electron captures and in β^- disintegrations of nuclei with A = 24, A = 28 and A = 32 respectively

	N = 12		N = 13		N = 14		N = 15		N = 16		N = 17		N = 18		N = 19
Z = 10					Ne^{24}										
				(2.5)											
Z = 11			Na^{24}								Na^{28}				
		(5.5)								(15.1)					
Z = 12	Mg^{24}								Mg^{28}						
								(1.8)							
Z = 13							Al^{28}								Al^{32}
						(4.6)								(13.7)	
Z = 14					Si^{28}								Si^{32}		
												(0.1)			
Z = 15											P^{32}				
										(1.7)					
Z = 16									S^{32}						

The figures of the second matrix of Table 6.3, confirm numerically, in the case of the neutronization transforming Fe^{56} into Mn^{56}, that the rest mass of Fe^{56} increased by the rest mass of an electron is not sufficient to give the rest mass of Mn^{56}, since $(873.84 + 0.51) < 878.55$. A similar situation exists for the neutronization of Mn^{56} since $(878.55 + 0.51) < 880.66$. In both cases, an additional energy, brought by the *kinetic* energy of the incident electron, is necessary. This increment, equal to 4.20 MeV for Fe^{56} and to 1.60 MeV for Mn^{56}, is given by Table 6.4. These figures are respectively equal, as expected, to $[878.55 - (873.84 + 0.51)]$ and $[880.66 - (878.55 + 0.51)]$. (Note that Salpeter (1961) gives a value of 3.7 MeV, instead of the exact value of 4.2 MeV for $(\epsilon_{k,e})_{\text{thr}}$ of neutronization of Fe^{56}. This error is reproduced by Zel'dovich-Novikov (1971).

Let us now consider a stellar mixture corresponding roughly to a mass density $\rho \approx 2.16 \times 10^{10}\,\text{g cm}^{-3}$, that is a density slightly above the mean density of massive white dwarfs, (see Table 5.3), in a very advanced stage of thermonuclear evolution. Such a star is still composed essentially of iron nuclei (A = 56; Z = 26) and free electrons (and practically no free neutrons). Thus, according to Chapter 4, Section 7, the parameters μ_e and μ_e' will both be equal to A/Z = 2.16 (this explains the choice of the coefficient of 10^{10} in the value of ρ taken as example, in order to simplify the numerical calculations).

Eq. (104) of Ch. 4 shows that the corresponding value of the numerical electron density n_e is equal to

$$n_e = (\rho/\mu_e)N_A \approx 6 \times 10^{33}\,\text{cm}^{-3}, \tag{20}$$

so that, by Eq. (51) of Ch. 4, the electron gas will be in a state of an (almost) complete degeneracy provided that the temperature T does not exceed the value of the limiting temperature $T_{\text{lim}}^{\text{deg}}(\text{UR})$ equal to

$$T_{\text{lim}}^{\text{deg}}(\text{UR}) = (0.70)n_e^{1/3} \approx 1.3 \times 10^{11}\,\text{K}. \tag{21}$$

Since this condition is usually realized, we can consider that the electron gas is actually in a state of a complete degeneracy, and that the ultra-relativistic approximation (UR) is permitted since n_e is far above the critical numerical density $n_{\text{cr},e}$ equal to 5.9×10^{29} cm^{-3} according to Eq. (36, e) of Ch. 4.

Now, Eqs. (112) and (43) of Ch. 4 immediately show that a mass density $\rho = 2.16 \times 10^{10}\,\text{g cm}^{-3}$ corresponds to a value of the energy parameter $x_{F,e}$ of the Fermi level of the electron gas equal to

$$x_{F,e} \approx (10^{-2})(\rho/\mu_e)^{1/3} \approx 21.5, \tag{22}$$

and to a value of the kinetic energy of electrons at the Fermi level $(\epsilon_{k,e})_F$ equal to

$$(\epsilon_{k,e})_F \approx \epsilon_{0,e} x_{F,e} \approx (0.51) \times (21.5) \approx 11\,\text{MeV}. \tag{23}$$

In a state of complete degeneracy of the electron gas, this system contains electrons of all kinetic energies between zero and $(\epsilon_{k,e})_F$, that is, in our example, of all kinetic energies between zero and 11 MeV. Thus, in our mixture there will be plenty of free electrons with kinetic energies *far above* the threshold values of 4.2 and 1.6 MeV required to allow a neutronization of Fe^{56} nuclei (transformed into Mg^{56} nuclei) and a neutronization of Mg^{56} transformed into Cr^{56} nuclei. The nuclei are becoming more and more neutron-rich and proton-poor.

As to the free electrons of the stellar mixture with kinetic energies *below* 4.2 MeV, the complete degeneracy of the electron gas combined with the Pauli exclusion principle (see Chapter 4, Section 2) will assign to them the following role. By their presence they will prevent the compensation by inverse β^- disintegrations of the electron captures transforming Fe^{56} into Mg^{56}. Indeed, according to explanations given in the beginning of this section, all electrons eventually produced by such β^- disintegrations would leave the Mg^{56} nuclei with a kinetic energy *below* the kinetic energy $\epsilon_{k,e}(56, 26, -1)$ equal to 4.2 MeV. But in a completely degenerate system of electrons, *all* virtual available energy states (two quantum states of different spin for each energy state) are already *occupied* up to the corresponding Fermi level, and those of these electrons (of kinetic energy below 4.2 MeV), which do not participate in neutronization of Fe^{56}, do not leave any 'place' for electrons which could be produced by β^- disintegrations, since the exclusion principle excludes the possibility of an occupancy of virtual available quantum states by more than one electron. Therefore, free electrons with kinetic energies below 4.2 MeV will transform neutronizations of Fe^{56} (by electrons with kinetic energies above 4.2 MeV) into an *irreversible* process, or, more precisely, displace the equilibrium between electron captures (5) and the β^- disintegrations (6) to favour neutronization.

Similarly, all free electrons with kinetic energies below 1.6 MeV will block reactions transforming Cr^{56} nuclei (formed by neutronization of Mg^{56} nuclei) back to Mg^{56} by possible β^- disintegations.

On the whole, in a mixture of Fe^{56} nuclei and of free electrons, of a mass density of the order of 2.16×10^{10} g cm^{-3}, the Fe^{56} will be transformed by a succession of electron captures into Cr^{56} nuclei (A = 56; Z = 24) or [N = 32; Z = 24], with *two more neutrons* (and two less protons) than Fe^{56} nucleus [N = 30; Z = 26].

Note that, as indicated by Table 6.4, in transformations of the type $Fe^{56} \rightarrow Mg^{56} \rightarrow Cr^{56}$ the first threshold (4.2) is *higher* than the second (1.6). Therefore the first neutronization, that of Fe^{56}, is necessarily followed by the second, that of Mg^{56}. They are not necessarily followed by a third neutronization, unless the mass density of the mixture becomes greater, and $(x_{F,e})$ higher, since the third threshold can become higher than the first two, as in the case of the chain $Si^{28} \rightarrow Al^{28} \rightarrow Mg^{28} \rightarrow Na^{28}$ or of the chain $S^{32} \rightarrow P^{32} \rightarrow Si^{32} \rightarrow Al^{32}$.

On the other hand, we see in the last matrix of the Table 6.3 that the binding energies B^+[N, Z] of nuclei (A, Z) = [N, Z] *decrease*, for a given value of A, when N increases, in the chain $Fe^{56} \rightarrow Mg^{56} \rightarrow Cr^{56}$, and it will be shown later, by taking the derivative of B^+[N, Z] given by Equation (4), (with numerical values of the coefficients C_1, C_2, C_3, C_4 given by Table 6.2), with respect to N, that this property is quite general for a continuation of this chain beyond N = 32. This decrease of the binding energy B^+ ends by a zero binding energy, that is, by a dissociation of the corresponding nuclei into a mixture of *free* neutrons.

3. (R_A) Reactions Leading to an Increase of Atomic Weight of Nuclei

As indicated in Section 1, (R_Z) reactions are isobaric: they do not modify the atomic weight A of nuclei present in the stellar mixture.

However, it is generally assumed that beside (R_Z) reactions (electron captures and β^- disintegrations), other types of nuclear reactions can take place, for instance (R_A) reactions leading to nuclei of increased atomic weight.

These reactions, which can also be called *reactions of 'rearrangement'* are summarized (if, to begin with, we leave the energy balance, and neutrinos, out of account), by Equation:

$$(A + 1)(A, Z) + Ze^- \rightleftarrows A(A + 1, Z). \tag{24}$$

Here (A, Z) represents a nucleus composed of A baryons (N neutrons and Z protons) and (A + 1, Z) represents a nucleus composed of (A + 1) baryons [(N + 1) neutrons and Z protons], whereas (A + 1) multiplying (A, Z), Z multiplying e^-, and A multiplying (A + 1, Z) are purely numerical coefficients. Indeed, if we let e represent the absolute value of the charge of an electron, we see that the charge of the left-hand side is equal to $[(A + 1)Ze - Ze]$ i.e. to the electric charge AZe of the right-hand side. On the other hand, the number of baryons of the left-hand side is (A + 1) A like the number of baryons of the right-hand side. Thus Equation (24), which is a (R_A) reaction, since it corresponds to an increase by one unit of A, satisfies both the conservation of charge and the conservation of the number of baryons.

Actually (R_A) reactions are not as simple as could be suggested by Equation (24). High temperatures are required, and it is necessary first to break up the (A, Z) nuclei into component parts, that is mainly into α-particles and neutrons, and then to rearrange them in a different way, in order to produce (A + 1, Z) nuclei. They are discussed in Section 6.

4. The Different Domains of Mass Density

As already suggested by the discussion at the end of Section 2, the composition of a totally ionized stellar mixture (initially composed of *neutral* atoms of one *single* 'species' [N, Z]) essentially depends on *the degree of compression* of the mixture, that is on its mass density ρ.

In the domain of densities considered in this Chapter, i.e. densities larger than $10^6\,\mathrm{g\,cm^{-3}}$, considerations similar to those leading to Equations (20) and (21), or more precisely Eqs. (104) and (50) of Ch. 4, show that temperatures higher than $10^9\,\mathrm{K}$ would be required to prevent an (almost) complete degeneracy of the free electrons even at the lower limit of $\rho = 10^6\,\mathrm{g\,cm^{-3}}$. (We have already found at the end of the same section that this limiting temperature would rise up to $10^{11}\,\mathrm{K}$ for $\rho \approx 10^{10}\,\mathrm{g\,cm^{-3}}$.) Therefore, we can take for granted that the complete degeneracy of the free electron gas is realized, as in white dwarfs, even at the lower end of the domain of ρ which we are considering.

According to the data published by Canuto (1974) one can roughly divide this domain into three parts, A, B and C, defined by the following values of the mass density.

Domain A

This domain corresponds to densities between 10^6 and $10^{12}\,\mathrm{g\,cm^{-3}}$. It can be considered as the domain of *white dwarfs in a process of neutronization*, by electron captures, similar to that described in Section 2 for Fe^{56} nuclei, but accompanied by a formation of more and more heavy nuclei (an increase of A). Starting with Fe^{56} nuclei [N = 30; Z = 26] for $\rho \approx 10^6\,\mathrm{g\,cm^{-3}}$ we end with Kr^{118} nuclei [N = 82; Z = 36], for $\rho \approx 4 \times 10^{11}\,\mathrm{g\,cm^{-3}}$. Thus we pass from A = 56 to A = 118 and from (N − Z) = 4 to (N − Z) = 46, that is to a strong increase in the 'neutron excess' (N − Z). This evolution also corresponds to a relatively strong increase of the number density n_{nuc} of (compound) nuclei from 2.8×10^{28} to 2.2×10^{33} cm^{-3}. On the other hand, the value of the parameter μ_e varies relatively little, from A/Z = 2.16 to A/Z = 3.30). (See Table 6.6, which summarizes the results concerning Domain A, as given by Baym *et al.* (1971), and reproduced by Canuto (1974).

TABLE 6.6

The values of A, N, Z of the nuclei corresponding to equilibrium conditions for different values of the mass density ρ of the stellar mixture in the domain A. The last three columns give the corresponding pressures P and the number densities n_e of free electrons and n_{nuc} of nuclei (A, Z) = [N, Z]

ρ g cm^{-3}	A	N	Z	P c.g.s.	n_e cm^{-3}	n_{nuc} cm^{-3}
2.62×10^{6}	56	30	26 (Fe)	9.8×10^{22}	7.3×10^{29}	2.8×10^{28}
1.31×10^{8}	62	34	28 (Ni)	2.6×10^{25}	3.6×10^{31}	1.3×10^{30}
1.04×10^{10}	82	50	32 (Ge)	7.7×10^{27}	2.5×10^{33}	7.8×10^{31}
4.27×10^{11}	118	82	36 (Kr)	7.8×10^{29}	7.8×10^{34}	2.2×10^{33}

When the mass density ρ of the stellar mixture exceeds a value of the order of 4×10^{11} g cm^{-3}, the nuclei suffer a kind of 'neutron evaporation', called *neutron drip*, and the result is an appearance of free neutrons, so that we enter the domain B characterized by a mixture of *nuclei*, free *electrons* and free *neutrons*.

Domain B

This domain corresponds to densities roughly between 4×10^{11} g cm^{-3} (beginning of the neutron drip) and 2×10^{14} g cm^{-3} (sudden disappearance of compound nuclei, which are totally decomposed).

In this domain, the evolution of the situation is clearly described by Table 6.7 (as given by Baym *et al.* (1971) and reproduced by Canuto (1974).

In the light of the comments made above, Table 6.7 is self-explanatory, except the last column, which represents the ratio $\bar{\omega}$ of the volume of the individual nucleus (A, Z) to the average 'available volume', this available volume being given in cm^3 by $(1/n_{\text{nuc}})$. We see that at the upper end of the domain B, the mass density ρ becomes so large that $\bar{\omega}$ tends to *one* which means that the corresponding nuclei are almost in contact with each other, and that they therefore lose a part of their individuality.

The limiting density corresponding to the decomposition of the last nuclei (the upper end of the domain B) computed by different authors is not exactly the same, but is always of the order of 10^{14} g cm^{-3} (the two extreme results being 1×10^{14} and 3×10^{14}).

Note that the number density of nuclei n_{nuc} increases monotonically with ρ up to a value of ρ equal to about 2.0×10^{14}, but begins to decrease for higher densities, and drops to zero for ρ larger than (roughly) 2.4×10^{14} g cm^{-3}. Note also, that beyond Ac we have 'theoretical nuclei' with Z equal to 120, 210 and 445 which, of course, have no names.

TABLE 6.7

The values of A, N, Z of the nuclei corresponding to equilibrium conditions for different values of the mass density ρ of the stellar mixture in the domain B. The last two columns give the number density n_n of free neutrons and the fraction ϖ of the volume of the mixture occupied by the nuclei (A, Z)

ρ g cm^{-3}	A	N	Z	P c.g.s.	n_e cm^{-3}	n_{nuc} cm^{-3}	n_n cm^{-3}	$\bar{\varpi}$
1.47×10^{12}	140	98	42 (Mo)	1.4×10^{30}	1.0×10^{35}	2.4×10^{33}	5.4×10^{35}	0.0022
2.67×10^{12}	149	105	44 (Ru)		1.2×10^{35}	2.7×10^{33}	1.2×10^{36}	0.0027
6.25×10^{12}	170	122	48 (Cd)	6.5×10^{30}	1.7×10^{35}	3.5×10^{33}	3.2×10^{36}	0.0040
1.50×10^{13}	211	157	54 (Xe)	2.2×10^{31}	2.6×10^{35}	4.8×10^{33}	8.0×10^{36}	0.0073
3.44×10^{13}	296	231	65 (Tb)		4.9×10^{35}	7.5×10^{33}	1.9×10^{37}	0.0170
8.01×10^{13}	548	459	89 (Ac)		1.2×10^{36}	1.3×10^{34}	4.3×10^{37}	0.0620
1.30×10^{14}	990	870	120 . . .	5.9×10^{32}	2.1×10^{36}	1.8×10^{34}	7.2×10^{37}	0.15
2.00×10^{14}	2500	2290	210 . . .	1.3×10^{33}	4.0×10^{36}	1.9×10^{34}	1.1×10^{38}	0.34
2.39×10^{14}	7840	7395	445 . . .		4.9×10^{36}	1.1×10^{34}	1.3×10^{38}	0.58

Domain C

After a complete dissolution of all compound nuclei, for a density roughly above $2 \times 10^{14}\,\mathrm{g\,cm^{-3}}$, we encounter a mixture of free neutrons, (some of which give rise to free protons by β^- disintegration) and free electrons. We are then in domain C discussed in Section 9.

This domain corresponds to a relatively narrow range of mass density between $2 \times 10^{14}\,\mathrm{g\,cm^{-3}}$ and $5 \times 10^{14}\,\mathrm{g\,cm^{-3}}$. For higher densities 'heavy elementary particles' (hyperons) should be taken into account, but their properties are not yet sufficiently well known (from the point of view of statistical physics) to be discussed in our relatively elementary exposition.

5. Different Forms of Equilibrium Equations for (R_Z) Reactions

5.1. Preliminary Remarks

For the relatively large values of Z considered below, a variation of Z by one unit – such as that encountered at the right-hand side of Equation (17′) corresponding to (R_Z) reactions – represents a variation sufficiently small in relative value to allow an assimilation of the *difference* $[m_{(A,Z)} - m_{(A,Z-1)}]$ to the *differential* of the *function* $m(A, Z) = m_{(A,Z)}$ corresponding to an increment of Z equal to $\Delta Z = +1$.

Therefore, we can write, doing an additional approximation assimilating the derivative at Z to the derivative at $(Z - 1)$, in a general way, for a function $F(A, Z)$:

$$F(A, Z) - F(A, Z-1) \approx \frac{\partial F(A, Z)}{\partial Z} \tag{25}$$

and, in a similar way:

$$F(A+1, Z) - F(A, Z) \approx \frac{\partial F(A, Z)}{\partial A}. \tag{25'}$$

Adopting approximations expressed by Equations (25) and (4) we can write Equation (16) as

$$\begin{aligned}\epsilon_{k,e}(A, Z, -1) &= \frac{\partial B^+(A, Z)}{\partial Z} + 0.773\ \mathrm{MeV}\\ &= \frac{4(A-2Z)}{A}a_T - 2a_c Z A^{-1/3} + 0.773\ \mathrm{MeV}\end{aligned} \tag{26}$$

or more explicitly, with Bethe's coefficients (of 1970) given by Table 6.2:

$$\epsilon_{k,e}(A, Z, -1) = \frac{96(A-2Z)}{A} - (1.44)\,Z A^{-1/3} + 0.773\ \mathrm{MeV}. \tag{26'}$$

Similarly, Equation (17′) can be approximated by

$$\epsilon_e(A, Z, -1) = -\frac{\partial m(A, Z)}{\partial Z}. \tag{26''}$$

5.2. THE DIRECT FORM OF THE EQUILIBRIUM EQUATION FOR (R_Z) REACTIONS

We have already used in Section 2 the fact that the value $(\epsilon_{k,e})_F$ of the kinetic energy at the Fermi level of the degenerate gas of electrons, is given, according to Equations (38, *e*), (43), (103) and (112) of Ch. 4 (in the domain A where the number density n_n of free neutrons is zero), by the approximate relations:

$$(\epsilon_{k,e})_F \approx \epsilon_{0,e}(1.2 \times 10^{-10})n_e^{1/3} \approx (0.51 \times 10^{-2})(\rho/\mathrm{A})^{1/3}\mathrm{Z}^{1/3}\,\mathrm{MeV} \qquad (27)$$

(where ρ is expressed in g cm^{-3}, and $\epsilon_{k,e}$ is in MeV).

This relation clearly shows that $(\epsilon_{k,e})_F$ is a function of three variables, A, Z and ρ, which can be symbolized by $(\epsilon_{k,e})_F(\mathrm{A, Z}, \rho)$. Note that according to Equation (27) this function is a quantity which *decreases* when (for a given value of ρ and A) Z *decreases*, or when n_e *decreases*.

On the other hand, according to Equation (26′), the partial derivative

$$\frac{\partial\epsilon_{k,e}(\mathrm{A, Z}, -1)}{\partial \mathrm{Z}} = -\frac{192}{\mathrm{A}} - \frac{(1.44)\,\mathrm{A}^{2/3}}{\mathrm{A}} \qquad (28)$$

is always *negative*, which means that, for a given value of A, the function $\epsilon_e(\mathrm{A, Z}, -1)$ *increases* for *decreasing* values of Z. At first sight this result seems in contradiction with the *observed* values of $\epsilon_{k,e}(\mathrm{A, Z}, -1)$ indicated in the Table 6.4 and by the lower part of Table 6.5. But we see that, in the Table 6.5, $\epsilon_{k,e}(28, 12, -1) = 15.1$ is indeed greater than both $\epsilon_{k,e}(28, 13, -1) = 1.8$ and $\epsilon_{k,e}(28, 14, -1) = 4.6$ and similarly $\epsilon_{k,e}(32, 14, -1) = 13.7$ is greater than the two other values of the same diagonal.

We must remember that Equation (28) is a consequence of Bethe–Weizsäcker semi-empirical analytical formula (4) which represents a *smooth interpolation* between the observed values. (One easily can verify, on a graphical representation of both the 'theoretical' $\epsilon_{k,e}(\mathrm{A, Z}, -1)$ given by Equation (26′) for A = 24 or 28 or 32, as a function of Z, by a *straight line* of *negative slope*, and of the approximate observed values of $\epsilon_{k,e}(\mathrm{A, Z}, -1)$ given by Table 6.5 for the same values of A, that the straight line represents a satisfactory interpolation between the points corresponding to observations.)

Now, in Section 2, we have explained the role played by the degenerate gas of electrons both as an agent of neutronization by electrons with kinetic energies $\epsilon_{k,e}$ *above* the threshold $\epsilon_{k,e}(\mathrm{A, Z}, -1)$, and as an agent preventing the inverse reactions by electrons with kinetic energies $\epsilon_{k,e}$ *below* the same threshold $\epsilon_{k,e}(\mathrm{A, Z}, -1)$. More precisely, we have explained that being given a value ρ of the mass density of the stellar mixture, and the type (A, Z) of the nuclei of the mixture, the value $(\epsilon_{k,e})_F = (\epsilon_{k,e})_F(\mathrm{A, Z}, \rho)$ of the kinetic energy of electrons at the Fermi level, given by Equation (27), can be sufficiently high (as was the case in the example of Section 2) to be *above* the corresponding threshold $\epsilon_{k,e}(\mathrm{A, Z}, -1)$ thus providing the necessary and sufficient conditions for *irreversible* neutronization transforming nuclei (A, Z) into nuclei (A, Z − 1). [In the example considered in Section 2, we had A = 56, Z = 26, $\rho = 2.16 \times 10^{10}$ g cm^{-3}, $(\epsilon_{k,e})_F \approx 11$ MeV and $\epsilon_{k,e}(\mathrm{A, Z}, -1) = 4.2$ MeV].

But in Section 2 we did not consider the 'feed back' effect of the *decrease* of the electron numerical density n_e due to electron captures transforming (A, Z) into (A, Z − 1) nuclei.

Each free electron captured by a nucleus (A, Z) represents a *loss* for the system of free electrons; each neutronization produces a decrease of n_e (without any modification of the mass density ρ). Therefore, if we suppose that we start from a situation of the type considered in Section 2, i.e. from a situation in which $(\epsilon_{k,e})_F > \epsilon_{k,e}(\mathrm{A}, \mathrm{Z}, -1)$, in which only electron captures (neutronizations) can take place, the decrease of n_e (or the corresponding decrease of Z) will produce, according to Equation (27), a lowering of $(\epsilon_{k,e})_F$ on the level diagram. At the same time the decrease of Z by neutronization will produce, according to Equation (28), a 'rise' of $\epsilon_{k,e}(\mathrm{A}, \mathrm{Z}, -1)$.

Thus a *convergence* towards a situation of *equality* between $(\epsilon_{k,e})_F$ and $\epsilon_{k,e}(\mathrm{A}, \mathrm{Z}, -1)$ would exist if this equality were not realized from the start.

Conversely, it is easily found that if one supposes that at the beginning $(\epsilon_{k,e})_F < \epsilon_{k,e}(\mathrm{A}, \mathrm{Z}, -1)$, the electrons with kinetic energies between $\epsilon_{k,e}(\mathrm{A}, \mathrm{Z}, -1)$ and $(\epsilon_{k,e})_F$, emitted by β^- disintegrations of nuclei (A, Z, − 1), will find 'places' unoccupied by the free electrons, so that β^- disintegrations will be possible. Hence both an increase of n_e and an increase of Z, i.e. again, according to Equations (27) and (28), a tendency to an equalization of $(\epsilon_{k,e})_F$ and $\epsilon_{k,e}(\mathrm{A}, \mathrm{Z}, -1)$.

Thus, whatever the initial situation, the mixture converges towards a *stable equilibrium* characterized by the following *'equation of equilibrium' with respect to* (R_Z) *reactions*:

$$\epsilon_{k,e}(\mathrm{A}, \mathrm{Z}, -1) = (\epsilon_{k,e})_F \tag{29}$$

or, in terms of total energies:

$$\boxed{\epsilon_e(\mathrm{A}, \mathrm{Z}, -1) = \epsilon_{F,e}.} \tag{29'}$$

This last equation can also, through the use of Equation (26″), take the form

$$\boxed{-\frac{\partial m(\mathrm{A}, \mathrm{Z})}{\partial \mathrm{Z}} = \epsilon_{F,e}.} \tag{29''}$$

More explicitly, taking into account Equations (26′) and (27), we can replace Equation (29) by the following equation which allows us to determine Z for given values of ρ and A:

$$0.773 + (96/\mathrm{A})(\mathrm{A} - 2\mathrm{Z}) - (1.44)\,\mathrm{ZA}^{-1/3} \approx (0.51 \times 10^{-2})(\rho/\mathrm{A})^{1/3}\mathrm{Z}^{1/3}. \tag{30}$$

It is easy to verify that the values of ρ, A and Z given by Table 6.6, roughly satisfy Equation (30), that is give a value of the same order to both sides of Equation (30) Thus, for instance, for $\rho = 1.04 \times 10^{10}$, A = 82 and Z = 32, the left-hand side is approximately equal to 11 and the right-hand side is approximately equal to 8. However, one cannot expect a precise verification. Indeed, Equation (30) is based (for pedagogical reasons) on the simplest choice of coefficients of Equation (4), whereas the Table 6.6 is established (on the same *principles*) by more refined calculations, with a different choice of the expression for the binding energy B^+ of nuclei as a function of A and Z.

Besides, Table 6.6 itself, though based on more refined data than Equation (30), must be considered as indicating only the order of magnitude of Z for a given value of ρ and A, since, as stated by Canuto (1974): "one can hardly hope to describe nuclei more and more neutron-rich by simply extrapolating low energy nuclear physics formulas such as the nuclear mass formula" (our Equation (4)). But the same caution must be exerted with respect to even more refined extrapolations used by Canuto himself.

5.3. THE THERMODYNAMIC FORMS OF THE EQUILIBRIUM EQUATION FOR (R_Z) REACTIONS

The fundamental Equation of equilibrium (29) with respect to (R_Z) reactions, can be established not only by the very explicit method used in Subsection 5.2, but also by a more general thermodynamic method, which represents a kind of short cut to the derivation of the final result without any study and *understanding* of intermediate stages.

Let us consider the sum W of all (total) *energy* densities of all particles (electrons and nuclei) of the stellar mixture. (Consideration of the (total) energies, instead of kinetic energies, as in Subsection 5.2, is not physically important, but simplifies somewhat the notation.)

Let n_e represent, as in Chapter 4, the number density of free electrons, and let u_e represent the (total) energy density of free electrons. Similarly, let n_{nuc} represent the number density of nuclei (A, Z) and $m(\text{A, Z})$ their rest mass (in energy units). Then, (neglecting the kinetic energy of nuclei with respect to their rest mass), we obtain:

$$W = u_e + m(\text{A, Z})\, n_{\text{nuc}}. \tag{31}$$

The mass, in grams, of each nucleus (A, Z) is equal to $\text{A}m_B$ (where m_B is the mass in grams of one baryon). Therefore, in the domain A, where free neutrons are still absent, the mass density ρ is equal to $(\text{A}m_B)\, n_{\text{nuc}}$, hence (if ρ is expressed in g cm^{-3}):

$$n_{\text{nuc}} = \rho/(\text{A}m_B). \tag{32}$$

On the other hand, the electric neutrality of the mixture demands that

$$n_e = \text{Z} n_{\text{nuc}}. \tag{33}$$

Equation (63) of Chapter 4, applied to the electron component of the mixture, and Equation (38, e) of Chapter 4, show that in the general (RR) case, the energy density u_e of the free electrons (supposed to be in a completely degenerate state) depends only on n_e:

$$u_e = u_e(n_e). \tag{34}$$

Thus, we see that, for a given value of ρ and A, n_{nuc}, given by Equation (32), is independent of Z, and u_e depends on Z only through n_e, given by Equation (33). Therefore, considering W as a function of three variables ρ, A and Z, the partial derivative of W with respect to Z will be given by

$$\frac{\partial W}{\partial \text{Z}} = \frac{du_e}{dn_e}\frac{\partial n_e}{\partial \text{Z}} + n_{\text{nuc}}\frac{\partial m(\text{A, Z})}{\partial \text{Z}}. \tag{35}$$

But, according to Equation (86′) of Ch. 4, applied to free electrons of the mixture,

$$\frac{du_e}{dn_e} = \epsilon_{F,e}, \tag{36}$$

where $\epsilon_{F,e}$ is the (total) energy of the Fermi level of the degenerate electron gas, and according to Equation (33)

$$\frac{\partial n_e}{\partial Z} = n_{\text{nuc}}. \tag{37}$$

Thus, replacing $\partial m/\partial Z$ by its value given by Equation (26″):

$$\frac{\partial W}{\partial Z} = [\epsilon_{F,e} - \epsilon_e(\text{A, Z}, -1)]\, n_{\text{nuc}}. \tag{38}$$

Since, in domain A, n_{nuc} cannot be zero, the necessary and sufficient condition for $W(\rho, \text{A, Z})$ to be *extremum* with respect to Z, i.e. the condition

$$\boxed{\frac{\partial W}{\partial Z} = 0,} \tag{39}$$

is equivalent to the equation of equilibrium (29′). Some authors introduce instead of $W(\rho, \text{A, Z})$ the 'mean (total) energy *per baryon*', $b(\rho, \text{A, Z})$, defined by

$$b(\rho, \text{A, Z}) = \frac{1}{n_B} W(\rho, \text{A, Z}), \tag{40}$$

where $n_B = \text{A}n_{\text{nuc}}$ is the number density of baryons. Since n_B is constant for (R_Z) reactions it is obvious that Equation (39) is equivalent to

$$\frac{\partial b}{\partial Z} = 0. \tag{39′}$$

Thus, imposing the condition that $W(\rho, \text{A, Z})$ or $b(\rho, \text{A, Z})$ be extremal with respect to Z, we find the same equation of equilibrium as by the direct method of Subsection 5.2.

5.3.1. Remark

It is interesting to note that instead of the thermodynamic condition (39) we could also use an even more general thermodynamic treatment in which the equilibrium is expressed by the extremum of the *thermodynamic potential* (or 'Gibbs free energy') defined by $(U - TS + PV)$ (see, for instance, Kittel, 1969, Ch. 19 and 21), with usual notations: (U = internal energy; T = Temperature; S = Entropy; P = Pressure; V = Volume). At constant pressure and at constant temperature, this leads to Equation (loc. cit. p. 347):

$$\tilde{\mu}_1 dN_1 + \tilde{\mu}_2 dN_2 + \tilde{\mu}_3 dN_3 = 0. \tag{41}$$

In this equation the chemical potentials (corresponding to the total energy) of the three components of the mixture are called $\tilde{\mu}_1$, $\tilde{\mu}_2$, $\tilde{\mu}_3$, instead of the notations usual in thermodynamics, in order to avoid confusion with the astrophysical use of the letter μ. Here dN_1, dN_2, dN_3 represent the change (for each component) in the number of particles due to the reactions under consideration.

In our case, (if as previously we neglect the neutrinos), the three components of the mixture corresponding to (R_Z) reactions (5′) or (6′) are the electrons, the (A, Z) and the (A, Z, − 1) nuclei.

The corresponding chemical potentials $\tilde{\mu}$ will be $\tilde{\mu}_e$, $\tilde{\mu}_{(A,Z)}$, $\tilde{\mu}_{(A,Z-1)}$ respectively. The corresponding values of dN_1, dN_2, dN_3 will be dN_e, $dN_{(A,Z)}$ and $dN_{(A,Z-1)}$ respectively. Since, according to Equation (5′), *each* creation of *one* nucleus (A, Z, − 1) demands a *decrease* of the number of (A, Z) nuclei and of the number of free electrons by one unit respectively, we have

$$dN_e = -dN_{(A,Z-1)}; \qquad dN_{(A,Z)} = -dN_{(A,Z-1)}. \tag{42}$$

Thus we obtain, by substitution into Equation (41) and elimination of $dN_{(A,Z-1)}$:

$$\boxed{\tilde{\mu}_{(A,Z-1)} - \tilde{\mu}_{(A,Z)} = \tilde{\mu}_e.} \tag{43}$$

A similar interpretation of Equation (6′) leads to the same result, so that Equation (43) holds for all (R_Z) reactions.

Now, we have seen in Chapter 4, Section 3 (Equation (30)) that the chemical potential $\tilde{\mu}$ corresponding to the *total* energy represents for a completely degenerate system the (total) energy ϵ_F of the Fermi level. Thus, for the electron component of the mixture

$$\tilde{\mu}_e = \epsilon_{F,e}. \tag{44}$$

On the other hand it is generally assumed (see for instance Landau and Lifshitz, 1958) that $\tilde{\mu}_{(A,Z)}$ is equal to $m(A, Z)$:

$$\tilde{\mu}_{(A,Z)} = m(A, Z). \tag{45}$$

Then, application of Equations (17′), (44) and (45) to Equation (43), yields a condition of equilibrium:

$$\epsilon_e(A, Z, -1) = \epsilon_{F,e} \tag{46}$$

identical to that obtained in Subsection 5.2 (Equation (29′)).

All these examples of thermodynamic treatment of the problem illustrate both the power and the dangers of such 'recipes'. The following comments, by Harrison *et al.*, 1965, p. 82, are quite appropriate:

"The idea has already become well established in chemical thermodynamics, though not without struggles at the time, decades ago, when it was first introduced, that one could analyze the possibilities for a reaction by comparing the free energies (. . .) without study or *even knowledge* [our italics!] of the perhaps dozens of intermediate stages, was in the end a principle too powerful to be overlooked. In the study of many important problems this principle short-circuited immense complexities of chemical-reaction-rate theory."

In other words, here we deal with a very powerful intellectual tool, very useful in research, but rather unsatisfactory from the pedagogical point of view, when the problem is precisely to *explain* all intermediate stages.

6. Equilibrium Equations for (R_{A}) Reactions

The study of the equilibrium with respect to the (R_{A}) reactions, defined in Section 3, provide an excellent opportunity of this kind of 'navigation without visibility'. We shall indeed apply first the thermodynamic method and emphasize both its elegance and its dangers.

6.1. THE THERMODYNAMIC TREATMENT

Let us assume, by analogy with the equilibrium (thermodynamic) condition (39) relative to (R_{Z}) reactions, namely $\partial W/\partial \mathrm{Z} = 0$, that the equilibrium condition for (R_{A}) reactions can be provided by Equation

$$\frac{\partial W(\rho, \mathrm{A}, \mathrm{Z})}{\partial \mathrm{A}} = 0, \tag{47}$$

where $W(\rho, \mathrm{A}, \mathrm{Z})$ is defined by the same Equation (31) as in Section 5, i.e.:

$$W(\rho, \mathrm{A}, \mathrm{Z}) = u_e(n_e) + m(\mathrm{A}, \mathrm{Z})\, n_{\mathrm{nuc}}. \tag{48}$$

Now, in the domain A considered here, there are no free neutrons, and all baryons (protons and neutrons) are in nuclei (A, Z). Therefore Equation (32) holds, viz.

$$n_{\mathrm{nuc}} = \rho/(\mathrm{A} m_B) = \frac{1}{\mathrm{A}} n_B, \tag{49}$$

where n_B is the number density of baryons, equal to (ρ/m_B).

In (R_{A}) reactions n_B remains constant, like the mass density ρ of the mixture.

On the other hand, Equation (33) also holds, so that n_e is given by

$$n_e = \mathrm{Z} n_{\mathrm{nuc}} = \frac{\mathrm{Z}}{\mathrm{A}} n_B. \tag{50}$$

Finally we obtain the following expression for $\partial W/\partial \mathrm{A}$:

$$\frac{\partial W}{\partial \mathrm{A}} = \frac{du_e}{dn_e}\frac{\partial n_e}{\partial \mathrm{A}} + \frac{\partial [m(\mathrm{A}, \mathrm{Z})\, n_{\mathrm{nuc}}]}{\partial \mathrm{A}} = \frac{du_e}{dn_e}\,\frac{\partial \left(\dfrac{\mathrm{Z}}{\mathrm{A}} n_B\right)}{\partial \mathrm{A}} + \frac{\partial [m(\mathrm{A}, \mathrm{Z})\, n_B/\mathrm{A}]}{\partial \mathrm{A}}. \tag{51}$$

Therefore, taking into account the fundamental equation (86′) of Ch. 4, and the constancy of n_B, $\partial W/\partial \mathrm{A}$ is given by

$$\frac{\partial W}{\partial \mathrm{A}} = \epsilon_{F,e}(-\mathrm{Z}\,\mathrm{A}^{-2} n_B) + \frac{1}{\mathrm{A}} n_B \frac{\partial m(\mathrm{A}, \mathrm{Z})}{\partial \mathrm{A}} - (\mathrm{A}^{-2} n_B)\, m(\mathrm{A}, \mathrm{Z}). \tag{52}$$

Thus, after multiplication by $(-\mathrm{A}^2/n_B)$, Equation (47) takes the form, similar to

Equation (29″):

$$\boxed{\frac{1}{Z}\left[A\,\frac{\partial m(A,Z)}{\partial A} - m(A,Z)\right] = \epsilon_{F,e}(\rho, A, Z),} \tag{53}$$

where we have written $\epsilon_{F,e}(\rho, A, Z)$ in order to make more explicit the dependence of $\epsilon_{F,e}$ on ρ, A and Z, already stressed in Subsection 5.2 (see Equation 27).

6.2. THE DIRECT TREATMENT

It is relatively easy to show that the thermodynamic equilibrium condition $\partial W/\partial A = 0$ corresponds to reactions (R_A) of the type described in Section 3 by Equation (24).

Indeed, let us apply the same type of discussion as the one used for (R_Z) reactions in previous sections.

We must then first replace Equation (24) by a more explicit *energy* balance, in which we denote by $Z\epsilon_e(A, 1, Z)$ the (total) threshold energy of the Z electrons transforming (A + 1) nuclei (A, Z) into A nuclei (A + 1, Z) by reactions (R_A) [or the maximum (total) energy of the Z electrons emitted by the corresponding inverse reaction]:

$$(A+1)\,m_{(A,Z)} + Z\epsilon_e(A, 1, Z) = A m_{(A+1,Z)}. \tag{54}$$

This equation is analogous, for (R_A) reactions, to Equation (17′) relative to reactions (R_Z).

It can be written in the form

$$Z\epsilon_e(A, 1, Z) = A[m_{(A+1,Z)} - m_{(A,Z)}] - m_{(A,Z)}, \tag{54′}$$

and, using the approximation expressed by Equation (25′), in the form:

$$\epsilon_e(A, 1, Z) = \frac{1}{Z}\left[A\,\frac{\partial m(A,Z)}{\partial A} - m(A,Z)\right]. \tag{54″}$$

This equation is analogous to Equation (26″).

For (R_Z) reactions, the convergence towards equilibrium was due (see Subsection 5.2) to the following circumstances:

(a) According to Equation (27) the kinetic energy of the Fermi level $(\epsilon_{k,e})_F$ and the total energy of the Fermi level $\epsilon_{F,e}$, were *decreasing* when Z or n_e were *decreasing*.

(b) According to Equation (28), the derivative of $\epsilon_{k,e}(A, Z, -1)$, and consequently the derivative of $\epsilon_e(A, Z, -1)$, with respect to Z was always negative. Thus when Z was *decreasing* (neutronization!) $\epsilon_{k,e}(A, Z, -1)$ was *increasing*.

Now, for (R_A) reactions transforming (A, Z) nuclei into (A + 1, Z) nuclei, A is *increasing*, but, according to Equation (50), n_e is decreasing.

Thus, according to Equation (27), $\epsilon_{F,e}$ is (again) *decreasing*.

On the other hand, a very tedious (but quite elementary) calculation, based upon Equations (54″), (3) and (4), (with Bethe's numerical values of a_s, a_c and a_τ indicated in Table 6.2), yields:

$$\frac{\partial \epsilon_e(\mathrm{A}, 1, \mathrm{Z})}{\partial \mathrm{A}} = \alpha(1 + \lambda), \tag{55}$$

where

$$\alpha = \frac{192\,\mathrm{Z}}{\mathrm{A}^2}; \qquad \lambda \approx \frac{1}{600}\,\mathrm{A}^{2/3}\left(1 - \frac{(\mathrm{A}/\mathrm{Z}^2)}{0.08}\right). \tag{56}$$

For all values of A and Z in the Table 6.6 (domain A) $|\lambda|$ remains between (0.001) and (0.010) and α is always positive. This means that $\epsilon_e(\mathrm{A}, 1, \mathrm{Z})$ *increases* when A *increases*.

As in the case of reactions (R_Z), the *opposite* trend of the variations of $\epsilon_{F,e}$ and of $\epsilon_e(\mathrm{A}, 1, \mathrm{Z})$ for increasing (or decreasing) of the independent variable (here A), produces a *convergence* towards a stable equilibrium corresponding to the equilibrium equation:

$$\boxed{\epsilon_e(\mathrm{A}, 1, \mathrm{Z}) = \epsilon_{F,e}.} \tag{57}$$

This equation, which expresses the equilibrium with respect to (R_A) reactions, is similar to Equation (29′) which expresses the equilibrium with respect to (R_Z) reactions.

Now taking into account Equation (54″), we immediately find that the condition of equilibrium (57) *is identical to Equation* (53), obtained much more easily (but with less physical insight) with the help of thermodynamic method. Note especially that in the thermodynamic treatment we do not make any use of Equation (24), i.e. of the actual transformations involved in (R_A) reactions. This emphasizes the dangers of the blind thermodynamic methods.

Indeed, the real nature of the reactions corresponding to the equilibrium condition $\partial W/\partial \mathrm{A} = 0$, disclosed by the direct treatment and described by Equation (24), justifies the following comments by Zel'dovich and Novikov (1971, p. 178):

"This [thermodynamic] analysis, while theoretically correct (. . .) is nevertheless an example of excessive rigor or even pedanticism. The truth of the matter is that at low temperature there does not exist a real mechanism which could transform iron into strontium within any reasonable time.

Let us reformulate this problem in a somewhat different way: in order to transform Fe^{56} to Sr^{120} within a reasonable time, high temperatures are required; it is necessary to break up the iron nuclei into component parts, i.e. α-particles and neutrons, and reshuffle them in a different way. But at the appropriately high temperature, the equilibrium will also be different."

7. The Domain A: Determination of A and Z Corresponding to an Equilibrium with Respect to Reactions (R_Z) and (R_A).

In a first introductory approach one can disregard the objections expressed above by Zel'dovich and Novikov concerning reactions (R_A). Then, considering A and Z as two independent variables, and assuming a convergence towards equilibrium with respect to reactions (R_Z) *and* reactions (R_A) (for a given value of the mass density ρ of the mixture), one has a system of two equations, (29″) and (53), for *two unknown* quantities A and Z.

Equation (29″) is equivalent to the more explicit Equation (30), and Equation (53) can now be replaced, through elimination of $\epsilon_{F,e}$ by means of Equation (29″), by a more symmetric and elegant equation:

$$\mathrm{A}\frac{\partial m(\mathrm{A},\mathrm{Z})}{\partial \mathrm{A}}+\mathrm{Z}\frac{\partial m(\mathrm{A},\mathrm{Z})}{\partial \mathrm{Z}} = m(\mathrm{A},\mathrm{Z}), \tag{58}$$

equivalent, through the use of Equation (3″), to

$$\boxed{\mathrm{A}\frac{\partial B^{+}(\mathrm{A},\mathrm{Z})}{\partial \mathrm{A}}+\mathrm{Z}\frac{\partial B^{+}(\mathrm{A},\mathrm{Z})}{\partial \mathrm{Z}} = B^{+}(\mathrm{A},\mathrm{Z}).} \tag{58'}$$

A very long and tedious (but quite elementary) calculation shows that Equation (58′) finally reduces when the expression (4) for B^{+} is used, to a very simple relation between A and Z, viz.

$$\mathrm{A} = (2a_c/a_s)\,\mathrm{Z}^2. \tag{59}$$

With Bethe's numerical values of a_c and a_s (see Table 6.2) Equation (59) takes the form:

$$\mathrm{A} = (0.080)\,\mathrm{Z}^2. \tag{59'}$$

It is very easy to verify that this equation is well satisfied by the values of A and Z given in Table 6.6. On the other hand, we have already shown in Subsection 5.2 that the same values of A and Z also satisfy Equation (30) reasonably well.

Therefore, we can consider that Table 6.6, relative to the domain A, represents the solution of the system describing an equilibrium with respect to reactions (R_{Z}) *and* (R_{A}), within the limits of uncertainty concerning the expression (4) for $B^{+}(\mathrm{A},\mathrm{Z})$.

7.1. Remark

Since $\mathrm{A} = \mathrm{N} + \mathrm{Z}$ one can consider the function $B^{+}(\mathrm{A},\mathrm{Z})$ given by Equation (4) as a function $B^{+}[\mathrm{N},\mathrm{Z}]$ of N and Z:

$$B^{+}[\mathrm{N},\mathrm{Z}] = B^{+}(\mathrm{N}+\mathrm{Z},\mathrm{Z}) = a_v(\mathrm{N}+\mathrm{Z}) - a_s(\mathrm{N}+\mathrm{Z})^{2/3} - \\ - a_c\mathrm{Z}^2(\mathrm{N}+\mathrm{Z})^{-1/3} - a_\tau(\mathrm{N}-\mathrm{Z})^2(\mathrm{N}+\mathrm{Z})^{-1}. \tag{60}$$

Conversely, since $\mathrm{N} = \mathrm{A} - \mathrm{Z}$, the function $B^{+}(\mathrm{A},\mathrm{Z})$ of A and Z can be expressed through the function $B^{+}[\mathrm{N},\mathrm{Z}]$ by

$$B^{+}(\mathrm{A},\mathrm{Z}) = B^{+}[(\mathrm{A}-\mathrm{Z}),\mathrm{Z}]. \tag{61}$$

Therefore, the partial derivatives of $B^{+}(\mathrm{A},\mathrm{Z})$ with respect to A or Z can be expressed in terms of partial derivatives of $B^{+}[\mathrm{N},\mathrm{Z}]$ by

$$\frac{\partial B^{+}(\mathrm{A},\mathrm{Z})}{\partial \mathrm{A}} = \frac{\partial B^{+}[\mathrm{N},\mathrm{Z}]}{\partial \mathrm{N}}\frac{\partial(\mathrm{A}-\mathrm{Z})}{\partial \mathrm{A}} = \frac{\partial B^{+}[\mathrm{N},\mathrm{Z}]}{\partial \mathrm{N}} \tag{62}$$

and

$$\frac{\partial B^{+}(\mathrm{A},\mathrm{Z})}{\partial \mathrm{Z}} = \frac{\partial B^{+}[\mathrm{N},\mathrm{Z}]}{\partial \mathrm{N}}\frac{\partial(\mathrm{A}-\mathrm{Z})}{\partial \mathrm{Z}} + \frac{\partial B^{+}[\mathrm{N},\mathrm{Z}]}{\partial \mathrm{Z}} \tag{62'}$$

hence

$$\frac{\partial B^{+}(\mathrm{A},\mathrm{Z})}{\partial \mathrm{Z}} = \frac{\partial B^{+}[\mathrm{N},\mathrm{Z}]}{\partial \mathrm{N}}(-1) + \frac{\partial B^{+}[\mathrm{N},\mathrm{Z}]}{\partial \mathrm{Z}}. \tag{63}$$

Taking into account Equations (62) and (63) we can transform Equation (58′) into

$$\mathrm{N}\frac{\partial B^{+}[\mathrm{N},\mathrm{Z}]}{\partial \mathrm{N}} + \mathrm{Z}\frac{\partial B^{+}[\mathrm{N},\mathrm{Z}]}{\partial \mathrm{N}} = B^{+}[\mathrm{N},\mathrm{Z}]. \tag{64}$$

8. The Domain B: a Mixture of Free Electrons, Free Neutrons and Nuclei (A, Z)

8.1. THE EQUILIBRIUM EQUATIONS

When free neutrons are present (domain B, defined in Section 4) the problem becomes so complex that the only convenient method of investigation of an equilibrium between different components of the mixture is the one based on thermodynamic conditions for equilibrium among reacting species.

An obvious generalization of thermodynamical treatment already used in Subsections 5.3 and 6.1 leads us to consider the sum W of all (total) energy densities of all particles of the stellar mixture: free electrons, nuclei (A, Z), and free neutrons,

$$W = u_e + u_{\mathrm{nuc}} + u_n, \tag{65}$$

where u_n is the (total) energy density of free neutrons.

Once more the electric neutrality of the mixture demands that

$$n_e = \mathrm{Z} n_{\mathrm{nuc}}, \tag{66}$$

and we have already explained in Section 5.3 why the energy density u_e of the free electrons (supposed to be in a completely degenerate state) depends only on the number density n_e of free electrons:

$$u_e = u_e(n_e) = u_e(\mathrm{Z} n_{\mathrm{nuc}}). \tag{67}$$

On the other hand, if n_n represents the number density of free neutrons, and n_{nuc} represents the number density of nuclei (A, Z), we can introduce (as in Equation (102) of Chapter 4) a number density n_B defined by

$$n_B = n_n + \mathrm{A} n_{\mathrm{nuc}}. \tag{68}$$

Physically, n_B represents the total number density of baryons: both free, as the free neutrons, *and* those (protons and neutrons) contained *in* nuclei (A, Z).

Solving Equation (68) with respect to n_n we can write

$$n_n = n_B - \mathrm{A} n_{\mathrm{nuc}}. \tag{68′}$$

In all reactions concerning the domain B, i.e. (R_{Z}), (R_{A}) reactions and reactions of '*neutron drip*' (emission of free neutrons by nuclei too neutron-rich) n_B *remains constant*. It is related to the constant mass density ρ of the mixture by the obvious equality:

$$n_B = \rho N_A, \tag{69}$$

where N_A is the Avogadro's number, so that Equation (68′) becomes:

$$n_n = \rho N_A - A n_{\text{nuc}}. \tag{68''}$$

Now, Equation (67) shows that u_e depends only on Z and n_{nuc}. On the other hand, $m(\mathrm{A, Z})$ being the rest mass of nuclei (A, Z) expressed in energy units, we have

$$u_{\text{nuc}} = m(\mathrm{A, Z})\, n_{\text{nuc}}. \tag{70}$$

This shows that u_{nuc} depends only on A, Z and n_{nuc}. Finally, it is easy to verify that in the domain B the free neutron component of the mixture is in a completely degenerate state. This results from an application of the criteria of degeneracy (see Chapter 4) to the values of n_n of Table 6.7, with reasonable values of the temperature.

Thus, u_n depends only on n_n, and through Equation (68′), on A and n_{nuc} only (for a given value of ρ or n_B):

$$u_n = u_n(n_n) = u_n(n_B - A n_{\text{nuc}}). \tag{71}$$

Therefore, taking into account the definition (65) of W, we see that W depends only on *three independent variables*: A, Z and n_{nuc}:

$$W = W(\mathrm{A, Z}, n_{\text{nuc}}) = u_e(\mathrm{Z} n_{\text{nuc}}) + m(\mathrm{A, Z})\, n_{\text{nuc}} + u_n(n_B - \mathrm{A} n_{\text{nuc}}). \tag{72}$$

By analogy with the conditions $\partial W/\partial \mathrm{Z} = 0$ and $\partial W/\partial \mathrm{A} = 0$ of Section 7, our present conditions will be

$$\frac{\partial W}{\partial \mathrm{Z}} = 0, \qquad \frac{\partial W}{\partial \mathrm{A}} = 0 \quad \text{and} \quad \frac{\partial W}{\partial n_{\text{nuc}}} = 0.$$

These three equations allow us to determine the three unknowns A, Z and n_{nuc}.

By application of the fundamental relation (86′) of Chapter 4 to du_e/dn_e and to du_n/dn_n these three equations form the system:

$$\epsilon_{F,e} n_{\text{nuc}} + \frac{\partial m}{\partial \mathrm{Z}} n_{\text{nuc}} = 0; \tag{73}$$

$$\frac{\partial m}{\partial \mathrm{A}} n_{\text{nuc}} + \epsilon_{F,n}(-n_{\text{nuc}}) = 0; \tag{74}$$

$$\epsilon_{F,e}\mathrm{Z} + m + \epsilon_{F,n}(-\mathrm{A}) = 0. \tag{75}$$

Dividing Equations (73) and (74) by n_{nuc}, and eliminating $\epsilon_{F,e}$ and $\epsilon_{F,n}$ from Equation (75), we obtain the system:

$$\epsilon_{F,e} = -\frac{\partial m(\mathrm{A, Z})}{\partial \mathrm{Z}}; \tag{76}$$

$$\epsilon_{F,n} = +\frac{\partial m(\mathrm{A, Z})}{\partial \mathrm{A}}; \tag{77}$$

$$\mathrm{A}\frac{\partial m(\mathrm{A, Z})}{\partial \mathrm{A}} + \mathrm{Z}\frac{\partial m(\mathrm{A, Z})}{\partial \mathrm{Z}} = m(\mathrm{A, Z}). \tag{78}$$

8.2. Solution of the Equilibrium Equations

Equation (76) is identical to Equation (29″) for reactions (R_Z) in the domain A. Equation (78) is identical to Equation (58) for reactions (R_A) in the domain. The only new equation is Equation (77) which involves the Fermi energy $\epsilon_{F,n}$ of the degenerate system of free neutrons. Since $\epsilon_{F,n}$ is a 'total' energy including the rest energy m_n of a neutron, the domain B is characterized by the condition $\epsilon_{F,n} > m_n$, or according to Equation (77) by the condition:

$$\frac{\partial m(\mathrm{A}, \mathrm{Z})}{\partial \mathrm{A}} > m_n, \tag{79}$$

where, of course, m_n must be expressed in energy units.

Taking successively into account Equations (39), (applied to electrons) and (38, e) of Ch. 4 and Equation (66), $\epsilon_{F,e}$ can be considered, quite generally, as a known function $\epsilon_{F,e}(\mathrm{Z}, n_{\mathrm{nuc}})$ of Z and n_{nuc}.

Similarly, taking successively into account Equations (39) (applied to neutrons) and (38, n) of Ch. 4 together with Equation (68″), $\epsilon_{F,n}$ can be considered, quite generally, as a known function $\epsilon_{F,n}(\rho, \mathrm{A}, n_{\mathrm{nuc}})$ of ρ, A and n_{nuc}.

Thus it is clear that for a given value of ρ a numerical solution of the system of Equations (76), (77) and (78) will give A, Z and n_{nuc}. Then, Equation (68″) will give the value of n_n.

In a more refined treatment, one must take into account the reduction of the *volume* available to free neutrons by the volume occupied by the huge nuclei formed at the upper end of the domain B. If $\bar{\omega}$ represents the fraction of the volume of the mixture occupied by these nuclei, the volume available to free neutrons will be *reduced* by a factor $(1 - \bar{\omega})$ and the number density of free neutrons *increased* by a factor $1/1 - \bar{\omega}$. Therefore Equations (68′) and (68″) must be replaced by

$$n_n = \frac{1}{(1-\bar{\omega})}(n_B - \mathrm{A} n_{\mathrm{nuc}}) = \frac{1}{(1-\bar{\omega})}(\rho N_A - \mathrm{A} n_{\mathrm{nuc}}), \tag{80}$$

and this modification must be taken into account in calculations leading to equations similar to (76), (77) and (78).

The values given in Table 6.7 have been obtained by this more refined method, but we do not reproduce the corresponding details, since the thermodynamic principles remain the same.

9. The Domain C: a Mixture of Free Electrons, Free Protons and Free Neutrons

9.1. The Equilibrium Equation

In our notation, a neutron n is represented by [1, 0] or (1, 0) and a proton p is represented by [0, 1] or (1, 1). Therefore, in the particular case N = 0, Z = 1, (A = 1), the (R_Z) reactions (5) and (6) become

$$p + e^- \rightarrow n + \nu_e \tag{81}$$

and

$$n \to p + e^- + \bar{\nu}_e \tag{82}$$

respectively.

Equation (81) describes the capture of an electron e^- by a proton; Equation (82) describes the β^- disintegration of a neutron.

A mixture of free electrons, free protons and free neutrons in which all components form degenerate systems (high mass density and relatively low temperature) converge towards an equilibrium between reactions (81) and (82), which are both of type (R_Z).

However, we cannot consider this equilibrium as a particular case of the equilibrium studied in Section 5 (in connection with the domain A) since, in that case for each value of the mass density ρ of the mixture, we had, beside the free electrons, only *one* type of nuclei: (A, Z). In the present case, the equilibrium mixture will be composed, beside the free electrons, of *two* types of nuclei: the protons and the neutrons. Therefore, the domains C must be considered as a particular case of the domain B with free *protons* playing the role of nuclei (A, Z) of Section 8, and a constant value of A equal to 1.

The electric neutrality of the mixture will now be expressed by

$$n_e = n_p, \tag{83}$$

where n_p is the number density of protons. This equation is a particular case of Equation (66) with $n_p = n_{\text{nuc}}$.

The *constant* number density of baryons, n_B, will now be given by

$$n_B = n_n + n_p = \rho N_A, \tag{84}$$

where, as before, n_n is the number density of free neutrons. This equation is a particular case of Equation (68) with A = 1 and $n_{\text{nuc}} = n_p$, and can be written like (68′), i.e.:

$$n_n = n_B - n_p. \tag{84′}$$

Once more the energy density u_e of electrons (in a degenerate state) will depend only on n_e, and Equation (67) will take the form

$$u_e = u_e(n_e) = u_e(n_p), \tag{85}$$

where we have taken into account Equation (83).

Similarly the energy density u_n of neutrons (in a degenerate state) will depend only on n_n, and Equation (71) will take the form

$$u_n = u_n(n_n) = u_n(n_B - n_p), \tag{86}$$

where we have taken into account Equation (84′).

However, we can no longer express the energy density u_p of protons as a particular case of Equation (70), with $m(\text{A, Z}) = m_p$. Indeed, in Equation (70) we could neglect the kinetic energy of the *heavy* nuclei (A, Z) with respect to their rest energy $m(\text{A, Z})$, but we have no right to neglect the kinetic energy of protons with respect to their rest energy m_p. Therefore, we must replace Equation (70) by an equation similar to Equation (85), and (considering that the energy density u_p of protons in a degenerate state depends only on n_p) write simply:

$$u_p = u_p(n_p). \tag{87}$$

Thus we see that in the domain C the sum W of all energy densities of all particles of the mixture will be given by

$$W = W(n_p) = u_e + u_p + u_n = u_e(n_p) + u_p(n_p) + u_n(n_B - n_p). \qquad (88)$$

This equation is more rigorous than Equation (72) and, at the same time, more simple, since it shows that in the present case W depends on a *single* variable n_p. This allows us to express the equilibrium by a single equation:

$$\frac{\partial W}{\partial n_p} = 0. \qquad (89)$$

By application of the fundamental relation (86′) of Chapter 4 to du_e/dn_e, du_p/dn_p and du_n/dn_n we immediately obtain the equilibrium condition:

$$\boxed{\epsilon_{F,e} + \epsilon_{F,p} - \epsilon_{F,n} = 0.} \qquad (90)$$

9.1.1. *Remark*

To obtain Equation (90) of equilibrium it is also possible to apply the thermodynamic recipe, already mentioned in Subsection 5.3.1 which consists of writing the 'chemical' Equation (81) as an equation among the corresponding chemical potentials $\tilde{\mu}$:

$$\tilde{\mu}_p + \tilde{\mu}_e = \tilde{\mu}_n. \qquad (91)$$

But, according to Equation (30) of Chapter 4, the chemical potential $\tilde{\mu}$ of a degenerate system is equal to the corresponding Fermi (total) energy ϵ_F. Hence Equation (90).

9.2. SOLUTION OF THE EQUILIBRIUM EQUATION

On using the expression, given for ϵ_F by Eq. (30) of Ch. 4, Equation (90) takes the form:

$$\epsilon_{0,e}(1 + x_{F,e}^2)^{1/2} + \epsilon_{0,p}(1 + x_{F,p}^2)^{1/2} = \epsilon_{0,n}(1 + x_{F,n}^2)^{1/2}. \qquad (92)$$

We write

$$\lambda_e = \frac{\epsilon_{0,e}}{\epsilon_{0,n}} = \frac{m_e}{m_n} \approx \frac{1}{1840}; \qquad \lambda_p = \frac{\epsilon_{0,p}}{\epsilon_{0,n}} = \frac{m_p}{m_n} \approx 1 - \frac{m_n - m_p}{m_n}, \qquad (93)$$

and, taking into account Equation (83),

$$\eta = x_{F,n}^2; \qquad X = (n_p/n_n)^{2/3} = (n_e/n_n)^{2/3}. \qquad (94)$$

Then we find, according to Equations (36) and (37) of Ch. 4,

$$X^{3/2} = \frac{n_e}{n_n} = \frac{m_e^3 x_{F,e}^3}{m_n^3 x_{F,n}^3} = \lambda_e^3 x_{F,e}^3 \eta^{-3/2} \qquad (95)$$

or

$$x_{F,e}^2 \lambda_e^2 = X\eta, \qquad (95')$$

and, similarly:

$$x_{F,p}^2 \lambda_p^2 = X\eta. \qquad (96)$$

Thus, dividing Equation (92) by $\epsilon_{0,n}$, it takes the form:

$$(X\eta + \lambda_e^2)^{1/2} + (X\eta + \lambda_p^2)^{1/2} = (1 + \eta)^{1/2}. \qquad (97)$$

Solving this equation with respect to X, one finds (after some elementary, but tedious, calculations):

$$X = \frac{\eta^2 + \alpha\eta + \beta}{4\eta(\eta + 1)}, \tag{98}$$

where

$$\alpha \approx 4\frac{(m_n - m_p)}{m_n} \approx 0.0055, \tag{99}$$

and

$$\beta \approx \frac{\alpha^2}{4} - 4\lambda_e^2 \approx 6.38 \times 10^{-6}. \tag{100}$$

From Equations (98) and (94) we immediately obtain the ratio z of equilibrium abundances (equal to the ratio of the partial mass densities ρ_p of protons and ρ_n of neutrons) as a function of η:

$$\boxed{z = \frac{n_p}{n_n} = \frac{n_e}{n_n} = \frac{\rho_p}{\rho_n} = X^{3/2} = \frac{1}{8}\left[\frac{\eta^2 + \alpha\eta + \beta}{\eta(\eta + 1)}\right]^{3/2}.} \tag{101}$$

Taking into account the orders of magnitude of α and β, given by Equations (99) and (100), it is easy to verify that z is *minimum* for $\eta_{\min} \approx \beta^{1/2} \approx 2.5 \times 10^{-3}$, since in the neighbourhood of this value of η we can neglect η with respect to 1 and consider that z varies as $(\eta + \alpha + \beta/\eta)$. The corresponding value, $z_{\min}$, of z is equal to

$$z_{\min} \approx 1.3 \times 10^{-4}. \tag{102}$$

Now, η is given as a function of ρ_n by Equation (113) of Chapter 4

$$\eta = x_{F,n}^2 = (2.9 \times 10^{-11})\rho_n^{2/3}, \tag{103}$$

hence

$$\rho_n = (6.4 \times 10^{15})\eta^{3/2}. \tag{103'}$$

Thus the value of ρ_n corresponding to $\eta_{\min}$ is of the order of 8×10^{11} g cm^{-3}.

On the other hand, when η increases beyond $\eta_{\min}$ z *increases monotonically* up to its asymptotic value, for $\eta \to \infty$ (and $\rho_n \to \infty$), equal to (1/8).

In the domain $2 \times 10^{14} < \rho_n < 5 \times 10^{14}$ g cm^{-3}, η varies, according to Equation (103), between 0.098 and 0.182, and z varies, according to Equation (101), between 0.0036 and 0.0076. This means that in this domain the number of protons never exceeds 1% of the number of neutrons, and the partial mass density ρ_p never exceeds 1% of the partial mass density ρ_n. Thus, in this domain $\rho = \rho_n + \rho_p \approx \rho_n$ and $n_B = n_n + n_p \approx n_n$. In other words, we are precisely in the domain C defined by $2 \times 10^{14} < \rho < 5 \times 10^{14}$ g cm^{-3}, and we can disregard both n_p and n_e (equal to n_p) in comparison with the number density n_n of free neutrons. In the domain C we deal with genuine '*neutron stars*'.

10. The Structure of Neutron Stars. Their Radius as a Function of their Mass

We have already explained in Chapter 2 that the mechanical equilibrium of a star demands a compensation of the gravitational force per unit volume, acting on each element of the star, by the force per unit volume produced by the *pressure gradient*.

No mechanical equilibrium is possible when the pressure gradient is too weak, and the star suffers a *gravitational collapse*.

Detailed calculations show that this happens over a rather large range of mean mass densities $\bar{\rho}$, which extends from the most dense of white dwarfs ($\bar{\rho} \approx 10^9\,\mathrm{g\,cm^{-3}}$) to the least dense of neutron stars (those of domain C) ($\bar{\rho} \approx 10^{14}\,\mathrm{g\,cm^{-3}}$).

Even near the upper limit $\rho \approx \rho_n \approx 5 \times 10^{14}\,\mathrm{g\,cm^{-3}}$ of the domain C, where $n_n \approx 300 \times 10^{36}$ and $n_p \approx n_e \approx 3 \times 10^{36}\,\mathrm{cm^{-3}}$, the number density n_n of neutrons is below the critical number density $n_{\mathrm{cr},n}$ equal to $3.7 \times 10^{39}\,\mathrm{cm^{-3}}$ (see Equation (36, n) of Ch. 4). The corresponding pressure of the degenerate neutron gas is therefore given (when interaction between neutrons is neglected) by Equations (135, n) or (141) of Ch. 4, viz.

$$P_n(\mathrm{NR}) = (1.3 \times 10^{-30}) n_n^{5/3} = (5.6 \times 10^9) \rho_n^{5/3} \approx (5.6 \times 10^9) \rho^{5/3}. \quad (104)$$

Similarly, we have, for protons

$$P_p(\mathrm{NR}) = (1.3 \times 10^{-30}) n_p^{5/3}. \quad (105)$$

On the other hand, the number density n_e of electrons is far above the critical density $n_{\mathrm{cr},e}$ equal to $5.9 \times 10^{29}\,\mathrm{cm^{-3}}$ (see Equation (36, e) of Ch. 4), so that P_e is given by Equation (136, e) of Ch. 4, viz.

$$P_e(\mathrm{UR}) = (2.4 \times 10^{-17}) n_e^{4/3}. \quad (106)$$

Thus we find that at the upper limit of the domain C

$$P_n \approx 2 \times 10^{34}; \qquad P_p \approx 0.001 \times 10^{34}; \qquad P_e \approx 0.011 \times 10^{34}\ \text{c.g.s.} \quad (107)$$

We see that P_p and P_e are entirely negligible with respect to P_n, and that according to Equation (104) we can write, if we call P the total pressure (in c.g.s. units):

$$P = (5.6 \times 10^9) \rho^{5/3} = P_n(\mathrm{NR}) = K_n \rho^{5/3}. \quad (108)$$

(A more advanced treatment takes into account nuclear forces between neutrons, which are no longer negligible at these high densities. Then one finds a somewhat lower pressure: $P \approx 0.7 \times 10^{34}$ for $\rho = 5 \times 10^{14}$ c.g.s. – See Canuto, (1974)).

Applying the rough treatment explained in Chapter 5, Section 2 we can use an approximate evaluation of the terms of the virial theorem

$$2T = B_g^+. \quad (109)$$

Here T is the total kinetic energy of all neutrons of the star, given, as a function of the approximate value $\bar{P}_{\mathrm{app}}$ of $\bar{P}$ and volume V, by

$$T = \tfrac{3}{2} \int_0^R P\,dV_r = \tfrac{3}{2} \bar{P} V \approx \tfrac{3}{2} \bar{P}_{\mathrm{app}} V, \quad (110)$$

and B_g^+ is the gravitational binding energy of the star given approximately, for a mean value $\bar{\rho} = M/V$ of the mass density, by

$$B_g^+ \approx (3/5)(G/R) M^2. \quad (111)$$

Now, according to Equation (108), we have as approximate value for the mean pressure $\bar{P}$

$$\bar{P}_{\mathrm{app}} = (5.6 \times 10^9)(\bar{\rho})^{5/3} = K_n \bar{\rho}^{5/3}. \quad (112)$$

In Equations (108) and (112) the K_n occurring in the expression for the pressure produced by neutrons in a neutron star, plays the same role as the coefficient $K_1 = K(\text{NR})$ occurring in the pressure produced by non-relativistic electrons in a white dwarf. But, of course, the corresponding *numerical values* are quite different.

Thus we do not need to repeat the tedious but elementary calculations leading to Equations similar (with K_1 replaced by K_n) to Equations (31) and (32) of Ch. 5, viz.

$$R \approx Q_n M^{-1/3} \tag{113}$$

and

$$Q_n \approx (1.92)(K_n/G), \tag{114}$$

which result from elimination of $\bar{\rho} = M/V$, $V = (4\pi/3)R^3$, and $\bar{P}_{\text{app}}$ among Equations (109), (110), (111) and (112).

Inserting in Equation (114) the numerical value of K_n given by Equation (108), and taking the constant of gravitation G in c.g.s. units ($G = 6.67 \times 10^{-8}$), we obtain

$$Q_n \approx 1.6 \times 10^{17} \text{ c.g.s.}. \tag{114'}$$

Finally, Equation (113) takes the form:

$$R \approx (1.6 \times 10^{17})M^{-1/3}, \tag{115}$$

where R is expressed in cm and M in g.

Expressing R in km and M by its ratio $M/M_\odot$ to the solar mass $M_\odot \approx 2 \times 10^{33}$ g, we obtain

$$\boxed{R^{\text{km}} \approx (12.7)\left(\frac{M}{M_\odot}\right)^{-1/3}.} \tag{115'}$$

Thus the radius R of a neutron star of the same mass as the Sun is of the order of 12.7 km.

Expressing R by its ratio $R/R_\odot$ to the solar radius $R_\odot \approx 7 \times 10^5$ km, Equation (115′) takes the form

$$\frac{R}{R_\odot} \approx (1.8 \times 10^{-5})\left(\frac{M}{M_\odot}\right)^{-1/3}, \tag{115''}$$

which, compared with Equation (33) of Ch. 5, shows that, for the same mass M, the radius R of a neutron star is *about 500 times smaller than the radius of a white dwarf.*

Eliminating R among Equation (115) and Equation (116):

$$\bar{\rho} = \frac{M}{V} = \frac{M}{\frac{4\pi}{3}R^3} \tag{116}$$

one easily finds that for a neutron star the mass M is given, as a function of $\bar{\rho}$, in c.g.s. units, by

$$M \approx (13.1 \times 10^{25})(\bar{\rho})^{1/2}, \tag{117}$$

which is equivalent, again in c.g.s. units, to

$$\frac{M}{M_\odot} \approx (6.55 \times 10^{-8})(\bar{\rho})^{1/2}. \qquad (117')$$

Thus, the values of $M/M_\odot$ corresponding to the extreme values $\bar{\rho} = 2 \times 10^{14}$ and $\bar{\rho} = 5 \times 10^{14}$ g cm^{-3} of the domain C, are $M/M_\odot \approx 0.92$ and $M/M_\odot \approx 1.5$. We conclude, from these figures, that a *neutron star must have a mass of the order of the mass* $M_\odot$ *of the Sun.*

The corresponding extreme values for the radius R are given by Equation (115′). They are 13 km and 11 km respectively.

Thus, as for white dwarfs, the radius R of neutron stars decreases when the mass M increases.

Equation (117) shows that, also as for white dwarfs, the mass is an increasing function of $\bar{\rho}$. However, contrary to what happens for white dwarfs, when $\bar{\rho}$ increases, no phenomenon of a 'limiting mass' occurs.

Indeed, when $\bar{\rho}$ increases beyond some 'critical density' the degeneracy of the free electron gas in white dwarfs becomes ultra-relativistic, and the non-relativistic polytropic index $n'(\text{NR}) = 1.5$ is replaced by the ultra-relativistic one $n'(\text{UR}) = 3.0$, hence the elimination of the radius R from the virial theorem (see Chapter 5, Section 2) and the existence of a limiting mass M_3.

Now, in the case of neutron stars, when $\bar{\rho}$ increases beyond some critical density (which obviously is not the same as for white dwarfs) the degeneracy of the free neutron component cannot become ultra-relativistic, with a polytropic index equal to 3.0, and cannot correspond to a limiting mass for the following reasons.

The phenomenon of limiting mass is associated with the properties of systems composed of *independent* particles, i.e. of particles with no other mutual interaction than elastic collisions. But for number densities of neutrons of the order of $n_n = 10^{38}$–10^{39} cm^{-3}, corresponding to mass densities $\rho \approx \rho_n$ of the order of 10^{14}–10^{15} g cm^{-3}, the neutrons become sufficiently closely packed to suffer short-range nuclear interactions.

As a consequence of this nuclear interaction, an increase of the density in this domain corresponds to a *decrease* of the mass M. (See, for instance, Cox and Giuli, 1968; pp. 1009–1015, where this phenomenon is rather clearly explained.)

Thus, instead of a monotonic increase up to a limiting mass, as for white dwarfs, we encounter here a *maximum mass* M_{OV}, called the '*Oppenheimer–Volkoff mass*' (hence the subscript OV). Because of uncertainties concerning nuclear forces the value of M_{OV} is rather uncertain. It is situated by different authors between $0.6M_\odot$ and $2M_\odot$ and corresponds to central densities between 2×10^{15} and 4×10^{15} g cm^{-3} (see Figure 6.2).

For even higher densities the theory of *general relativity* must be taken into account and leads to a *zero* limiting mass (see Weinberg (1972, p. 320) or Cox and Giuli (1968, pp. 1015)).

Let us note, finally, returning to normal neutron stars, that according to Börner (1973) the internal structure of a neutron star corresponds to the schematic stratification indicated in Figure 6.3.

11. Conclusion

The very strange theoretical properties of neutron stars are almost incredible. Fortunately the discovery of *pulsars*, studied in Chapter 7, provides at least a proof of the existence of

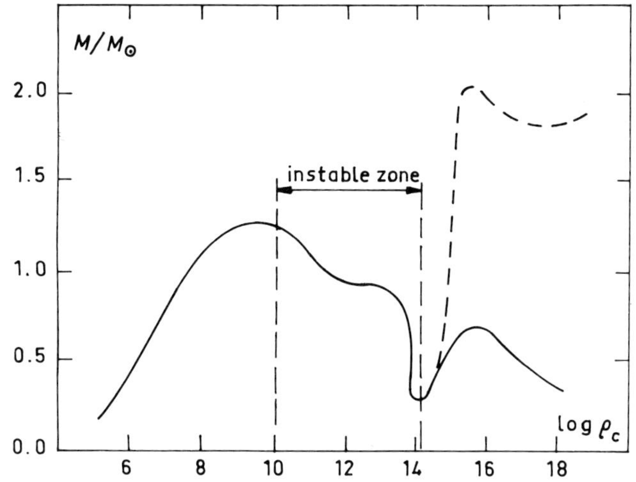

Fig. 6.2. The variations of the equilibrium mass of a star of high and ultrahigh density as a function of the log of the central density ρ_c (in g cm^{-3}). Full line, as given by Oppenheimer and Volkoff; dotted line, as given by Cameron.

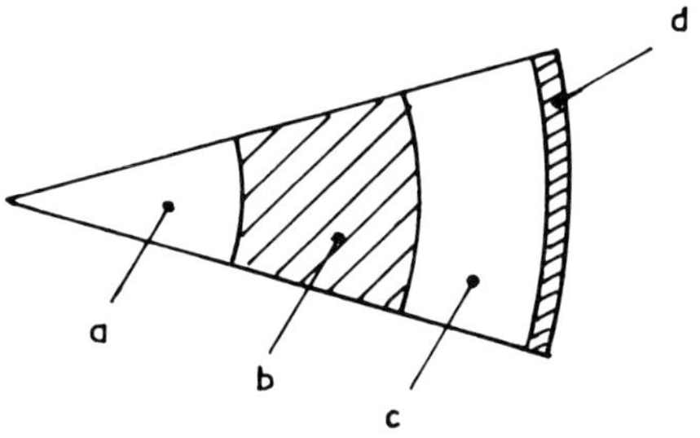

Fig. 6.3. (a) The hyperon core; (b) the 'liquid' part; (c) the 'crust' formed by a mixture of electrons, neutrons and nuclei; (d) the atmosphere.

these strange objects. And this existence is confirmed by the study of binary X-ray sources presented in Chapter 9, where the 'observational' values of the masses of neutron stars are indicated (Table 9.3).

References

1. WORKS CITED IN THE TEXT

(Works that can be used for further study are marked by an asterisk.)

Blatt, J. M. and Weisskopf, V. F.: 1952, *Theoretical Nuclear Physics*, Wiley, London.

Bethe, H. and Morisson, P.: 1956, *Elementary Nuclear Theory*, Wiley, London.

Baym, G. *et al.*: 1971, *Nuclear Physics* **A175**, 225.

* Börner, G.: 1973, 'On the Properties of Matter in Neutron Stars', *Springer Tracts in Modern Physics* **69**, Berlin.

Cameron, A. G. W.: 1959, *Astrophys. J.* **130**, 884.
* Canuto, V.: 1974, *Ann. Rev. Astron. Astrophys.* **12**, 190, 205.
* Cox, J. P. and Giuli, R. T.: 1968, *Principles of Stellar Structure*, Vol. II. (Ch. 26.5), Gordon and Breach, New York.
* Harrison, B. H. *et al.*: 1965, *Gravitation Theory and Gravitational Collapse* (Mainly Ch. 9 and 10), Univ. of Chicago Press, Chicago.
Kittel, C.: 1969, *Thermal Physics*, Wiley, New York.
* Landau, L. and Lifshiftz, E.: 1958, *Statistical Physics* (Ch. 11), Pergamon Press, London.
Oppenheimer, J. R. and Volkoff, G. M.: 1939, *Phys. Rev.* **55**, 374.
Salpeter, E. E.: 1961, *Astrophys. J.* **134**, 679 (Table I).
Tondeur, F.: 1971, *Astron. Astrophys.* **14**, 451.
* Weinberg, S.: 1972, *Gravitation and Cosmology* (Ch. 11. 4, pp. 317–325), Wiley, New York. [Equation (11.4.28) is incorrect: delete the coefficient $\frac{1}{2}$ and replace 0.002 by $(0.0026)^{3/2} = 0.000\,135$.]
* Zel'dovich, Ya. B. and Novikov, I. D.: 1971, *Stars and Relativity* (Relativistic Astrophysics, Vol. I, Ch. 5, 6, 10, 11) Univ. of Chicago Press, Chicago.

2. REFERENCES FOR FURTHER STUDY

(See first the works cited in 1 marked by an asterisk.)

A. *Review Articles and Monographs*

Cameron, A. G. W.: 1970, 'Neutron Stars', *Ann. Rev. Astron. Astron. Astrophys.* **8**, 179.
Canuto, V.: 1974, 'Equation of State at Ultrahigh Densities', in *Ann. Rev. Astron. Astrophys.* **12**, 168.
Chiu, H. Y.: 1968, *Stellar Physics* (Mainly Chapter 3), Blaisdell, Waltham, Mass. [To read with extreme caution because of very many misprints!]
Misner, C. W. *et al.*: 1973, *Gravitation* (Ch. 24 and 36.4), Freeman, San Francisco.
Ruderman, M.: 1972, 'Pulsars: Structure and Dynamics', *Ann. Rev. Astron. Astrophys.* **10**, 427.

B. *Articles*

Baym, G., Pethick, C., and Sutherland, P. P.: 1971, *Astrophys. J.* **170**, 299.
Cameron, A. G. W. and Cohen, J.: 1969. 'The Instability of Stellar Structures Intermediate between White Dwarfs and Neutron Stars', *Astrophys. Letters* **3**, 3.
Langer, W. D. and Cameron, A. G. W.: 1969, 'An Equation of State at Subnuclear Densities', *Astrophys. Space Sci.* **5**, 259.
Bethe, H., Börner, G., and Sato, K.: 1970, 'Nuclei in Neutron Matter', *Astron. Astrophys.* **7**, 279.

N.B. – Note, as a source of possible confusion, that in Canuto, p. 177; Zel'dovich *et al.*, Ch. 8, 2; Salpeter; Cox *et al.*, p. 822; Chiu, p. 120, the symbol x denotes our x_F or, more specifically, our $x_{F,e}$ but in Canuto, p. 182; Börner, p. 15; Bether, Börner, Sato, p. 261 the symbol x denotes our μ_e' (or Z/A), whereas in Cameron, p. 314; Salpeter, p. 670, the symbol μ denotes our $(\mu_e')^{-1}$. Conversely, in Canuto, p. 170; Börner, p. 15; Bethe, Börner, Sato, p. 281, the symbol μ_e represents our $\epsilon_{F,e}$.

CHAPTER 7

PULSARS

1. The Discovery of Pulsars

1.1. A PARTLY CONTESTED NOBEL PRIZE

The 1974 Nobel Prize in physics was awarded to Martin Ryle, the leader of the Cambridge (G.B.) radio astronomy team, and Antony Hewish, under whose supervision pulsars were discovered.

Martin Ryle was not directly involved in the discovery of pulsars and he was awarded the Nobel Prize mainly as the creator of the Cambridge Radio Observatory and, more specifically, as creator of radio telescopes of large resolving power.

Hewish's citation was 'for his decisive role in the discovery of pulsars'. This somewhat vague qualification hides (according to the very thorough and fascinating report of Wade (1975)) the contribution of Miss J. Bell (now Mrs Burnell) a graduate student working in the Cambridge group.

Some astronomers have described the award of the prize to Hewish without any mention of Miss Bell's name as a scandal. Indeed, as stated later by Manchester and Taylor (1977)*, in the dedication of their monograph on pulsars, 'without her perceptiveness and persistence the discovery of pulsars might have been retarded for an indefinite time'.

1.2. AN 'IMPOSSIBLE' PHENOMENON

In order fully to appreciate Miss Bell's contribution to the discovery of pulsars, it can be useful to show how unexpected was this phenomenon for most competent astronomers.

For this purpose, let us indulge in a long quotation of Sir Bernard Lovell (1973) who described, with a typically British humour, some of the first reactions to the announcement of this remarkable discovery in his book *Out of the Zenith*. His comments merit particular attention because the discovery of pulsars was made at the rival Cambridge observatory and not at the radio astronomical observatory of Jodrell Bank (Manchester, G.B.) created by Lovell.

"The discovery of pulsars came to me and my colleagues [Lovell writes] as an abrupt and almost incredible surprise."

On 21 February 1968, at a meeting of the British Science Research Council, Lovell met Fred Hoyle who had returned from America a few days earlier.

* This excellent monograph will be quoted very often below, both for observational data and for the corresponding diagrams. For the sake of brevity it will be indicated by the abbreviation MT, followed by the number of the quoted page.

"Any news?" Lovell whispered, meaning was there any fresh news about quasar identification.

"Not much", Hoyle replied. Then, almost as an afterthought, he added, "but last night, at a colloquium in Cambridge, Hewish said that he had discovered some radio sources which emitted in pulses with intervals of about a second."

"Impossible", Lovell said, "he must be picking up some odd form of interference."

"No," Hoyle added, "the evidence for extraterrestrial origin seems convincing."

Then Hoyle told him that Hewish had been investigating these pulsed emissions for months, that four had been discovered and that their motion in the sky and other characteristics pointed conclusively to an origin in the Milky Way. An account in *Nature* was to be published on 24 February 1968.

"On the journey home [Lovell writes] I could think of little but this fantastic news. Some of my people at Jodrell must surely have known about this. Why hadn't I been told and why weren't we checking these objects with the telescope? In fact, it turned out that they had no information at all."

The truth is that Hewish and the whole Cambridge group had, for several months, achieved a screen of security and secrecy which, in itself, was almost as much of an accomplishment as the discovery itself.

"The story of the discovery of the pulsating radio sources [adds Lovell] is like the story of the discovery of the radio emissions from space and of many other events in science, one of an accidental and unintentional observation. The research programme which brought evidence of their existence to human eyes was not even the main research interest of Martin Ryle and his group."

Actually, the discovery of pulsars came about as an unexpected result of putting into operation a large radio telescope array designed to the study of the phenomenon of the '*interplanetary scintillation*' of compact radio sources, discovered by Hewish and his group in 1964.

This mention of interplanetary scintillation deserves a short historical and physical comment. When a radio telescope receiving signals in the meter wave band is used to record the radio emission from sources of small angular diameter such as distant radio galaxies or quasars, the record obtained is normally steady.

However, non-correlated fluctuations in intensity, with time scales of the order of several seconds or minutes, are observed when two receivers, separated by a large distance, are used. It was soon realized that these variable signals were observed because the *steady* emission from a source of small angular diameter was *diffracted* by the irregularities of the ionization in the upper regions of the Earth's ionosphere (the variability of the brightness of quasars, on a much larger time scale, is not involved here).

This effect is analogous to the twinkling of an ordinary star. In the case of optical scintillation the diffraction of the light is produced by irregularities in the density of relatively low atmospheric layers. The diffraction of radio waves in the ionosphere takes place at an altitude of about 200 km. As in the optical domain, only radio sources of sufficiently small angular diameter show this effect: the planets do not scintillate.

This interplanetary scintillation represents a somewhat similar effect. Only radio sources of small angular diameter 'twinkle' but when the time scale of this scintillation is only *of the order of seconds* the fluctuations arise when the radio waves from the source traverse *the irregularities of ionized interplanetary space.*

It was soon realized that this discovery provided a new method for setting upper limits to the angular diameters of the radio galaxies and quasars (which had not yet been resolved by the 'long baseline intercontinental interferometry') and had several other interesting applications.

The telescope constructed by Hewish and his group, for that purpose, was operating at 81.5 MHz ($\approx$ 3.7 m) because the plasma effects involved were more pronounced at longer wavelengths. To attain the maximum of sensitivity the telescope had a large collecting area and consisted of a rectangular array of 2048 dipoles spread over 4.5 acres (see the photographic picture of this telescope, and all technical details in MT). The construction started in March 1965, and was completed in 1967. Routine recordings began in July of that year.

1.3. MISS BELL'S CONTRIBUTION

Now we revert to Wade's report, already mentioned above. Having helped to build the telescope, it was Miss Bell's task to operate it single-handed and analyze its data until she had enough material for a thesis.

This analysis was rather arduous. The telescope churned out about 30 m of three-track paper every day. It took four days to cover the sky, so Miss Bell had about 130 m of paper chart to analyze for each complete coverage of the sky.

Her job was to scan the chart by eye, mapping the signals that were true twinkling radio sources and discarding many man-made sources of interference.

By October 1967, Miss Bell was 300 m of chart behind and, by the end of November, she was lagging a third of a mile. It was in October 1967 that she discovered the first pulsar.

Its signal, which she describes as 'a bit of scruff' occupied about 1 cm of the 130 m of chart (see, for instance, Figure 1.2 in MT). "The first thing I noticed," Miss Bell says, "was that sometimes within the record there were signals that I could not quite classify . . . I began to remember that I had seen this particular bit of scruff before, and from the same part of the sky. It seemed to be keeping pace with 23 hours 56 minutes, i.e. with the rotation of the stars."

As stated by Wade, *the nub of the discovery lay in that single instant of recognition.*

Again, in Miss Bell's words: "When it clicked that I had seen it before I did a double click. I remembered I had seen it from the same part of the sky before. This bit of scruff was something I didn't completely understand – my brain just hung on to it and I remembered I had seen it before."

She discussed the signals with Hewish and they jointly decided to look at them on the observatory's fast recorder, so as to get a clearer picture of the signal's structure.

But when this recorder, occupied with another task, became available, in mid-November, 1967, *for several weeks nothing happened.* The signal, at all times variable, failed to show at all.

"Hewish was thinking," Miss Bell says, "at that stage that it was a flare star and that we had missed it." (Don't forget that any other interpretation was 'impossible' for a competent astronomer!)

"Finally one day I managed to catch it, and I got a series of pulses coming out of the recorder. They were $1\frac{1}{3}$ seconds apart. That is a very sort of man-made record. Tony

Hewish had left the recording to me. I phoned him up to tell him about the pulses and he said: 'Oh that settles it, it must be man-made'. . ."

In some way Hewish was right (and victim of his competence in astrophysics): the fastest variable star then known had a period of *one third of a day*, and no one conceived of a star with a period of the order of *one second.*

The day before she left Cambridge for her Christmas holiday, she discovered a second 'source'.

"I saw," Miss Bell says, "something which looked remarkably like the bits of scruff we had been working with. . . That particular bit of sky was due to go through the beam at 1 a.m. It was a very cold night, and the telescope doesn't perform very well in cold weather. I breathed hot air on it, I kicked and swore at it, and I got it to work for just 5 minutes. It was the right 5 minutes, and at the right setting. The source gave a train of pulses, but with a different period, of about $1\frac{1}{4}$ seconds."

Hewish himself did a recording in the middle of the night in mid-January 1968 and confirmed the second source of pulses.

"That removed," Miss Bell says, "the worry about little green men, since there wouldn't be two lots signalling us at different frequencies. So obviously we were dealing with some sort of very rapid star. I threw up another two sometime in January 1968."

As rightly emphasized by Ward, Miss Bell seems to have been less ready than her elders to doubt that her 'bit of scruff' was a real signal from a real star. However, she confesses that she 'joined in the business of trying to explain it away as wholeheartedly as everybody else'.

1.4. UNEXPECTED BUT NOT ACCIDENTAL DISCOVERY

Thus we see that the discovery of pulsars was neither 'planned' nor purely 'accidental'.

Miss Bell's contribution has not been as routine as a hired scanner's. According to Thomas Gold, quoted by Ward, Miss Bell, unlike a photo scanner, understood the basis of what she was doing and, moreover, had not been told to look for pulsars: "She was told to plot scintillating radio sources, but she noted and pursued in her own way a different kind of signal (Gold remarks)."

On the whole, as Ward states it, the discovery of pulsars was a feat of unaided memory and observation, not the result of following a preset routine; in turning up the pulsars she was going beyond the precise letter of her instructions, even if acting within their general framework.

As to the contribution of Hewish, apart from the conception and construction of a rather special radio telescrope, the most important seems to have been the use of receivers having very short time constant of about 0.1 s.

As noted by Manchester and Taylor (1977), radio telescopes with sensitivity adequate for detecting pulsars had been in existence since the 1950s. However, because rapid time variations in emission from celestial sources were unknown (except for those within the solar system), receivers and recording devices were generally equipped with time constants of several seconds to smooth random noise fluctuations. The *mean* flux level from most pulsars is quite low, well below the detection threshold of early surveys made with such systems.

Finally, the most difficult was to prove that the strange, sporadic, signals observed by Miss Bell, had been emitted by natural astronomical objects.

Man-made signals transmitted from space probes or reflected from the moon or planets were finally ruled out because the absence of any parallax greater than about 2 arcmin showed that the source lay outside the solar system.

The possibility that the signals might have been transmitted by an extra-terrestrial civilization was briefly entertained. But when three more similar pulsating sources were detected, it become clear, as reported above by Miss Bell, that the sources had to be natural phenomena.

When the discovery of the first pulsar was published, astronomers everywhere were amazed.

"The idea that celestial sources could produce pulses of energy required a fundamental reorientation of thought," concludes Lovell.

Many well written and easily accessible reports dispense us to give more details about the discovery of pulsars. We refer the interested reader to Hewish's introduction in the reprint of the 115 papers published in *Nature* in 1968 and 1969 on pulsars (under the title *Pulsating Stars*), to Ward (1975) and to MT.

2. The First Investigations and the First General Properties

2.1. THE 'FIRST PULSAR', THE CRAB PULSAR AND THE VELA PULSAR

2.1.1. The Spinning Neutron Star Model

After a short period of theoretical interpretation of the pulsar phenomenon, as due to radial pulsations of white dwarfs, to orbital motions of white dwarfs or neutron stars, the phenomenon was attributed by Gold (1968) (following a suggestion by Pacini (1967)) to *rapid spinning of a neutron star*. The radiation observed as pulses is supposed to be confined to a single (or multiple) narrow beam, so that the pulse repetition period P is representing the *rotation period of the neutron star*.

2.1.2. The Names of Pulsars

The terminology used to define different pulsars is the following.

Formerly, the names of pulsars were based on a combination of their approximate position with the abbreviated reference to observatories where they were found.

Today, the pulsars are denoted by the prefix PSR (an obvious abbreviation of the word 'pulsar'). This prefix is followed by a four-digit number indicating right ascension (in 1950.0 coordinates) but, in addition, a sign and two digits indicate degrees of declination. When further resolution is required, declination is given in tenths of a degree by adding another digit.

Thus, for instance, the 'first pulsar', initially called CP 1919+21 (C for Cambridge, P for pulsar), is today called PSR 1919+21. And two other pulsars are distinguished by writing PSR 1913+16 for the first (remarkable by a very short period $P = 0.0590$ s) and PSR 1913+167 for the second, in order to distinguish it from the first.

2.1.3. *The 'Constancy' of a Slowly Increasing Period*

In their very first paper on pulsars, Hewish *et al.* (1968) announced that after suitable corrections taking into account the Doppler effect due to the orbital motion of the Earth, the pulse repetition period P_0 of the 'first pulsar' PSR 1919+21 was equal to

$$P_0 = 1.337\,279\,5 \pm 0.000\,002\,0\ \mathrm{s}, \tag{1}$$

and remained stable at this approximation, during an observation interval of about 15 days.

This result soon appeared as slightly incorrect. Indeed, two independent determinations, at Parkes (Australia) and at Arecibo (Puerto Rico), yielded a more precise period:

$$P_0 = 1.337\,301\,09 \pm 0.000\,000\,07\ \mathrm{s}, \tag{2}$$

showing that Hewish and his collaborators had made an error of (only) *one unit* in the count of 64 000 pulses observed in two weeks!

On the other hand, and more important was the fact that, in the meantime, it also appeared that Hewish *et al.* were also slightly wrong in stating that P_0 remains vigorously constant, because their observations covered a too short period.

Accurate observations extending over several months disclosed later that the periods of *all* pulsars were *regularly increasing.*

Observed rates of change are typically of about 10^{-15} seconds per second, i.e. of a few tens of nanoseconds per year. Thus, for instance, the period of the 'Crab pulsar' (considered in the next subsection) is lengthening at a rate of about one part in 2000 per year.

According to the model of a rotating neutron star, the gradual increase of P can be interpreted as due to a *slowing-down* of the rotation of the star responsible for pulses.

2.1.4. *Vela Pulsar and Crab Pulsar Discovered*

In late 1968, the Mongolo observatory (Australia) announced the discovery of a pulsar PSR 0833–45, now called the '*Vela pulsar*', located near the 'center' of the supernova remnant Vela X. Its period was very short (of only 0.0892 s).

The same year, Staelin and Reifenstein (1968) announced the detection, at Green Bank Observatory (U.S.A.), of two other pulsars, one with an even shorter period of 0.0331 s (the shortest of all pulsars known today, according to MT), and another with a very long period of 3.7454 s, both near the 'center' of the Crab Nebula, the most widely studied of all supernova remnants. These two pulsars are PSR 0531+21 (now called the '*Crab pulsar*') and PSR 0525+21.

It is well known that the Crab Nebula (NGC 1952) represents a diffuse emission nebula situated rather near the galactic plane (its galactic latitude b is about $-6°$) in a direction nearly opposed to the direction of the galactic center. Its distance, according to a determination of Trimble (1968) is of the order of 2 kpc. The 'center' of the Nebula – a rather vague concept because of the irregular form of this object – is situated approximately at $\alpha_{1950} = 05^{\mathrm{h}}31^{\mathrm{m}}30^{\mathrm{s}}$, $\delta_{1950} = +21°58'$.

The two pulsars found by Staelin and Reifenstein have (according to more recent measures based on the optical identification, see Subsection 2.2.1) practically the same declination, viz. $\delta_{1950} = +21°58'54''.8$ for the pulsar with the period of 0.0331 s and

$\delta_{1950} = +21°58'18''$ for the pulsar with the period of 3.7454 s. However, their right ascensions are respectively $\alpha_{1950} = 05^h31^m31\overset{s}{.}46$ and $\alpha_{1950} = 05^h25^m52^s$.

Thus we understand why only the first one is called the Crab pulsar, since it is the only one which corresponds to the center of the Crab Nebula (and, as will be shown later, the only one genetically bound to this nebula). Note that the Crab pulsar is often denoted by its provisional name: NP 0532.

2.1.5. Association of Some Pulsars with Supernova Remnants

Long before the discovery of pulsars it has been suggested by Baade and Zwicky (1934) that neutron stars would be formed as residues of supernova explosions.

The association of the Crab pulsar and of the Vela pulsar having the shortest periods (0.0331 and 0.0892 s), known at the time, with two supernova remnants gave support to this suggestion, in accordance with Gold's interpretation of the pulsar phenomenon as due to a rotating neutron star. Moreover, it fulfilled predictions made by Wheeler (1966) and Pacini (1967) before the discovery of pulsars, namely, that the energy source in the Crab Nebula could be a rotating neutron star (see Subsection 5.3.).

On associating the birth of some pulsars with known dates of the corresponding supernova explosion, it became possible to find the 'age' of these pulsars and to confirm the 'age' deduced from the rate of change of the period (the latter method will be described in Section 4).

2.1.6. The Glitches

A little later another sensational observation created great interest. In March, 1969, a large, and apparently discontinuous, decrease in the period of the Vela pulsar was detected. And since that time, several similar events have been observed for this pulsar. Thus, e.g. between June 21 and July 14 1978, one had $\Delta P/P \approx (-3.05) \times 10^{-6}$, with $\Delta\dot{P}/\dot{P} = (+8.8 \pm 0.1) \times 10^{-3}$, and later observations confirmed this jump.

Similarly, several discontinuities have been observed in the period of the Crab pulsar. (See, e.g., Figure 6-4 in MT, p. 114, which illustrates these discontinuous steps in the regular increase of the period of the Vela pulsar.)

These period discontinuities are often called the *'glitches'*. In each of the three events observed from 1968 to 1976 for the Vela pulsar, the period decreased by about 200 ns, a large change compared to the regular rate of increase of about 11 ns per day. In several cases the period decrease occurred between successive observations separated by about a week.

In each of these events the decrease in period was accompanied by an *increase* in $\dot{P}$ (as in June–July 1978). However, this increase in derivative decayed away after about a year, leaving the derivative close to its original value.

Shortly after the first Vela speed-up Baym *et al.*, (1969) proposed that a sudden cracking of the neutron star crust (a kind of 'starquake') results in a decrease of the moment of inertia and hence an increase in the angular velocity of rotation of the pulsar.

The starquake model seems however unable to account for the relatively short intervals between the Vela speed-ups. Other models have been proposed (*corequakes* and more sophisticated models) excellently summarized with recent references in MT, pp. 113–121, and, for less recent data, by Ruderman (1972).

2.2. MULTIFREQUENCY OBSERVATIONS AND THE CONCEPT OF 'MULTIPULSE'

2.2.1. The Optical Pulses. Identification of the Crab Pulsar

Many efforts were made after the discovery of pulsars to attempt to correlate radio astronomical observation with phenomena in other domains of frequency: optical, X, γ, etc.

After first deceptions, many astrophysicists became sceptical about the possibility of finding such correlations. Thus, in the Preface to the first volume of *'Pulsating Stars'*, F. G. Smith (1968) could write: "It would, of course, have been encouraging to find a visible object in one of the accurately determined positions, but the chance of finding the type of the pulsating stars through their visible spectra now seems very slight. As compensation, it may become that *the star is one which cannot be reached optically* [our italics], so that the radio pulses offer the only means of detection of an object the physical processes of which are of great interest."

However, in contradiction with these pessimistic views, as early as 16 January 1969 an optical observation of pulses from the Crab pulsar, with a period of 0.0331 s, was made at the Steward Observatory (U.S.A.) and confirmed a few days later by McDonald at Kitt Peak Observatories (U.S.A.).

The exact coincidence of the periods and the approximate coincidence of the positions of the sources of the optical and radio pulses, left no doubt that a definite 'identification' of a pulsar with an optical object was achieved.

Figure 7.1 shows the optical pulses, obtained by two different teams, whose form is very similar to that of radio pulses (with slight differences shown by Figure 7.2.)

A little later, extraordinary stroboscopic television photographs were obtained by chopping the light beam from the telescope by means of a rotating disk so that the 'open' periods were separated by a time close to that of a period of the optical flashes. The photographs obtained at Lick Observatory (U.S.A.) on 3 February 1969 of the optical pulses from the Crab pulsar were initially published by Miller and Wapler (1969). They are reproduced in the April 1969 issue of *Sky and Telescope*, in Lovell (1973, p. 88) and in MT, p. 69, so that we refrain from reproducing them once more.

One of these photographs shows, in the region near the center of the Crab Nebula, a stroboscopic selection of the maximum phase of optical brightness of the pulsar. It appears, by integration of light, as an ordinary star near normal ones. The other photograph of the same region, but corresponding to a stroboscopic selection of the *minimum* ('off') phase of the optical brightness, shows the disappearance of the pulsar image observable on the first picture, still leaving the images of the normal stars.

One immediately realizes that an optical identification of this kind yields a very precise position of the pulsar. A precise position is less accessible to radio telescopes, because of the great length of radio waves. We also encounter this fact in Chapter 13 dealing with quasars.

The extraordinary feature of the optical identification of the Crab pulsar was that the

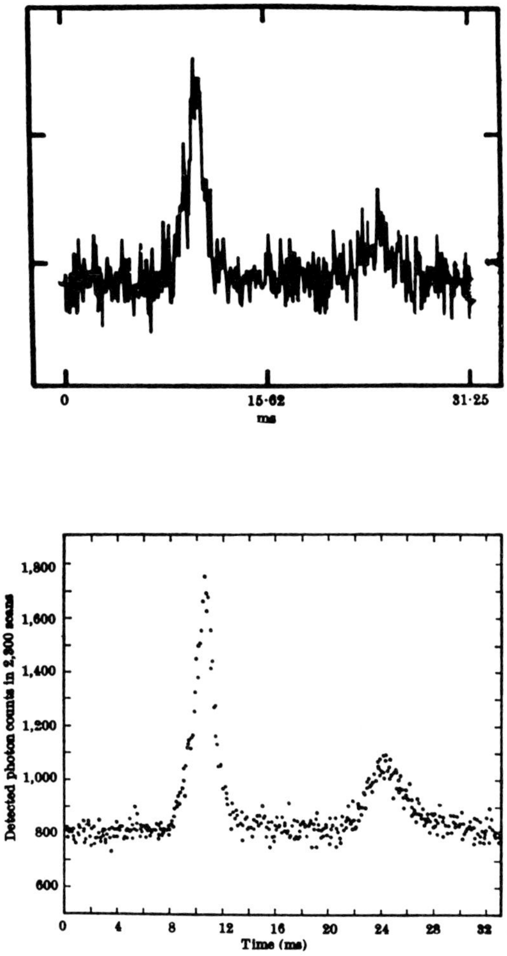

Fig. 7.1. Upper figure: the first optical pulse observed on the recording screen of the 36-inch telescope at the Steward Observatory (U.S.A.), from the Crab pulsar. The amplitude scale is arbitrary and represents the superposition of about 300 optical pulses. [From Cocke *et al.* (1969).] Lower figure: the effect of summing about 12 000 optical pulses from the Crab pulsar. [From Nather *et al.* (1969).] Note the structure of the pulse (similar on all frequencies) a *main* pulse followed by an '*interpulse*' inside the period of 0.0331 s.

light flashes coincided with the position of the star which, 27 years earlier, Baade and Minkowski had suspected to be the remnant of the supernova explosion producing the diffuse nebula. Apparently, the major part of the light responsible for the ordinary photographic image of this star is actually *the integrated effect of the optical flashes of the pulsar*.

As it often occurs in science, and more particularly in astronomy, Cambridge astronomers, who also were searching for optical pulses from the Crab pulsar, (but perhaps not very actively, if we judge by Smith's comments quoted above) have registered as early as

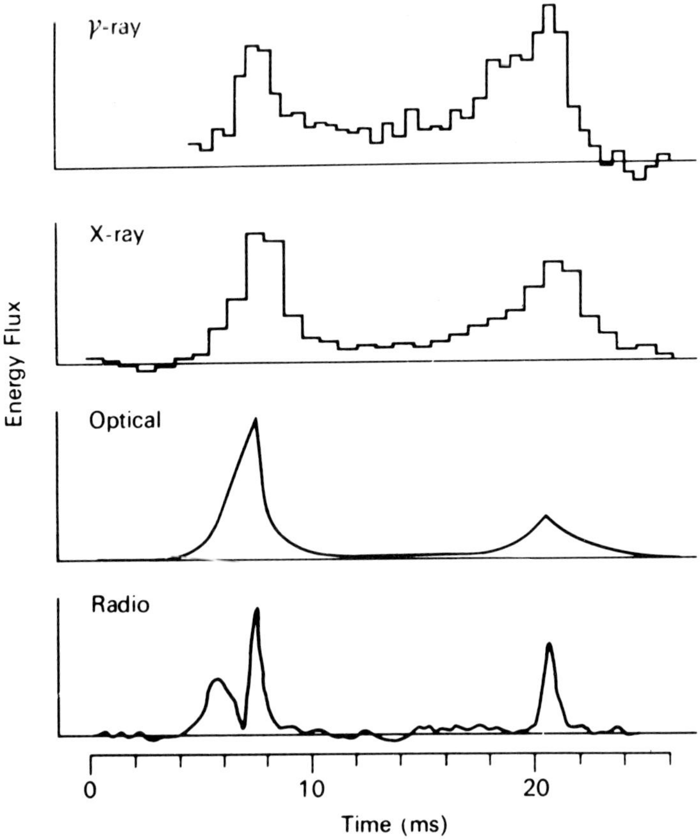

Fig. 7.2. Integrated profile shapes of the Crab pulsar from radio (410 MHz) to γ-ray frequencies. The radio frequency pulse components are, from the left, the 'precursor', the main pulse and the 'interpulse'. Only the main pulse and interpulse are present at higher frequencies. [From *Pulsars*, p.70, by R. N. Manchester and J. H. Taylor. Copyright 1977 by W. H. Freeman and Company. Reproduced by permission of the authors and the publishers]

24 November 1968, i.e. almost two months before the Americans, such flashes, but did not analyze the data previously. Willstrop, 1969, from Cambridge observatory, explains:

"Because no flashes had been detected from the many pulsars studied at different observatories with sensitive equipment and because the radio position available on 24 November was uncertain." This reminds us that the director of the Cambridge Observatory didn't trust enough the theoretical computations of W. Adams to look for Neptune and that about a similar misadventure occurred to LeVerrier at the Paris Observatory, before the observational discovery of Neptune, in 1846, at Berlin.

2.2.2. X-Ray and γ-Ray Pulses

However, 'the exciting saga about the Crab pulsar', as Lovell says, does not end with the discovery of optical pulses. Between 1969 and 1971, several rockets and balloons launched in U.S.A., equipped with X-ray and γ-ray detectors registered X-ray and γ-ray pulsations of the same period of 0.0331 s, in phase with optical and radio pulses. (See also Chap. 9 on X-ray pulses).

Figure 7.2 shows the remarkable correspondence of the main phase for all these pulses (for the corresponding references see MT, pp. 68–76).

The case of the Vela pulsar (PSR 0833–45) is very different from that of the Crab pulsar.

Indeed, optical pulsations of the Vela pulsar have been detected only quite recently, in 1977, at the same time as it was identified with a faint blue star (the magnitude of the pulsed emission is of the order of 24).

As the Crab pulsar, the Vela pulsar has been observed at γ-frequencies, but not yet positively in X-rays. Apparently in X-rays the pulsations of the Vela pulsar are $\sim 10^4$ times weaker than the Crab's, although the γ-ray pulsed energy of the two objects is roughly the same.

Another major difference between the Crab pulsar and the Vela pulsar is that the remarkable correspondence of phases shown by Figure 7.2 for the Crab, does not exist for the radio, optical and γ-ray pulses of the Vela pulsar.

On the other hand, the Crab and the Vela pulsars are for the moment (November 1978) the only *radio* pulsars observed *optically*. (This is not true for all pulsars: in Chapter 9 we will see that some of X-ray sources represent X-ray pulsars, and many of these do exhibit optical pulsations.)

As for the Crab Nebula and the Crab pulsar, the 'Vela pulsar source' contains both pulsed and steady components. The latter include the surrounding supernova remnant and a compact X-ray source.

N.B. One must take care to avoid confusion between the Vela pulsar and the X-ray source called Vela X-1.

2.2.3. The Concept of Multipulse

Multifrequency observations of pulsars show that the term '*pulse*' used to describe pulsar radiation is very unfortunate and *exceedingly misleading*, not only because the so-called 'pulsating stars' are actually *rotating* and not pulsating, but also because in physics this term is already used with a quite *different and precise meaning*.

Indeed, (see for instance Crawford, Jr. (1968) Sec. 6.3 of Vol. III of Berkeley Physics Course) in physics a '*pulse*' is a signal carried by an electromagnetic wave of *a single frequency* ν_0, whose amplitude is *adequately modulated*.

The use of the term 'pulse' for pulsar flashes is particularly misleading in connection with application of the physical theory of 'wave packets' to be developed in Subsection 3.1.3 in relation with one of the methods of measurement of the distance of pulsars.

Of course, as will be recalled in Subsection 3.1.1, the Fourier transform describing a physical pulse is a function $f(\nu)$ which only presents a maximum for $\nu = \nu_0$, and does not vanish for frequencies ν in a more or less wide range on both sides of this 'average' (or 'dominant') frequency ν_0.

Contrary to pulsar flashes, a true 'physical pulse' is, for instance, at work in radar equipment: a klystron working on a well defined *single* frequency is delivering physical pulses of several milliseconds. Similarly a commercial AM radio receiver must be tuned on a *single* frequency ν_0 corresponding to a single '*carrier frequency*' assigned to the emitting station, in the range between about 0.5 and 1.6 MHz.

In absolute contradiction with all these examples the 'pulsar pulses' are observed with equipment working on transmission frequencies ('carrier frequencies') extending from radio to γ-ray frequencies. Each flash of a pulsar thus corresponds to a continuous set of commercial radio stations working on different 'carrier frequencies'.

Thus, the so-called pulsar 'pulses' actually represent from the physical point of view a phenomenon which we propose to call a '*multipulse*', in order to emphasize that we deal in this case with a source which, like an ordinary star, emits *simultaneously* radiation over a wide range of frequencies, which form a continuum from at least 30 MHz to about 10^{14} MHz.

The only difference with the radiation of an ordinary star is that each 'component' of this multipulse has *a Fourier transform of a particular form*, which can be denoted by $f(\nu, \nu_0)$ in order to emphasize that each of these components can have a (slightly) different form according to the value of ν_0, as illustrated by Figure 7.2. The corresponding law of variation of intensity as a function of time depending both on the form of the function f and on the value of ν_0.

The most widely accepted theoretical model for a pulsar, that of the rotating neutron star emitting radiation in a relatively narrow beam (relatively fixed in orientation with respect to the star) explains the identity of the pulse repetition period P observed at different 'carrier frequencies' composing the multipulse.

Of course, as already stated above, this identity of P does not imply an absolute equality in shapes of intensity or polarization as a function of time of different frequency components of a multipulse of a given pulsar.

From the point of view of a simultaneous presence of different frequencies in a natural transient phenomenon, one can compare the multipulse of a pulsar to the atmospheric discharge producing *a lightning*. Here, also, we deal with a phenomenon of short duration associated with a continuous set of frequencies (from 500 to 10 000 MHz) observed, after the dispersion in ionosphere, as 'whistlers'. Of course, this analogy does not imply any similarity in emission mechanisms. Besides, in spite of various theoretical models hitherto proposed (see, for instance, MT, Ch. 10 and, on a more advanced level, Guinzburg and Zheleznyakov (1975) the emission mechanism of pulsar multipulses is still unknown.

3. Pulsar Distances

3.1. Introduction

In all discussions concerning pulsars it is very important to be able to derive their 'monochromatic' luminosities from their brightness observed in different domains of frequency. This implies the knowledge of their distances.

One of the most interesting and fruitful (though not the most precise) methods of determination of pulsar distances is based on the theoretical interpretation of the observed effects of interstellar scattering on pulsar's multipulse.

These effects involve both the theory of dispersion of electromagnetic waves in an ionized plasma and the theory of propagation of 'wave packets'.

The almost perfect presentation of both theories in Berkeley Physics Course, Vol. III

(Section 4.3, example 7; and Sections 6.1 to 6.4), i.e. in '*Waves*', by F. S. Crawford, Jr. (1968) supplemented by L. Brillouin's '*Wave Propagation and Group Velocity*' (1960) provide a sufficient preliminary knowledge of these theories. However, for the convenience of the reader we present below (in Subsections 3.1.1, 3.1.2 and 3.1.3) a short elementary summary of these texts. Advanced readers can start with Subsection 3.2.

3.1.1. The Physical Pulses

Consider a mechanical oscillating system, such as a simple pendulum. Its motion depends only on a single parameter x, representing the angular or the linear '*displacement*', in a given plane, from the equilibrium position.

The maximum displacement of the pendulum is supposed *to decrease gradually*, as in the case of oscillations damped by friction, before the state of harmonic 'small oscillations' is reached.

These anharmonic oscillations will be described by a *non-rigorously periodic* function $x = f(t)$ of time t.

However, it will be possible to generalize the concept of *amplitude* and of *period* for this motion and to define a 'pseudo-amplitude' and a 'pseudo-period' in the following way.

The absolute value of *successive* 'positive' and 'negative' maximum deflections $|x_{\text{max}}|$, numbered $1, 2, 3, \ldots, n$, form a discrete set $|x_{\text{max}}|_n$, which can be interpolated by a continuous function $A(t)$ of time, defining the variable '*pseudo-amplitude*' of oscillation.

Similarly we can define two variable '*pseudo-periods*' (very near each other), either $T(t)$ by continuous interpolation between the double of the intervals of time between two successive 'zeros' of $x(t)$, or $T'(t)$ by continuous interpolation between the double of the intervals of time between two successive $|x_{\text{max}}|$.

Thus, the plot of A or of $(-A)$ *versus* t will appear as a kind of *envelope* of the plot of x *versus* t, joining the 'summits' or the 'dips' of $x(t)$.

The diagram describing the function $x(t)$, i.e. the *kinematical properties* of the considered motion, has some geometrical form, and by extension (somewhat abusively) it is usual to speak about the 'form of the oscillation'. (In the case of 'local oscillations' considered for the moment, with no propagation, the concept of wave has not yet any place!)

Now, one can encounter in physics, and more particularly in connection with electromagnetic systems, a phenomenon represented *as a whole* by a form of oscillation called a (physical) '*pulse*'.

Generally, a pulse is characterized by an oscillation of a short overall duration of the non-negligible part of its rapidly decreasing, more or less regular, pseudo-amplitude.

The corresponding form of $x(t)$ can be very complicated. However, different theoretical simple 'models of pulse' are easily described by the form of their *Fourier transform*. The easiest to analyse, even if its does not represent the most typical physical pulse concerning the function $A(t)$, is the pulse corresponding to a 'square' (actually rectangular) Fourier transform.

3.1.1.1. The Pulse Corresponding to a 'Square' Spectrum. Let $\omega = 2\pi\nu$, measured in radians per second, denote the '*angular frequency*'. A plot of the Fourier transform *versus* ω is called the '*frequency spectrum*'.

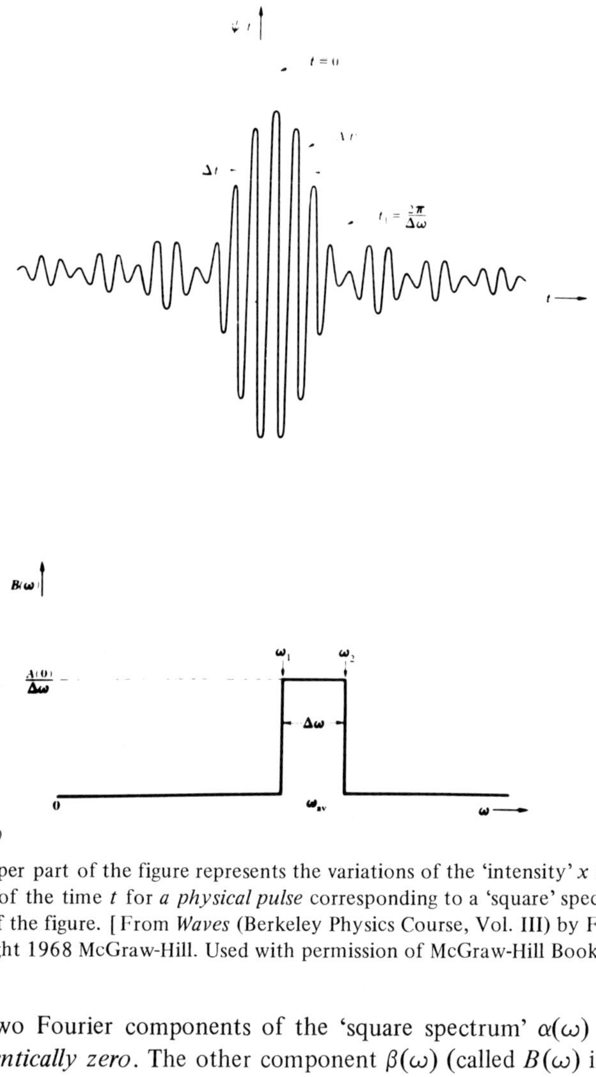

Fig. 7.3. The upper part of the figure represents the variations of the 'intensity' x of a phenomenon as a function $\psi(t)$ of the time t for *a physical pulse* corresponding to a 'square' spectrum represented in the lower part of the figure. [From *Waves* (Berkeley Physics Course, Vol. III) by F. S. Crawford, Jr. Copyright 1968 McGraw-Hill. Used with permission of McGraw-Hill Book Co.]

One of the two Fourier components of the 'square spectrum' $\alpha(\omega)$ (not shown in Figure 7.3) is *identically zero*. The other component $\beta(\omega)$ (called $B(\omega)$ in Figure 7.3) is 'flat', i.e. is *constant*, over a certain limited frequency band of width $\Delta\omega = 2(\delta\omega)$, between the frequencies ω_1 and ω_2, and it is zero elsewhere. (Thus $\delta\omega$ represents the *half-width* of the spectrum).

The corresponding form of the pulse itself, i.e. of $x(t)$ is represented by $\psi(t)$ in the upper part of Figure 7.3. It is characterized by a rigorously constant value T_0 of the pseudo-period T (thus, by a constant value of the corresponding frequency $\nu_0 = 1/T_0$) and by a pseudo-amplitude $A(t)$ corresponding to a rather rapidly 'damped sinusoid' (the 'envelope' of the plot of x *versus* t).

Analytically, this pulse is represented by the function $x(t)$ given by

$$x(t) = \int_{-\infty}^{+\infty} \alpha(\omega) \sin \omega t \, d\omega + \int_{-\infty}^{+\infty} \beta(\omega) \cos \omega t \, d\omega, \tag{3}$$

where $\alpha(\omega)$ and $\beta(\omega)$ are the Fourier transforms defined above for the 'square' spectrum. If β_0 denotes the constant value of $\beta(\omega)$ between ω_1 and ω_2, we successively find for $x(t)$:

$$x(t) = \beta_0 \int_{\omega_1}^{\omega_2} \cos \omega t \, d\omega = \frac{1}{t} \beta_0(\sin \omega_2 t - \sin \omega_1 t)$$

$$= \frac{2}{t} \beta_0 \sin \tfrac{1}{2}(\omega_2 - \omega_1) t \cos \tfrac{1}{2}(\omega_2 + \omega_1) t$$

$$= 2\beta_0 \delta\omega \left[\frac{\sin (t\,\delta\omega)}{t\,\delta\omega}\right] \cos \omega_0 t = A(t) \cos \omega_0 t, \tag{4}$$

where ω_0 is the 'average' (or 'dominant') frequency defined by

$$\omega_0 = \tfrac{1}{2}(\omega_2 + \omega_1), \tag{5}$$

and $A(t)$ is the pseudo-amplitude given by

$$A(t) = A(0) \frac{\sin (t\,\delta\omega)}{t\,\delta\omega}, \tag{6}$$

with

$$A(0) = x(0) = 2\beta_0 \delta\omega, \tag{6'}$$

since for $t \to 0$ the square bracket in Equation (4) tends to 1.

This analytical representation corresponds indeed to the announced general properties of this model of pulse. The oscillations of $x(t)$ with a constant period T_0 are described by the variations of the term $\cos \omega_0 t$. This period T_0 can be very small for sufficiently large values of the 'dominant frequency' $\omega_0 = 2\pi\nu_0$. On the other hand, the decrease of the function $A(t)/A(0)$ can be very rapid for sufficiently large values of the 'half-width' $\delta\omega$ of the spectrum.

Indeed, $\delta\omega$ determines the time $t_1 = \pi/\delta\omega$ of the first zero of $A(t)$, t_1 itself equal to the temporal half-width $\delta t = (\pi/\delta\omega)$ of $A(t)/A(0)$ at the level corresponding to $(2/\pi) \approx 0.64$ of the maximum (equal to 1). Thus the rate of the decrease of the 'super envelope' $F(t) = 1/t\,\delta\omega$ of the envelope $A(t)$ of $x(t)$ is entirely determined by the value of $\delta\omega$.

Figure 7.4 (overpage) illustrates this influence of $\delta\omega$, on the rate of decrease of the pseudoamplitude $A(t)$. It shows how a *broadening* of the spectrum is correlated with a *narrowing* of the pulse (with respect to time). In quantum mechanics, this corresponds to the 'Principle of Heisenberg' (Heisenberg's Principle of Uncertainty).

3.1.1.2. The Pulse Corresponding to a Gaussian-Shaped Spectrum. A physical pulse more typical than the pulse corresponding to a 'square' spectrum is represented by a pulse corresponding to a 'Gaussian-shaped' spectrum, because its envelope $A(t)$ is deprived of sinusoidal oscillations (see the dotted curves in Figure 7.4).

It is defined by

$$\alpha(\omega) = 0; \qquad \beta(\omega) = \frac{1}{\pi} \exp\left[-\left(\frac{\omega - \omega_0}{\delta\omega}\right)^2\right], \tag{7}$$

where, obviously, $\delta\omega$ represents the half-width of the Gaussian curve representing $\beta(\omega)$ at a level corresponding to $\beta(\omega)/\beta(\omega_0) = \beta(\omega)/\beta_{\max} = 1/e$.

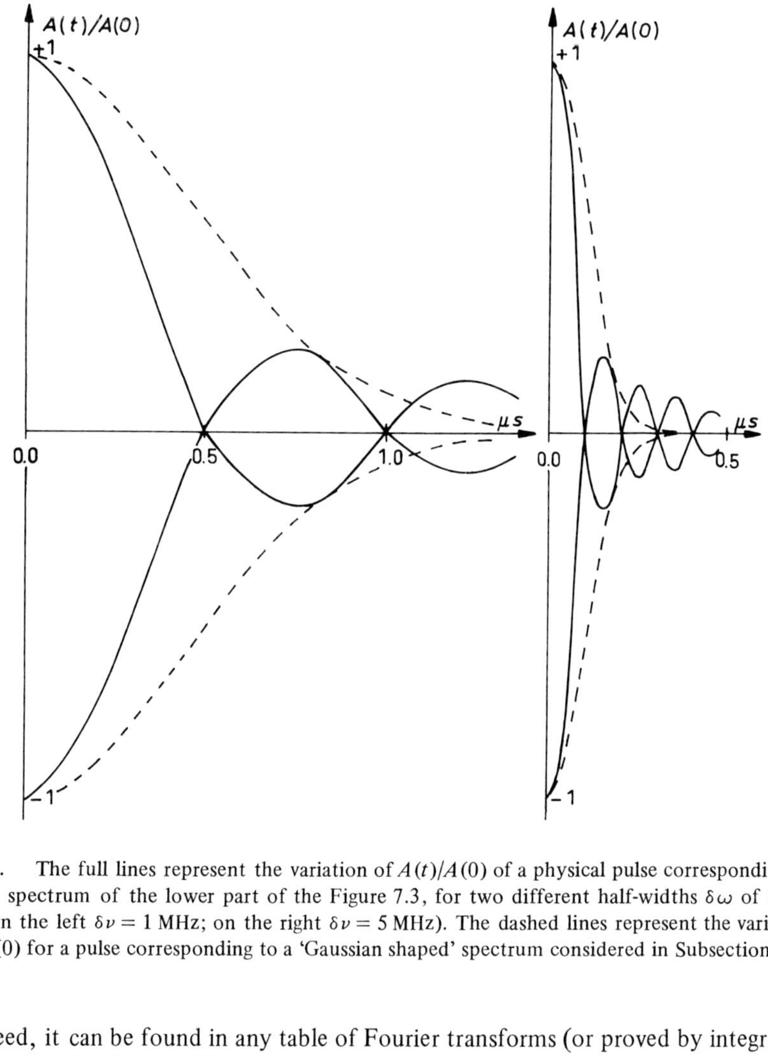

Fig. 7.4. The full lines represent the variation of $A(t)/A(0)$ of a physical pulse corresponding to the 'square' spectrum of the lower part of the Figure 7.3, for two different half-widths $\delta\omega$ of the spectrum (on the left $\delta\nu = 1$ MHz; on the right $\delta\nu = 5$ MHz). The dashed lines represent the variations of $A(t)/A(0)$ for a pulse corresponding to a 'Gaussian shaped' spectrum considered in Subsection 3.1.1.2.

Indeed, it can be found in any table of Fourier transforms (or proved by integration in the complex plane) that $x(t)$ corresponding to this spectrum is given by

$$x(t) = A(t) \cos \omega_0 t, \tag{8}$$

where

$$A(t) = A(0) \exp\left[-\left(\frac{t\delta\omega}{2}\right)^2\right]. \tag{9}$$

According to this expression of $A(t)$ the envelope of this pulse is itself of Gaussian shape, with a half-width (δt) equal to $2/\delta\omega$ at a level where $A(t)/A(0) = 1/e$.

Thus, as in the case of 'square' spectrum, the temporal half-width δt of the pulse is inversely proportional to the half-width $\delta\omega$ of its spectrum, in correspondence with Heisenberg's Principle. The dashed lines in Figure 7.4 show this narrowing of δt corresponding to the broadening of the spectrum.

3.1.2. Dispersion of Waves in a Plasma and Phase Velocity

Consider now, instead of local oscillations, the *propagation* of a monochromatic plane electromagnetic *wave* of angular frequency $\omega = 2\pi\nu$ in a macroscopically *homogeneous* and totally ionized *plasma*.

If we introduce a Cartesian reference system Oxyz such that the Oz axis represents the direction of propagation (normal to the plane of waves), the Ox axis can be chosen in such a way that the transverse linearly polarized oscillations of the electric field $\boldsymbol{E}$ are parallel to the axis Ox, and the oscillations of the magnetic field $\boldsymbol{B}$ are parallel to the axis Oy. In our discussion we shall consider only the propagation of $\boldsymbol{E}$.

The standard discussion of this propagation is given in Crawford (1968, Section 4.3, example 7), in Ginzburg (1970) and many other textbooks. We simply recall that the component E_x of $\boldsymbol{E}$ along the Ox axis (the only one in the case of linear polarization) at a point M of the Oz axis, depends on the coordinate z of M, on the time t, and, in a plasma on the angular frequency ω of the oscillations of $\boldsymbol{E}$, so that it can be denoted by $E(z, t; \omega)$. Thus E_x is a function of two variables z and t depending on the parameter ω.

When there is no static magnetic field imposed to the plasma, $E(z, t; \omega)$ is described by a function of the form:

$$E(z, t; \omega) = E_0 \cos(\omega t - kz), \tag{10}$$

where E_0 is a constant.

In Equation (10) k is a parameter which depends only on the electron density n_e of free electrons in the plasma. It is given by the *'dispersion equation'*

$$c^2k^2 = (\omega^2 - \omega_p^2), \tag{11}$$

where c is the speed of light in vacuum, and ω_p is defined (in Gaussian units) by

$$\omega_p^2 = 4\pi n_e e^2/m_e. \tag{12}$$

In Equation (12) e and m_e are the charge and the mass of the electron respectively, and n_e is expressed in cm^{-3}. For rapid evaluations of the order of magnitude of $\omega_p = 2\pi\nu_p$ the following approximate expression for ν_p in MHz (where n_e is again in cm^{-3}) can be useful:

$$\nu_p = 9 \times 10^{-3}(n_e)^{1/2}. \tag{13}$$

The physical meaning of k, called the *'angular wave number'*, is well known and very simple. If one considers *the whole* of the electric field (travelling as a sinusoidal wave) at a given moment t_0, the representation of $E(z, t_0; \omega)$ as a function of z, will have the form of a sinusoidal 'wave' of wavelength λ (λ is a kind of a 'space period'). The value of λ will be given by equation:

$$\omega t_0 - kz = [\omega t_0 - k(z + \lambda) + 2\pi], \tag{14}$$

hence

$$k = 2\pi/\lambda. \tag{15}$$

Thus k plays with respect to the 'space period' λ the same role as ω with respect to the 'time period' T. Another way of expressing the physical meaning of k consists to state that it represents the rate of increase of the 'phase angle', i.e. of the argument of cosines in Equation (10) per unit length.

At time t_0 and at the point $\text{M}(z_0)$ of coordinate z_0 this phase angle φ_0 is equal to

$(\omega t_0 - kz_0)$. The same value of E will reappear at the point $M(z_0 + \Delta z)$ at time $t_0 + \Delta t$ if the corresponding value of the phase angle is again φ_0, i.e. if

$$\omega(t_0 + \Delta t) - k(z_0 + \Delta z) = \omega t_0 - kz_0 = \varphi_0, \tag{16}$$

i.e. if

$$\omega \Delta t = k\,\Delta z. \tag{17}$$

Thus we can consider that any given phase (for instance a given wave 'crest', i.e. a maximum of E) 'propagates' with a '*phase velocity*' v_φ equal to $\Delta z/\Delta t$ where Δz and Δt are related by Equation (17). Hence.

$$v_\varphi = \frac{\omega}{k}. \tag{18}$$

According to Equations (11) and (18), in a given plasma characterized by a given value of ω_p, both k and v_φ depend only on ω, and can be denoted by $k(\omega)$ and $v_\varphi(\omega)$ respectively, so that Equation (18) can also be written:

$$\boxed{v_\varphi(\omega) = \frac{\omega}{k(\omega)}.} \tag{18'}$$

Some authors also consider the quantity $n_\varphi(\omega)$ defined by

$$n_\varphi(\omega) = \frac{c}{v_\varphi(\omega)} = \frac{c}{\omega}k(\omega), \tag{19}$$

called the 'phase index' (or the 'refraction index').

3.1.3. *The Wave Packets and the Group Velocity*

3.1.3.1. The Wave Packet Corresponding to a 'Square' Spectrum. Consider, as in Subsection 3.1.2, the propagation along the Oz axis of a Cartesian reference system, of a transverse sinusoidal wave, similar to a linearly polarized oscillating electric field $\boldsymbol{E}$, of angular frequency ω, i.e. of a field $E(z, t; \omega)$ reduced to its component along the Ox axis.

For more generality, and in order to emphasize the relation with the theory of physical pulses presented in Subsection 3.1.1, let us denote here $E(z, t; \omega)$ by $x(z, t; \omega)$.

Thus the function $x(z, t; \omega)$ will describe the 'displacement', along Ox, at time t, of a transverse oscillation of a given angular frequency ω, travelling along Oz, defined analytically by

$$x(z, t; \omega) = a_0 \cos(\omega t - kz). \tag{20}$$

The same oscillation, if taking place *locally*, at $z = 0$, without any propagation, would be described by the function $x(0, t; \omega)$:

$$x(0, t; \omega) = a_0 \cos \omega t. \tag{21}$$

This function corresponds to the function $x(t)$ considered in the theory of pulses (Subsection 3.1.1).

However, an essential difference is emphasized by the presence of the parameter ω in $x(0, t; \omega)$, whereas no argument indicating a variable frequency appears in $x(t)$ given by

$$x(t) = A(t) \cos \omega_0 t. \tag{22}$$

Indeed, if we limit our discussion to the case of a 'square' spectrum considered in Subsection 3.1.1.1, we can interpret the pulse $x(t)$ given by Equation (22), identical to Equation (4), and defined by

$$x(t) = \int_{\omega_1}^{\omega_2} \beta_0 \cos \omega t \, d\omega, \tag{22'}$$

as *generated physically by a superposition*, at a given time t, *of elementary oscillations*, of the same amplitude β_0, of *different angular frequencies* ω continuously distributed between ω_1 and ω_2, represented analytically by $\beta_0 \cos \omega t$.

Consider now a similar superposition, at time t, and at the point $M(z)$ of coordinate z, of elementary transverse oscillations $x(z, t; \omega) = a_0 \cos(\omega t - kz)$ of different angular frequencies ω continuously distributed between ω_1 and ω_2. The function $x(z, t)$ describing the physical result of this superposition, by analogy with Equation (22′), is

$$x(z, t) = \int_{\omega_1}^{\omega_2} a_0 \cos(\omega t - kz) d\omega. \tag{23}$$

In the case of the propagation in a plasma k depends on ω and Equation (23) takes the form:

$$x(z, t) = \int_{\omega_1}^{\omega_2} a_0 \cos[\omega t - k(\omega) z] \, d\omega. \tag{24}$$

3.1.3.2. The Group Velocity of a Wave Packet Corresponding to a 'Square' Spectrum in a Linear Approximation for $k(\omega)$. We have seen in Subsections 3.1.1.1 and 3.1.1.2 that in order to generate a typical pulse of *short* duration (i.e. of small temporal width) it is necessary to superpose elementary oscillations distributed over a *broad* range $(2\delta\omega)$ of frequencies.

Nevertheless, in order to facilitate our analysis, we shall assume here, as a first approximation, that the range (ω_1, ω_2) is sufficiently *narrow*, to permit, in the domain of integration of Equation (24), the replacement of $k(\omega)$ by its Taylor expansion, in the neighbourhood of the 'central frequency' $\omega_0 = \frac{1}{2}(\omega_2 + \omega_1)$, limited to the linear term proportional to $(\omega - \omega_0)$, viz.

$$k(\omega) \approx k_0 + k_0'(\omega - \omega_0), \tag{25}$$

where

$$k_0 = k(\omega_0), \tag{26}$$

and

$$k_0' = \left[\frac{dk(\omega)}{d\omega}\right]_{\omega = \omega_0} \tag{26'}$$

Let us now introduce the variable $\tilde{\omega}$ defined by

$$\omega = \omega_0 + \tilde{\omega}, \tag{27}$$

and, as above, the half-width $\delta\omega$ defined by

$$\delta\omega = \tfrac{1}{2}(\omega_2 - \omega_1). \tag{28}$$

Then ω_1 and ω_2 will correspond to $\tilde{\omega} = -\delta\omega$ and $\tilde{\omega} = +\delta\omega$ respectively, so that $x(z, t)$ defined by Equation (24) will take successively the form:

$$x(z,t) \approx a_0 \int_{\tilde{\omega}=-\delta\omega}^{\delta\omega} \cos\,[(\omega_0 + \tilde{\omega})t - (k_0 + k_0'\tilde{\omega})z]\, d\tilde{\omega}$$

$$\approx a_0 \int_{-\delta\omega}^{+\delta\omega} \cos\,[(\omega_0 t - k_0 z) + (t - k_0' z)\tilde{\omega}]\, d\tilde{\omega}$$

$$\approx a_0 \int_{-\delta\omega}^{+\delta\omega} \cos\,(\omega_0 t - k_0 z)\cos\,[(t - k_0' z)\tilde{\omega}]\, d\tilde{\omega}, \tag{29}$$

since the integral corresponding to $\sin\,[\tilde{\omega}(t - k_0' z)]$ vanishes because $\sin(-y) = -\sin y$.

Finally we find

$$x(z,t) \approx a_0 \cos\,(\omega_0 t - k_0 z) \int_{-\delta\omega}^{+\delta\omega} \cos\,[(t - k_0' z)\tilde{\omega}]\, d\tilde{\omega}$$

$$\approx 2a_0 \frac{\sin\,[(t - k_0' z)\delta\omega]}{(t - k_0' z)} \cos\,(\omega_0 t - k_0 z). \tag{30}$$

Thus $x(z, t)$ can be written as

$$x(z,t) \approx A(z,t)\cos\,(\omega_0 t - k_0 z), \tag{31}$$

where the pseudo-amplitude $A(z, t)$ is given by

$$A(z,t) = A(0,0)\frac{\sin\,[(t - k_0' z)\delta\omega]}{(t - k_0' z)\delta\omega}. \tag{31'}$$

For $z = 0$ Equation (31′) reduces to Equation (6).

Thus the function $x(z, t)$ represents *an amplitude-modulated carrier wave* of a very high frequency $\nu_0 = (1/2\pi)\omega_0$ (practically several thousands of MHz).

The carrier wave is described by the factor $\cos\,(\omega_0 t - k_0 z)$. It is modulated by the factor $A(z, t)$ (the envelope of the plot of $x(z, t)$ *versus* t for a given z).

The carrier wave propagates with a constant *phase velocity* given by Equation (18′) where ω is replaced by ω_0, viz. by

$$v_\varphi(\omega_0) = \omega_0/k(\omega_0) = \omega_0/k_0. \tag{32}$$

The pseudo-amplitude $A(z, t)$ has (in the particular model of a 'square' spectrum considered here) a sinusoidal form 'damped' by the term $1/(t - k_0' z)\delta\omega$.

The main maximum of $A(z, t)$, corresponding to $z = 0$, $t = 0$, *is not fixed*, as in the case of a local pulse, but *propagates* with a velocity

$$v_g = \frac{dz}{dt}, \tag{33}$$

such that the factor $\sin\,[(t - k_0' z)\delta\omega]$ keeps a constant value, i.e. such that

$$(t + dt) - k_0'(z + dz) = t - k_0' z. \tag{34}$$

Hence,

$$v_g = 1/k_0', \tag{35}$$

and a more usual expression:

$$\frac{1}{v_g} = \left[\frac{dk(\omega)}{d\omega}\right]_{\omega=\omega_0} \tag{36}$$

The velocity v_g is called the '*group velocity*', and the oscillation described by the function $x(z, t)$ is called the '*wave packet*'. It corresponds to a kind of a moving pulse.

Similar results are obtained for more complicated spectra, in particular for the 'Gaussian-shaped' spectrum.

Figure 7.5 (overleaf) illustrates the case of a wave packet in which the phase velocity v_φ of the carrier wave is twice the group velocity v_g.

3.1.3.3. General Wave Packets and their Group Velocity. More generally, all wave packets are characterized by their carrier wave of high frequency ν_0 (corresponding to the 'wavelets' of a local physical pulse) and an envelope of pseudo-amplitude *modulating* the carrier wave. The latter *moves as a whole*, but is *slowly deformed* because of the difference between the exact function $k(\omega)$ and its linear approximation.

The amplitude-modulated carrier wave always presents some 'reference feature' (the crest of the damped sinusoid, the maximum of a Gaussian-shaped distribution, etc.). The velocity of displacement of this reference feature, characteristic of the velocity of displacement of the whole wave packet represents the 'group velocity' v_g of the wave packet.

3.2. PULSAR DISTANCES DERIVED FROM DIFFERENCES IN ARRIVAL TIMES OBSERVED AT DIFFERENT FREQUENCIES OF A MULTIPULSE

3.2.1. Dispersion Relation and 'Dispersion Measure'

Consider a multipulse observed at the carrier angular frequency $\omega_1 = 2\pi\nu_1$ and let t_1 denote the interval of time between the departure of a 'crest' (maximum) of the corresponding wave packet from a pulsar and its arrival to the observer's receiver.

If ds' denotes the element of distance s between the pulsar and the observer's receiver, and if $v_g(\omega_1)$ denotes the group velocity corresponding to the frequency ω_1, the interval of time dt_1' necessary for the crest of the wave packet to move along ds' will be given, according to the definition (33) of v_g, by

$$dt_1' = \frac{1}{v_g(\omega_1)} ds'. \tag{37}$$

When the dispersion is produced by the free electrons (of a number density n_e cm^{-3}) of an interstellar ionized plasma, the function $k(\omega)$ is given, in Gaussian units, according to Equations (11) and (12), by

$$k(\omega) = \frac{\omega}{c}\,[1-(\omega_p/\omega)^2]^{1/2}, \tag{38}$$

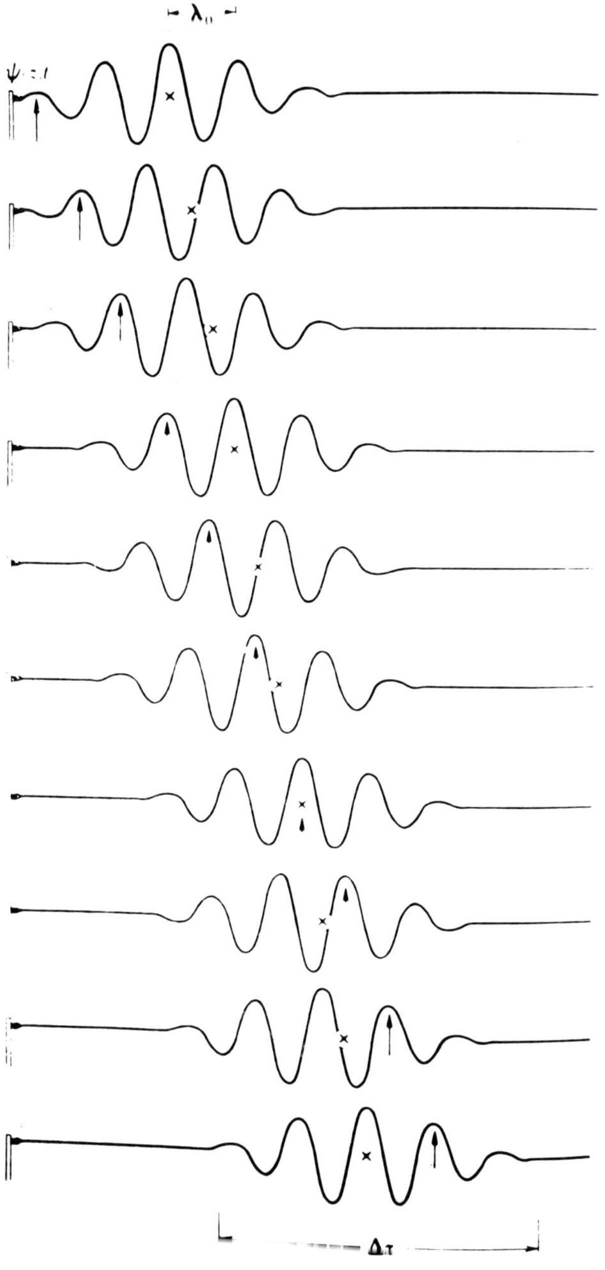

Fig. 7.5. A wave packet in which the phase velocity v_φ is twice the group velocity v_g. One well determined wave crest (maximum) of the carrier wave is marked by a vertical arrow. The crest (maximum) of the envelope $A(z, t)$ is marked by a cross. We see that the arrow overtakes the cross because $v_\varphi = 2v_g$. [From *Waves* (Berkeley Physics Course, Vol. III) by F. S. Crawford, Jr. Copyright 1968 McGraw-Hill. Used with the permission of McGraw-Hill Book Co.]

where

$$\omega_p^2 = 4\pi n_e e^2/m_e, \tag{39}$$

and

$$e \approx 4.80 \times 10^{-10}; \qquad m_e \approx 9.11 \times 10^{-28}; \qquad c \approx 3 \times 10^{10}. \tag{40}$$

Generally the interstellar medium is inhomogeneous so that n_e depends on the distance s' of the considered element ds' from the receiver, and is described by some function $n_e(s')$.

The values of $\nu_p = (1/2\pi)\,\omega_p$ in MHz are given as a function of $n_e\,\text{cm}^{-3}$ by the approximate relation (13), viz.

$$\nu_p \approx 9 \times 10^{-3}(n_e)^{1/2}. \tag{41}$$

Therefore, for the interstellar medium, where n_e seldom exceeds 0.1, ν_p is usually *smaller* than 3×10^{-3} MHz, whereas the receivers used to observe the multipulses of pulsars are, even in the range of radio frequencies, always tuned at frequencies higher than 10 MHz.

Thus, $\omega_p/\omega = \nu_p/\nu$ is certainly much smaller than 1, so that $k(\omega)$ given by Equation (38) can be approximated by the limited expansion:

$$k(\omega) \approx \frac{\omega}{c}\left[1 - \tfrac{1}{2}(\omega_p/\omega)^2\right] \approx \frac{\omega}{c} - \frac{1}{2c\omega}\,\omega_p^2. \tag{42}$$

The derivative $dk(\omega)/d\omega$ will then be given by

$$\frac{dk(\omega)}{d\omega} \approx \frac{1}{c} + \frac{1}{2c}\,\omega_p^2\omega^{-2}, \tag{43}$$

hence, according to Equation (36):

$$\frac{1}{v_g(\omega)} \approx \frac{1}{c} + \frac{1}{2c}\,\omega_p^2\omega^{-2}. \tag{44}$$

If we substitute this expression for $1/v_g$ into Equation (37) we obtain, according to Equations (39) and (40):

$$dt_1' \approx \frac{ds'}{c} + \frac{n_e(s')\,ds'}{\alpha\nu_1^2}, \tag{45}$$

where the frequency ν_1 is in MHz and α is a constant defined by

$$\alpha = 2\pi c m_e e^{-2} \approx 745 \text{ c.g.s.}. \tag{46}$$

On integrating with respect to s', from $s' = 0$ to $s' = s$, we find the travel time t_1 in seconds for the wave packet carried at frequency ν_1 (and observed at frequency ν_1):

$$t_1 \approx \frac{s}{c} + \frac{1}{\alpha\nu_1^2}\int_0^s n_e(s')\,ds'. \tag{47}$$

If we denote by $\bar{n}_e$ the average value of $n_e(s')$ between the pulsar and the receiver, defined by

$$\bar{n}_e = \frac{1}{s}\int_0^s n_e(s')\,ds', \tag{48}$$

Equation (47) takes the form:

$$t_1 \approx \frac{s}{c} + \frac{\bar{n}_e s}{\alpha \nu_1^2}. \tag{49}$$

Since t_1 depends on frequency only through the divisor ν_1^2 it is obvious that for a different frequency of receipt ν_2 of a given multipulse, the travel time t_2 will be given by

$$t_2 \approx \frac{s}{c} + \frac{\bar{n}_e s}{\alpha \nu_2^2}. \tag{50}$$

If $\nu_1 < \nu_2$ observations at two different frequencies will present a difference $\Delta t_{12} = t_1 - t_2$ in arrival times of the 'maximum' of a pulsar pulse given by

$$\Delta t_{12} = t_1 - t_2 \approx \frac{1}{\alpha} \bar{n}_e s(\nu_1^{-2} - \nu_2^{-2}), \tag{51}$$

hence the '*dispersion relation*'

$$D = \bar{n}_e s \approx \alpha\, \Delta t_{12}(\nu_1^{-2} - \nu_2^{-2})^{-1}, \tag{52}$$

where all quantities, and in particular the '*dispersion constant*' $D = \bar{n}_e s$, are in c.g.s. units, and α is given by Equation (46).

When $\bar{n}_e$ is expressed in cm^{-3} whereas s is expressed in parsecs, the quantity $\bar{n}_e s$ expressed in these hybrid units is called the '*dispersion measure*' and is denoted by *DM*. Since $\alpha \approx 745$ c.g.s. and $\text{pc} \approx 3.09 \times 10^{18}$ cm, the dispersion relation (52) can also be written, with Δt_{12} in seconds, and ν_1, ν_2 in MHz, as

$$DM = (\bar{n}_e\, \text{cm}^{-3}) \times (s\, \text{pc}) \approx \frac{2.410 \times 10^{-4}\, \Delta t_{12}}{\nu_1^{-2} - \nu_2^{-2}}\, \text{cm}^{-3}\, \text{pc}, \tag{53}$$

or, when ν_1 and ν_2 are expressed in $\text{GHz} = 10^9$ Hz, as

$$DM \approx \frac{241.0\, \Delta t_{12}}{(\nu_1\, \text{GHz})^{-2} - (\nu_2\, \text{GHz})^{-2}}\, \text{cm}^{-3}\, \text{pc}. \tag{54}$$

3.2.1.1. The Case of Small Values of $\Delta\nu = \nu_2 - \nu_1$. When the difference $\Delta\nu$ between the frequencies ν_2 and ν_1 is sufficiently small, one can neglect $\Delta\nu$ with respect to ν_1; replace $(\nu_1 + \Delta\nu)$ by ν_1 and $(\nu_1 + \nu_2)$ by $2\nu_1$.

Then Equation (54), where all frequencies are expressed in GHz, and Δt_{12} in seconds, will take the form:

$$DM = (\bar{n}_e\, \text{cm}^{-3}) \times (s\, \text{pc}) \approx (120.5)(\nu_1\, \text{GHz})^3 (\Delta t_{12})/(\Delta\nu\, \text{GHz}). \tag{54$'$}$$

The distance of the 'first pulsar' PSR 1919+21 was derived by Hewish *et al.*, 1968, in their first paper, from measure of Δt_{12} corresponding to observations at $\nu_1 = 0.0805$ GHz and $\nu_2 = 0.0815$ GHz. They have used Equation (54′) and have found $DM \approx 13\ \text{cm}^{-3}\,\text{pc}$, which is very near the value $DM \approx 12.43\ \text{cm}^{-3}$ pc resulting from more recent observations.

3.2.1.2. Verification of the Dispersion Relation by Observation of Multipulses Over a Wide Range of Frequencies. Since each signal from a given pulsar represents a multipulse (see Subsection 2.2.3) observable over a wide range of frequencies, one can generalize the dispersion relation (54), replacing ν_1 by ν_i and ν_2 by ν_j, where i and j can take different values $1, 2, 3, \ldots, n$.

Hence

$$\Delta t_{ij} = t_i - t_j \approx \frac{DM}{241.0} \Delta_{ij}, \tag{54''}$$

where t_i and t_j, in seconds, correspond to frequencies ν_i and ν_j, in GHz, respectively, and where Δ_{ij} represents

$$\Delta_{ij} = (\nu_i \, \mathrm{GHz})^{-2} - (\nu_j \, \mathrm{GHz})^{-2}. \tag{55}$$

Observations over frequencies as widely spaced as 0.151; 0.408; 0.922; 1.412 GHz, by Rickett and Lyne (1968), combined in different ways, yield the values of Δ_{ij} and Δt_{ij} which very accurately obey the relation, (54″), and yield very precise values of *DM*, with an uncertainty of about 0.01 cm^{-3} pc, i.e. about a part in 10^4. (See Figure 7.6).

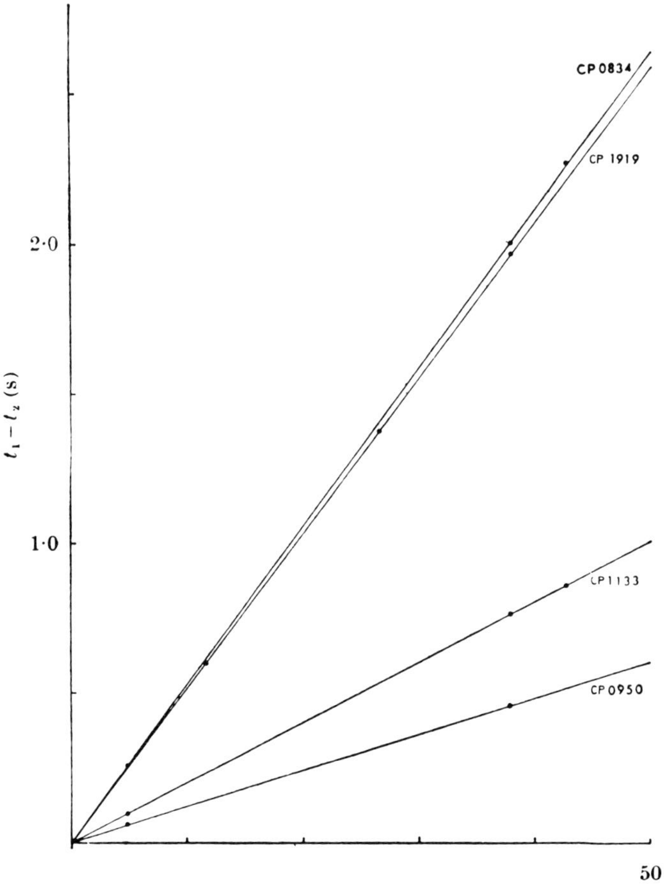

Fig. 7.6. The dispersion Δt_{ij} in arrival time of the multipulses for four pulsars as a function of Δ_{ij} defined by Equation (55). (CP 0834 = PSR 0834+06; CP 1919 = PSR 1919+21; CP 1133 = PSR 1133+16; CP 0950 = PSR 0950+08. The respective values of *DM* are: 12.8; 12.4; 4.8; 3.0). [From Lyne and Rickett (1968).]

A more thorough discussion of 'interstellar scattering' can be found in MT, pp. 128–146.

3.2.1.3. The Distribution of Pulsar Dispersion Measures with Respect to Galactic Latitude. The distribution of pulsar dispersion measures *DM* with respect to galactic latitude is given in Figure 7.7 which shows that moderate to large values of *DM* ($DM \gtrsim 40$) are found only at galactic latitudes smaller than about 10° in absolute value.

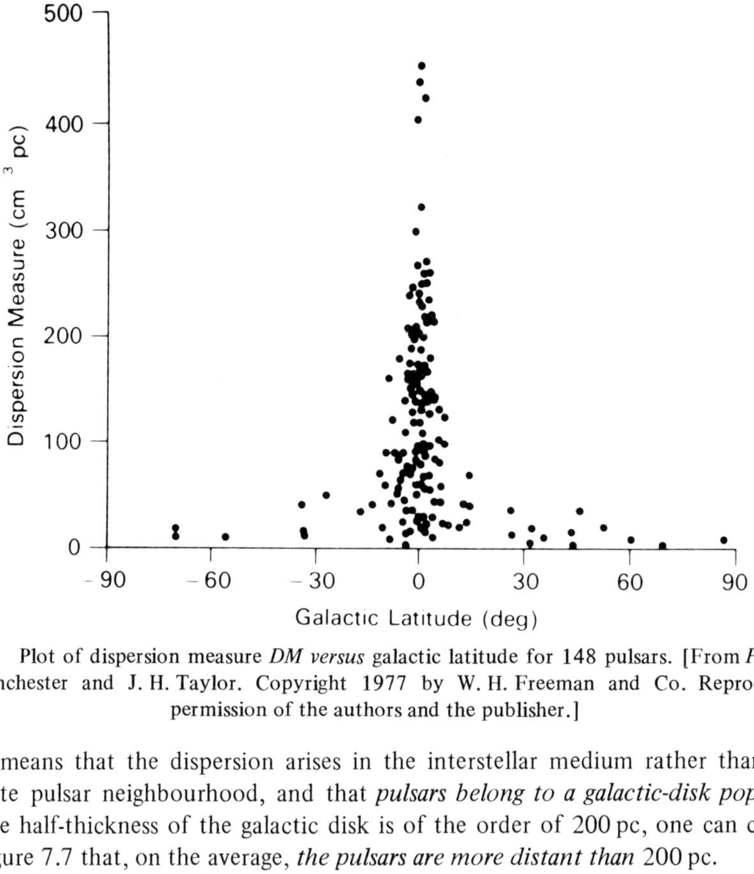

Fig. 7.7. Plot of dispersion measure *DM versus* galactic latitude for 148 pulsars. [From *Pulsars*, by R. N. Manchester and J. H. Taylor. Copyright 1977 by W. H. Freeman and Co. Reproduced by permission of the authors and the publisher.]

This means that the dispersion arises in the interstellar medium rather than in the immediate pulsar neighbourhood, and that *pulsars belong to a galactic-disk population.* Since the half-thickness of the galactic disk is of the order of 200 pc, one can conclude from Figure 7.7 that, on the average, *the pulsars are more distant than* 200 pc.

3.2.1.4. Pulsar Distance Derived from a Combination of the Observed DM with Assumptions Concerning the Mean Values of the Electron Density of the Interstellar Medium. The mean density of the dispersing electrons in the galactic disk can be estimated because the distances to some pulsars are known from their association with known supernova remnants, and from neutral H absorption data (discussed in Subsection 3.3).

A recent discussion (MT, pp. 131 f) confirms the 'standard value' 0.03 cm^{-3} of $\bar{n}_e$, already given by Terzian, as soon as 1970 (but it also shows that this standard value can be in error within a factor of about two) over a large portion of the Galaxy.

This standard value of $\bar{n}_e$ corresponds to the rather well known distance (~ 2 kpc) of the Crab Nebula and to the very well known dispersion measure *DM* (~ 56.7) of the Crab pulsar PSR 0531+21.

For Vela pulsar, PSR 0833–45 the corresponding supernova remnant seems, from optical measures, to be at a distance of about 0.5 kpc, whereas the *DM* of the pulsar is equal to 69.0. The corresponding value of $\bar{n}_e$ is 0.14, about five times the 'standard value' of $\bar{n}_e$. This alone already shows that the adoption of a standard value for $\bar{n}_e$ is a somewhat dangerous procedure.

Nevertheless, it can be statistically interesting to derive pulsar distances from their *DM* by application of the standard value of $\bar{n}_e$. The result is (according to the data given by MT, Appendix) that these distances range between 0.1 kpc for the high galactic latitude ($b = 43.7°$) pulsar PSR 0590+08 and 15 kpc for the low galactic latitude ($b = -0.6°$) pulsar PSR 1353–62 (see also the end of Subsection 3.3.4).

3.3. Pulsar Distances Derived from Observations of Neutral Hydrogen Line near 1420 MHz

3.3.1. *Introduction*

Estimates of pulsar distances from measurements of absorption of multipulses near 1420 MHz by neutral H atoms in the galactic disk are based on a combination of the theory of radiative transfer with the data concerning the structure and the 'differential rotation' of our Galaxy.

We start with a brief summary of these two bases.

3.3.2. *Position and Differential Rotation of the Spiral Arms of Our Galaxy*

According to the data compiled by Allen (1973) the position of spiral arms of our Galaxy in the neighbourhood of the Sun S can be schematically represented as in Figure 7.8.

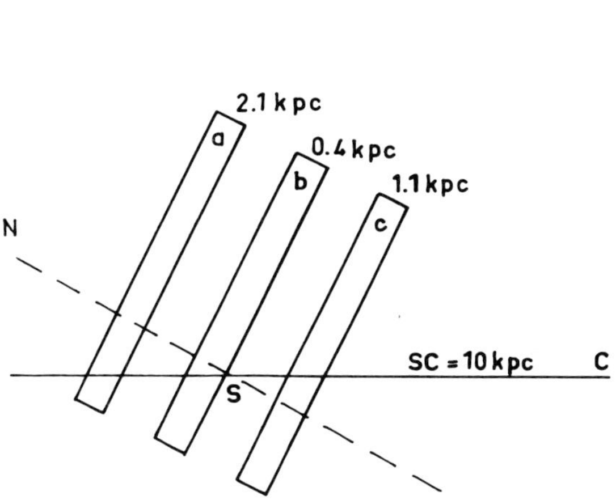

Fig. 7.8. A schematic representation of three spiral arms of the Galaxy, in the neighbourhood of the Sun. (a) The Perseus arm; (b) the 'local' (Cygnus–Orionis) arm; (c) the Sagittarius arm. The dotted line indicates the direction of the normal SN to the arms. The distances of the middle of each arm from the Sun, along SN, are indicated near each 'rectangle' representing a portion of an arm.

The width of each arm is, in a first approximation, of the order of 0.6 kpc. The center C of the Galaxy is at an approximate distance of 10 kpc from the Sun. The angle between the normal SN to the three arms represented in Figure 7.8 (considered, in a first approximation, as parallel strips, and represented, for the sake of clarity, as rectangles) and the direction SC of the center C, is approximately equal to 27°.

Galactic longitudes l are usually reckoned positively counterclockwise, around S, with the direction from S to C as origin, in a figure like Figure 7.8.

In a second approximation the two 'edges' of the Perseus arm should be placed, between 2.3 and 2.9 kpc, in the direction $l \approx 145°$ of the pulsar PSR 0329+54 considered in Subsection 3.3.4.

3.3.3. *Application of the Theory of Radiative Transfer to the Neutral H Line near 1420 MHz* ($\lambda \approx 21$ cm)

The line of central frequency $\nu \approx 1420$ MHz ($\lambda \approx 21$ cm) of neutral H corresponds to a 'forbidden transition' between two sublevels of hyperfine structure (F = 1 and F = 0) of the fundamental state of neutral H atom.

The precise central frequency ν_r of this line (r) is equal to 1420.4058 MHz.

In the theory of radiative transfer (Chapter 1, Section 2) the lower level is denoted by the subscript (1), and the upper level by the subscript (2).

Let us denote the number densities of neutral H atoms in states (1) and (2) by N_1 and N_2 respectively.

The probability A_{21} of a spontaneous transition (see Chapter 1, Subsection 2.5) from the state (2) to the state (1) is

$$A_{21} = 2.84 \times 10^{-15}\ \mathrm{s}^{-1}. \tag{56}$$

This transition can be considered as 'forbidden', and the sublevel (2) can be considered as 'metastable', since to this value of A_{21} corresponds a mean life time $\tau_2 = 1/A_{21}$ (with respect to a radiative transition) particularly long, viz.

$$\tau_2 = \frac{1}{A_{21}} = 35 \times 10^{13}\ \mathrm{s} \approx 11 \times 10^6\ \mathrm{yr}. \tag{56'}$$

Actually this life time τ_2 must be appreciated by comparison with the mean interval of time $\bar{\tau}_c$ between two successive collisions of neutral H atoms of the interstellar gas (in the so-called HI regions). Therefore, let us determine the order of magnitude of $\bar{\tau}_c$.

In a first approximation, the frequency ν_c of collisions is given, as a function of the number density n of a gas of identical particles (assimilated to spheres of radius R) by

$$\nu_c = nV, \tag{57}$$

where V is the volume swept by the cross section $\sigma = 4\pi R^2$ of each atom in one second.

If the atoms move with a mean velocity $\bar{v}$, the volume V will be equal to $\sigma\bar{v}$, hence

$$\nu_c = n\sigma\bar{v}. \tag{58}$$

For a perfect monoatomic gas, the mean velocity $\bar{v}$ is determined by the well known expression $\frac{3}{2}kT$ for the mean kinetic energy $\frac{1}{2}m\bar{v}^2$ of each particle, i.e. by

$$\tfrac{1}{2}m\bar{v}^2 = \tfrac{3}{2}kT_{\text{kin}}, \tag{59}$$

where m is the mass of each particle and T_{kin} the kinetic temperature of the gas (see, for instance, Chapter 4, subsection 9.2).

Let us take for R of the neutral H atoms the radius of the first Bohr's orbit:

$$R \approx \tfrac{1}{2} \times 10^{-8}\ \text{cm}. \tag{60}$$

On taking as a mean number density $\bar{n}$ of the interstellar medium the value

$$\bar{n} \approx 1\ \text{cm}^{-3}, \tag{61}$$

and adopting as a mean kinetic temperature $\bar{T}_{\text{kin}}$ the value

$$\bar{T}_{\text{kin}} \approx 100\,\text{K}, \tag{62}$$

we find, (with $m = m_{\text{H}} \approx m_B \approx 1.66 \times 10^{-24}$ g), for the mean value $\bar{\nu}_c$ of ν_c:

$$\bar{\nu}_c \approx 4\pi R^2 \bar{n}(3k/m)^{1/2}(\bar{T}_{\text{kin}})^{1/2} \approx 5 \times 10^{-11}\,\text{s}^{-1}. \tag{63}$$

Hence the mean interval of time $\bar{\tau}_c$ between collisions:

$$\bar{\tau}_c = (1/\bar{\nu}_c) \approx 2 \times 10^{10}\ \text{s}. \tag{63$'$}$$

Thus we see the $\bar{\tau}_c$ is about 17 000 times smaller than τ_2.

Therefore, most of (upward and downward) transitions between the states (1) and (2) are due to collisions; only a few downward transitions can be (spontaneous or stimulated) *radiative* transitions.

The population of the excited level No. 2 is thus, in a steady state, maintained by collisions, at the expense of the internal energy of the interstellar gas. In other words, the excitation of the level No. 2 can be considered as essentially 'thermal', (see Chapter 1, subsection 2.5).

The corresponding excitation temperature $T_{21} = T_{\text{exc}}$ is defined by Eq. (33) of Ch. 1, viz.

$$N_2/N_1 = (g_2/g_1) \exp(-h\nu_r/kT_{\text{exc}}), \tag{64}$$

where the central frequency ν_r of the line must be expressed in Hz (when c.g.s. units are used for h and k):

$$\nu_r \approx 1.42 \times 10^9\ \text{Hz}. \tag{65}$$

The factor $h\nu/k$ is approximately equal to

$$h\nu/k \approx \frac{(7 \times 10^{-27}) \times (1.42 \times 10^9)}{1.4 \times 10^{-16}} \approx 0.07\,\text{K}, \tag{66}$$

(it is expressed in K since it has the dimension of a temperature).

It is usually assumed, in a first approximation, that in 'normal' conditions (i.e. when the number density of the gas is not too small and when the radiation field is not intense, which is the case far from bright sources) T_{exc} can be assimilated with the mean kinetic temperature $\bar{T}_k$ of the interstellar gas, given by Equation (62).

Now, for a given spectral line (r), the total coefficient of removal Σ_ν (taking into account stimulated emissions) is given by Eq. (37) of Ch. 1, viz. with $j = j' = 1$; $k = k' = 2$; and $B_{12}(\nu) = B_{12}\psi_r(\nu)$:

$$\Sigma_\nu = \frac{h\nu}{c} N_1 B_{12} \psi_r(\nu)[1 - \exp(-h\nu/kT_{\text{exc}})]. \tag{67}$$

In this Equation the function $\psi_r(\nu)$ describes the variation, within the domain of the line (r), of the coefficient of removal Σ_r of the whole of the line, as a function of the frequency ν (it is normalized to unity by Eq. 26 of Ch. 1). The other symbols have their usual meaning.

In the domain to the line at 1420 MHz the ratio

$$\epsilon = h\nu/kT_{\text{exc}} \approx h\nu/k\bar{T}_{\text{kin}} \approx \frac{0.07}{100} \approx 7 \times 10^{-4}, \tag{68}$$

is so small that one can, in a first approximation, replace $\exp(-\epsilon)$ by $(1-\epsilon)$, so that the square bracket in Equation (67) reduces to ϵ.

On the other hand, it is well known (and proved in the Appendix to Chapter 1) that

$$B_{12} = B_{21}(g_2/g_1), \tag{69}$$

and

$$B_{21} = A_{21} c^3/(8\pi h \nu_r^3). \tag{70}$$

For the state No. 2 (F = 1) and the state No. 1 (F = 0) spectroscopic tables give:

$$g_2/g_1 = 3, \tag{71}$$

hence

$$B_{12} = 3c^3 A_{21}/(8\pi h\nu_r^3). \tag{72}$$

On taking into account Equation (68) and on assimilating to 1 the factor $\exp(-\epsilon)$ in Equation (64), we thus find:

$$N_2/N_1 \approx g_2/g_1 \approx 3. \tag{73}$$

If we introduce the total number density n_{H} of neutral H atoms in the ground state, and assume that $n_{\text{H}} = N_1 + N_2$, Equation (73) yields:

$$N_1 = N_1 \frac{n_{\text{H}}}{N_1 + N_2} = \frac{n_{\text{H}}}{1 + (N_2/N_1)} \approx \tfrac{1}{4} n_{\text{H}}. \tag{74}$$

On replacing in Equation (67) N_1 by its expression (74), B_{12} by its expression (72), A_{21} by its value (56), ν_r by its value (65), the constants c, h, k by their values in c.g.s. units, and the square bracket by ϵ given by Equation (68), one finds, after all reductions:

$$\Sigma_\nu \approx C(n_{\text{H}}/\bar{T}_{\text{kin}})\, \psi_r(\nu), \tag{75}$$

where

$$C \approx \frac{3c^2 h}{32\pi k}(A_{21}/\nu_r) \approx 2.58 \times 10^{-15} \text{ c.g.s.}. \tag{76}$$

The 'natural width' of the line $\lambda \approx 21$ cm is very small and the collisional broadening is equally very small even in 'dense' HI regions of relatively small number density of neutral H (practically all in the fundamental state), n_{H}.

In these conditions the large and complex profile of the 21 cm line observed in radio astronomy is generated by the Doppler effect corresponding to *macrosopic motions* of cold 'dense' isolated HI regions of the spiral arms of the Galaxy, participating to the 'differential rotation' of these arms.

For small radial velocities V_{rad} of these HI regions the first order Doppler effect, described by

$$\nu = \nu_0 \left(1 - \frac{1}{c} V_{\text{rad}}\right), \tag{77}$$

where ν_0 is the 'laboratory' frequency, is involved.

Let us now introduce the element of 'optical' (or, better, of 'physical') displacement (see Chapter 1, Subsection 3.2) $d\tau_\nu$

$$d\tau_\nu = \Sigma_\nu \, ds', \tag{78}$$

where we have called ds' the element of geometrical displacement, in order to reserve s to denote the *total distance* travelled by the radio waves, and where we have dropped the subscript ω indicating the direction, since we consider here only the propagation along the line of sight.

For a macroscopic geometrical distance s we have

$$\tau_\nu = \tau_\nu(s) = \int_0^s \Sigma_\nu(s') \, ds', \tag{79}$$

where Σ_ν is considered as a function of s' because in Equation (75), where T_{kin} is assumed to be equal to $\bar{T}_{\text{kin}}$, i.e. constant, n_{H} is a function $n_{\text{H}}(s')$, (non-homogeneous medium!)

Thus, on replacing Σ_ν by its value (75) we find:

$$\tau_\nu(s) \approx C\psi_r(\nu)(1/\bar{T}_{\text{kin}}) \int_0^s n_{\text{H}}(s') \, ds'. \tag{80}$$

Let N_{H} denote the total number of neutral H atoms (assumed to be all in the ground state) per cm^2 of the column between the source and the receiver, given by

$$N_{\text{H}} = \int_0^s n_{\text{H}}(s') \, ds'. \tag{81}$$

Then Equation (80) takes the form

$$\tau_\nu(s) \approx C\psi_r(\nu)(1/\bar{T}_{\text{kin}}) N_{\text{H}}. \tag{82}$$

When a pulsar is observed through one or several spiral arms of the Galaxy one must distinguish, according to the instant of observation, the phase when the pulsar is 'off' and the phase when the pulsar is 'on'.

When the pulsar is 'off' one observes the (partially self-absorbed) radiation *emitted* by the HI regions of the spiral arms situated along the line of sight of the pulsar. Observation shows that the result of emissions exceeds the result of absorptions, since an *emission line* (broadened by the Doppler effect) is observed. The shape of this emission line can, of course, be discussed theoretically. However the procedure used in the study of pulsars, described in Subsection 3.3.4, makes use of the *observed profile* of the emission line corresponding to the phase 'off' of the pulsar, and is not based on the *theory* of this emission line.

When the pulsar is 'on' one observes a superposition of the above mentioned emission line and of pulsar radiation affected by removals ('absorptions') along the path travelled by this radiation. Now, as explained in Subsection 3.3.4, on subtracting from this observed superposition the part corresponding to the emission line (phase 'off') one is left with the sole effect of removals upon the pulsar radiation. The latter can be discussed theoretically, on using the transfer equation with no emission term.

On reducing the transfer equation for the stationary case (Eq. 63 of Ch. 1) to the sole effect of removals, i.e. on neglecting the 'source function', we obtain:

$$dI_\nu/d\tau_\nu = -I_\nu. \tag{83}$$

If $I_\nu(s)$ denotes the observed and $I_\nu(0)$ the 'emitted' intensity:

$$I_\nu(s) = I_\nu(0)\,e^{-\tau_\nu(s)}. \tag{84}$$

Thus we see that on a plot of $I_\nu(s)$ *versus* ν, the phenomenon of removal will be represented by a more or less dip 'trough' ('absorption line').

However, if the source (pulsar) emits a *continuous spectrum* (which is the case for the carrier waves of a multipulse) its intensity $I_{\nu,c}(0)$ emitted at frequencies (ν, c) on both 'sides' of the range of frequencies corresponding to this trough (dip) will be very near the intensity $I_\nu(0)$ of the radiation emitted at frequencies within the domain of the line (r):

$$I_\nu(0) \approx I_{\nu,c}(0). \tag{85}$$

On the other hand, since there is no 'absorption' (removal) outside the domain (r) of frequencies, in the domain (ν, c) we have

$$I_{\nu,c}(0) = I_{\nu,c}(s). \tag{86}$$

Thus, we finally obtain:

$$I_\nu(s) \approx I_{\nu,c}(s)\,e^{-\tau_\nu(s)}, \tag{87}$$

a relation which permits to test our assumptions on comparing $I_\nu(s)$ observed on both sides of the domain (r).

On forming the ratio of 'observable quantities' $I_\nu(s)$ and $I_{\nu,c}(s)$ we can find $\tau_\nu(s)$, given by

$$\tau_\nu(s) = \log_e\,[I_{\nu,c}(s)/I_\nu(s)]. \tag{88}$$

Generally in the plot of $I_\nu(s)$ versus ν the frequency ν is expressed, through Equation (77), in terms of radial velocities V_{rad} (expressed in km s^{-1}). Thus $\tau_\nu(s)$ becomes a function $\tau(V_{\text{rad}})$ of V_{rad}. Of course, before operating the subtraction mentioned above, the intensity observed during the phase 'off' of the pulsar is also represented as a function of V_r.

If we denote the speed of light in vacuum expressed in km s^{-1} by $c_0 = 3 \times 10^5$, Equation (77) will give:

$$\int_{(r)} \tau_\nu\,d\nu = \int_{(r)} \tau(V_{\text{rad}})\,\nu_0 \frac{1}{c_0}\,dV_{\text{rad}} \approx \frac{1.42 \times 10^9}{3 \times 10^5} \int_{(r)} \tau(V_{\text{rad}})\,dV_{\text{rad}}$$

$$\approx 0.473 \times 10^4 \int_{(r)} \tau(V_{\text{rad}})\,dV_{\text{rad}}. \tag{89}$$

On replacing, in Equation (89), τ_ν by its expression (82), we find:

$$C(1/\bar{T}_{\text{kin}})N_{\text{H}} \int_{(r)} \psi_r(\nu)\,d\nu \approx 0.473 \times 10^4 \int_{(r)} \tau(V_{\text{rad}})\,dV_{\text{rad}}. \tag{90}$$

On taking into account the value 1 of the integral at the left-hand side of Equation (90), and the numerical value of C given by Equation (76), we obtain:

$$N_{\text{H}} = (1.82 \times 10^{18})\,\bar{T}_{\text{kin}} \int_{(r)} \tau(V_{\text{rad}})\,dV_{\text{rad}}, \tag{91}$$

where N_{H} is in cm^{-2}, whereas V_{rad} is in km s^{-1} and $\tau(V_{\text{rad}})$ is given by

$$\tau(V_{\text{rad}}) = \log_e \left[I_{\text{continuum}}/I_{\text{line}}(V_{\text{rad}})\right]. \tag{92}$$

Application of Equation (91) to numerical integration of the absorption profiles of PSR 0329+54 and PSR 1749–28 by Gordon *et al.* (1969) together with an assumed temperature $\bar{T}_{\text{kin}} = 100\,\text{K}$, yields $2.8 \times 10^{21}\,\text{cm}^{-2}$ and $0.7 \times 10^{21}\,\text{cm}^{-2}$ respectively.

On assuming further the mean value $\bar{n}_{\text{H}}$ of n_{H} equal to $0.4\,\text{cm}^{-3}$ one finds for PSR 0329+54 a distance of 2.3 kpc and with $\bar{n}_{\text{H}}$ equal to $0.3\,\text{cm}^{-3}$ a distance of 3.0 kpc. Similarly, with $\bar{n}_{\text{H}} \approx 0.35\,\text{cm}^{-3}$ one finds for PSR 1749–28 a distance of about (2/3) kpc, certainly smaller than 1 kpc.

3.3.4. *Pulsar Distances with Respect to Spiral Arms of the Galaxy from Absorption Line Observations Near 1420 MHz*

The absolute value $|b|$ of the galactic latitude of most pulsars is so low (see Figure 7.7) that, if they belong to our Galaxy (which is sure from other evidence) they must be situated within the galactic disk.

The galactic latitude b of one of them, PSR 0329+54, (formerly called CP 0328), is equal to ($-1.2°$), and its direction is almost parallel to the normal SN to spiral arms (see Figure 7.8), since its galactic longitude is equal to $145.0°$.

This means that its signals must cross:

(1) Either a part of the 'local arm' (containing the Sun);
(2) Or a part of the 'local arm' and the 'Perseus arm';
(3) Or even more distant arms (but we show below that this possibility is excluded by observations).

A similar possibility exists for a few other pulsars whose $|b|$ is smaller than $10°$, whose signals must cross (more or less obliquely) a part of the local arm and one or more other arms, when their longitudes l lie approximately between the limits $27° < l < 207°$ or $207° < l < 360° + 27°$. In the latter case the signal must cross a part of the local arm, the Sagittarius arm, etc.

Now, the *differential rotation* of the Galaxy manifests itself by differences in radial velocity V_{rad} of HI regions situated in different parts of different arms (i.e. at different distances in different directions).

These radial velocities are near zero for HI regions in directions near $l = 0°$ or $l = 180°$ parallel to SC (towards the galactic center and anticenter), but are already perceptible for $l = 145°$, especially for the elements lying sufficiently far from the Sun.

When an absorption line of a pulsar is detected near 1420 MHz, *the range of possible*

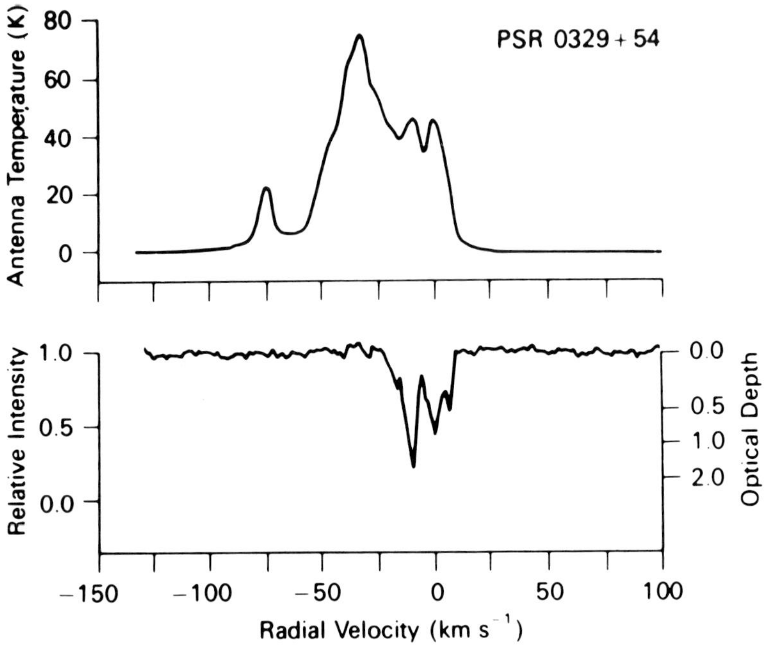

Fig. 7.9. Profiles of emission (upper) and absorption (lower) by neutral hydrogen in directions of the pulsar PSR 0329+54. The radial velocities are with respect to that local standard of rest. [From Gordon and Gordon (1973).]

distances can be inferred from the radial velocities of the main 'features' of the profile of the absorption line, as illustrated by the example of Figure 7.9.

The intensities of the *emission* spectrum of the HI regions in the direction of PSR 0328+54 near $\nu_0 = 1420\,\text{MHz}$ (usually expressed in terms of the 'antenna temperature') present a profile corresponding to frequencies displaced by Doppler effect generated by the radial velocities V_{rad} of the HI regions situated in the line of sight.

The two small emission features at about -2 and $-12\,\text{km s}^{-1}$ are generally ascribed to the local spiral arm ($d \leqslant 0.6\,\text{kpc}$); the 'high' (strong) emission feature at about $V_{\text{rad}} = -37\,\text{km s}^{-1}$ is generally ascribed to HI regions situated in the Perseus arm ($2.3 < d < 2.9\,\text{kpc}$); and the rather 'low' (faint) emission feature at about $V_r = -77\,\text{km s}^{-1}$ is ascribed to the 'outer arm' (the first arm more distant than the Perseus arm).

The 'relative intensity' (absorption profile obtained by the procedure explained below) represents the result of 'absorption' by *some* of the HI regions producing the emission line, on the pulsar's multipulse continuous spectrum, *only those situated between the pulsar and the receiver*. Figure 7.9 shows that in the 'absorption line' both the features of PSR 0329+54 corresponding to $V_r = -2\,\text{km s}^{-1}$ and to $V_r = -12\,\text{km s}^{-1}$ are present (absorption in the 'local arm'), whereas the feature at $V_r = -77\,\text{km s}^{-1}$ is definitely absent (no absorption in the 'outer arm'). The very small and questionable feature near $V_r = -37\,\text{m s}^{-1}$ is considered as real by competent investigators, hence the conclusion that the pulsar lies *in* the Perseus arm, i.e. that its distance is between 2.3 and 2.9 kpc.

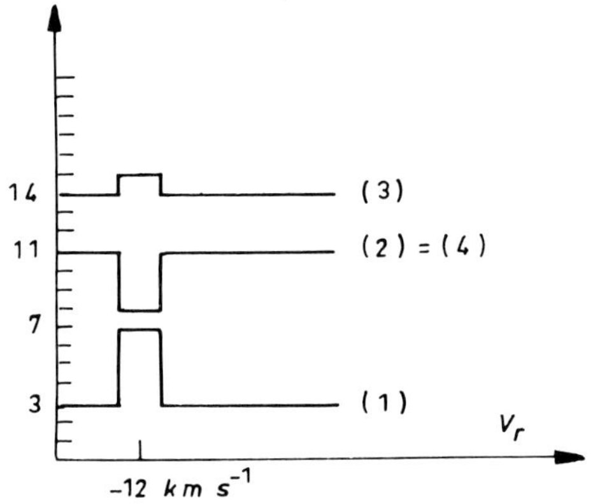

Fig. 7.10. Schematic profiles relative to the features at $V_r \approx -12\,\mathrm{km\,s^{-1}}$ in Figure 7.9. (1) Intensity observed when the pulsar is 'off'; (2) Relative intensity obtained by subtracting (3) from (1); (3) Intensity observed when the pulsar is 'on'. (See the detailed explanations in the text.)

The 'relative intensity' (absorption profile) is obtained in the following way. Let us represent schematically in Figure 7.10, in arbitrary units, the part of Figure 7.9 (PSR 0329+54) relative uniquely to the features near $V_r = -12\,\mathrm{km\,s^{-1}}$, and neglect the slope of the profile outside these features.

The line marked (1) represents the emission intensity (emission profile) observed during the portion of the pulsar period when the pulsar is 'off'. It represents the 'ordinary' emission from the HI regions, situated in the local arm, whose approximate radial velocity is about $(-12)\,\mathrm{km\,s^{-1}}$ in the direction of PSR 0329+54.

The line marked (2) represents the 'trough' which, in the absence of the emission (1), would be produced in the continuous spectrum of the carrier waves of the pulsar multipulse by the same HI regions, near 1420 MHz, displaced by the Doppler effect.

The line marked (3) represents the superposition [(1)+(2)] of phenomena (1) and (2), observed during the portion of the pulsar period when the multipulse of the pulsar is 'on'.

The curve marked 'relative intensity' in Figure 7.9 corresponds to the difference, marked (4) in Figure 7.10, between the line (3) (pulsar 'on') and the line (1) (pulsar 'off'): (4) = (3)−(1).

It is obvious that (4) = (2), since (3)−(1) = [(1)+(2)]−(1)=(2). By a similar method, upper limits can be inferred from the absence of absorption features at more negative (or more positive – according to the quadrant of galactic longitude) radial velocities, in correspondence with emission features. The derived distances are most reliable, as noted by MT, for pulsars of large *DM* and away from directions $l = 0^\circ$ and $l = 180^\circ$.

MT (p. 127) gives a compilation of the distances of 26 pulsars obtained by this method. They range between 0.16 ± 0.01 kpc for the high latitude ($b = +26.1^\circ$) pulsar PSR 1642–03 (Graham *et al.*, 1974) and 20 ± 2 kpc for the low latitude ($b = -0.6^\circ$) pulsar PSR 1859+03 (Ables and Manchester, 1976). The latter distance is the largest attributed to any pulsar.

For other methods of determination of pulsar distances we refer the reader to the fundamental work of MT.

3.3.5. *Application of the Determinations of Pulsar Distances to a Rough Analysis of the Distribution of Pulsars in the Galaxy*

We have already seen, in Subsection 3.2.1.3, that the plot of the dispersion measure *DM* versus galactic latitude (see Figure 7.7) shows that pulsars belong to a galactic disk population.

Another method of investigation of the distribution of pulsars in the Galaxy is provided by an analysis of the number $N(d)$ of pulsars whose distance is smaller than some given distance d. In this analysis the distances given by *all* available methods (interstellar dispersion, interstellar absorption, pulsar-supernova associations, etc.) are used.

If the number of pulsars per unit volume of the Galaxy were constant, with a number density of n pulsars per pc^3, the number $N(d)$ of pulsars within a sphere of radius d pc centered on the Sun, would be equal to

$$N(d) = n\frac{4\pi}{3}d^3, \tag{93}$$

so that one would have

$$\log_{10} N(d) = \text{const.} + 3 \log_{10} d, \tag{94}$$

and the line representing (after interpolation) $\log_{10} N(d)$ *versus* $\log_{10} d$ would be a straight line with a slope equal to 3.

With the same assumption of a constant value of n and all pulsars confined to the galactic disk of a half-width h pc, the number $N(d)$ of pulsars inside a cylinder based on a circle of radius d pc centered on the Sun (above and below the galactic plane) would be equal to

$$N(d) = n2\pi d^2 h, \tag{95}$$

so that one would have

$$\log_{10} N(d) = \text{const.} + 2 \log_{10} d, \tag{96}$$

and the line representing (after interpolation) $\log_{10} N(d)$ *versus* $\log_{10} d$ would be a straight line with a slope equal to 2.

This means that if pulsars were uniformly distributed through the whole of the galactic disk, of half-thickness equal to about 200 pc, one should expect the slope of the plot of $\log_{10} N(d)$ *versus* $\log_{10} d$ to remain equal to 3 for $d < 200$ pc and to become equal to 2 for $d > 200$ pc.

Figure 7.11 shows that the results concerning pulsar distances smaller than 1500 pc are in satisfactory agreement with this prediction up to $d \approx 500$ pc. At greater distances the sample becomes noticeably incomplete. (For more information about the galactic distribution of pulsars see MT, pp. 153–160.)

4. Pulsar Ages

4.1. THE CONSTANCY OF $P\dot{P}$?

Let us list the pulsars, with *known values of the period derivatives* $\dot{P}$, by order of increasing periods P, and assign a number N to the (provisional) place of each of these pulsars in

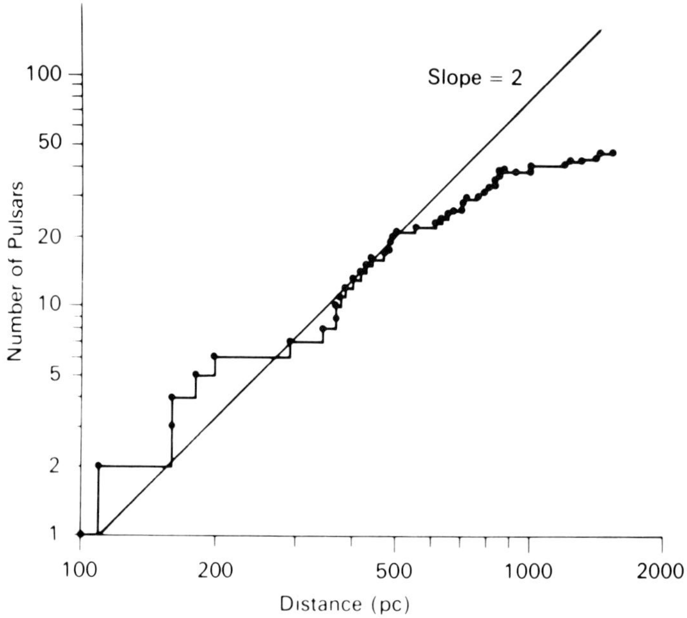

Fig. 7.11. Number $N(d)$ of pulsars with distances less than d pc, as a function of d, (at a logarithmic scale). [From *Pulsars* by R. N. Manchester and J. H. Taylor. Copyright 1977 by W. H. Freeman and Co. Reproduced by permission of the authors and the publisher.]

this list (which, of course, will be modified when more pulsars will be discovered). The corresponding period will be denoted by P_N. We thus obtain, first a Table, such as Table 7.1, where we consider only 10 of these pulsars with the shortest periods (known for the moment) and next Table 7.2, where we consider, for the sake of illustration, the

TABLE 7.1

The periods P (in s), the period derivatives $\dot{P}$ (in 10^{-15} s s^{-1}), the values of $P\dot{P}$ (in s), and the age $T(t_0)$ (in 10^6 yr) given by Equation (108). Only pulsars with known values of $\dot{P}$ are listed, and ordered with respect to the value of P, for the 10 pulsars with the *shortest* known period.
[After MT, Appendix.]

No.	PSR	P (s)	$\dot{P}$ (10^{-15} s s^{-1})	$P\dot{P}$ (10^{-15} s)	$T(t_0)$ (10^6 yr)	Notes
1	0531+21	0.0331	422.69	14.1	0.00124	Crab pulsar
2	1913+16	0.0590	0.0088	0.00052	106	Binary pulsar
3	0833–45	0.0892	125.03	11.1	0.011	Vela pulsar
4	1930+22	0.1444	63	9.1	0.036	
5	0355+54	0.1563	4.39	0.685	0.56	
6	0740–28	0.1667	16.80	2.80	0.016	
7	1557–50	0.1925	5.06	0.98	0.60	
8	1915+13	0.1946	7.20	1.40	0.43	
9	1929+10	0.2265	1.16	0.0263	3.1	
10	2324+60	0.2336	0.36	0.0845	10.0	

TABLE 7.2

The same data as in Table 7.1, for the 10 pulsars, with known $\dot{P}$ and with the *longest* known period (X-ray pulsars are not considered here). [After MT, Appendix.]

No.	PSR	P (s)	$\dot{P}$ (10^{-15} s s^{-1})	$P\dot{P}$ (10^{-15} s)	$T(t_0)$ (10^6 yr)
77	2154+40	1.5253	2.9	4.4	8.3
78	2303+30	1.5758	2.91	4.6	8.6
79	1112+50	1.6564	2.6	4.3	10.0
80	1819–22	1.8742	0.6	1.1	49
81	2045–16	1.9615	10.96	21.7	2.8
82	2002+31	2.1112	74.58	157	0.45
83	1910+20	2.2329	10.18	22.8	3.5
84	2319+60	2.2564	7.6	17.1	4.7
85	0153+61	2.3516	189	444	0.20
86	0525+21	3.7454	40.06	150	1.5

10 of these pulsars with the longest known periods (binary X-ray sources are excluded, see Chapter 10). (The total number of pulsars listed in the Appendix of MT is 149; the number of pulsars with known $\dot{P}$ is equal to 86; about more 20 pulsars have been found recently.)

It was already indicated in Subsection 2.1.3 that the periods P of all pulsars *increase* (when the 'glitches' are left out of account). Let us denote by $P_N(t)$ the function describing the time variation of P_N. And let us denote by t_0 some given 'standard instant', in some chronolgy, *near the present time*. The periods listed in Tables 7.1 and 7.2 (and later in Table 7.3) represent $P_N(t_0)$ and $\dot{P}$ represents $[dP_N/dt]_{t=t_0}$. The precise value of t_0 does not need to be given in such Tables where periods are given only to four decimal digits (although, in most cases, they are known to much greater accuracy) since near the present time these digits remain invariable. A similar remark holds for $\dot{P}$ given in our Tables.

When in a diagram the product $P_N(t_0)\dot{P}_N(t_0)$ is represented as a function of N, it is found that, in spite of a very considerable dispersion, this product can be considered as constant (in a very rough approximation) and equal to 2×10^{-15} s.

A more convenient representation of all data concerning $P_N(t_0)$ and $\dot{P}_N(t_0)$ is the one found in MT (p. 110) and reproduced in Figure 7.12, where $\dot{P}(t_0)$ is represented as a function of $P(t_0)$, without taking into account N, *on a logarithmic scale*.

If, omitting the subscript N for the sake of brevity, we assume that

$$P\dot{P} = D = \text{const.}, \tag{97}$$

we find

$$\log_{10} \dot{P} = \log_{10} D - \log_{10} P. \tag{98}$$

Thus the approximate relation (97) is represented in Figure 7.12 by straight lines of negative slope, differing by the value of the constant D.

We see in Figure 7.12 that the 'cloud' of points lies roughly between the solid lines of negative slope corresponding to the values 10^{-14} s and 10^{-16} s of the constant D (a more detailed investigation shows that the average D is equal to 2×10^{-15} s). Note that the Crab pulsar PSR 0531+21 lies almost exactly on the line $D = 10^{-14}$ s, whereas the binary pulsar

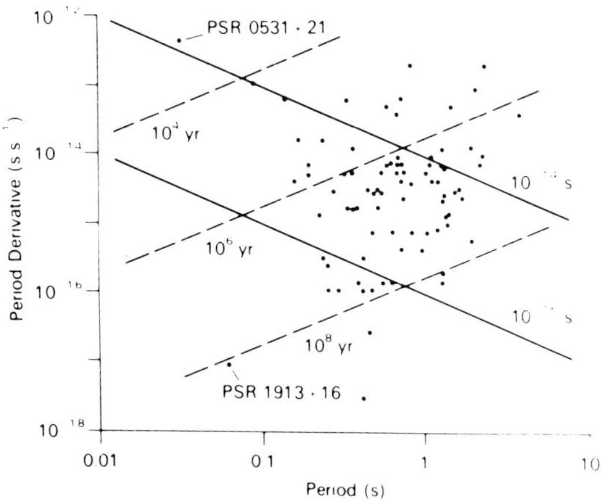

Fig. 7.12. Period derivatives $\dot{P}$ plotted (at a logarithmic scale) against periods P for 83 pulsars. The upper solid line and the lower solid line correspond to a constant value D of $P\dot{P}$ equal to 10^{-14} and 10^{-16} s respectively. The dashed lines correspond to constant values of $T(t_0)$ given by Equation (108). [From *Pulsars* by R. N. Manchester and J. H. Taylor. Copyright 1977 by W. H. Freeman and Co. Reproduced by permission of the authors and the publisher.]

PSR 1913+16 lies much below the line $D = 2 \times 10^{-15}$ s, since the corresponding value of $P\dot{P}$ is roughly 5×10^{-19} s.

Nevertheless, it is usual to assume, in a very first approximation, that all pulsars obey the empirical relation:

$$P_N(t_0)\dot{P}_N(t_0) \;=\; 2 \times 10^{-15}\,\text{s}, \tag{99}$$

whatever the value of N. (However, it will appear later that the *constancy* of D is more important than its numerical value!)

4.2. THE CONCEPT OF 'STANDARD PULSAR' AND THE ESTIMATE OF THE PRESENT AGE OF A PULSAR BASED ON THE VALUES OF ITS P AND $\dot{P}$

Let now $T_N(t)$ denote the interval of time separating the moment t from the moment $t_{f,N}$ of *formation* of the Nth pulsar:

$$T_N(t) \;=\; t - t_{f,N}. \tag{100}$$

It is natural to consider $T_N(t_0)$ as the present 'age' of the Nth pulsar.

A rough estimate of $T_N(t_0)$ based on the observed values of P and of $\dot{P}$ is possible if the following assumptions are accepted:

Assumption No. 1

All pulsars are born with an almost infinite angular velocity of rotation Ω.

Since P is inversely proportional to Ω this assumption means that $P(t_{f,N})$ is assumed to be zero for all pulsars.

Assumption No. 2

All pulsars behave in the same way, like a '*standard pulsar*' (S), concerning the decrease of their angular velocity Ω as a function time. Or, equivalently, concerning the increase of their period P.

Thus if t_f represents the time of formation of (S) and $T(t)$ its age at time t, so that

$$T(t) = t - t_f, \tag{101}$$

the law of increase $P(t)$ of the period P of (S), considered as a function of $T(t)$, will represent the law of variation

$$P = P(t) = \Phi(T), \tag{102}$$

of the period P_N of any pulsar as a function of its age T_N.

Assumption No. 3

The empirical relation (97), (whatever the value of the constant D) holds not only for the present time t_0 but for any time t, i.e. for any value of $T(t)$.

Then according to assumptions No. 1 and No. 2, applied to the standard pulsar, we shall have

$$\Phi(0) = 0, \tag{103}$$

and according to assumptions No. 2 and No. 3, applied to the standard pulsar, we shall have

$$\frac{d\Phi}{dT} = \frac{D}{\Phi(T)}. \tag{104}$$

On integrating this differential equation we find

$$\Phi^2 = 2DT + \text{const.}. \tag{105}$$

Hence, on taking into account the condition (103):

$$\Phi^2 = 2DT. \tag{106}$$

Elimination of the constant D between Equations (104) and (106) yields:

$$T(t) = \tfrac{1}{2}\left[\Phi(T)\Big/\frac{d\Phi}{dT}\right] = \tfrac{1}{2}[P(t)/\dot{P}(t)] \tag{107}$$

or, on applying Equation (107) to the present time t_0:

$$\boxed{T(t_0) = \tfrac{1}{2}[P(t_0)/\dot{P}(t_0)].} \tag{108}$$

If, according to assumption No. 2 we apply this 'standard law' to all pulsars with known P and $\dot{P}$ we obtain a quantity which can be considered (in the frame of all three assumptions) as an estimate of the age of the corresponding pulsar (indicated in Tables 7.1 and 7.2).

It is remarkable that Manchester and Taylor's ingenious representation of $\dot{P}$ against P, at a logarithmic scale, given by Figure 7.12, also clearly shows the values (and dispersion) of the ages $T(t_0)$ given by Equation (108). Indeed, this Equation is equivalent to

$$\log_{10} \dot{P} = + \log_{10} P - \log_{10} [2T(t_0)], \tag{109}$$

so that, in Figure 7.12, the dashed straight lines of a constant positive slope correspond to constant values of the age $T(t_0)$, indicated in the figure.

We see, in Figure 7.12 and in Tables 7.1 and 7.2, that these ages $T(t_0)$ are roughly distributed between 10^4 and 10^8 yr, with two extreme cases: that of the Crab pulsar (PSR 0531+21), for which $T(t_0)$ is equal to 1240 yr and the pulsar PSR 1952+29 for which $T(t_0)$ is equal to 3600×10^6 yr.

The age $T(t_0)$ of the Crab pulsar is somewhat larger but in reasonable agreement with the actual age (deduced from the date of the supernova explosion), viz. 920 yr (for $t_0 =$ 1974).

Similarly the age $T(t_0)$ of the Vela pulsar, of about 1.1×10^4 yr is consistent with the estimates of $(1 - 3) \times 10^4$ yr for the age of the Vela supernova remnant by Clark and Caswell (1976).

However, no conclusion can be drawn from these two particular cases concerning the validity of assumptions No. 1, No. 2 and No. 3, and consequently concerning the assumption that $T(t_0)$ really represents the true age of a pulsar.

These reservations are supported by statistics concerning the kinematics and the $|z|$-distribution (z being the 'height' above the galactic plane), of pulsars which show that pulsars velocities (proper motions) and $|z|$-distribution are usually in contradiction with the assumption that they are born near the galactic plane where lie their genitors (O and B stars) and an age equal to $T(t_0)$.

From such arguments concerning about 10 pulsars whose proper motion, distance d, and distance $|z|$ to the galactic plane are rather well known, their true age must be less than $T(t_0)$ *by a factor of about four*. For more details see MT, pp. 160–168.

Similarly, if our assumptions correspond to reality, we would have, for all pulsars, a relation resulting from Equation (106), viz.

$$T(t_0) = \frac{1}{2D} P^2(t_0), \tag{110}$$

i.e. the age $T(t_0)$ would be a monotonously increasing function of the period P, and consequently of the number N in Tables 7.1 and 7.2. The case of the binary pulsar PSR 1913+16 is in contradiction with such a law, since it has the second shortest period (No. 2) of only 0.0590 s and, instead of a small age, the largest age of 106×10^6 yr after the pulsars PSR 1952+29 and PSR 1944+17 whose $T(t_0)$ are equal to 3600×10^6 and 286×10^6 yr respectively, in spite of relatively small periods of about 0.43–0.44 s.

It is thus quite possible that assumption No. 1 is quite wrong, and that the angular velocity Ω_f of pulsars, at the time of their formation, is far from being infinite. In this case, the true age would be smaller than the one given by Equation (108), in conformity with the results of the discussion concerning the $|z|$-distribution.

Indeed, on replacing the boundary condition (103) by

$$\Phi(0) = \Phi f \neq 0, \tag{111}$$

one easily finds that Equation (108) is replaced by

$$T(t_0) = [\tfrac{1}{2}P(t_0)/\dot{P}(t_0)] - [\tfrac{1}{2}P_f^2/P(t_0)\dot{P}(t_0)]. \qquad (112)$$

On the other hand, assumption No. 2, typical of argument currently used in astrophysics, has already been criticized in the following way (in particular by the late Russian astronomer Parenago).

When, at present time, one observes bassets, spaniels, or great danes, one can be tempted to construct an 'evolutive' theory of the 'standard dog', and assume that bassets simply represent an early stage whereas the great danes represent a late stage of the evolution of the 'standard dog'. Such is the danger of considering as belonging to different evolutionary stages of a standard representative for objects of different races.

Finally, one can note that large values of $T(t_0)$ generally correspond to small values of $\dot{P}$, difficult to measure in limited time intervals. Moreover, extrapolation to large values of $T(t_0)$ obviously should imply the knowledge of derivatives of P of an order higher than the first.

5. Luminosity Problems and the Pacini Model

5.1. THE ROTATIONAL LUMINOSITY OF PULSARS

Suppose, in a first approximation, that the main source of energy radiated by a pulsar (at the totality of all frequencies) can be attributed to the loss of its rotational energy, associated with the decrease in its angular velocity $\Omega = 2\pi/P$.

If $E_{\rm rot}(t)$ denote this rotational energy at time t, with this assumption the corresponding *'rotational luminosity'* $L_{\rm rot}$ will be given by

$$L_{\rm rot}(t) = -dE_{\rm rot}(t)/dt. \qquad (113)$$

As an additional assumption, suppose that the moment of inertia I of the spinning neutron star is constant, and, as a further first approximation, suppose that it is given by the moment of inertia of a *homogeneous* sphere of mass M and radius R, viz.

$$I = \tfrac{2}{5}MR^2. \qquad (114)$$

The neutron star typical radius R (see Chapter 6, Section 10) is of the order of 10 km, and its mass M is of the order the mass $M_\odot$ of the Sun, i.e. about 2×10^{33} g. Thus I given by Equation (114) would be of the order of $8 \times 10^{44}\,\mathrm{g\,cm^2}$.

However, as noted by Ruderman (1972, p. 430), a star is never homogeneous, and its mass is generally concentrated towards its center (see Chapter 3, Section 8). This explains why more precise estimates ascribe to I a value between 7×10^{43} and $7 \times 10^{44}\,\mathrm{g\,cm^2}$. Therefore, it is usual to adopt the standard value $I_s = 1.6 \times 10^{44}\,\mathrm{g\,cm^2}$, (though MT, p. 177, adopts the value $I_s' = 10 \times 10^{44}\,\mathrm{g\,cm^2}$).

With all these assumptions $E_{\rm rot}(t)$ will be given by

$$E_{\rm rot} \approx \tfrac{1}{2}I_s\Omega^2 \approx \tfrac{1}{2}I_s\left(\frac{2\pi}{P}\right)^2, \qquad (115)$$

hence, according to Equation (113):

$$L_{\rm rot}(t) \approx 4\pi^2 I_s (P\dot{P})P^{-4}. \qquad (116)$$

For the Crab pulsar PSR 0531+21 we have $(P\dot{P}) \approx 14 \times 10^{-15}$ s, so that (with I equal to I_s):

$$[L_{\mathrm{rot}}(t)]_{\mathrm{Crab}} \approx 0.7 \times 10^{38} \text{ erg s}^{-1}. \tag{117}$$

(With I equal to I_s' we would find 4.7×10^{38} erg s^{-1}.)

More generally, for all pulsars, in a very rough first approximation we can take for $P\dot{P}$ the standard value given by Equation (99). Then we find, with $I = I_s$:

$$L_{\mathrm{rot}}(t) \approx 1.3 \times 10^{31} P^{-4} \text{ erg s}^{-1}. \tag{118}$$

5.1.1. *Remark*

It was shown in Chapter 5 (Appendix) that the gravitational binding energy B_g^+ of an homogeneous star of mass M and radius R is given by

$$B_g^+ = \tfrac{3}{5} GM^2/R, \tag{119}$$

where G is the gravitational constant. Thus, for a homogeneous standard neutron star considered above ($M \approx M_\odot \approx 2 \times 10^{33}$ g and $R \approx 10^6$ cm) B_g^+ would be of the order of 1.6×10^{53} erg, corresponding to a potential gravitational energy equal to $(-B_g^+)_n \approx -1.6 \times 10^{53}$ erg.

On the other hand, the same relation (119) yields for a star (S) of M' solar masses and of a radius equal to R' solar radii ($R_\odot \approx 7 \times 10^{10}$ cm) a value of B_g^+ of the order of $2.3 \times 10^{48} M'^2/R'$ erg, corresponding to a potential gravitational energy equal to $(-B_g^+)_s \approx -2.3 \times 10^{48} M'^2/R'$ erg.

Thus we see that for all reasonable values of M' and R' (see Chapter 8, Section 3) proper to O and B stars generating supernovae and (as residue) neutron stars, $|(-B_g^+)_s| \ll |(-B_g^+)_n|$.

Now, the energy liberated by the explosion of a supernova, leaving as residue a neutron star, must be equal to the difference between the initial potential energy (of the star S) and the final potential energy (of the neutron star N), i.e. to

$$[(-B_g^+)_s - (-B_g^+)_n] \approx (B_g^+)_n \approx 1.6 \times 10^{53} \text{ erg}. \tag{120}$$

However, according to observations, the energy liberated in the formation of a nebula such as the Crab Nebula, by explosion of a supernova represents only 3×10^{49} ergs.

The difference between 1.6×10^{53} and 3×10^{49} ergs must represent the initial rotational energy of the rotating neutron star (plus eventually the kinetic energy of translation of the neutron star). This difference, of the order of 10^{53} erg, is quite sufficient to account for a loss of $L_{\mathrm{rot}}(t)$ erg per second of the order given by Equation (117), i.e. of the order of 10^{38} erg s^{-1}, during a time of 10^{15} s $\approx 10^8$ yr much longer than the 'age' of the Crab pulsar.

5.2. The Radio Luminosities of Pulsars

If W_e denotes the 'mean' duration (defined as a temporal *equivalent width*) of the pulse, this interval of time will determine, in a first approximation, the very small angular diameter θ of the corresponding beam, in radians or degrees, (with W_e and P in seconds) by

$$\theta = 2\pi(W_e/P) \text{ radian} = 360^\circ(W_e/P), \tag{121}$$

hence (this relation is proved in Subsection 9.1 of Chapter 12) the solid angle Ω' of the beam (with the same units for W_e and P):

$$\Omega' \approx \pi \left(\frac{\theta}{2}\right)^2 \approx \pi^3 (W_e/P)^2 \text{ steradian.} \tag{122}$$

At a distance d cm from the pulsar the total area A covered by the beam will be given by

$$A = d^2\Omega' \approx \pi^3 d^2 (W_e/P)^2 \text{ cm}^2. \tag{123}$$

If the energy (in ergs) received per cm^2, per Hz, in W_e seconds of mean duration of a pulse, near the carrier wave frequency of 400 MHz is denoted by E_{400} the *'monochromatic radio brightness'* B_{400}, i.e. the corresponding energy *per second* will be given by

$$B_{400} = E_{400}/W_e. \tag{124}$$

Let finally $\Delta\nu$ denote the (more or less arbitrary) range of frequencies considered as the *radio*-range of carrier frequencies, in Hz.

Then the corresponding *'radio luminosity'* of pulses L_{radio} will be given, in erg s^{-1}, by

$$L_{\text{radio}} \approx A\,\Delta\nu B_{400} \approx \pi^3 d^2 W_e P^{-2} E_{400} \Delta\nu \text{ erg s}^{-1}. \tag{125}$$

On the other hand, if we introduce the *'flux density'* S_{400} defined as the mean energy (in ergs) received in form of pulses, *per second*, per cm^2 and per Hz, near 400 MHz, i.e. the pulse energy E_{400} divided by the period P, we have

$$S_{400} = E_{400}/P. \tag{126}$$

On eliminating E_{400} between Equations (125) and (126) we obtain (in c.g.s. units):

$$L_{\text{radio}} \approx \pi^3 d^2 (W_e/P) S_{400} \Delta\nu \text{ erg s}^{-1}. \tag{127}$$

Some of the largest values of S_{400} are given in Table 7.3 (in units of milliJanskys: 1 mJy $= 10^{-29}$ W m^{-2} Hz^{-1}).

If we adopt with MT (p. 158) a relatively small value of $\Delta\nu$ equal to 400 MHz, and express d in kpc and S_{400} in mJy, Equation (127) will take the form:

$$L_{\text{radio}} \approx 1.2 \times 10^{27} (d \text{ kpc})^2 (S_{400} \text{ mJy})(W_e/P) \text{ erg s}^{-1}. \tag{127$'$}$$

If, in addition, we adopt, for statistical purposes, a standard value of (W_e/P) equal to 1/36 (corresponding to a conical radiation beam of width 10°) L_{rad} will be given in erg s^{-1} by

$$L_{\text{radio}} \approx 3.4 \times 10^{25} (d \text{ kpc})^2 (S_{400} \text{ mJy}) \text{ erg s}^{-1}. \tag{127$''$}$$

For the Crab pulsar we have, according to Table 7.3:

$$d \approx 2.0 \text{ kpc}; \qquad W_e/P \approx 0.057; \qquad S_{400} \approx 480 \text{ mJy}. \tag{128}$$

Hence, on applying Equation (127′):

$$(L_{\text{radio}})_{\text{Crab}} \approx 1.3 \times 10^{29} \text{ erg s}^{-1}. \tag{129}$$

Most pulsars have, according to MT, p. 203, radio luminosities (with $\Delta\nu = 400$ MHz) given by Equation (127′) in the range 10^{26}–10^{30} erg s^{-1}, thus generally considerably smaller than the radio luminosity of the Crab pulsar. (Note that, in a such comparison, the value adopted for $\Delta\nu$ has no importance.)

It is interesting to compare radio luminosities of pulsars to their luminosities in other domains of frequency. According to MT (pp. 72 and 202), at infrared and higher frequencies the luminosity of the Crab pulsar certainly exceeds 7.5×10^{34} erg s^{-1} and the luminosity of the optical and X-ray emission from the Crab pulsar is about 10^{35} erg s^{-1}.

TABLE 7.3

The value of the flux density S at 400 MHz S_{400} (in mJy), for the ten 'brightest' (in S_{400}) pulsars. Distances marked by an asterisk (in kpc) are obtained by HI measures or by identification with a supernova remnant. The column W_e gives pulse 'equivalent widths' (mean durations), in ms, defined as pulse energy (obtained by integration in time of each pulse profile) divided by peak flux density. The column (W_e/P) shows that the 'standard' value of $1/36 \approx 0.028$ adopted in Equation (127″) is quite reasonable.

PSR	S_{400} (mJy)	W_e (ms)	d (kpc)	W_e/P	P (s)	Notes
0833–45	2800	1.71	2.4*	0.019	0.0892	Vela pulsar
0329+54	2270	8.7	2.6*	0.012	0.7145	
1641–45	1300	9	4.9*	0.02	0.4550	
1749–28	1070	7	1.0*	0.01	0.5625	
1451–68	840	13	0.3	0.049	0.2633	
1929+10	610	5.5	0.1	0.024	0.2265	
0531+21	480	1.9	2.0*	0.057	0.0331	Crab pulsar
1933+16	360	6.5	6.0*	0.018	0.3587	
0740–28	360	8	2.0*	0.05	0.1667	
1642–03	350	4.0	0.2*	0.01	0.3876	

5.3. Pacini's Model

The 0.7×10^{38} erg s^{-1} supplied (with $I = I_s$, see Subsection 5.1) by the slowdown of the rotation of the Crab pulsar are, according to the figures quoted above, obviously much larger than the sum of pulsar luminosities at all observed frequencies.

More generally, as noted by MT (p. 110), except for the recently observed γ-ray emission from PSR 1747–46 and PSR 1818–04, the observed pulses constitute a very small fraction of the total rotational luminosity $L_{\rm rot}$ (energy loss) of a pulsar.

It is now, therefore, usually assumed that the Crab pulsar, more or less directly, accelerates particles to ultra-relativistic energies required to produce the optical and X-ray emission of the Crab *Nebula* (by synchrotron process).

The loss of rotational energy is just of the proper order to account for the nebular luminosity, especially if the value I_s' is used for I instead of I_s.

As early as in 1967, prior to the discovery of pulsars, the Italian astronomer Pacini pointed out that a rapidly rotating magnetized neutron star would radiate significant amounts of energy in form of electromagnetic waves, at the rotation frequency. The observed effects in this interpretation would result from the non-alignment of the magnetic and the rotation axes. On assuming that these axes are orthogonal, Pacini

(1968) and later Ostriker and Gunn (1969), found a relation between the radius R, the moment of inertia I, the product $P\dot{P}$ of a pulsar, and the surface magnetic field strength B_0.

For a neutron star with a radius R of 10^6 cm and a moment of inertia I_s' of 10×10^{44} g cm^2 they obtain, in Gaussian units:

$$B_0 \approx 3.2 \times 10^{19}(P\dot{P})^{1/2}. \tag{130}$$

The values of B_0 given by Equation (130) range between about 2×10^{10} and 2×10^{13} G, with fields about 10^{12} G being typical. For the Crab pulsar Equation (130) yields a value of B_0 of the order of 4×10^{12} G.

The same model shows that the reaction torque N transmitted to the star by the magnetic field is proportional to $\Omega^3 = (2\pi/P)^3$. Hence a rate of variation of Ω of the form

$$\dot{\Omega} = \frac{d\Omega}{dt} = -K\Omega^3, \tag{131}$$

where K is a constant.

5.4. The Braking Index and the General 'Characteristic Age'

More generally, in most theoretical models the braking torque is proportional to some power n of Ω (n is called the '*braking index*'):

$$\dot{\Omega} = \frac{d\Omega}{dt} = -K\Omega^n, \tag{132}$$

which is equivalent to

$$\dot{P} = \frac{dP}{dt} = DP^{(2-n)}, \tag{132'}$$

where D is a constant.

In the Pacini model $n = 3$ and Equation (132′) becomes

$$P\dot{P} = D, \tag{133}$$

in agreement with the rough empirical relation (97).

If we return to the theory of the age of the 'standard pulsar', discussed in Section 4, we can replace, in the general Equation (132′), the function $P(t)$ by $\Phi(T)$ and write

$$\frac{d\Phi}{dT} = D\Phi^{(2-n)}. \tag{134}$$

On integrating this differential equation with the boundary condition (103) corresponding to the assumption No. 1 (see Subsection 4.2), viz.

$$\Phi(0) = 0, \tag{135}$$

we obtain

$$T = \frac{\Phi^{(n-1)}}{(n-1)D}. \tag{136}$$

Elimination of D between Equations (134) and (136) yields:

$$T = \frac{1}{(n-1)}\left[\Phi(T) \Big/ \frac{d\Phi}{dT}\right]. \tag{137}$$

Hence, by application of the same assumptions as in Subsection 4.2, the 'characteristic age' $T_n(t_0)$:

$$T_n(t_0) = \frac{1}{(n-1)}[P(t_0)/\dot{P}(t_0)], \tag{138}$$

which represents a generalization of Equation (108) for the case when n is different from 3.

Pacini's value $n = 3$ of the braking index is generally used in Tables giving the 'characteristic age' of pulsars, because of its agreement with the empirical relation (97). (For more sophisticated models of rotating neutron stars and of pulse emission mechanisms see MT, Chapters 9 and 10.)

6. The Problem of Association of Pulsars with Supernovae

We have already mentioned in Subsection 2.1.5 the remarkable association of the Crab pulsar and of the Vela pulsar with remnants of supernovae. Such associations suggest that pulsars are formed, as rotating neutron stars, through explosions, observed as supernova phenomenon, of massive O and B stars.

However, among about 150 pulsars listed in MT, and approximately 120 known galactic supernova remnants (SNR) the evidence for associations on a one-for-one basis is not very strong.

Thus, for instance, it seems that there is no association between the brightest of radio sources Cas A (whose flux density S_{400}, at 400 MHz, is about five times larger than that of the Crab Nebula) and any pulsar. The nearest known pulsar lies about 1.6° from the 'center' of Cas A. Nevertheless, because of the very brightness of Cas A, any eventual pulsar inside Cas A would be undetectable unless much more luminous than the Crab pulsar (which itself is exceptionally bright).

The associations such as those of PSR 0611+22 with SNR IC 443 must be regarded according to MT (p. 165) as 'tentative at best'. A number of other possible pulsar-SNR associations have been proposed, but in no case is the evidence compelling (see MT, p. 165, f).

Because of the great age of most pulsars it is quite possible that the corresponding supernova remnant already dissipated to such an extent as to become undetectable if this dissipation is more rapid than the braking of pulsar's rotation.

However, the problem remains why the vast majority of supernova remnants do not have detectable young pulsars embedded in them. For most of the SNRs listed by MT as 'questionable associations' or 'no association' an undetected pulsar of *low luminosity* ($L_{\text{radio}} \leqslant 10^{29}$ erg s^{-1}) is not excluded by the observations.

According to MT (p. 168) a comparison between the birth rate of observable SNR with the observed number and lifetimes of pulsars, suggests that "some, perhaps the majority, of pulsars are born in less spectacular circumstances than the explosive detonation of a high-mass star".

Nevertheless, in spite of all these difficulties, it remains noteworthy that *the space distribution* of pulsars, near the galactic plane (see Subsection 3.2.1.3) corresponds to the galactic distribution of Population I 'young stars'. Many of these stars are very massive and *statistically* correspond to the distribution of supernovae belonging to the Population I (called, in a misleading terminology, supernovae of type II).

A similar statistical correspondence exists concerning the 'concentration towards the galactic center' in projection over the galactic plane. (For more details see MT, pp. 153–160.)

7. The 'Celibacy' of Radio Pulsars (and Binary Character of the 'X-ray' Pulsars)

The few known 'X-ray' pulsars discussed in Chapter 9, apart for a somewhat longer period, are physically very similar to the 'radio-pulsars' discussed hitherto.

However, all X-ray pulsars are members of *binary systems*, whereas the radio-pulsars are (with just one exception: PSR 1913+16) *solitary* objects. This is extremely surprising since about half of all stars in the Galaxy seems to be members of binary or even multiple systems.

If, in spite of the difficulties discussed in Section 6, one assumes that neutron stars and in particular pulsars, are formed by supernova explosions, two problems arise: are the radio-pulsars 'divorced' members of previously binary systems? Can a fraction of neutron stars created in close binary systems remain orbitally bound to their unexploded companions, and why are they observed as X-ray pulsars?

A partial answer was given to the second question by Wheeler *et al.* (1975) who didn't explain why (except 1913+16) only X-ray pulsars are members of binary systems, but found that neutron stars *can* be formed in close binary systems with a resulting *orbit of finite eccentricity*.

But this result deepens the mystery of the 'solitude' of radio-pulsars, and makes the very low incidence of the majority of pulsars (radio-pulsars) in orbiting systems more surprising.

An attempt to explain, in a very rough way, the solitude of radio pulsars as the result of a 'divorce' of a binary system is provided by the following considerations of Van den Heuvel, showing that a massive component of a binary system, on losing a great part of its mass in a supernova explosion, becomes unable to 'retain' gravitationally its companion (just as a ruined husband becomes sometimes unable to 'retain' his wife, formerly attracted by his wealth).

Consider a massive star A, 'married' to a companion B. Initially, the two components A and B of the binary system rotate about the center of gravity G of the system with linear velocities V_A and V_B inversely proportional to the respective masses M_A and M_B. (See Chapter 8).

This immediately results from the derivation with respect to time t of the relation

$$M_A \mathbf{r}_A + M_B \mathbf{r}_B = 0, \tag{139}$$

expressing that the radius vector $\mathbf{r}_G$ of the center of gravity G is zero when the motion is referred to an inertial system with origin at G.

Let α denote the fraction of M_A 'volatilized' and transformed into a diffuse nebula, by the supernova explosion of the component A. Then the mass of the residual neutron star N will be $(1-\alpha)M_A$.

Now, it is well known that the escape velocity V_e of any mass m placed in a gravitation field of a spherical homogeneous mass M_A at a distance R from the center O_A of a star A is given by

$$V_e = \left(\frac{2GM_A}{R}\right)^{1/2} \approx (1.4)(GM_A/R)^{1/2}, \qquad (140)$$

where G is the gravitational constant. Indeed, the gain in potential energy in removing m from the distance R (from O_A) to infinity equal to $GM_A m/R$ must be equal to the loss in kinetic energy $\frac{1}{2}mV_e^2$ from launching to zero velocity at infinity.

On the other hand, the linear orbital velocity $V_{orb,B}$ of a mass M_B gravitating in the field of M_A, in a relative circular orbit, is easily deduced from the Newtonian form of Kepler's third law, simplified by the assumption that $M_B \ll M_A$ (see Chapter 8, Section 1), viz.

$$R^3\omega^2 = G(M_A + M_B) \approx GM_A, \qquad (141)$$

where R is the distance between A and B and where ω is the angular relative velocity.

Indeed, we have

$$V_{orb,B} = R\omega. \qquad (142)$$

Hence:

$$V_{orb,B} \approx (GM_A/R)^{1/2}. \qquad (143)$$

If, for the sake of clarity, we denote by $(V_e)_{B/A}$ the escape velocity of B in the field of A, we obtain on comparing Equations (140) and (143):

$$(V_e)_{B/A} \approx 2^{1/2}V_{orb,B} \approx (1.4)(GM_A/R)^{1/2}. \qquad (144)$$

This relation is illustrated by the well known fact that, in order to escape the solar gravitational attraction, a cosmic missile must be launched from the Earth with a velocity of about 42 km s^{-1}, whereas the orbital velocity of the Earth is equal to about 30 km s^{-1}.

It is physically obvious, and confirmed by Equation (144), that before the explosion of the star A, the orbital velocity $V_{orb,B}$ of the star B, is *lower* than the escape velocity $(V_e)_{B/A}$:

$$V_{orb,B} < (V_e)_{B/A}. \qquad (145)$$

Now, if a sufficiently important fraction α, for instance $\alpha = \frac{3}{4}$, of the mass M_A of A is 'volatilized' by an explosion reducing the rest of A to the state of a collapsed neutron star N of a mass equal to $\frac{1}{4}M_A$, the new escape velocity of B in the field of its new companion N will be given, according to Equation (140) by

$$(V_e)_{B/N} \approx (1.4)(G\tfrac{1}{4}M_A/R)^{1/2} \approx (0.7)(GM_A/R)^{1/2}, \qquad (146)$$

hence, according to Equation (143):

$$(V_e)_{B/N} \approx (0.7)\, V_{orb,B}, \qquad (147)$$

equivalent to

$$V_{orb,B} \approx (1.4)(V_e)_{B/N}. \qquad (147')$$

If one makes the additional assumption that after the explosion of the star A the linear velocity V'_B remains equal to its previous value $V_{orb,B}$, one finds, in the new situation, according to Equation (147′):

$$V_B' \approx V_{orb,B} \approx (1.4)(V_e)_{B/N}, \tag{148}$$

which means that just after the explosion the velocity V_B' will exceed the escape velocity of B in the new system (B, N).

After the reduction of A to the state of a neutron star N, the gravitational attraction of N upon B will not be sufficient to bind B to N, since the (unchanged) remainder V_B' of the former orbital velocity of B is now larger than the escape velocity of B. Thus, the component B would escape and leave the pulsar (neutron star N) in a state of 'divorced' (or 'abandoned') object.

This theory is very ingenious, and attractive by its simplicity. However, it does not seem to explain in a quite satisfactory way the enigma of radio-pulsar's 'solitude'. The problem must be considered as not yet solved.

7.1. Binary Radio Pulsar PSR 1913+16

As already mentioned above, the non-binary character of the *radio* pulsars suffers only one exception: PSR 1913+16.

This binary pulsar was discovered, in July 1974, by Hulse and Taylor, during a systematic search of the galactic plane for new pulsars (Hulse and Taylor, 1975a, b).

As reported by MT (p. 91): "This pulsar was of immediate interest because its period of 0.059 s was less than that of any known pulsar except the one in the Crab Nebula. It soon became clear that the large cyclic variations observed in its period could be easily understood if the pulsar was in orbit about another massive object."

Indeed, instead of remaining constant (except for a slow secular increase of the order of 10^{-15} s s^{-1}) as for other radio pulsars, the period of this pulsar varies systematically between 0.058 967 and 0.059 045 s, within a cycle of 7.75 h ($\sim$ 0.3230 d). A velocity curve derived from an interpretation of these variations of period as a Doppler effect due to an orbital motion (after correction for the Doppler effect of the orbital motion of the Earth) was shown to be consistent with an orbit of eccentricity $e = 0.62$. No eclipses are observable (contrary to what happens for most binary X-ray pulsars studied in Chapter 9). At the present time PSR 1913+16 is detectable only at radio wavelengths, and *its companion has not been directly observed*. However it is quite certain that the unseen companion is also a compact object, with a mass comparable to that of the pulsar.

The high rate of advance $d\upsilon_0/dt$ of the periastron (see Chapter 8, Subsection 2.7.1) ($d\upsilon_0/dt = 4.22 \pm 0.04$ deg yr^{-1}), a phenomenon analogous to the famous advance of perihelion of Mercury, suggests that through further observations of this binary system a number of interesting gravitational and relativistic phenomena will permit to test different gravitational theories. (For a more complete discussion the reader is referred to MT, pp. 91–96.)

References

1. Works Cited in the Text

(Works that can be used for further study are marked by an asterisk.)

Ables, J. G. and Manchester, R. N.: 1976, *Astron. Astrophys.* **50**, 177.

Allen, C. W.: 1973, *Astrophysical Quantities* (3rd edn.), Athlone Press, London.

Baade, W. and Zwicky, F.: 1934, *Proc. Nat. Acad. Sci.* **20**, 254.
Baym, W., Penthick, C., Pines, D., and Ruderman, M.: 1969, *Nature* **224**, 872.
Brillouin, L.: 1960, *Wave Propagation and Group Velocity*, Academic Press, New York.
Clark, D. H. and Caswell, J. L.: 1976, *Mon. Not. Roy. Astron. Soc.* **174**, 267.
Cocke, W. J., Disney, M. J., and Taylor, D. J.: 1969, *Nature* **221**, 525 and 527.
Crawford, Jr, F. S.: 1968, *Waves* (Berkeley Physics Course, Vol. III), McGraw-Hill, New York.
Ginzburg, V. L.: 1970, *The Propagation of Electromagnetic Waves in Plasmas* (2nd edn.), Pergamon, New York.
* Ginzburg, V. L. and Zheleznyakov, V. V.: 1975, 'On the Pulsar Emission Mechanisms', *Ann. Rev. Astron. Astrophys.* **13**, 511.
Gold, T.: 1968, *Nature* **218**, 731.
Gordon, C. P., Gordon, K. L., and Shalloway, A. M.: 1969, *Nature* **222**, 129.
Gordon, K. J. and Gordon, C. P.: 1973, *Astron. Astrophys.* **27**, 119.
Graham, D. A., Mebold, U., Hesse, K. H., Hills, D. L., and Wielebinski, R.: 1974, *Astron. Astrophys.* **37**, 405.
Hewish, A., Bell, S. J., Pilkington, J. D. H., Scott, P. F., and Collins, R. A.: 1968, 'Observation of a Rapidly Pulsating Radio Source', *Nature* **217**, 709.
Hulse, R. A. and Taylor, J. H.: 1975a, *Astrophys. J.* (*Letters*), **195**, L51; Hulse, R. A. and Taylor, J. H.: 1975b, *Astrophys. J.* (*Letters*), **201**, L55.
* Lovell, B.: *Out of the Zenith*, Oxford Univ. Press, London.
* Manchester, R. N. and Taylor, J. H.: 1977, *Pulsars*, Freeman, San Francisco.
Miller, J. S. and Wampler, E. J.: 1969, *Nature* **221**, 1037.
MT – See Manchester and Taylor, 1977.
Nather, R. E., Warner, B., and Macfarlane, M.: 1969, *Nature* **221**, 527.
Ostriker, J. P. and Gunn, J. E.: 1969, *Astrophys. J.* **157**, 1391.
Pacini, F.: 1967, *Nature* **216**, 567.
Pacini, F.: 1968, *Nature* **221**, 454.
Rickett, B. J. and Lyne, A. G.: 1968, *Nature* **218**, 326.
Ruderman, M. A.: 1972, *Ann. Rev. Astron. Astrophys.* **10**, 427.
Smith, F. G.: 1968, in *Pulsating Stars*, Vol. I, MacMillan, London (p. VI).
Staelin, C. H. and Reifenstein, E. C.: 1968, *Science* **162**, 1481.
Terzian, Y.: 1970, in A. M. Lenchek (ed.), *The Physics of Pulsars*, Gordon and Breach, N.Y. (pp. 85–109).
Trimble, V.: 1968, *Astron. J.* **73**, 535.
Wade, N.: 1975, 'Discovery of Pulsars: A Graduate Student's Story', *Science* **189**, 358.
Wheeler, J. C., Lecar, M., and McKee, C. F.: 1975, *Astrophys. J.* **200**, 145.
Wheeler, J. A.: 1966, *Ann. Rev. Astron. Astrophys.* **4**, 393.
Willstrop, R. V.: 1969, *Nature* **221**, 1023.

2. REFERENCES PARTLY USED IN THE TEXT AND WHICH CAN BE USED FOR FURTHER STUDY

(See first the works cited in (1) marked by an asterisk.)
MN = *Monthly Notices of Roy. Astron. Soc.*; *AA* = *Astron. Astrophys.*; *ApJ* = *Astrophys. J.*; *IAUC* = *Int. Astron. Union Circulars*; *Nat* = *Nature*

General Monographs

Smith, F. G.: 1977, *Pulsars*, Cambridge Univ. Press. Cambridge, G.B.
Taylor, J. H. and Manchester, R. M.: 1977, *Ann. Rev. Astron. Astrophys.* **15**, 19.

General Theories

1978, *ApJ*, **220**, 1101; **222**, 197; **222**, 675; **222**, 1006; **224**, 988, **225**, 226; **225**, 557; **225**, 574; **225**, 582.
1978, *MN*, **182**, 157; **182**, 735.
1978, *AA*, **65**, 173; **65**, 179; **66**, 325; **68**, 289.

Discontinuities in the Period Derivative

1978, *MN*, **184**, 35P.

Pulsating White Dwarfs
1978, *ApJ*, **220**, 614.

The Total Number of Known Pulsars
1978, *ApJ* (*Letters*), **225**, L31.

The Problem of the Scarcity of Binary Radio Pulsars
1977, *Nat*, **268**, 606.
1978, *Nat*, **275**, 725.
1978, *IAUC*, No. 3242.

Crab Pulsar and Vela Pulsar
1977, *ApJ*, **216**, 560; **216**, 865.
1977, *AA*, **61**, 279.
1977, *Nat*, **265**, 121; **266**, 692.
1978, *ApJ*, **221**, 268.
1978, *MN*, **182**, 39P; **184**, 159.
1978, *AA*, **69**, 141.
1978, *IAUC*, No. 3274; 3724.

PART III

NEWTON'S LAW, BINARY SYSTEMS AND GALACTIC X-RAY SOURCES

CHAPTER 8

THEORY OF SPECTROSCOPIC AND ECLIPSING BINARIES. STELLAR MASSES

1. The Newtonian Form of Kepler's Third Law

Let us consider a system of two celestial bodies revolving about their center of mass under the influence of their mutual gravitational attraction. Such a system can take various forms: the Sun and one of the planets of the solar system; Earth and Moon; a planet and one of its satellites; a binary star; and a double galaxy.

As will be shown below, the main property of such a system is that it behaves *as if* one of the components of the system were *fixed*, the mass M of this component being simply replaced by the *sum* of masses $(M + m)$ of the two components.

We know, indeed, that the dynamical principle expressed by the classical relation $F = m\gamma$ is valid only when the motion of the 'point mass' m is referred to an inertial reference system (a reference system with respect to which the motion of m, in the absence of the force F, would be rectilinear and uniform).

Now, taking a concrete example, suppose that one of the components of our system is the Sun S of mass M, and the other a planet P of smaller mass m. A reference system attached to the Sun cannot be inertial since the Sun is subject to the gravitational attraction of the planet P (even if we neglect, as we shall, in this 'two-body problem', the action of the other planets).

We shall thus intially consider the equations of motion *in a truly inertial reference system*.

If we call **S** and **P** the 'radius vectors' of S and P respectively, the motion of S under the gravitational attraction of P (with a force F') can be described, in a truly inertial reference system, by the vector equation:

$$\mathbf{S} = \mathbf{S}(t), \tag{1}$$

where $\mathbf{S}(t)$ is a *vectorial function* satisfying the equation

$$\frac{d^2\mathbf{S}}{dt^2} = \gamma_{\mathrm{S}} = \frac{F'}{M}. \tag{2}$$

Putting r for the distance between S and P, Newton's law of gravitation yields (G is the gravitational constant):

$$\frac{d^2\mathbf{S}}{dt^2} = \frac{GmM}{r^2}\frac{1}{M}\frac{(\mathbf{P}-\mathbf{S})}{r} = \frac{Gm}{r^2}\frac{(\mathbf{P}-\mathbf{S})}{r}. \tag{3}$$

Similarly, since the planet P is acted upon by a force F, equal to $(-F')$, its motion will be described by

$$\mathbf{P} = \mathbf{P}(t), \tag{4}$$

with

$$\frac{d^2\mathbf{P}}{dt^2} = \frac{GMm}{r^2}\frac{1}{m}\frac{(\mathbf{S}-\mathbf{P})}{r} = -\frac{GM}{r^2}\frac{(\mathbf{P}-\mathbf{S})}{r}. \tag{5}$$

Subtracting Equation (3) from Equation (5) we obtain:

$$\frac{d^2(\mathbf{P}-\mathbf{S})}{dt^2} = -\frac{G(M+m)}{r^2}\frac{(\mathbf{P}-\mathbf{S})}{r}. \tag{6}$$

Thus, referred to a system whose radius vectors have their origin at S, and whose cartesian axes are parallel to those of the truly inertial system of reference, the *relative motion* of P satisfies an equation of the same form as Equation (7):

$$\frac{d^2\mathbf{P}}{dt^2} = -\frac{GM}{r^2}\frac{\mathbf{P}}{r}. \tag{7}$$

This is precisely the equation which would describe the motion of P if the Sun were *fixed* at the origin of a truly inertial reference system, but with M replaced by $(M+m)$.

According to Equations (3) and (5), added after multiplying by M and m respectively, we also have:

$$M\frac{d^2\mathbf{S}}{dt^2} + m\frac{d^2\mathbf{P}}{dt^2} = \frac{d^2(M\mathbf{S}+m\mathbf{P})}{dt^2} = 0. \tag{8}$$

Therefore, defining (in the truly inertial reference system) a radius vector **G** by

$$(M+m)\mathbf{G} = M\mathbf{S} + m\mathbf{P}, \tag{9}$$

we find that

$$\frac{d^2\mathbf{G}}{dt^2} = 0. \tag{10}$$

Thus, since G is the end of the radius vector **G** (of course, this G is not to be confused with the constant of universal gravitation), *the motion of* G *is inertial*, and *any system of reference attached to* G, with its cartesian axes parallel to those of a truly inertial reference system, *is itself inertial.*

Now, according to the definition (9) of **G**, we have

$$M\mathbf{G} + m\mathbf{G} = M\mathbf{S} + m\mathbf{P}, \tag{11}$$

hence

$$M\mathbf{G} - M\mathbf{S} = m\mathbf{P} - m\mathbf{G}, \tag{12}$$

or

$$M(\mathbf{G}-\mathbf{S}) = m(\mathbf{P}-\mathbf{G}). \tag{13}$$

Thus, G is simply the *center of mass* (or the *center of gravity*) of the system formed by the two bodies S and P. Writing the distances of S and P from G as |GS| and |GP| we have the well known relation:

$$|\mathrm{GS}|M = |\mathrm{GP}|m. \tag{14}$$

We can consider the reference system attached to the center of mass to be truly intertial.

Consider now the simplest case in which the *relative orbit* of P, in the *non-inertial* reference system attached to S, is *circular*. We can then apply Huyghens' formula for the modulus of the centripetal acceleration $|\gamma|$ of a body (a mass point) moving with constant *angular velocity* ω on a circle of radius r, viz.

$$|\gamma| = \omega^2 r = \left| \frac{d^2(\mathbf{P} - \mathbf{S})}{dt^2} \right|. \tag{15}$$

Combined with Equation (6), this gives the *Newtonian form of Kepler's third law* for the circular case, with $|\mathrm{SP}| = r$:

$$\frac{G(M+m)}{r^2} = \omega^2 r \quad \text{or} \quad \boxed{\omega^2 r^3 = G(M+m)}\,. \tag{16}$$

It can be shown, by a relatively elementary analysis (see Appendix A of this Chapter), that the relation (16) remains valid in a more general case of an *elliptical* orbit when ω is replaced by $2\pi/T$, (where T is *the orbital period*) and r by the *semimajor axis* a of the ellipse described by P in its *relative* motion around S. (Note that a is the semimajor axis of the *relative* orbit!)

Thus we have the general law (*Kepler's third law*):

$$\boxed{\frac{a^3}{T^2} = \frac{G(M+m)}{4\pi^2}}\,. \tag{17}$$

In the case of a system of binary stars, Equation (17) can take different forms according to the choice of the units for the quantities involved. To find the corresponding relations it is sufficient to remember that

$$\begin{aligned}
&M = \text{The mass of the Sun} \approx 2 \times 10^{33}\,\text{g};\\
&R = \text{The radius of the Sun} \approx 7 \times 10^{10}\,\text{cm};\\
&\text{One year} \approx 30 \times 10^{6}\,\text{s};\\
&\text{One parsec} = 1\,\text{pc} \approx 3 \times 10^{18}\,\text{cm};\\
&\text{One Astronomical unit} = \text{AU} \approx 15 \times 10^{12}\,\text{cm}.
\end{aligned} \tag{18}$$

2. Elementary Interpretation of Observations

2.1. Introduction

The *physical* aspects of different situations must not be drowned in a too general and abstract formalism. Therefore, we shall consider the different aspects of the problem in an order of *growing complexity*.

2.2. The 'Velocity Curves' in the Case of Circular Orbits and an Observer in the Orbital Plane

Let S be the more massive component of mass M and P be the less massive component of mass m in a binary system of *stars*. Generally M is supposed to be different from m. In the case of binary stars the ratio $\alpha' = m/M$ usually lies between 0.1 and 1.0. Then Equation (14) shows that the radius a_S of the circular orbit described by S around the center of gravity G will be *smaller* than the radius a_P of the circular orbit described by P around G.

The upper part of the Figure 8.1 shows four typical situations I, II, III, IV, for an observer situated in the orbital plane at the bottom of the page.

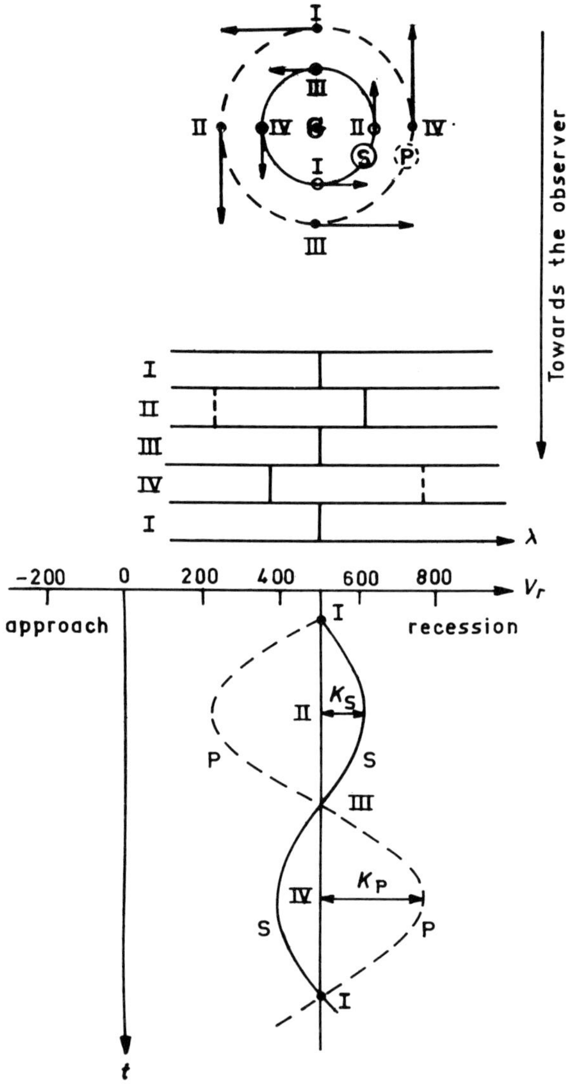

Fig. 8.1. The upper part of the figure shows four typical situations I, II, III, and IV, of the components P and S with respect to the observer. The middle part of the figure indicates the corresponding positions of *one* spectral line affected by the Doppler effect. The lower part of the figure describes the continuous displacement of the same spectral line as a function of time, in terms of radial velocities V_r deduced from the Doppler shifts $\Delta\lambda$ of the line.

The middle part of the figure indicates schematically (by full bars for S and dotted bars for P, unless they overlap) the corresponding positions of *one* spectral line (i.e. a line of a given laboratory wavelength) affected by the Doppler effect.

The lower part in Figure 8.1 describes the continuous displacement of the same spectral line as a function of time t, in terms of radial velocities V_r (expressed in km s^{-1}) deduced from Doppler shifts $\Delta\lambda$ by the classical relation:

$$\frac{\Delta\lambda}{\lambda} = \frac{1}{c} V_r, \tag{19}$$

where $\Delta\lambda$ is the difference $(\lambda_{\text{obs}} - \lambda)$ between the observed wave-length λ_{obs} and the rest (laboratory) wave length λ of the spectral line (e.g. the line H_γ of H). The *'astrophysical radial velocity'* V_r, i.e. the projection of the velocity vector upon the line of sight, is considered as *positive* when $\Delta\lambda$ is positive (i.e. in the case when S *or* P are *receding* with respect to the observer.)

The corresponding curves which represent the variations of V_r as a function of t (full line for S and dotted line for P) are called *'velocity curves'* or more precisely 'curves of radial velocities'.

It is assumed in Figure 8.1 that the mass M of S is double the mass m of P. Thus, the distance $|GP| = a_P$ is double the distance $|GS| = a_S$. The *linear* velocity of P in its uniform circular motion around G is double the linear velocity of S (whereas the *angular* velocities have *the same* value ω).

It is also assumed that the center of gravity G is *receding* from the observer with a *'gamma velocity'* $(V_r)_G = (V_r)_\gamma$ of $500\,\text{km s}^{-1}$. Radial velocities of circular motions around G are subtracted from, or added to, $(V_r)_G$ in the actual observations. Thus the common (overlapping) positions of lines at phases (I) or (III) do *not* correspond to λ but are affected by the gamma velocity.

It is customary to denote by K_S the semiamplitude of the velocity curve relative to S and by K_P the semiamplitude of the velocity curve relative to P. These two quantities are usually expressed in km s^{-1}; the corresponding linear velocities V_S and V_P of circular motions are expressed in the same units.

Obviously $K_S = V_S$ and $K_P = V_P$. If T is the *period*, i.e. the interval of time of two successive identical phases (for instance phases I) we have:

$$V_S = \omega a_S = \frac{2\pi}{T} a_S = K_S, \tag{20}$$

and

$$V_P = \omega a_P = \frac{2\pi}{T} a_P = K_P. \tag{21}$$

Therefore from measures of T, K_S and K_P on the velocity curves of S and P we obtain:

$$a_S = |GS| = \frac{T}{2\pi} K_S, \tag{22}$$

and

$$a_P = |GP| = \frac{T}{2\pi} K_P. \tag{23}$$

According to Equation (17), where a represents $|SP| = a_S + a_P$, we can use Equations (22) and (23) and first write:

$$a = a_S + a_P = \frac{T}{2\pi} K, \tag{24}$$

with

$$K = K_S + K_P, \tag{24'}$$

and then

$$M + m = \frac{4\pi^2}{GT^2} a^3 = \frac{T}{2\pi G} K^3. \tag{25}$$

According to Equation (14) we also have the relations:

$$\frac{M}{(1/a_S)} = \frac{m}{(1/a_P)} = \frac{M+m}{(1/a_S)+(1/a_P)} = \frac{(M+m)a_S a_P}{a} = \frac{K^2 a_S a_P}{G}, \tag{26}$$

which give the individual masses M and m by

$$M = \frac{1}{G} K^2 a_P = \frac{T}{2\pi G} K^2 K_P, \tag{27}$$

and

$$m = \frac{1}{G} K^2 a_S = \frac{T}{2\pi G} K^2 K_S, \tag{28}$$

as a function of observable quantities T, K_S, K_P and $K = K_S + K_P$ (when the reality corresponds to the theoretical model considered up to now: circular orbits in a plane containing the observer).

In this particular case, the velocity vector V_S of S in its circular motion aroung G always remains perpendicular to GS. Therefore, if we call t_I the time corresponding to the phase (I), the (astrophysical) radial velocity (corrected for the radial motion of G) $(V_r)_S$ will be given as a function of t, by

$$(V_r)_S = V_S \sin \omega(t - t_I), \tag{29}$$

where $V_S = |\mathbf{V}_S|$.

Similarly, one easily finds:

$$(V_r)_P = -V_P \sin \omega(t - t_I). \tag{30}$$

Since V_S, V_P and $\omega = 2\pi/T$ are constant, these equations show that, in the particular case considered here, the velocity curves of S and P are *sinusoidal*, with opposite phases, as in Figure 8.1.

2.3. The Use of Velocity Curves in the Case of Circular Orbits and an Observer outside the Orbital Plane

When the observer O is no longer in the orbital plane (R) containing the center of gravity G, we can introduce the unit vector **u** corresponding to the direction from O to G. Let us next introduce a system of cartesian axes $\mathbf{x}_G$, $\mathbf{y}_G$, $\mathbf{z}_G$ (with its origin at G), chosen in the following way (see Figure 8.2).

Let $\mathbf{z}_G$ be the *unit* vector normal to the orbital plane (R), directed in such a way that the (common) direction of circulation of S and P around $\mathbf{z}_G$ is counter-clockwise.

Let $\mathbf{y}_G$ be the *unit* vector in the plane (R) directed along the projection of $\mathbf{u}$ upon (R).

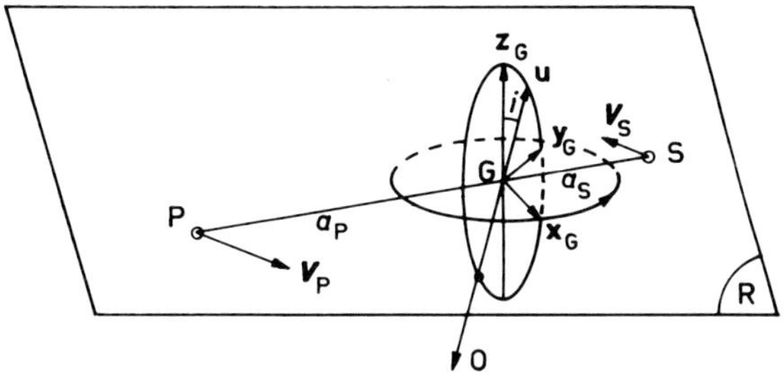

Fig. 8.2. The reference system in the case of an observer outside the orbital plane.

Finally, let us choose the unit vector $\mathbf{x}_G$ normal to $\mathbf{y}_G$ in the (R) plane, in such a way that $(\mathbf{x}_G, \mathbf{y}_G, \mathbf{z}_G)$ form a *direct* system of axes.

It is customary to denote by i the angle between $\mathbf{u}$ and $\mathbf{z}_G$ $(0 \leqslant i \leqslant \pi)$; then the cartesian components of $\mathbf{u}$ are $(0; \sin i; \cos i)$.

The astrophysical radial velocity V_r (corrected for the radial motion of G) will be given by the algebraic measure of the projection of the linear velocity V of the circular motion of S or P upon the unit vector $\mathbf{u}$ i.e. by $(V \cdot \mathbf{u})$.

If we denote by $[(V_x)_S; (V_y)_S, 0]$ and $[(V_x)_P; (V_y)_P; 0]$ the cartesian components of the velocity vectors V_S and V_P, and take the same origin of times t_I (phase I) as in Subsection 2.2, we find:

$$(V_r)_S = (V_S \cdot \mathbf{u}) = (V_y)_S \sin i = V_S \sin i \sin \omega(t - t_I)$$
$$= K_S \sin \omega(t - t_I), \qquad (29')$$

and

$$(V_r)_P = (V_P \cdot \mathbf{u}) = (V_y)_P \sin i = - V_P \sin i \sin \omega(t_I - t)$$
$$= - K_P \sin \omega(t - t_I), \qquad (30')$$

where $V_S = |V_S|$ and $V_P = |V_P|$.

Thus the velocity curves of S and P are again *sinusoidal*, with opposite phases, as in Figure 8.1.

Equations (29′) and (30) show that, in the present case, relations (22) and (23) are replaced by

$$a_S = \frac{1}{\omega} V_S = \frac{T}{2\pi \sin i} K_S, \qquad (31)$$

and

$$a_P = \frac{1}{\omega} V_P = \frac{T}{2\pi \sin i} K_P. \qquad (32)$$

Equation (24) becomes

$$a = \frac{T}{2\pi \sin i} K, \qquad (33)$$

with a still representing $(a_S + a_P)$ and K still representing, as in Equation (24′), the sum $(K_S + K_P)$.

Thus, the computation of Subsection 2.2 can be repeated replacing everywhere K_S, K_P and K by $(K_S/\sin i)$, $(K_P/\sin i)$ and $(K/\sin i)$ respectively.

Then Equations (25), (27) and (28) will take the form:

$$(M+m)\sin^3 i = \frac{T}{2\pi G}K^3, \tag{34}$$

$$M\sin^3 i = \frac{T}{2\pi G}K^2 K_P, \tag{35}$$

$$m\sin^3 i = \frac{T}{2\pi G}K^2 K_S. \tag{36}$$

A determination of the angle i is generally very difficult or impossible, but the *ratio* M/m equal to

$$\frac{M}{m} = \frac{K_P}{K_S} \tag{37}$$

does not depend on i and can be deduced, in the case considered here, directly from observations.

Moreover, since $\sin i$ is always less than 1 (or equal to 1) we have

$$(M+m) \geqslant \frac{T}{2\pi G}K^3. \tag{38}$$

Thus we can obtain a *minimum* value of $(M+m)$ and similarly minimum values of the individual masses M and m.

2.4. Utilization of a Single Curve of Radial Velocities in the Case of Circular Orbits. Mass Function

When the second component of a binary system is too faint, or presents a spectrum deprived of spectral lines, the only observable displacements of spectral lines (or more generally the only observable Doppler shifts) are those of *one* of the two components. This happens, for instance, in the study of X-rays sources considered in Chapter 9.

In such a case, observation yields only *one* of the two curves of Figure 8.1, this single curve giving, for instance, only K_P. Then, using relations (34) and (35) we can form the ratio:

$$\frac{M^3\sin^3 i}{(M+m)^2} = \frac{T}{2\pi G}(K_P)^3 = f(K_P), \tag{39}$$

which is called '*mass function*' and which depends only on T and K_P.

Fortunately, one sometimes knows (or can guess) the ratio M/m. Then, introducing the notation

$$\alpha = \frac{M}{m}, \tag{40}$$

one finds, after some elementary calculations:

$$m = \frac{(1+\alpha)^2}{\alpha^3 \sin^3 i} f(K_{\rm P}), \tag{41}$$

where as above m and $K_{\rm P}$ correspond to the *less* massive component.

However, the single curve of radial velocities usually corresponds to the more luminous and more massive component (i.e., in our notation, the S component of mass M), which provides the semiamplitude $K_{\rm S}$. In this case one introduces the *mass function* $f(K_{\rm S})$ defined by

$$f(K_{\rm S}) = \frac{T}{2\pi G} K_{\rm S}^3 = \frac{m^3 \sin^3 i}{(M+m)^2}, \tag{42}$$

and the mass M of S is given by

$$M = \frac{\alpha(1+\alpha)^2}{\sin^3 i} f(K_{\rm S}) = \frac{\alpha(1+\alpha)^2}{\sin^3 i} \frac{T}{2\pi G} K_{\rm S}^3. \tag{43}$$

Some authors introduce, instead of the ratio M/m, the ratio:

$$\alpha' = \frac{m}{M}. \tag{44}$$

Then, using Equation (43), the expression for M takes the form:

$$M = \frac{(1+\alpha')^2}{\alpha'^3 \sin^3 i} f(K_{\rm S}) = \frac{(1+\alpha')^2}{\alpha'^3 \sin^3 i} \frac{T}{2\pi G} K_{\rm S}^3. \tag{45}$$

(Note that some authors use α to denote m/M.)

2.5. Utilization of the Curves of Radial Velocities in the Case of Elliptical Orbits and an Observer in the Orbital Plane

Let us now consider a more general case of *elliptical* orbits, and assume again that the *observer is situated in the orbital plane* ($i = 90°$).

We can fix our attention on the component S alone, in order to make the corresponding Figure 8.3 clearer, since the same method of analysis remains valid for the component P.

The plane of Figure 8.3 is the common *orbital plane* (R) of S and P. It contains the direction O_G of the observer.

It is shown in Appendix A of the present Chapter that (under assumption of certain initial conditions favoring elliptical rather than parabolic or hyperbolic motions) the most general motion of *each component* of a binary system, under the influence of their mutual gravitational attraction, is *elliptical*. In this case, the center of gravity G represents the *focus* of each orbit (Kepler's first law), and each radius vector GS or GP *sweeps areas proportional to the interval of time* (Kepler's second law). (The system of reference being the intertial system bound to G.)

Let A be the apex *nearest* to G of the ellipse described by S, and let $\mathbf{y}_G$ be the *unit* vector originating at G in a direction *opposite* to that of the observer O, in the orbital plane (R). Finally, introduce the unit vector $\mathbf{x}_G$, in the plane (R), normal to $\mathbf{y}_G$, in such a way that S crosses, *first* the cartesian axis $G\mathbf{x}_G$ and then (after a *positive* rotation of 90°) the cartesian axis $G\mathbf{y}_G$.

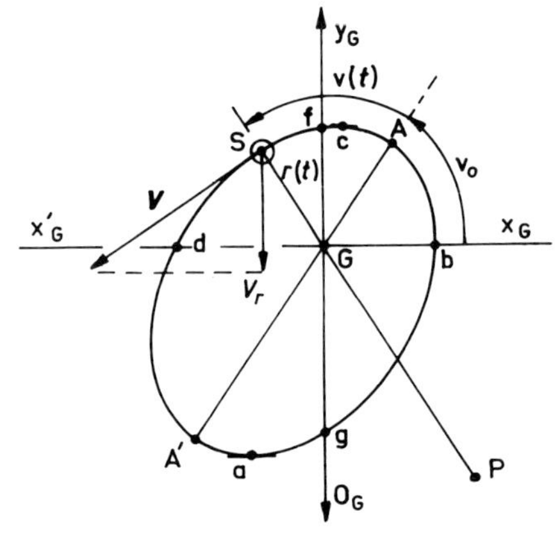

Fig. 8.3. The elliptical orbit of the component S. The corresponding notations are explained in the text.

Consider the *polar coordinates*, referred to a pole at G, and to the axis GA used as origin of the polar angles. The angle $\widehat{AGS}$, called in astronomy '*true anomaly*' (reckoned positively in the direction of the motion), is denoted by v. The distance GS is called, as usual, *r*.

Sometimes the angle $\widehat{xGA}$ is denoted in astronomy by ω, but in order to avoid confusion with the average angular velocity ($\omega = 2\pi/T$) we shall denote $\widehat{xGA}$ by v_0, (see Figure 8.3). The angle v_0 is constant, whereas $r = r(t)$ and $v = v(t)$ are functions of time t.

Let $x(t)$ and $y(t)$ be the functions describing the variations of cartesian coordinates (x, y) of S in the system of axes defined above. The cartesian components V_x, V_y of the velocity vector V of S, (we drop the subscript S since here we deal only with S), will be given by $(dx/dt; dy/dt)$.

The unit vector $\mathbf{y}_G$ defining the axis Gy_G is, by construction, *opposite* to the direction of the observer. Therefore V_y (positive in the case of a recession with respect to the observer) will represent, in conformity with its definition, the astrophysical radial velocity V_r. In Figure 8.3 we have represented a phase of the motion in which S is approaching the observer and in which V_y and V_r are *negative*. One easily verifies that, in the case considered here ($i = 90°$), the unit vector **u** introduced in Subsection 2.3, whose components are (0; 1; 0), is identical to $\mathbf{y}_G$. Thus the general relation $V_r = (V \cdot \mathbf{u})$ also gives here $V_r = V_y = dy/dt$. (One must take care to avoid confusion between the *astrophysical radial velocity* introduced above and the component of V along GS, which in *kinematics* represents the '*radial velocity*' of V).

It is obvious, in Figure 8.3, that

$$y(t) = r(t) \sin [v_0 + v(t)], \tag{46}$$

hence

$$V_r = \frac{dy}{dt} = \frac{dr}{dt} \sin (v_0 + v) + r \frac{dv}{dt} \cos (v_0 + v). \tag{47}$$

Now, it is well known (see, if necessary, Section A1 of Appendix A) that in polar coordinates (r, v) the equation of an ellipse (when the pole is placed at the focus G situated *near* the apex A from which one reckons the polar angle v) is:

$$r = \frac{p}{1 + e \cos \mathrm{v}}, \tag{48}$$

where a is the semimajor axis, e the eccentricity, b the semiminor axis, and p the parameter defined by

$$p = \frac{b^2}{a} = a(1 - e^2). \tag{48'}$$

The gravitational force acting upon S is directed towards the 'center' G. Therefore, according to a well known theorem of mechanics, the moment of the momentum MV about G must be constant. The corresponding moment of V is equal to $V_n r$, where V_n is the component of V normal to the radius vector GS. In polar coordinates, the component V_n is given by $(r\,d\mathrm{v})/dt$, so that we finally can write:

$$r^2 \frac{d\mathrm{v}}{dt} = r\left(r \frac{d\mathrm{v}}{dt}\right) = C = \text{const.}. \tag{49}$$

Now, the area $d\tilde{A}$ of the infinitesimal sector swept by the radius vector GS during the interval of time dt is given (if we neglect the terms of second order) by

$$d\tilde{A} = \tfrac{1}{2}r(r\,d\mathrm{v}) = \tfrac{1}{2}r^2\,d\mathrm{v}, \tag{50}$$

where $d\mathrm{v} = (d\mathrm{v}/dt)\,dt$ is the variation of the true anomaly v during the interval of time dt.

Combining Equations (49) and (50) we get the '*Newtonian form of Kepler's second law*':

$$2\frac{d\tilde{A}}{dt} = r^2 \frac{d\mathrm{v}}{dt} = C = \text{const.}. \tag{51}$$

Thus, integrating Equation (51), and reckoning the area $\tilde{A}$ swept by the radius vector, and the time t, from the passage of S through the Apex A, we obtain

$$\tilde{A} = \frac{C}{2}t + 0 = \frac{C}{2}t. \tag{52}$$

Then let T be the 'period', that is the duration of a complete revolution of S (the time taken for GS to sweep out the whole of the ellipse, whose area is known to be given by πab). We have, according to Equation (52), with t equal to T and $\tilde{A}$ equal to πab:

$$\pi ab = \frac{C}{2}T. \tag{53}$$

However, (see Appendix A, Equation (A5))

$$b = a(1 - e^2)^{1/2}. \tag{53'}$$

Hence, according to Equations (49), (53) and (53'):

$$\boxed{C = r^2\frac{d\mathrm{v}}{dt} = \frac{2\pi}{T}a^2(1 - e^2)^{1/2}}\,. \tag{54}$$

Equation (54) gives the value of the '*constant of areas*' C in terms of a, e and T. Differentiation of Equation (48) with respect to t, and its re-utilization gives:

$$\frac{dr}{dt} = \frac{-p}{(1 + e \cos \mathrm{v})^2}(-e \sin \mathrm{v})\frac{d\mathrm{v}}{dt} = \frac{e}{p}\sin \mathrm{v}\left(r^2\frac{d\mathrm{v}}{dt}\right), \tag{55}$$

but $(r^2(d\mathrm{v}/dt))$ can be replaced, according to Equation (54), by C, hence

$$\frac{dr}{dt} = \frac{C}{p}(e \sin \mathrm{v}). \tag{56}$$

Now, according to Equation (54) we also have $r(d\mathrm{v}/dt) = C/r$, and the substitution of these expressions for dr/dt and $r(d\mathrm{v}/dt)$ into Equation (47) gives:

$$V_r = \frac{C}{p}e \sin \mathrm{v} \sin (\mathrm{v}_0 + \mathrm{v}) + \frac{C}{r}\cos (\mathrm{v}_0 + \mathrm{v}). \tag{57}$$

Elimination of r by means of Equation (48) yields:

$$\begin{aligned} V_r &= \frac{C}{p}\left[e \sin \mathrm{v} \sin (\mathrm{v}_0 + \mathrm{v}) + (1 + e \cos \mathrm{v}) \cos (\mathrm{v}_0 + \mathrm{v})\right] \\ &= \frac{C}{p}\left[\cos (\mathrm{v}_0 + \mathrm{v}) + e \cos \mathrm{v}_0\right]; \end{aligned} \tag{58}$$

but, according to Equation (54), divided by Equation (48′):

$$\frac{C}{p} = \frac{2\pi a}{T}(1 - e^2)^{-1/2}. \tag{59}$$

Thus, we finally obtain (assuming that the observations have already been corrected for the motion of the center of gravity G):

$$\boxed{V_r = c\frac{\Delta\lambda}{\lambda} = K_\mathrm{S}[\cos (\mathrm{v}_0 + \mathrm{v}) + e \cos \mathrm{v}_0]}\,, \tag{60}$$

where $\mathrm{v} = \mathrm{v}(t)$, $\mathrm{v}_0 = const.$ and $K_\mathrm{S} = \mathrm{const.}$

Replacing a by a_S, in order to recall that here we deal with the component S, we see that, according to Equations (58) and (59), K_S in Equation (60) is equal to:

$$K_\mathrm{S} = \frac{2\pi}{T(1 - e^2)^{1/2}}a_\mathrm{S}. \tag{61}$$

Obviously, a similar Equation (62) can be written for the component P, since P describes, according to Equation (14), an ellipse homothetic to the ellipse described by S, with respect to G, with a scaling factor equal to $\mathrm{GP/GS} = M/m$:

$$K_\mathrm{P} = \frac{2\pi}{T(1 - e^2)^{1/2}}a_\mathrm{P}. \tag{62}$$

Note that Equations (61) and (62) become identical to Equations (22) and (23) respectively for circular orbits ($e = 0$).

Equation (60) shows that V_r is a *sinusoidal function of* v, through the term $\cos(\mathrm{v}_0 + \mathrm{v})$. However, since $\mathrm{v}(t)$ is, in elliptical motion, a complicated function of t, in this case V_r is no longer *a sinusoidal function of time t* (but remains *periodical*).

Depending on the respective values of e and v_0, that is, depending on the *form* and the *situation* of the elliptical orbit, the curve representing the variations of the radial velocity V_r as a function of time t can take the typical forms (including the circular case) represented in Figure 8.4, due to Hynek (1951, p. 458).

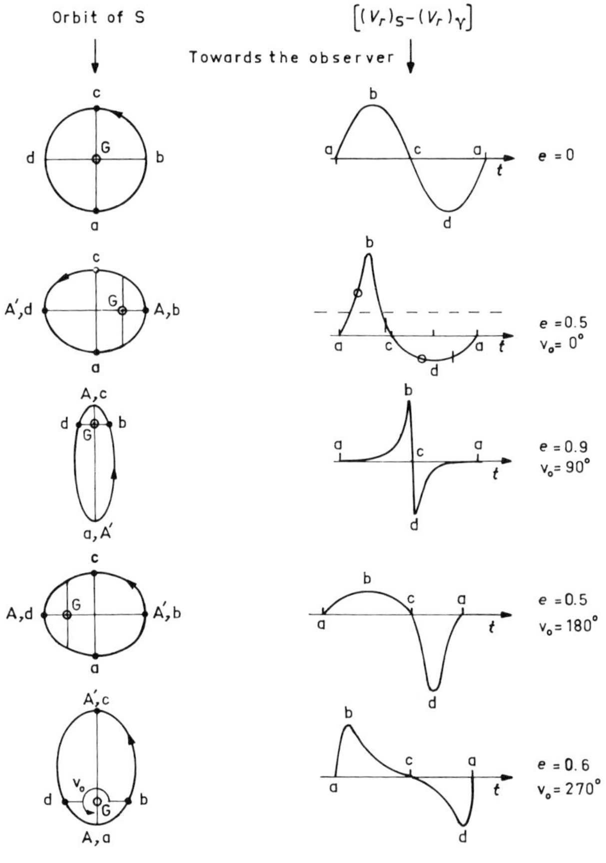

Fig. 8.4. The variations of the radial velocity V_r as a function of time, for different *forms* (value of e) and for different situations of the elliptical orbit with respect to the observer (value of v_0).

We have already mentioned in Section 1 the fact (proved in Appendix A, Section A.2), that the Newtonian form of Kepler's third law, given by Equation (17), can be applied to elliptical motion.

Let us put aside for the moment the problem of the determination of K_S and K_P with the help of the *observed* curves $V_r(t)$ of the general type represented in Figure 8.4. We see that, generalizing Equation (24′) for the case of elliptical motion, and writing

$$K = K_S + K_P, \tag{63}$$

we can add Equations (61) and (62), obtaining

$$a = a_S + a_P = \frac{T}{2\pi}(1-e^2)^{1/2}K. \tag{64}$$

Substitution into Equation (17) then yields an expression for the sum $(M+m)$ of the masses of S and P:

$$M+m = \frac{T}{2\pi G}(1-e^2)^{3/2}K^3. \tag{65}$$

This represents a generalization of Equation (25) for the case of elliptical orbits.

Applying the mathematical procedure based on Equation (14), already used in Equation (26), we obtain the individual masses:

$$\boxed{M = \frac{T}{2\pi G}(1-e^2)^{3/2}K^2K_P}\ , \tag{66}$$

and

$$\boxed{m = \frac{T}{2\pi G}(1-e^2)^{3/2}K^2K_S}\ . \tag{67}$$

Equations (66) and (67) are obvious generalizations of Equations (27) and (28).

Now, let us come back to the observational determination of K_S and K_P.

Note first that in the general case considered here, K_S still represents *one half of the total amplitude* of the periodic variations of V_r of S with t, but is no longer equal to the corresponding $(V_r)_{max}$ or $|(V_r)_{min}|$ since, in the general case, these quantities are *unequal*.

Indeed, the slope of the tangents to the curve representing $V_r(t)$ is given by dV_r/dt, i.e., according to Equation (60), by

$$dV_r/dt = -K_S \sin(\mathrm{v}_0 + \mathrm{v})\frac{d\mathrm{v}}{dt}. \tag{68}$$

The angular velocity, $d\mathrm{v}/dt$, of S is physically always positive. Therefore Equation (68) shows that the *maximum* of V_r will be reached when $\sin(\mathrm{v}_0 + \mathrm{v})$ changes from *minus* to *plus*, for $(\mathrm{v}_0 + \mathrm{v}) = 0°$, whereas the minimum of V_r will be reached for $(\mathrm{v}_0 + \mathrm{v}) = 180°$.

Thus the maximum of V_r is reached when S crosses at b the axis Gx (see Figures 8.3 and 8.4), whereas the minimum of V_r is reached when S crosses at d the axis Gx'_G opposite to Gx_G.

In other words, $\mathrm{v}_{max} = -\mathrm{v}_0$ and $\mathrm{v}_{min} = 180° - \mathrm{v}_0$, and substitution of these values into Equation (60) gives, on introducing the *absolute value* of the negative quantity $(V_r)_{min}$:

$$(V_r)_{max} = K_S(1 + e\cos \mathrm{v}_0), \tag{69}$$

and

$$|(V_r)_{\mathrm{min}}| = K_S(1 - e \cos v_0) = -(V_r)_{\mathrm{min}}. \tag{69'}$$

From Equations (68) and (69′) we deduce that

$$K_S = \tfrac{1}{2}[(V_r)_{\mathrm{max}} + |(V_r)_{\mathrm{min}}|]. \tag{70}$$

These equations show that, as announced above, K_S still represents *one half of the total amplitude* of V_r, whereas (except for $v_0 = 90°$ or $v_0 = 270°$) in the case $e \neq 0$, $(V_r)_{\mathrm{max}}$ and $|(V_r)_{\mathrm{min}}|$ are *unequal*, each of them being different from K_S.

Figure 8.4 exhibits clearly this asymmetry in the case $e = 0.5$ and $v_0 = 0°$, since in this case $(V_r)_{\mathrm{max}} = \tfrac{3}{2}K_S$ and $|(V_r)_{\mathrm{min}}| = \tfrac{1}{2}K_S$. This asymmetry disappears in the particular cases where $v_0 = 90°$ or $v_0 = 270°$.

Note that, with points a, b, c, d having the positions indicated in Figures 8.3 and 8.4, (and with a and c corresponding to a *zero* value of V_r), a very general property of the curves representing $V_r(t)$ is the following. Let (abc) be the area enclosed (from a to c via b) between the curve $V_r(t)$ and the t axis, and let |(cda)| be the absolute value of the negative area (cda) enclosed (from c to a via d) between the curve and the t axis. It is easy to prove that (abc) = |(cda)| $= y_c - y_a$.

Indeed, we have:

$$(\mathrm{abc}) = \int_{\mathrm{abc}} V_r\, dt = \int_{\mathrm{abc}} \frac{dy}{dt}\, dt = \int_{\mathrm{abc}} dy = y_c - y_a, \tag{71}$$

and

$$(\mathrm{cda}) = \int_{\mathrm{cda}} dy = y_a - y_c; \quad \text{thus } |(\mathrm{cda})| = y_c - y_a. \tag{72}$$

If the curve of $V_r(t)$ were not yet corrected for the velocity of the center of gravity, this property would enable one (by tracing an axis on such a level that both areas mentioned above are equal) to find the gamma velocity from the level of the new axis.

Solving Equations (69) and (69′) for $e \cos v_0$ we obtain:

$$e \cos v_0 = \frac{(V_r)_{\mathrm{max}} - |(V_r)_{\mathrm{min}}|}{(V_r)_{\mathrm{max}} + |(V_r)_{\mathrm{min}}|}. \tag{73}$$

On the other hand (see Figure 8.3), if we denote by A′ the apex corresponding to $v = 180°$ (apex A corresponding to $v = 0°$), then (still concentrating our attention on the component S) Equation (60) will give, with obvious notation:

$$\begin{cases} (V_r)_A = K_S e \cos v_0 + K_S \cos v_0; \\ (V_r)_{A'} = K_S e \cos v_0 - K_S \cos v_0. \end{cases} \tag{74}$$

On subtracting Equation (69′) from Equation (69) we obtain:

$$K_S\, e \cos v_0 = \tfrac{1}{2}[(V_r)_{\mathrm{max}} - |(V_r)_{\mathrm{min}}|]. \tag{75}$$

This relation shows that the straight line $V_r = K_S\, e \cos v_0$, parallel to the t axis, is situated (as the dotted line relative to the case $e = 0.5$ and $v_0 = 0°$ in Figure 8.4) *half-way* between the maximum and the minimum.

We advise the reader, in order to clarify the present analysis, to draw a schematic curve analogous to that of Figure 8.4, for instance for the case: $T = 12$; $t_b = t_{max} = 2$; $t_d = t_{min} = 8$; $(V_r)_{max} = 11$; $(V_r)_{min} = -5$; $t_c = 4$; $t_A = 3$; $(V_r)_A = 8$; $t_{A'} = 9$; $(V_r)_{A'} = -2$. As explained below, t_A and $t_{A'}$ are *not immediately given by observation*, but can be determined, on the observational curve, by the following procedure.

We note first that according to Equation (74) the points A and A′ must lie at the same *'distance'* (in ΔV_r) *from the line* $V_r = K_S e \cos v_0$ (which in our example will be at the level $V_r = 3$). Quite generally, an ellipse is *symmetric* with respect to its major axis AA′ and, according to Kepler's second law, *equal areas* are swept by the radius vector in *equal times*. Therefore, the time interval between A and A′ must be equal to *one half of the period T*. [Thus, in our example $(t_{A'} - t_A)$ must be equal to 6.]

Now, *generally* (see, for instance, in the case $e = 0.5$; $v_0 = 0°$, the points denoted in Figure 8.4 by open circles, or by small vertical bars) on the curve representing $V_r(t)$, the points corresponding to an interval of time of half a period *are not* lying at the same *'distance'* from the line $V_r = K_S\, e \cos v_0$, (i.e. the dotted line in Figure 8.4, or the line $V_r = 3$ of the numerical example).

Only *two particular* pairs, corresponding to the points b and d or to the points A and A′ possesses this *double* property. Note that in the case $e = 0.5$; $V_0 = 0°$ the number of such pairs is reduced to *one*, since in this case A = b and A′ = d.

Thus, quite generally, the points A and A′ can be found on the curve $V_r(t)$ by *trial and error* (for instance by the use of a transparent sheet of paper suitably graduated).

The knowledge of the positions of the points A and A′ on the curve of radial velocities of S, will give $K_S \cos v_0$ by their 'distance' to the line $V_r = K_S\, e \cos v_0$ (or to the line $V_r = 3$ of the numerical example in which the value of $K_S \cos v_0$ is, by construction, equal to 5). Since we already know the value of $K_S\, e \cos v_0$ the knowledge of $K_S \cos v_0$ will give us the observational value of the eccentricity e of the orbit of S. (In the numerical example, $K_S\, e \cos v_0 = 3$, and $K_S \cos v_0 = 5$, hence $e = \frac{3}{5}$.)

The orbit of the component P being, through Equation (14), homothetic to the orbit of S, both orbits have, as already pointed out, the same eccentricity e. On the other hand, Equation (70) gives us K_S, and a similar relation for the curve $V_r(t)$ relative to P gives us K_P.

Thus, one can find all of the observational values T, e, K_S, K_P, and $K = K_S + K_P$ on which depend the values of M and m, in relations (66) and (67).

Another way of attacking the same problem is based on the fact that, according to Equation (60), the *form* of the curve $V_r(t)/K_S = f(t)$ depends only on the pair of values taken by e and v_0. Therefore, on drawing *once and for all* on transparent paper a network of curves representing $f(t)$ for different combinations of values of e and v_0, one can read off the values of e and v_0 from the curve of the network which corresponds to that observed. (K_S and K_P are given directly by the curves representing $V_r(t)$ for S and P.)

2.6. Utilization of the Curves of Radial Velocities in the Case of Elliptical Orbits and an Observer Outside the Orbital Plane

On combining the results of Subsections 2.3 and 2.5, and on using the same systems of axes and of unit vectors as in Figures 8.2 and 8.3, we immediately find that $y(t)$ is still given by Equation (46), but that Equation (47) is replaced by

$$V_r = (V \cdot \mathbf{u}) = V_y \sin i = \sin i \frac{dy}{dt}$$

$$= \sin i \left[\frac{dr}{dt} \sin (\mathrm{v}_0 + \mathrm{v}) + r \frac{d\mathrm{v}}{dt} \cos (\mathrm{v}_0 + \mathrm{v}) \right], \tag{76}$$

(where the subscript S is omitted).

This yields, by the same transformations as those leading from Equation (47) to Equation (60), a result which can still be written as Equation (60), but with a value of K_S, different from Equation (61), given by

$$K_S = \frac{2\pi}{T(1-e^2)^{1/2}} a_S \sin i. \tag{77}$$

Similarly K_P will now be given by

$$K_P = \frac{2\pi}{T(1-e^2)^{1/2}} a_P \sin i. \tag{78}$$

Hence, with K still defined as $(K_S + K_P)$, and a as $(a_S + a_P)$:

$$a \sin i = a_S \sin i + a_P \sin i = \frac{T}{2\pi} (1-e^2)^{1/2} K, \tag{79}$$

which substituted in Equation (17) gives

$$(M + m) \sin^3 i = \frac{T}{2\pi G} (1-e^2)^{3/2} K^3, \tag{80}$$

which is a generalization of Equations (34) and (65). Next, the use of Equation (14) yields the following generalizations of Equations (35), (36), and (66) and (67):

$$M \sin^3 i = \frac{T}{2\pi G} (1-e^2)^{3/2} K^2 K_P, \tag{81}$$

and

$$m \sin^3 i = \frac{T}{2\pi G} (1-e^2)^{3/2} K^2 K_S. \tag{82}$$

There is no change, as compared with Subsection 2.5, in the necessary determination of e. As already stated in Subsection 2.3, a precise and direct determination of the angle i is generally very difficult or impossible. However, we see from Equations (81) and (82) that $\sin i$ disappears from the ratio M/m, still equal to K_P/K_S as in Equation (37), if the Doppler shifts of the spectral lines of both S and P can be observed.

On the other hand, a statistical treatment is possible if a great number of binary systems is observed, and if the values of $\sin i$ are assumed to be distributed *at random*.

One also can select, by the study of 'light curves' considered in Subsection 2.7, *either* systems which undergo *total eclipse* (since the value of i in this case is certainly 90°) *or* systems undergoing *partial eclipse* (since the value of i in these cases is sufficiently near 90° and $\sin i$ is very near 1).

2.6.1. Remark.

In the case, analogous to that considered in Subsection 2.4, of a single observable velocity curve (for instance that of S), Equations (42) and (43) are obviously replaced by

$$\frac{m^3 \sin^3 i}{(M+m)^2} = F(K_S) = \frac{T}{2\pi G} K_S^3 (1-e^2)^{3/2}, \tag{83}$$

and

$$M = \frac{\alpha(1+\alpha)^2}{\sin^3 i} F(K_S). \tag{84}$$

Hence $M \geqslant \alpha(1+\alpha)^2 F(K_S)$.

2.7. Some Elementary Properties of Eclipsing Binaries, in the Case of an Observer in the Orbital Plane

When the observer is situated in the orbital plane of a system of close double stars the total energy received by his instruments decreases every time one of the stars partly or completely occults the other.

It is customary in the study of these '*eclipsing binaries*' to disregard the distinction between the *luminosity* (i.e. the energy *emitted* per second) and the *brightness* (i.e. the corresponding energy *received* by the observer's instruments per second and per cm^2 normal to the line of sight).

Indeed, distance and possible interstellar extinction effects are the same for the two stars since they are so close that they cannot be observed individually.

Therefore, it is usual to call '*variations of the total luminosity*' the observed variations of the total brightness of the system due to the effects of mutual screening. The corresponding representation of this total luminosity as a function of time is called the *light curve*.

The form of the light curve depends on very many factors. Thus, for instance, the apparent stellar disks may be darkened toward the limb as is the Sun's disk. However, to a first approximation, adopted henceforth, it can be assumed that the '*luminosity*' *of each disk is uniformly distributed over its area*. Then the terminology introduced above allows us to reserve the term 'surface brightness', or simply '*brightness*', for the ratio between the luminosity of a disk and its area. In this way the theory of eclipsing binaries is reduced to the study of the screening effects of two circular disks, of different radii and different brightness, in relative motion.

The decrease in total luminosity can correspond, during an orbital period T, to two different configurations of the system with respect to the observer, the component S being either behind or in front of the component P. These two decreases being generally of unequal 'depth', the deeper one is called *primary minimum*, the other one is called *secondary minimum*.

The minimum corresponding to a situation where the small star is behind the big one represents a real *eclipse*; the minimum corresponding to a situation where the small star is in front of the big one (hiding a part of the disk of the latter) represents an *occultation*. However, the difference between eclipse and occultations is often disregarded, and both phenomena are called 'eclipse'.

When the disks partly overlap, the situation is called *the phase of partial eclipse*. When no overlapping occurs, the situation is that of the *phase outside the eclipse*; then the total luminosity of the system conserves a constant *maximum* value L_{max}.

Each of the minima (the primary and the secondary) generally includes a *phase of totality* (or 'total phase') during which the total luminosity conserves a *constant* value L_{prim} for the primary *or* L_{sec} for the secondary minimum (both L_{prim} and L_{sec} being less than L_{max}, and by definition L_{sec} being greater than L_{prim}).

Both phases of totality present some respective durations, Δt_{prim} and Δt_{sec}. These must not be confused with the corresponding durations D_{prim} and D_{sec} of the primary and secondary minima, which include, by definition, the partial phases surrounding the total ones.

It is relatively easy to prove that, beside other factors, the durations D_{prim} and D_{sec} depend on the *sum* of radii r_S and r_P of the components, whereas the durations Δt_{prim} and Δt_{sec} (under the same conditions with respect to the other factors) depend on the *difference* of radii of the components.

Let us suppose that the area A_S of the disk of the component S is greater than the area A_P of the disk of the component P. Obviously if we keep the distinction between eclipses and occultations, the eclipse will correspond to the passage of P *behind* S, whereas occultation will correspond to the passage of P *in front of* S.

In this case ($A_S > A_P$) the loss in total luminosity during the total phase of the eclipse will be equal to the individual luminosity L_P of the disk of P, whereas (B_S being the brightness of the component S) the loss in total luminosity during the total phase of the occultation will be equal to $B_S A_P$. The minimum corresponding to the total phase of the eclipse will be a *primary minimum* if $L_P > B_S A_P$ which is equivalent to $B_P > B_S$ (if B_P represents the brightness of the component P, since B_P is equal to L_P divided by A_P).

Thus, *an eclipse* (if we keep in mind the distinction between eclipses and occultations) *will correspond to a primary minimum if the (surface) brightness of the eclipsed disk is greater than the brightness of the eclipsing disk*; but *the corresponding depth of the primary minimum* depends only *on the individual luminosity* (L_P in the case considered above) *of the* disk of the *eclipsed component*, whereas the 'depth' of the secondary minimum is equal to $B_S A_P = L_S(A_P/A_S)$.

Exercise: In order to familiarize himself with all these properties, the reader should compute the respective depths of minima for the two following systems, which present the same ratio of radii, with the radius r_S of S equal to 6 (arbitrary) units of length and the radius r_P of P equal to 2 units of length.

In the *System I* the respective individual luminosities of S and P are, for instance, $L_S = 18$ and $L_P = 4$ in arbitrary units of luminosity, which means that the brightness, in corresponding units, are proportional to 18/36 and 4/4 respectively. Since $B_P > B_S$ the primary minimum corresponds to an *eclipse* of P by S. The depth of this primary minimum is equal to $L_P = 4$ units of luminosity. The depth of the secondary minimum (corresponding to an occultation of a part of S by P), which is generally equal to $L_S(A_P/A_S) = L_S(r_P/r_S)^2$, is here equal to 2 units of luminosity.

In *System II* the respective individual luminosities of S and P are, for instance, $L_S = 18$ and $L_P = 19$, and the corresponding values of B_S and B_P are proportional to 18/36 and 19/4. Since $B_P > B_S$, the primary minimum corresponds to an eclipse of P by S. The depth of the secondary minimum is again equal to 2 units (since L_S is the same as in System I), but the depth of the primary minimum is considerably greater than for System I, since it is now equal to 19 units. The difference is obviously due to the choice of a much greater luminosity of the smaller component P. (*End of the Exercise.*)

The relative orbit of one of the components, e.g. P, with respect to an inertial reference system bound to the component S, can be an *ellipse* of eccentricity e. Then, if the orbital period is denoted by T, the interval of time T between the midpoints of two successive *primary* minima is *not* separated into two equal parts by the middle of the intermediate secondary minimum.

Let us consider the *relative* motion of the component P around the component S, (Figure 8.5).

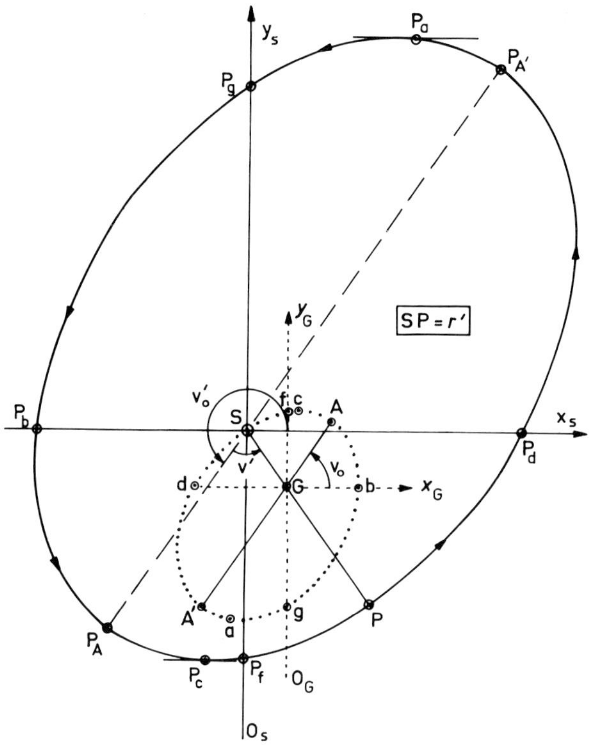

Fig. 8.5. The relative motion of the component P around the component S. The corresponding notations are explained in the text.

In this figure, S is the center of the star S and P is the center of the star P. Moreover, as an example, it is supposed that the ratio of the masses is such that $|GP| = 2|GS|$, G being the center of gravity of the system. Then the orbit of S relative to G (represented by small dots) is the same as in Figure 8.3. The physically remarkable points of the orbit of S around G being again called a; b; c; d; A; A′; f; g, the corresponding points of the relative orbit of P around S are called P_a; P_b; P_c; P_d; P_A; $P_{A'}$; P_f; P_g respectively. The cartesian axes parallel to $\mathbf{x}_G$ and $\mathbf{y}_G$, with origin at S, will be defined by the unit vectors $\mathbf{x}_S$ and $\mathbf{y}_S$, and the positive rotation around S from $\mathbf{x}_S$ to SP_A will be defined by the angle

v_0' ($0 \leqslant v_0' \leqslant 360°$). It is obvious that v_0 being the angle defined in Subsection 2.5, we have $v_0' = v_0 + 180°$. Let us call v' the corresponding true anomaly and r' the radius vector SP of P.

We suppose again $r_S > r_P$ and $B_P > B_S$. Then an *eclipse* of P by S corresponds to a *primary* minimum. The *midpoint of the total phase of this primary minimum* corresponds to the position P_g of P, at time t_g, just behind the center of S with respect to the observer O_S. Similarly, the *midpoint of the total phase of the first secondary minimum* after t_g (occultation of S by P), corresponds to the position P_f of P, at time t_f, just in front of the center of S with respect to the observer O_S.

The difference between $(t_f - t_g)$ and $T/2$ is easy to find in the particular case of a sufficiently small eccentricity e. We can then apply the approximate relation (A32) established in the Appendix A to the present Chapter. This relation gives the interval of time t between the passage through the nearest apex and the first passage through the point of true anomaly v' (called v in Equation (A32)), as a function of T, e and v'. This relation can be written in the present case (v' being here measured in *radians*) as:

$$t \approx \frac{T}{2\pi}(v' - 2e \sin v'). \tag{85}$$

In Figure 8.5 where v_g' denotes the value of v' corresponding to P_g and v_f' the value of v' corresponding to P_f, we see that: $v_g' = (\pi/2) - v_0'$ and $v_f' = (3\pi/2) - v_0'$. Hence

$$\sin v_g' = \cos v_0'; \quad \sin v_f' = -\cos v_0'; \quad \cos v_g' = \sin v_0'; \quad \cos v_f' = -\sin v_0'. \tag{86}$$

On applying Equation (85) we obtain:

$$\begin{aligned} t_f &\approx \frac{T}{2\pi}\left[\left(\frac{3\pi}{2} - v_0'\right) + 2e \cos v_0'\right]; \\ t_g &\approx \frac{T}{2\pi}\left[\left(\frac{\pi}{2} - v_0'\right) - 2e \cos v_0'\right]. \end{aligned} \tag{87}$$

Finally, the difference between $(t_f - t_g)$ and $T/2$ is given by

$$(t_f - t_g) - \frac{T}{2} \approx \left(\frac{2e}{\pi} \cos v_0'\right) T. \tag{88}$$

The maximum value of this difference, for a given small value of e, corresponds to $v_0' = 0$. The left-hand side of Equation (88) can be expressed (but at the price of more complicated computations) as a function of v_0' for any value of e.

The left-hand side of Equation (88) for $e \neq 0$ is *zero* for $v_0' = 90°$ or $270°$. This is physically obvious because of the geometrical and dynamical symmetry of the corresponding situations with respect to the observer O_S.

Figure 8.6 (overleaf) illustrates in the case $v_0' = 0$ the correspondence between the curve of radial velocities of the component P, and the light curve.

The apparent discrepancy between the curve of radial velocities of Figure 8.4 (case where $e = 0.5$ and $v_0 = 180°$) and Figure 8.6 is due to the fact that Figure 8.4 relates to the component S, whereas Figure 8.6 relates to the component P. This difference produces, by symmetry with respect to G, a reversal of the sign of V_r for the corresponding situations. The same symmetry explains why the point b, defined for S as the crossing of the axis Gx, does *not* correspond for P_b to the crossing by P of the homologous axis Sx; it corresponds to the crossing by P of the axis Sx′ opposite to Sx.

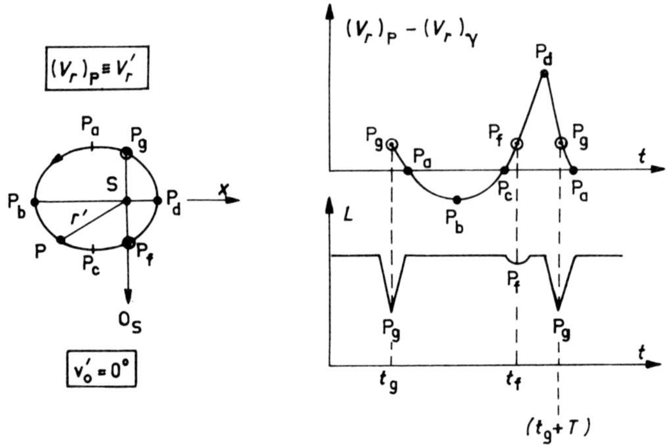

Fig. 8.6. The correspondence between the curve of radial velocities and the luminosity curve, in the case $v_0' = 0$.

In Figure 8.6 the points of the velocity curve corresponding to the middle of the primary minimum (points P_g of the orbit) are indicated by open circles like the point corresponding to the middle of the secondary minimum (point P_f of the orbit). On the corresponding light curve the durations of the total phases are neglected in order to emphasize the times t_g and t_f of the respective midpoints of primary and secondary minima.

Let us consider now the ratio of the duration D_{sec} of the secondary minimum to the duration D_{prim} of the primary minimum, (both include the partial phases!). A drawing shows that both D_{prim} and D_{sec} correspond respectively to an interval of time taken by the center of P to move the *same distance*, equal to $(r_P + 2r_S + r_P) = d$, either *behind* S or *in front of* S. Therefore, the ratio D_{sec}/D_{prim} will depend only on the ratio of the *moduli* $|V_n'|_f$ and $|V_n'|_g$ of the component V_n' of the velocity V' of P normal to the radius vector $SP = r'$ (in its *relative* motion around S) near P_f and near P_g. As a reasonable approximation we assume that both normal components are *constant* during the corresponding complete eclipse *or* occultation. This will be acceptable if r_P and r_S are small in comparison with the distance between S and P. We have:

$$d = 2(r_P + r_S) = |V_n'|_f D_{sec} = |V_n'|_g D_{prim}, \quad (89)$$

whence

$$D_{sec}/D_{prim} = |V_n'|_g / |V_n'|_f. \quad (89')$$

Since for $v_0' = 0°$ (see Figure 8.6) or for $v_0' = 180°$, $|V_n'|_g$ and $|V_n'|_f$ are equal by symmetry, the corresponding durations D_{sec} and D_{prim} will be equal. But it is relatively easy to find the value of D_{sec}/D_{prim} for *any* value of v_0' (and $e < 1$).

Let us apply Equation (51) to the motion of $P(r', v')$ (Kepler's second law), and take into account the well known expression for V_n' in polar coordinates: $V_n' = r'(dv'/dt)$; in our case dv'/dt is always positive, and therefore $V_n' = |V_n'|$. This gives us

$$r'|V'_n| = C = \text{const.} \tag{90}$$

Hence, according to Equation (A4) of Appendix A applied to $P(r', v')$:

$$\frac{|V'_n|_g}{|V'_n|_f} = \frac{r'_f}{r'_g} = \frac{1 + e \cos v'_g}{1 + e \cos v'_f}. \tag{91}$$

Finally Equations (86), (89′) and (91) give us the fundamental *general* relation (for any $e < 1$):

$$\frac{D_{\text{sec}}}{D_{\text{prim}}} = \frac{1 + e \sin v'_0}{1 - e \sin v'_0}. \tag{92}$$

For a given value of e, the ratio $D_{\text{sec}}/D_{\text{prim}}$ takes the maximum value $(1 + e)/(1 - e)$ for $v'_0 = 90°$. This case is considered in Figure 8.7.

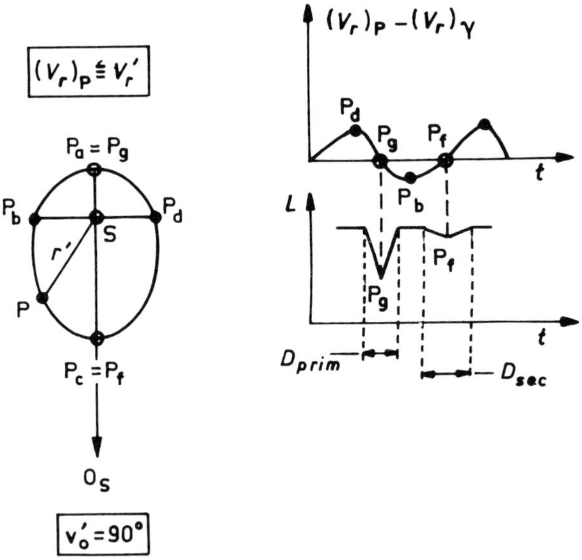

Fig. 8.7. The correspondence between the curve of radial velocities and the luminosity curve, in the case $v'_0 = 90°$.

According to Equation (88) in the case $v'_0 = 90°$, for small values of e, $(t_f - t_g) \approx T/2$, but by symmetry, obvious in Figure 8.7, in this case $(t_f - t_g)$ is rigorously equal to $T/2$ for any value of $e < 1$.

The main application of the properties expressed by Equations (88) and (92), is that they permit the deduction of both $e \cos v'_0$ and $e \sin v'_0$ (whence e and v'_0) from the *observable* quantities defined by the left-hand sides of these two Equations. (Equation (88) very often does not need any extension since eclipses are mainly observed for binary systems of small eccentricity. However the method can be extended to any value of e.)

2.7.1. Remark.

It is shown in advanced Monographs that *tidal effects* in very close binaries produce a *deformation* of both components: instead of being spherical they become approximately *ellipsoidal*. Newton's theorem cannot be applied to such stars (i.e. one cannot suppose the mass of each star to be concentrated at a single point) in the determination of the gravitational field created by each star.

The main consequence of this situation is that the angle v_0 (for S) or v_0' (for P) of the major axis of the *relative* orbit with a fixed direction (normal to that of the observer) does not remain constant, but *varies* slowly as a function of time t.

It is shown in these Monographs that dv_0/dt and dv_0'/dt depend on the *distribution of masses inside* each component: when the stars are strongly 'concentrated' towards their centers (*strong gradient of the mass density* ρ) the derivatives such as dv_0/dt are smaller since one is nearer to the validity of Newton's theorem, thus nearer to the constancy of v_0.

Therefore, the observation of variations of v_0 or v_0' in a binary system in a sense reveals the interior of a star.

3. The Values of Stellar Masses. Relations Between Masses, Luminosities and Spectral Classes

The application to the appropriate observations of the theories discussed above, adequately improved, yields the stellar masses indicated in Table 8.1 (as given by Allen, (1973)). Each mass is a function of two variables: the spectral type Sp and the 'luminosity class'.

TABLE 8.1

The stellar masses as a function of spectral type and 'luminosity class'

Type Sp	log M			
	I	III	MS	V
05	+ 2.2	.....	+ 1.60	.....
B0	+ 1.7	.....	+ 1.25	.....
B5	+ 1.4	.....	+ 0.81	.....
A0	+ 1.2	.....	+ 0.51	.....
A5	+ 1.1	.....	+ 0.32	.....
F0	+ 1.1	.....	+ 0.23	.....
F5	+ 1.0	.....	+ 0.11	.....
G0	+ 1.0	+ 0.4	......	+ 0.04
G5	+ 1.1	+ 0.5	......	− 0.03
K0	+ 1.1	+ 0.6	......	− 0.11
K5	+ 1.2	+ 0.7	......	− 0.16
M0	+ 1.2	+ 0.8	......	− 0.33
M2	+ 1.3	.....	......	0.41
M5	.....	.....	......	− 0.67
M8	.....	.....	......	− 1.00

In this Table, stellar masses are expressed in solar units ($M_\odot = 2 \times 10^{33}$ g). The Table gives the logarithm (to the base 10) of the mass M. The luminosity class I corresponds to 'supergiants'; the luminosity class III corresponds to 'giants'; the luminosity class V corresponds to 'dwarfs' (white dwarfs excluded). A distinction between class III and class V is meaningless for spectral types earlier than G0, therefore between O5 and F5 only MS ('main sequence') stars are considered.

Table 8.2 indicates the values of luminosities L and radii R expressed in solar units ($L_\odot = 4 \times 10^{33}\,\mathrm{erg\,s^{-1}}$ and $R_\odot = 7 \times 10^{10}\,\mathrm{cm}$) as a function of the mass $M/M_\odot$ (again the white dwarfs are not considered in this Table).

TABLE 8.2
The values of luminosities L and radii R of normal stars, as a function of mass M

$M/M_\odot$	$L/L_\odot$	$R/R_\odot$
0.10	0.001	0.13
0.16	0.003	0.20
0.25	0.01	0.32
0.40	0.03	0.50
0.63	0.16	0.72
1.00	1.00	1.00
1.6	6.31	1.3
2.5	4.0×10^1	2.1
4.0	2.0×10^2	3.1
6.3	1.0×10^3	3.8
10.0	5.0×10^3	5.2
15.8	2.8×10^4	7.2
25.1	7.9×10^4	10.0
39.8	2.5×10^5	13.0
63.1	1.0×10^6	20.0

As a first rough approximation the *observed* relation between L and M, can be expressed by the '*mass-luminosity relation*' $L \propto M^{3.45}$, but more refined relations between L and M depend on the luminosity class.

Appendix A. On Keplerian Motion

A1. Equation of an Ellipse in Polar Coordinates

Let us use the same notations as in Figure 8.3, but to be more general let us call F and F′ the foci of our ellipse, (see Figure 8.A1 overleaf).

As usual, the distance FF′ is called $2c$, the distance AA′ (the major axis) is called $2a$, and the minor axis is called $2b$.

Let M be some point of the ellipse. We call r the distance FM and r' the distance F′M. The apex A is the one close to the focus F taken as origin of the polar coordinates (r, v); the polar angle v is defined by the positive rotation from FA to FM.

The ellipse is defined by the relation

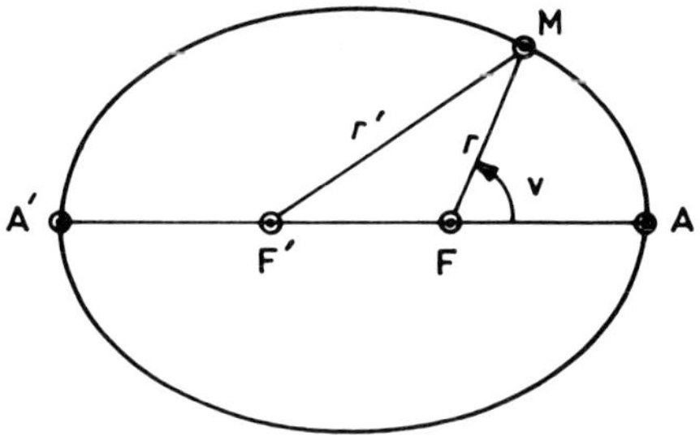

Fig. 8.A1. The notations used in Equation (A4).

$$r + r' = 2a = \text{const.} \tag{A1}$$

and it is assumed that $c/a < 1$.

In the triangle FF′M we have the classical relation

$$r'^2 = (2c)^2 + r^2 - 2(2c)r \cos(180° - \mathrm{v}). \tag{A2}$$

Replacing r' by its value $(2a - r)$ given by Equation (A1) we obtain, after elementary simplifications:

$$ra\left(1 + \frac{c}{a}\cos \mathrm{v}\right) = a^2 - c^2. \tag{A3}$$

On introducing the eccentricity $e = c/a$ of the ellipse, we find:

$$r = \frac{p}{1 + e \cos \mathrm{v}}, \tag{A4}$$

where p is defined by

$$p = a(1 - e^2) = \frac{a^2 - c^2}{a} = \frac{b^2}{a}. \tag{A5}$$

A2. THE PROOF OF THE NEWTONIAN FORM OF KEPLER'S THIRD LAW FOR AN ELLIPTICAL MOTION

Consider Equation (6) describing the relative motion of a heavenly body P of mass m, around a heavenly body S of mass M, referred to a reference system attached to S (with axes parallel to those attached to an inertial reference system), under the mutual gravitation according to Newton's law: i.e.

$$\frac{d^2(\mathbf{P} - \mathbf{S})}{dt^2} = -\frac{G(M + m)}{r^2}\frac{(\mathbf{P} - \mathbf{S})}{r}. \tag{A6}$$

Here G is the constant of gravitation, t the time, and r the distance between P and S, respective centers of the two bodies (the masses m and M are supposed to obey 'Newton's theorem', see Chapter 2, Subsection 2.1).

Interpreting this equation as that of a material point of unit mass, acted upon by an attractive radial (central) force of intensity $G(M + m)/r^2$ (in an *inertial* system of reference) we could immediately apply (V being the modulus of the velocity vector V of P) the theorem of kinetic energy, and write:

$$\tfrac{1}{2}d(V^2) = -\frac{G(M+m)}{r^2}dr. \tag{A7}$$

However, it is almost as easy to start directly from Equation (A6). Let us, indeed, call (V_x, V_y) the cartesian components of the velocity V of the relative motion of P referred to a system of cartesian axes (Sx, Sy), with (x, y) as coordinates of P. The radius vector r and the 'true anomaly' v represent the polar coordinates of P in the orbital plane, with pole at S and Sx as origin of the polar angle v. On writing μ instead of $G(M+m)$, the cartesian equivalent of Equation (A6) is given by equations:

$$\frac{dV_x}{dt} = -\frac{\mu}{r^2}\cos \mathrm{v} = -\frac{\mu}{r^2}\frac{x}{r}, \tag{A8}$$

$$\frac{dV_y}{dt} = -\frac{\mu}{r^2}\sin \mathrm{v} = -\frac{\mu}{r^2}\frac{y}{r}. \tag{A8$'$}$$

Now

$$V^2 = V_x^2 + V_y^2 \tag{A9}$$

hence

$$\frac{1}{2}\frac{d(V^2)}{dt} = V_x\frac{dV_x}{dt} + V_y\frac{dV_y}{dt}, \tag{A9$'$}$$

and

$$r^2 = x^2 + y^2, \tag{A10}$$

hence

$$r\frac{dr}{dt} = x\frac{dx}{dt} + y\frac{dy}{dt} = xV_x + yV_y. \tag{A10$'$}$$

On multiplying Equation (A8) by V_x and Equation (A8$'$) by V_y, and on taking the sum, we obtain

$$\frac{1}{2}\frac{d(V^2)}{dt} = -\frac{\mu}{r^2}\frac{dr}{dt}, \tag{A11}$$

which is equivalent to Equation (A7), and which can be transformed, through multiplication by $dt/d\mathrm{v}$, into

$$\frac{1}{2}\frac{d(V^2)}{d\mathrm{v}} = -\frac{\mu}{r^2}\frac{dr}{d\mathrm{v}}. \tag{A12}$$

Since the gravitational force acting upon P is directed towards the 'center' S, the moment of the momentum $(M+m)V$ about S must be constant. But the corresponding moment of V is given by a multiplication of V_n (the component of V *normal* to the radius vector SP) by the distance r. In polar coordinates, the component V_n is given by $(r\,d\mathrm{v})/dt$, so that we finally can write:

$$r^2\frac{d\mathrm{v}}{dt} = C = \text{const}. \tag{A13}$$

In order to avoid confusion with the astrophysical radial velocity V_r defined in Subsection 2.2, we denote the kinematical radial velocity dr/dt by $\bar{V}_r$, and for more homogeneity in notation we write $\overline{V}_n$ instead of V_n. Then, taking into account the value of dt given by Equation (A13) and the value of $r(d\mathrm{v}/dt) = C/r$ given by Equation (A13), we find

$$\overline{V}_r = \frac{dr}{dt} = \frac{C}{r^2}\frac{dr}{d\mathrm{v}} = -C\frac{d(1/r)}{d\mathrm{v}} \tag{A14}$$

and

$$\overline{V}_n = r\frac{d\mathrm{v}}{dt} = \frac{C}{r} = C\left(\frac{1}{r}\right). \tag{A15}$$

It thus appears that it can be useful to introduce for a while, instead of the function $r(\mathrm{v})$, the function $u(\mathrm{v})$ defined by

$$u(\mathrm{v}) = \frac{1}{r(\mathrm{v})}. \tag{A16}$$

This allows us to write Equations (A14) and (A15) as

$$\overline{V}_r = -C\frac{du}{d\mathrm{v}} \tag{A14$'$}$$

and

$$\overline{V}_n = +Cu. \tag{A15$'$}$$

Hence the expression of V^2 as a function of u and of $du/d\mathrm{v}$:

$$V^2 = \overline{V}_r^2 + \overline{V}_n^2 = C^2\left[\left(\frac{du}{d\mathrm{v}}\right)^2 + u^2\right]. \tag{A17}$$

According to Equation (A16) we have

$$\frac{du}{d\mathrm{v}} = -\frac{1}{r^2}\frac{dr}{d\mathrm{v}} = -u^2\frac{dr}{d\mathrm{v}}. \tag{A18}$$

Consequently, on replacing the right-hand side of Equation (A12) by $+\mu(du/d\mathrm{v})$ and on replacing in the left-hand side of Equation (A12) V^2 by its expression (A17), we finally obtain (after division of both sides by $C^2(du/d\mathrm{v})$):

$$\frac{d^2u}{d\mathrm{v}^2} + u = \frac{1}{C^2}\mu. \tag{A19}$$

It is relatively easy to show that this differential equation can be satisfied by Equation (A4) describing an elliptical orbit in polar coordinates (after replacement of $1/r$ by u), i.e. by the function

$$u(\mathrm{v}) = \frac{1}{p}(1 + e\cos\mathrm{v}), \tag{A20}$$

provided that the parameter p in Equation (A20) satisfies the relation

$$\frac{1}{p} = \mu/C^2. \tag{A21}$$

Indeed, if $u(\mathrm{v})$ is defined by Equation (A20), $d^2u/d\mathrm{v}^2$ will be given by

$$\frac{d^2u}{dv^2} = -\frac{1}{p} e \cos v. \tag{A22}$$

A substitution into Equation (A19) immediately shows that Equation (A21) represents a sufficient condition for Equation (A20) to be a solution of Equation (A19) for any value of e.

Thus the orbit of P is indeed an ellipse if e is less than 1.

Now, let $d\tilde{A}$ represent, as in Subsection 2.5, the area, equal to $\frac{1}{2}r(r\,dv)$, of the infinitesimal sector swept by the radius vector SP during the interval of time dt. Then Equation (A13) shows that the *finite* area $\Delta\tilde{A}$ swept by the radius vector SP is equal to $\frac{1}{2}C\Delta t$, where Δt is the corresponding finite interval of time (this is *Kepler's second law* or the *law of areas*).

If T represents the time of a complete revolution, the corresponding area $\tilde{A}_{\text{tot}}$ swept by SP will be the *total* area $\pi ab = \pi a^2(1-e^2)^{1/2}$ of an ellipse, and C will be given by:

$$C = 2\tilde{A}_{\text{tot}}/T = \frac{2\pi}{T} a^2(1-e^2)^{1/2} = \frac{2\pi}{T} ab. \tag{A23}$$

Since μ denotes $G(M+m)$, Equation (A21) can be written, according to Equation (A5):

$$G(M+m) = C^2/p = \frac{4\pi^2}{T^2} a^2b^2 \Big/ \left(\frac{b^2}{a}\right) = \frac{4\pi^2}{T^2} a^3 \tag{A24}$$

or

$$\boxed{\frac{a^3}{T^2} = \frac{G(M+m)}{4\pi^2}}\,. \tag{A25}$$

Equation (A25) represents the *Newtonian form of Kepler's empirical third law* $a^3/T^2 = \text{const.} = C_k$.

Its advantage on the Keplerian empirical formulation is double: first it extends a purely planetary law (proper to the solar system) to any binary system; next it shows how the constant C_k of the empirical law depends on the masses M and m. It also shows that for a given value of M, like the mass $M_\odot$ of the Sun in the solar system, the empirical constant C_k is not really constant, but depends on the mass m of each planet, thus opening a way to the determination of masses by observation of quantities such as a and T.

In the particular case of the solar system, the existence of satellites revolving round some planets can be used for a determination of the ratio of the mass of a planet to the mass of the Sun $M_\odot$. Let us illustrate this in the case of a satellite of Jupiter.

We call a' the semimajor axis of the ellipse described by a given satellite of mass m' with an orbital period equal to T'. We call a_j and T_j the corresponding parameters of the motion of Jupiter itself (of mass m_j). Then, the fundamental relation (A25) applied to the systems Sun–Jupiter and Jupiter–Satellite will give:

$$a_j^3/T_j^2 = \frac{G}{4\pi^2}(M_\odot + m_j); \qquad a'^3/T'^2 = \frac{G}{4\pi^2}(m_j + m'). \tag{A26}$$

On eliminating G by division we obtain:

$$\frac{m_j + m'}{M_\odot + m_j} = \frac{m_j}{M_\odot}\left(1 - \frac{m_j}{M_\odot} + \frac{m'}{m_j} + \ldots\right) = (a'/a_j)^3 (T_j/T')^2. \tag{A27}$$

The observation of Jupiter's satellite No XII yields $a' \approx 0.142$ astronomical units (AU) and $T' \approx 630$ days (d); that of Jupiter $a_j \approx 5.203$ AU and $T_j \approx 4332.6$ d.

Thus, neglecting m'/m_j (for the largest satellite of Jupiter, Ganymede, this ratio is of the order of 1/12 300), an iterative solution of Equation (A27) gives $m_j/M_\odot \approx 1/1047$.

Among all planets of the solar system Jupiter's mass is the greatest, but it represents less than 0.1% of $M_\odot$. This explains why in Kepler's empirical law the *constancy* of the 'constant' C_k was not significantly affected by the presence of m at the right-hand side of Equation (A25).

A3. AN APPROXIMATE EXPRESSION FOR THE TRUE ANOMALY AS A FUNCTION OF TIME (FOR SMALL VALUES OF ECCENTRICITY)

Let us assume that the eccentricity e of a binary system is sufficiently small to allow the neglect of the terms proportional to e^2, e^3, etc. in the series expansions in powers of e. Then, limiting these expansions to the terms of first order in e, Equations (A4) and (A5) will give:

$$r \approx a(1 - e \cos \mathrm{v}), \tag{A28}$$

hence,

$$r^2 \approx a^2(1 - 2e \cos \mathrm{v}). \tag{A29}$$

According to Equations (A13) and (A23)

$$r^2 d\mathrm{v} \approx \frac{2\pi}{T} a^2 dt. \tag{A30}$$

On combining Equations (A29) and (A30) we find:

$$(1 - 2e \cos \mathrm{v})\, d\mathrm{v} \approx \frac{2\pi}{T} dt. \tag{A31}$$

On integrating and taking into account a choice of the origin of times t corresponding to the *zero* value of v, we obtain:

$$\boxed{\mathrm{v} - 2e \sin \mathrm{v} \approx \frac{2\pi}{T} t}\ . \tag{A32}$$

On introducing the '*mean anomaly*' M defined by

$$\mathrm{M} = \frac{2\pi}{T} t, \tag{A33}$$

we deduce from Equation (A32) that:

$$\mathrm{v} \approx \mathrm{M} + e(2 \sin \mathrm{v}). \tag{A34}$$

In the term proportional to e we can replace v by M, hence we finally obtain:

$$\mathrm{v} \approx \mathrm{M} + e(2 \sin \mathrm{M}), \tag{A35}$$

or

$$\mathrm{v} \approx \frac{2\pi}{T} t + e\left(2 \sin \frac{2\pi}{T} t\right). \tag{A36}$$

A4. DETERMINATION OF THE RIGOROUS VALUE OF THE TRUE ANOMALY AS A FUNCTION OF TIME, FOR ANY VALUE OF THE ECCENTRICITY

Returning to the general case of an eccentricity $e < 1$ let us introduce an auxiliary parameter ξ, called '*eccentric anomaly*', defined by the relation:

$$r = a(1 - e \cos \xi). \tag{A37}$$

On substituting this expression of r into the relation (A4) between r and v, and solving for cos v, we obtain after elimination of p by Equation (A5):

$$\cos \mathrm{v} = \frac{-e + \cos \xi}{1 - e \cos \xi}, \tag{A38}$$

hence

$$\sin \mathrm{v} = (1 - \cos^2 \mathrm{v})^{1/2} = \frac{(1 - e^2)^{1/2} \sin \xi}{1 - e \cos \xi}. \tag{A38'}$$

On introducing this expression for sin v into Equation (56), we find:

$$\frac{dr}{dt} = \frac{Ce}{p} \sin \mathrm{v} = \frac{Ce(1 - e^2)^{1/2} \sin \xi}{p(1 - e \cos \xi)}. \tag{A39}$$

On combining this equation with Equations (A5) and (A23) we obtain:

$$\frac{dr}{dt} = a \frac{2\pi}{T} \frac{e \sin \xi}{(1 - e \cos \xi)}. \tag{A40}$$

By differentiation of Equation (A37) we find:

$$\frac{dr}{dt} = ae \sin \xi \frac{d\xi}{dt}. \tag{A41}$$

On comparing Equations (A40) and (A41), and on taking into account the definition (A33) of M we obtain:

$$(1 - e \cos \xi)\, d\xi = d\mathrm{M} = \frac{2\pi}{T} dt. \tag{A42}$$

Hence by integration (with obvious initial conditions):

$$\boxed{\xi - e \sin \xi = \mathrm{M} = \frac{2\pi}{T} t}. \tag{A43}$$

This Equation, called '*Kepler's Equation*', can be solved numerically for ξ. Thus one obtains numerical tables for ξ (for different values of e) as a function of time t. Then Equations (A38) and (A38′) give v as a function of t.

The absence of proportionality in the general case between v and t is evident from Kepler's second law (A13) written in the form $dv/dt = C/r^2$, since r is variable on an elliptical orbit. This explains the complicated forms of the curves in Figure 8.4.

Elimination of $(1 - e \cos \xi)$ between Equations (A37) and (A38), or between (A37) and (A38′) yields with b given by Equation (A5):

$$r \cos v = a \cos \xi - c \qquad \text{(A44)}$$

and

$$r \sin v = b \sin \xi = \frac{b}{a}(a \sin \xi). \qquad \text{(A44′)}$$

Hence an interesting geometrical interpretation of ξ illustrated by Figure 8.A2:

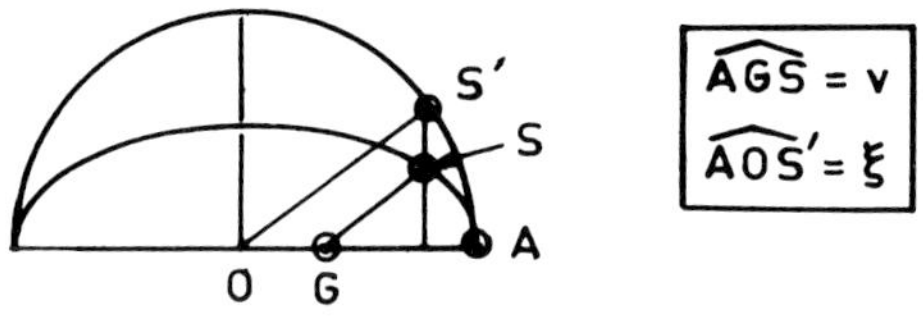

Fig. 8.A2. A geometrical interpretation of the parameter ξ.

Appendix B. Inductions Leading from Kepler's Empirical Laws to Newton's Law of Gravitation

B1. Introduction

The deduction of Kepler's laws from Newton's law of gravitation, presented in the Appendix A, should not let us forget the role played by Kepler's laws in Newton's construction of his gravitational theory.

Therefore, it can be worth attempting (without any claim at rigorous historical accuracy) an analysis of *the logical structure of inductions* which led Newton, in 1687, to his law of universal gravitation. On doing this we try to describe the shortest way to this discovery. The actual historical way was certainly much more tortuous.

B2. The Concept of Continuous Fall of a Planet towards the 'Attracting' Central Body

In 1609 Kepler formulated his first empirical law: planetary orbits are *elliptical* and the Sun is situated at one of the foci of these ellipses.

However, according to the *law of inertia*, introduced in dynamics by Galileo and Descartes, such as it was understood and formulated by Newton, a body not acted upon by any external force should remain at rest or move along a *straight line* with a constant speed with respect to a reference system at rest in the 'absolute space' (and in the concrete 'attached' to fixed stars). (Today, of course, we do not believe any more in the existence of a rigid absolute space and we define the 'interial reference systems' as those with respect to which the law of inertia is satisfied, but this does not play any role in the arguments presented below.)

Since, according to Kepler's first law, the path of a planet P is *curved*, with its concavity turned towards the Sun, Newton was led to admit that the planet P is *acted upon by a force F* responsible for a transformation of the straight inertial motion into a motion curved towards the Sun.

However, it was not at all obvious that this force could be directly related to some 'solar gravity' analogous to the terrestrial gravity attracting material terrestrial bodies towards the center of the Earth, and involved in both their *free fall* and their weight.

It was, therefore, important to start by showing that this curvature of the planetary orbit could be interpreted as a result of a kind of continuous *'free fall'* of the planet towards the Sun.

This was achieved by Newton through a genial combination of the law of inertia with Kepler's second law, also formulated in 1609, according to which the areas swept by the line connecting the planet to the Sun are proportional to the interval of time.

More or less schematically Newton's 'geometrical' proof can be summarized in the following way (see Figure 8.B1).

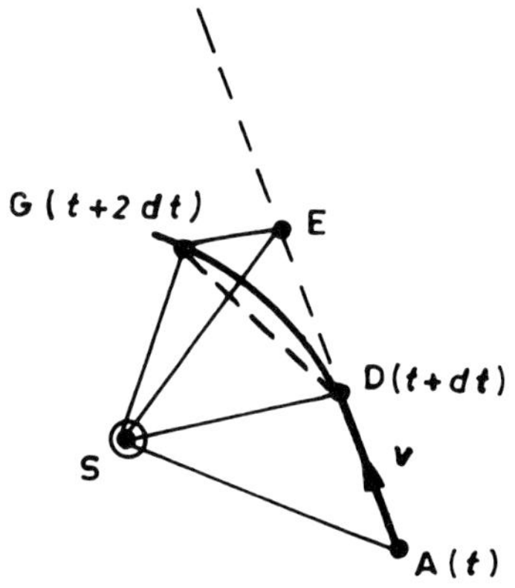

Fig. 8.B1. The real trajectory ADG, and the theoretical inertial trajectory ADE. The 'fall' EG, parallel to DS, is attributed to the attractive gravitational force of the Sun S.

Consider the motion of a planet P describing an elliptical orbit round the Sun S. At time t the planet is at the point A. Its vector velocity $\boldsymbol{v}$ is tangent to the trajectory at A. After a very short interval of time dt, i.e. at time $(t + dt)$, the planet is at the point D such that the *arc* AD has a length equal to $(dt)\,|\boldsymbol{v}|$.

At time $(t + 2dt)$ the planet, on its actual trajectory, is at the point G.

Now, if dt is sufficiently small one can assimilate the *arcs* AD and DG with *straight lines* AD and DG and, applying Kepler's second law, write:

$$\text{The area of the triangle SAD} = \text{The area of the triangle SDG.} \qquad \text{(B1)}$$

If at time t the force responsible for the curvature of the orbit suddenly vanished, the planet would escape along the tangent at the point A to the trajectory. The inertial motion being rectilinear and uniform, and the speed at A being equal to $|\boldsymbol{v}|$, the planet, at time ($t + dt$), would come at the point D of the orbit (because of the assumption made above about the smallness of dt allowing the assimilation of the arc AD with the straight line AD). Thus, at time ($t + 2dt$) the planet would come at the point E of the straight line AD such that AD = DE.

Since the triangles SAD and SDE would have equal 'bases' AD and DE and the same height (the distance from S to the line ADE), we would have

$$\text{The area of the triangle SAD} = \text{The area of the triangle SDE.} \tag{B2}$$

Hence, by comparison of Equations (B1) and (B2):

$$\text{The area of the triangle SDE} = \text{The area of the triangle SDG.} \tag{B3}$$

Since the triangles SDE and SDG have the same 'base' SD and the same area, their heights must be equal, which means that EG *must be parallel to* DS.

Thus, near D the planet P must move under the action of a 'central' force $(F_s)_p$ directed *along the line going from* P *to* S, a force responsible for the fall, equivalent to EG, from its virtual inertial position E to its actual position G.

This new concept of '*equivalent fall*', introduced by Newton, showed itself very fruitful.

B3. The Dependence of the Force Curving the Path of a Planet on its Mass and on its Distance from the Sun

B3.1. The Gravitational Field of the Earth, and the Proportionality of the Weight of a Body to its Mass

It is well known that Newton performed the following experiment. A vacuum, as perfect as possible, was created in a vertical tube, containing different material bodies, in order to minimize the resistance of the air. A lead ball proved to fall with the same vertical vector acceleration g_0 as a feather, when the tube was turned over.

The vector g_0 represents, by definition, the '*acceleration of gravity*' at the Earth surface at the see leavel.

Since two bodies, as different by their mass as a ball and a feather, fall with the same acceleration g_0, this means, according to the fundamental principle of Newtonian dynamics

$$F = mg_0, \tag{B4}$$

that the mass m of each body *cancels out* in Equation (B4) when the force F_e acting upon the body is its *weight*. This cancellation demands the proportionality between F_e and m: the force must be of the form:

$$F_e = mE, \tag{B5}$$

where E is the force of the Earth's 'attractive power' upon the unit mass near the surface of the Earth.

In modern terminology (used below for more clarity), E represents the gravitational field created by the Earth (supposed to be spherical) at a distance r_0, equal to Earth's radius, from its center. This terminology and notation will later prove very convenient, and no confusion with an electric field is, of course, possible!

On comparing Equations (B4) and (B5) one finds

$$E = E(r_0) = g_0. \tag{B6}$$

On the other hand, the static weight **W**, i.e. the weight measured by the lengthening of a vertical spring stretched by a hanging mass m, is a force proportional to the mass m, conceived as a quantity of matter. The force **W** is vertical and directed downwards.

On identifying the force $m\mathbf{E}$ responsible for the dynamical acceleration $\mathbf{g}_0$ of the free fall of the mass m with the force **W** responsible for the lengthening of the spring, one obtains:

$$W = mE = mg_0.. \tag{B7}$$

This allows one to use the symbol **W** and the word 'weight' to denote the force $m\mathbf{E}$ responsible for the acceleration $\mathbf{g}_0$ in the free fall.

B3.1.1. Remark

Note incidentally that the same mass m thus occurs in two phenomena of apparently different physical nature: a kind of '*resistance to acceleration*' such as the mass m appears in Equation (B4) when written in the form

$$\gamma = \frac{F}{m}, \tag{B8}$$

and the '*grasp offered to the attraction by the Earth*' such as it appears in Equation

$$W = mE. \tag{B9}$$

It is well known that this remark represents a keystone in the elaboration of Einstein's relativistic theory of gravitation.

B3.2. The Force of Solar Gravity: Its Inverse Proportionality to the Square of the Distance and its Direct Proportionality to the Mass of the Attracted Planet

Let us consider now a planet P and let us assume (for pedagogical reasons, i.e. in order to simplify the calculations) that its orbit is a circle of radius R. Of course, the Sun is supposed to be at the center of the orbit. It follows then from the Kepler's second law that the motion is uniform with a constant angular velocity ω and a constant linear speed $|\mathbf{v}|$.

Because of the curvature of the trajectory and in spite of the constancy of the angular velocity the vector velocity $\mathbf{v}$ does not remain constant in direction and its variations per unit time $\Delta\mathbf{v}/\Delta t$ define in the limit (when $\Delta t \to 0$) a non-zero vector acceleration:

$$\gamma = \frac{\Delta \mathbf{v}}{\Delta t}. \tag{B10}$$

The nowadays classical expression for the modulus γ of $\boldsymbol{\gamma}$, obtained for the first time by C. Huyghens in 1673, is

$$\gamma = \omega^2 R, \tag{B11}$$

and it was already known by Newton that $\boldsymbol{\gamma}$ is directed towards the center of the circular orbit, i.e. towards the Sun.

On replacing ω by $2\pi/T$, where T is the orbital period, one obtains

$$\gamma = 4\pi^2 R/T^2. \tag{B12}$$

On comparing the orbital motion of the planet Earth, characterized by the parameters (γ_e, R_e, T_e) to the orbital motion of one of the other planets P characterized by (γ_p, R_p, T_p), and on applying Kepler's empirical general third law, viz.

$$\frac{T^2}{R^3} = \text{const. (for any planet)}, \tag{B13}$$

Newton could obtain:

$$T_e^2/T_p^2 = R_e^3/R_p^3. \tag{B14}$$

Then on combining this consequence of Kepler's *observational* third law with Huyghens' *theoretical* relation (B12) Newton was able to obtain the following chain of relations:

$$\frac{\gamma_p}{\gamma_e} = \frac{R_p/T_p^2}{R_e/T_e^2} = \frac{R_p\,T_e^2}{R_e\,T_p^2} = \frac{R_p R_e^3}{R_e R_p^3} = \frac{R_e^2}{R_p^2} = \frac{1/R_p^2}{1/R_e^2}. \tag{B15}$$

Now Newton introduces two new and bold ideas. He assumes that the attractive force $(\boldsymbol{F}_s)_e$ of the Sun, responsible for the curvature of the terrestrial (Earth's) orbit and the attractive force $(\boldsymbol{F}_s)_p$ of the Sun responsible for the curvature of the orbit of the considered planet P, i.e. the two forces which constrain the Earth and the planet to 'fall' towards the Sun, represent a kind of '*solar gravity*' (solar weight).

If so, these forces must both possess the property of the terrestrial gravity, expressed by Equation (B5), to be proportional to the mass m of the body acted upon. Thus, m_p being the mass of the planet P, and $\boldsymbol{S}_p$ being the force of Sun's action upon a unit mass at the distance R_p of P from the Sun (i.e., in modern terminology, $\boldsymbol{S}_p$ being the gravitational field of the Sun at the distance R_p) Newton could write

$$(\boldsymbol{F}_s)_p = m_p \boldsymbol{S}_p \tag{B16}$$

for the planet P and similarly write

$$(\boldsymbol{F}_s)_e = m_e \boldsymbol{S}_e \tag{B16$'$}$$

for the planet Earth of mass m_e.

On the other hand, application of the Newtonian principle (B4) to the motion of the planet P and to the motion of the Earth respectively, yields

$$(\boldsymbol{F}_s)_p = m_p \boldsymbol{\gamma}_p, \tag{B17}$$

and

$$(\boldsymbol{F}_s)_e = m_e \boldsymbol{\gamma}_e. \tag{B17$'$}$$

Hence, on comparing Equations (B16) and (B17) or Equations (B16′) and (B17′), and on writing the moduli of $\boldsymbol{S}_p$ and $\boldsymbol{S}_e$ in a way which recalls the respective distances R_p and R_e from the Sun:

$$\text{S}_p = \text{S}(R_p) = \gamma_p, \tag{B18}$$

and

$$\text{S}_e = \text{S}(R_e) = \gamma_e. \tag{B18$'$}$$

On dividing Equation (B18) by Equation (B18′) one obtains:

$$\mathrm{S}_p/\mathrm{S}_e = \gamma_p/\gamma_e. \tag{B19}$$

This, combined with Equation (B15), gives the final result, viz.

$$\frac{\mathrm{S}_p}{\mathrm{S}_e} = \frac{1/R_p^2}{1/R_e^2}, \tag{B20}$$

which can also be written as

$$\frac{\mathrm{S}_p}{1/R_p^2} = \frac{\mathrm{S}_e}{1/R_e^2}. \tag{B20$'$}$$

Since the right-hand side of Equation (B20′) keeps the same value for different planets, Newton finally obtains, as the expression for the gravitational field $S(R_p)$ of the Sun, at the distance R_p:

$$\mathrm{S}(R_p) = \frac{C_s}{R_p^2}, \tag{B21}$$

where the constant C_s, *characteristic of the Sun's 'attractive power'*, keeps the same value for all planets revolving round the Sun.

On taking into account Equation (B16), one obtains:

$$(F_s)_p = \frac{m_p}{R_p^2}C_s. \tag{B22}$$

Thus Newton already finds that the force $(F_s)_p$ of the solar gravity (solar attraction upon some planet P) is proportional to the mass m_p of the planet and inversely proportional to the square R_p^2 of the distance R_p of the planet P from the Sun (the constant of proportionality C_s remaining the same for all planets of the solar system).

B4. A VERIFICATION BY APPLICATION TO THE MOTION OF THE MOON

The interpretation of the curved motion of a planet round the Sun as a continuous fall of the planet towards the Sun, under the force of solar gravity, and the extension to this force of the property of terrestrial gravity to be a force proportional to the mass m of the falling body (in addition to all theoretical and empirical ingredients used in Sections B2 and B3) enabled Newton to discover the property expressed by Equation (B21), viz. (in modern terminology) that the gravitational field $S(R)$ of the Sun at the distance R from its centre is inversely proportional to R^2.

Now, guided by his genial idea of the similarity between the properties of solar and terrestrial 'attractive powers' (gravities) Newton will make a vertiginous conceptual jump and assume the validity of the same law, of inverse proportionality to the square of the distance, for the gravitational field $\boldsymbol{E}$ of the Earth.

Considering, henceforth, the gravitational field $\boldsymbol{E}$ as a function of the distance r to the center of the Earth, we can denote it by $\boldsymbol{E}(r)$. Thus, Newton's new assumption will be expressed by

$$\frac{E(r)}{E(r_0)} = \frac{1/r^2}{1/r_0^2}, \tag{B23}$$

where r_0 denotes the radius of the Earth (supposed to be spherical).

On the other hand, Equation (B6) reduced to the corresponding moduli takes the form

$$E(r_0) = g_0, \tag{B24}$$

where g_0 represents the modulus of the acceleration $\mathbf{g}_0$ of gravity observed at the sea level of the surface of the Earth.

On comparing Equations (B23) and (B24) Newton obtains

$$E(r) = \frac{g_0 r_0^2}{r^2} = \frac{C_e}{r^2}, \tag{B25}$$

where C_e is a constant, equal to $g_0 r_0^2$, characteristic of the 'attractive power' of the Earth.

Of course, by an extension of the law expressed by Equation (B5), the gravitational force $(F_e)_m$ exerted by the Earth on a body of mass m, at a distance r from its center, will be given by

$$(F_e)_m = mE(r) = \frac{mC_e}{r^2}. \tag{B26}$$

Now, Newton interprets the curved path of the Moon round Earth in the same way as the curved path of a planet round the Sun: i.e. like a continuous fall towards the central 'attracting heavenly body': the Earth.

Since, on the other hand, Newton knows that the Moon obeys (at least at the circular approximation) the two first Kepler's laws, he is able to apply the same type of argument as in Section B4 to lunar motion round Earth.

Hence, on distinguishing henceforth by the subscript m all quantities relative to the Moon, one can write for the lunar motion (assumed to be circular and uniform) by analogy with Huyghens' relation (B12):

$$\gamma_m = 4\pi^2 \frac{r_m}{T_m^2}, \tag{B27}$$

where obviously r_m is the distance of the Moon to the center of the Earth, and T_m is the duration of a complete revolution of the Moon round Earth.

As to the modulus γ_m of the vector acceleration of the lunar motion, Newton can equate it, by application of the arguments similar to those leading to Equation (B18), with the modulus $E(r_m)$ of the gravitational field of the Earth near the Moon, and write

$$E(r_m) = \gamma_m. \tag{B28}$$

On the other hand, Equation (B25) applied with $r = r_m$ becomes

$$E(r_m) = \frac{g_0 r_0^2}{r_m^2} = \frac{C_e}{r_m^2}. \tag{B29}$$

The combination of Equations (B29), (B27) and (B28) finally yields:

$$4\pi^2 \frac{r_m}{T_m^2} = \frac{g_0 r_0^2}{r_m^2} = \frac{C_e}{r_m^2}, \tag{B30}$$

a relation which can also be written:

$$\frac{T_m^2}{r_m^3} = \frac{4\pi^2}{C_e} = \frac{4\pi^2}{g_0 r_0^2}, \tag{B30'}$$

or, on introducing the average angular velocity $\omega_m = 2\pi/T_m$ of the Moon:

$$\omega_m^2 r_m^3 = C_e = g_0 r_0^2. \tag{B30''}$$

Since all quantities in Equations (B30), (B30′) or (B30″) are measurable by observations ($r_0 \approx 6.4 \times 10^8$ cm; $r_m \approx 3.8 \times 10^{10}$ cm; $T_m \approx 2.4 \times 10^6$ s; $g_0 \approx 9.8 \times 10^2$ c.g.s.; $\omega_m \approx 2.6 \times 10^{-6}$ s^{-1}), Newton was thus able to perform, within the precision allowed by the measures of his time, a first *observational test* of his gravitational theory.

B4.1. Remarks

B4.1.1. A wide-spread and lasting legend states that Newton could not perform this test because of his ignorance of a sufficiently precise value of r_0.

Actually it seems quite sure nowadays that the book *Mesure de la Terre*, published in 1671, and containing Jean Picard's sufficiently precise measures of r_0, was known to Newton eight years before the publication of his *Philosophiae Naturalis Principia Mathematica* in 1687.

B4.1.2. Considering the system of artificial satellites in addition to the natural satellite of the Earth (the Moon) Equation (B30″) represents an extension of Kepler's third law to the 'moons' travelling round Earth, in a form intermediate between the one where $\omega_m^2 r_m^3$ is simply assigned a constant value for all 'moons', and the one called now the 'Newtonian form of Kepler's third law', where C_e is expressed in terms of Earth's mass m_e (see Section 1 of this Chapter).

Indeed, here the constant C_e is already expressed in terms of measurable quantities g_0 and r_0, though the mass of the Earth does not yet appear in this relation.

B5. DEPENDENCE OF THE 'ATTRACTIVE POWER' OF THE CENTRAL BODY ON ITS OWN MASS

We have explained in Sections B3 and B4 how Newton succeeded in showing that both the Sun and the Earth were able to 'generate' what would be called today a *gravitational field*, each of these two 'central bodies' being characterized, with respect to its 'attractive power', by a specific constant: C_s in Equation (B21) for the Sun and C_e in Equation (B25) for the Earth.

A new induction will enable Newton to discover that C_s is proportional to the mass $M_s = M_\odot$ of the Sun, and that C_e is proportional to the mass m_e of the Earth.

Indeed, Newton considers that the gravitational field of modulus $E(r)$ of the Earth, given by Equation (B25), viz.

$$E(r) = \frac{C_e}{r^2}, \tag{B31}$$

where r is the distance to the center of the Earth, can extend its influence as far as the Sun. If so, the distance between the Earth and the Sun being denoted by a, the Earth field's intensity $E(a)$ near the Sun will be given by

$$E(a) = \frac{C_e}{a^2}. \tag{B32}$$

On the other hand, the gravitational field represents only the attractive force *per unit mass* of the *attracted* body. Therefore, since the attractive force exerted by the central body is proportional to the mass of the attracted body, the gravitational force $(F_e)_s$ exerted by the Earth upon the Sun will be given, in absolute value, by

$$(F_e)_s = \frac{C_e M_s}{a^2}. \tag{B33}$$

By a similar argument, and in accordance with Equation (B16′), viz.

$$(F_s)_e = m_e S_e = m_e \mathrm{S}(R_e) = m_e \mathrm{S}(a), \tag{B34}$$

and Equation (B21) applied to the planet Earth, viz.

$$\mathrm{S}(a) = \mathrm{S}(R_e) = \frac{C_s}{R_e^2} = \frac{C_s}{a^2}, \tag{B35}$$

the absolute value of the gravitational force $(F_s)_e$ exerted by the Sun upon the Earth will be given by

$$(F_s)_e = \frac{m_e C_s}{a^2}. \tag{B36}$$

The roles of $(F_s)_e$ and $(F_e)_s$ are perfectly reciprocal, since they both represent the *mutual attraction* between the Sun and the Earth, or, in virtue of the principle of equality between action and reaction. Therefore Newton sets:

$$(F_s)_e = (F_e)_s. \tag{B37}$$

Hence, on taking into account Equations (B33) and (B36), the announced result:

$$M_s C_e = m_e C_s, \tag{B38}$$

or

$$\frac{C_s}{M_s} = \frac{C_e}{m_e} = G, \tag{B38′}$$

where G is a constant, called the '*gravitational constant*'.

From Equation (B38′) we deduce that

$$C_s = GM_s, \tag{B39}$$

and

$$C_e = Gm_e. \tag{B40}$$

Finally Equations (B33) and (B36) take the form

$$(F_s)_e = (F_e)_s = \frac{Gm_e M_s}{a^2}, \tag{B41}$$

and Equations (B22), (B26) take the forms:

$$(F_s)_p = \frac{Gm_p M_s}{R_p^2}, \tag{B42}$$

and

$$(F_e)_m = \frac{Gmm_e}{r^2} \,. \tag{B43}$$

B6. The Universal Law of Gravitation

By a last induction Newton assumes that the relations like (B41), (B42) and (B43) relative to the gravitational interactions between the Earth and the Sun, between the Sun and a planet of mass m_p, or between the Earth and a body of mass m, respectively, can be applied to the force F of gravitational interaction between two microscopic (or punctual macroscopic) masses m and m' separated by a distance d, with the same gravitational constant G.

Hence the law

$$\boxed{F = G\frac{mm'}{d^2} \,.} \tag{B44}$$

This universal law will take the name of '*Law of universal gravitation*' when one wants to emphasize the universality of the phenomenon of gravitation, or the name '*Universal law of gravitation*' when one wants to emphasize the universality of Equation (B44).

The constant G cannot be measured by astronomical observations. Laboratory experiments (first performed by Cavendish in 1798) yield its numerical value, which according to our present knowledge is equal to

$$G = 6.67 \times 10^{-8}\,\text{c.g.s.} \,. \tag{B45}$$

References

WORKS CITED IN THE TEXT

(Work that can be used for further study is marked by an asterisk.)

Allen, C. W.: 1973, *Astrophysical Quantities* (3rd Ed.), Athlone Press, London.

*Hynek, J. A.: 1951, *Astrophysics* (A Topical Symposium), McGraw-Hill, New York.

CHAPTER 9

GALACTIC X-RAY SOURCES

1. Introduction

By 'galactic' X-ray sources we mean below all sources situated either in the galactic disk or in the galactic 'bulge', but also those associated with some of the globular clusters surrounding our galaxy, and even those situated in such satellites of our galaxy as the Small and the Large Magellanic clouds.

The X-rays emitted by some of these sources are definitely known to originate in *binary systems* in which a collapsed object (such as a white dwarf, a neutron star or, perhaps, a 'black hole' defined in Subsection 3.3.1) is associated with a star still burning its nuclear fuel.

It will appear that the theoretical interpretations of spectroscopic binaries (Chapter 8) can be applied to some binary X-ray sources, along with an elementary interpretation of eventual eclipses.

However, in addition to the Doppler effect already at work in spectroscopic binaries, some binary X-ray sources behave like X-ray pulsars, and all present a much more complicated interaction with their companion than in the case of ordinary, even close, double stars.

Several specific effects are observed in X-ray sources, both binary, nonbinary and non surely binary, which will be described below, after an introduction of the basic physical concepts used in the presentation of observations, and a brief historical remark.

The subject of X-ray sources deserves a special word of warning. The subject is so complex, is changing so rapidly; the wealth of observational data, complicated and uncertain theories are so overwhelming, that we do not attempt to present a complete and thorough picture of all present (November 1978) knowledge in this field. We try simply to describe systematically the main phenomena and illustrate them by most typical examples. The basic physical mechanisms at work (theoretical 'models') will be simply suggested in a qualitative way. Any future lack of 'up-to-date-ness' will mainly prove the vitality of astrophysics!

1.1. Historical Remarks

The study of cosmic X-ray sources (other than the Sun) started on 18 June 1962, the day of the launch by a team of the Massachusetts Institute of Technology (MIT) of a rocket whose X-ray equipment disclosed an X-ray source in the constellation of Scorpio. This rocket reached a relatively small altitude of about 150 km, but this was already sufficient to avoid the atmospheric extinction of X-rays.

However, a real advance in the knowledge of cosmic X-ray sources (Giacconi, 1973) started only through the launch, on 12 December 1970, of the first small astronomy satellite ('orbiting observatory') *Uhuru* entirely devoted to X-ray observations. This artificial satellite was launched from a site in Kenya, and was named *Uhuru* after the Swahili word for 'freedom' because the day of launching was that of the seventh anniversary of the independence of Kenya. The satellite was built for NASA by R. Giacconi and his colleagues at American Science & Engineering, Inc.

Uhuru's orbit was almost circular: its distance to the Earth was between 520 and 560 km. The inclination of the plane of the orbit with respect to the terrestrial equator was about 3°. The orbital period was of 96 minutes.

The use of *Uhuru* brought about several very important discoveries: already in 1974 more than 160 X-ray sources were discovered.

More recent X-ray satellites can detect sources much weaker than could *Uhuru*: the British *Ariel* V, the Dutch ANS, the American SAS 3, OSO 8, etc. The HEAO A (High Energy Astronomy Observatory), launched by NASA in 1977, was able to detect sources more than 10 times fainter than the limits of *Uhuru* and SAS-3. The HEAO B launched in 1978 carries the first image-forming X-ray telescope. The instrument has an angular resolution of 2 arcsec and achieves a thousandfold increase in sensitivity.

Due to all these developments, X-ray astronomy has been extremely rich in new and often quite unexpected important results, both in galactic and extragalactic (see Chapter 13) astronomy.

1.2. TERMINOLOGY

According to the constellation in which they are located and the sequence in which they were discovered, X-ray sources were formerly given names such as Sco X-1 (in Scorpio), Cyg X-1, Cyg X-2, Cyg X-3 (in Cygnus) and so on. This terminology is still very usual for the brightest of X-ray sources.

However, since the publication of the 'Third *Uhuru* Catalog of X-ray Sources' (Giacconi *et al.*, 1974), and of the 'Fourth *Uhuru* Catalog' (Forman *et al.*, 1978) the same type of terminology as for pulsars is also used, with the prefix 3U or 4U replacing the prefix PSR. The source designation is given as α_{1950} (i.e. the right ascension) with four digits (hours and minutes of α for 1950.0) and δ_{1950} (i.e. the declination) with two or three digits (degrees and eventually tens of degree of δ).

Unfortunately, many observers prefer to use a terminology which recalls the name of their satellite or their team, still followed by the indication of (α, δ) of the source. Moreover, the progress in precision of location introduces very often a (slight) change of (α, δ).

Thus, for instance, the source Cyg X-1 has also the following 'name': 3U 1956 + 35. And 3U 1809 + 50 = 4U 1813 + 50.

Similarly the same source can be called 4U 1538 − 52 and A 1540 − 53 (where the minus sign indicates a negative, i.e. southern, declination).

A last complication is introduced by the fact that many observers (in particular those observing with hard X-ray or γ-ray equipment) use instead of (α, δ) coordinates, the $(l^{\mathrm{II}}, b^{\mathrm{II}})$ coordinates, where l^{II} is the galactic longitude and b^{II} is the galactic latitude. In this case the letter G, preceded or followed by the letter X, or by a letter indicating the specific satellite, can be followed by one, two or three digits expressing the value of l^{II}, and by one or (exceptionally) two digits (with a *plus* or a *minus* sign) expressing the value of b^{II}.

Thus, for instance, the source 4U 1658 − 48 is sometimes named GX 339 − 4, and the source GX 17 + 2 is the same as 3U 1813 − 14. Similarly, when observed with the γ-rays satellite COS B, the Crab pulsar PSR 0531 + 21 (often called NP 0532 by radio astronomers), is also named 3U 0531 + 21 as X-ray source, and CG 185 − 5 as γ-ray source (C stands for COS B, the European γ-ray satellite).

In order to help the non-specialst to find his way in this confusing array of names given to X-ray and γ-ray sources, we give in the Appendix to this chapter, a 'dictionary of abbreviations' and a 'Table of Names' with (α, β), GX, HD, and NGC entries, and 'constellation entries' for galactic sources most often encountered in scientific literature.

1.3. Units and the Corresponding Physical Concepts

In the X-ray domain, it is usual to define the frequency ν of a photon through its energy $\epsilon = h\nu$ (where, of course, h is the Planck's constant). In astrophysics this energy ϵ is usually expressed in keV (kiloelectron volts). Between ϵ, ν and the corresponding wavelength λ one has the following fundamental relations:

$$\epsilon\ \text{keV} = \frac{12.4}{\lambda\ \text{Å}} = (4.13 \times 10^{-18})(\nu\ \text{Hz}). \tag{1}$$

For some X-ray sources the energry dE received per cm^2 normal to the line of sight, per second, per unit of 'spectral interval' of one keV, in the neighbourhood of some energy ϵ, is sufficiently large to be measurable, and represents the '*intensity*' I_ϵ of the source. This intensity I_ϵ is very similar to the intensity defined in the theory of radiative transfer (see Chapter 1), with two major and one minor differences. In the present case, the direction $\boldsymbol{\omega}$ has not to be specified (since it is simply the direction of the source), and the solid angle $d\omega$ has not to be considered (since the sources are, hitherto, considered as punctual). Moreover the 'spectral interval' is expressed in units of energy (keV) and not in units of frequency (Hz). Finally, I_ϵ will be expressed in keV (of dE) cm^{-2} s^{-1} keV^{-1} (of $\Delta\epsilon$).

Thus, for instance, the continuous spectrum of the source Sco X-1, could be described, at some time t_0, (in units just defined above) by the following table:

TABLE 9.1
Intensity I_ϵ of the continuous X-ray spectrum of the source Sco X-1 (in keV cm^{-2} s^{-1} keV^{-1}) at the time t_0 (this spectrum is variable)

ϵ keV	1	2	5	10	20	50
I_ϵ	63	40	20	7.9	1.3	0.1

The time t_0 is not specified because Sco X-1 is a variable source, and the purpose of the Table 9.1 is simply to show that I_ϵ depends on ϵ.

However, for many sources, the sensitivity of the available receivers is not sufficient to permit the determination of I_ϵ and one is obliged to measure the energy ΔE received in a 'spectral interval' (ϵ_1, ϵ_2) of *several* $\Delta\epsilon$ keV per cm^2 normal to the line of sight and per second. In this case observation simply yields some sort of '*X-brightness*' of the source (which is not rigorously monochromatic, but corresponds to a relatively *extended* range of frequencies $\Delta\nu$ or energies $\Delta\epsilon$). Most authors call this brightness ΔE the *flux* (a few call it 'the intensity'!). This 'flux' does not correspond to the concept of flux used in the theory of radiative transfer, but corresponds to the concept of flux used by the radio astronomers. Of course, the X-brightness depends on (ϵ_1, ϵ_2) and not only on $(\epsilon_2 - \epsilon_1) = \Delta\epsilon$.

Very often the energy ΔE, instead of being expressed in keV (as the energy dE considered in the definition of I_ϵ), is expressed in *ergs* (per cm^2, per second). Thus, for instance, the '*X-brightness*' of Sco X-1 between $\epsilon_1 = 2$ keV and $\epsilon_2 = 6$ keV was, at time t_0, equal to 2×10^{-7} erg cm^{-2} s^{-1}, and at the same time was equal only to 4×10^{-7} erg cm^{-2} s^{-1} between 1 keV and 20 keV, (because of the decrease of I_ϵ for increasing ϵ shown by Table 9.1).

Many authors, and in particular those working with *Uhuru*, express the X-brightness of the observed sources in terms of the *number of photons* registered per second by their receivers, of known effective *area* and known spectral sensitivity limited to the interval (ϵ_1, ϵ_2).

Taking into account the particularities of the instruments borne by *Uhuru*, the authors of the 3U-Catalog assume that the corresponding unit of X-brightness called the '*one count per second*' (or simply '*one count*') corresponds (with a margin of possible error equal to about 40%) to an X-brightness (in the 2–7 keV energy band) equal to 1.7×10^{-11} erg cm^{-2} s^{-1}. Expressed in '*Uhuru counts*' the (variable) X-brightness of Sco X-1, at its maximum between 1970 and 1974, was of about 17 000 counts.

Some observers also use the '*Crab*' unit of X-brightness, i.e. the brightness (in some specified band ϵ_1, ϵ_2 of energies) of the Crab *Nebula*, the strongest permanent X-ray source.

A parameter similar to the usual colour index of the optical astronomy is provided by the *ratio* of the X-brightness in a band of a relatively high energy to the X-brightness in a band of low energy.

Just as it is usual to speak of stars of large colour index as 'red stars' and of stars of small (or negative colour index) as 'blue stars', one uses in the X-ray domain of wavelengths the rough expressions of *hard* or *soft* spectrum, with a somewhat different meaning defined below.

Quite generally, for most galactic X-ray sources the X-brightness of the source is not proportional to the extent $\Delta\epsilon$ of the corresponding spectral interval, since generally I_ϵ is not constant with respect to ϵ.

Conversely, when the function $f(\epsilon)$ which represents the variation of I_ϵ with respect to ϵ is known, the X-brightness (or 'flux') corresponding to (ϵ_1, ϵ_2) can be computed and will be given by

$$[\text{X-brightness in the range } (\epsilon_1, \epsilon_2)] = \int_{\epsilon_1}^{\epsilon_2} f(\epsilon)\, d\epsilon, \tag{2}$$

where, of course, one must take care to use the same units everywhere. Thus, for instance, the spectrum of Sco X 1 can be represented, in a first approximation (for $\epsilon < 20\,\text{keV}$) by the 'exponential law':

$$f(\epsilon) \approx K_0 \exp(-\epsilon/\epsilon_0), \tag{3}$$

where $\epsilon_0 = 4.4\,\text{keV}$ and $K_0 = 70\,\text{keV}\,\text{cm}^{-2}\,\text{s}^{-1}\,\text{keV}^{-1}$. This permits the application of Equation (2) to any interval (ϵ_1, ϵ_2) covering the validity of Equation (3).

One often writes kT instead of ϵ_0, with k representing the Boltzmann constant, and T representing the 'equivalent temperature' of the source. This temperature is normally expressed in degrees K, but very often it is also expressed in keV, with 1 keV of ϵ_0 equivalent to 1.14×10^7K. This last procedure amounts to call T the corresponding value of ϵ_0 in keV.

The spectrum is said to be 'soft' when ϵ_0 is large and 'hard' when ϵ_0 is small (because in the latter case it extends up to the region of hard X-rays).

Very often, instead of the 'exponential law' of the form (3) one fits the observed values of I_ϵ by a 'power law', written in the form

$$dN/d\epsilon = C(\epsilon/\epsilon_0)^{-\alpha}, \tag{4}$$

where dN is the number of photons received, per cm^2 and per second, in an energy interval of $d\epsilon$ keV, near the energy ϵ keV. In Equation (4), the parameter ϵ_0, usually different from the one in Equation (3), represents some constant energy (in keV) which, with the proper choice of the constant parameter α, yields the best fit for the observed X-ray spectrum, in a given energy band (ϵ_1, ϵ_2). The parameter α is sometimes called the 'power law number index'.

Thus, for instance, the high-energy X-ray spectrum of the Crab Nebula is sufficiently well fitted, in the energy band 23–513 keV, by a power law of the form (4) with $\epsilon_0 \approx 39.1\,\text{keV}$ and $\alpha = 2.00 \pm 0.06$. It seems that no long-term variability affects these values of ϵ_0 and of α.

Similarly, the X-ray spectrum of the source associated with the variable star AM Her can be represented in the $\sim 2\,\text{keV}$ to 1000 keV energy band by a power law of the form (4) with $\alpha = 0.84$.

Sometimes an extra factor of the form $\exp[-(\epsilon_\alpha/\epsilon)^{8/3}]$ is introduced at the right-hand side of Equation (3) or (4) in order to take into account the existence of a *cut off* in I_ϵ at *low energy*, the parameter ϵ_α, called the 'cut off energy', being determined by the best fit of the complete empirical law in observations.

In the usual case where the function $f(\epsilon)$ relative to a particular given source is not known, the relation between the [X-brightness in (ϵ_1, ϵ_2)] and the [X-brightness in $(\epsilon_1', \epsilon_2')$] is generally obtained on assuming that the $f(\epsilon)$ is of the form (3) or (4), but with an average 'standard' value of ϵ_0 different from the value 4.4 keV proper to Sco X-1. As to K_0 or C for each [X-brightness in (ϵ_1, ϵ_2)] it is given precisely by Equation (2) as soon as the value of ϵ_0 is given. Such procedures can of course only indicate the orders of magnitude.

1.4. THE BASIC PHYSICAL AND ASTROPHYSICAL PROCESSES

The main astrophysical process generally assumed to be responsible for the observed X-ray emission is the phenomenon of '*accretion*'.

In this process, the matter is supposed to 'fall' onto a compact (collapsed) object: a white dwarf, a neutron star, or a black hole. Thus, from a purely kinematical point of view, the accreting compact object behaves as a 'vacuum sweeper'. The very strong gravitational field in the neighbourhood of the compact object obliges the surrounding gas to flow over to the object. Indeed, the gravitational field at one radius (i.e. $\sim 10\,\text{km}$) above the surface of a neutron star, is about 10^{10} times as strong as that at one radius above the surface of the Sun.

The accreted matter can be supplied in very different ways. In a binary system it can come from a non-collapsed companion of the compact object. This (more or less) 'normal' companion, often called the 'primary', still burning its nuclear fuel, can emit a 'stellar wind', similar to the solar wind. On the other hand, the primary might also be unable to retain its outer layers against the attraction of the collapsed secondary. This can happen for two different reasons.

First, the two components of the binary system can be very close and the primary can have a very extended atmosphere. Second, and more frequently, the primary and the companion can be surrounded by an extended gaseous envelope or halo filling two pear-shaped critical surfaces (called the 'Roche lobes') on which the gravitational forces due to both components, acting on the gas surrounding both stars, are approximately in equilibrium with the centrifugal forces due to the orbital motion of the system. (For the best presentation of the theory of these critical surfaces see Plavec and Krotochvil (1964).)

Of course, the presence of a binary companion is not a necessary condition for an accretion process. Indeed, the necessary gas can be supplied to an isolated compact object occasionally passing through the stellar winds of remote red giant stars, or passing through a sufficiently dense interstellar cloud.

Whatever the origin of the accreted matter, in the process of accretion the gravitational potential energy of the system formed by the collapsed object and the infalling gas is converted into other forms of energy, and produces the observed X-rays in several ways.

The strong gravitational field of the collapsed object causes large accelerations of the infalling particles. This generates high velocities and kinetic energies. On the other hand, when these high velocity particles encounter dense regions of the collapsed object they are strongly decelerated, like the electrons which are stopped by the anticathode of an X-ray tube. Both acceleration or deceleration along a straight line, or a circulation with a constant velocity along a curved path, both of electrons or of ions, generate electromagnetic waves. In particular, a strong deceleration generates the so called *Bremsstrahlung* ('Braking radiation' in German), which in astrophysical conditions consists of X and γ-rays.

Alternatively, the incoming matter moving into regions of increasing density can become highly compressed and sufficiently hot ($\sim 10^8\,\text{K}$) to generate a blackbody (non polarized) 'thermal emission' already described in Chapter 1. At such temperatures the maximum intensity of $B_\nu(T)$ lies in the X-ray band of frequencies. The temperature can also rise by 'shock heating' if the infalling matter generates shock waves.

When the force of gravity is balanced by radiation pressure, it was shown by Eddington that the luminosity L resulting from accretion, for an accreting object of mass M, cannot exceed the 'Eddington limit' $L_E \sim 10^{38}(M/M_\odot)\,\text{erg s}^{-1}$.

Finally, the collapsed object, if it is a white dwarf or a neutron star, can be surrounded by a magnetic field. The action of such a magnetic field on the infalling charged particles is similar to that of the Earth's magnetic field on the flow of particles coming from the Sun: these particles are trapped inside the 'magnetosphere' and circulate more or less rapidly, spiralling around the magnetic lines.

Just as the particles producing auroral phenomena in polar regions of the Earth, the accreted particles can be forced by the magnetic field of the compact object to funnel down at each pole of the white dwarf or of the neutron star, with a spread over a region of about one tenth of the star's diameter.

On the other hand, because of the rapid orbital motion in most binary systems, the gas accreted by the compact component does not fall into the compact object, but often spirals inward to form a spinning 'accretion disk'.

Moreover, at a small distance from the accreting star the accretion disk can become highly turbulent. Then frictional heating can also raise the temperature of the gas to 10^7–10^8 K.

According to the velocity of the ionized particles, spiralling around the magnetic lines produces a 'cyclotron' or a 'synchrotron' radiation.

Synchrotron radiation is a very common physical process in astrophysical objects. It is well established that the Crab Nebula radiates by synchrotron process generated by high-energy, 'relativistic' electrons. The frequency of the corresponding radiation depends (in addition to other parameters) on the strength of the magnetic field and on the energy of the spiralling particles. Electrons of very high energy, in strong fields, emit γ-rays or X-rays. However, in certain conditions optical or radio waves can be emitted. Synchrotron radiation is polarized because the electronic paths are curved only around the magnetic lines.

The intensity of the synchrotron radiation depends in a very specific way on its frequency, so that a synchrotron source can be recognized by the spectral distribution of the corresponding continuum.

When a cloud of 'relativistic electrons' is injected into a medium where a magnetic field is present, the radiation is mostly of high frequency. As the electrons lose a part of their energy (both by radiation and through the expansion of the cloud) the radiation of a lower frequency (e.g. radio frequency) is emitted.

For about the same reason as that explaining their polarization the synchrotron photons are emitted in a fairly narrow beam, in a given direction.

When low frequency (low energy) photons encounter high energy electrons they can gain some energy from electrons (by a process inverse of the Compton effect). Thus, by this 'Inverse Compton Scattering', low energy blackbody photons can become X-ray or γ-ray photons. Thus, arrival of high energy electrons in a *cold* cloud can generate X-rays.

Contrary to the synchrotron process the *cyclotron* process generates spectral *lines*. Observations of such lines permits a determination of the approximate strength of the corresponding magnetic field. Thus, for instance, the first more or less direct measurement of a neutron star's magnetic field was recently performed (by Trümper *et al.*, 1977) through observation of a line at about 53 keV in the spectrum of the source Her X-1. This observation indicates a field strength of about 10^{12} G, in agreement with some of pulsar models.

Similarly, a line at 0.85 keV was observed in the X-ray spectrum of Capella (see below, Subsection 3.4).

1.5. Observational Effects and Identification Criteria

One of the most important and difficult problems in X-ray astronomy is, as for pulsars, the problem of 'identification' of each source with astronomical objects observed in other domains of frequency, and first of all, in the optical domain. The optical, radio or infrared 'counterpart' can be a star, a radio source, a globular cluster (and in extragalactic X-ray astronomy, a galaxy, a cluster of galaxies, a quasar, etc.).

We have written the word identification in inverted commas because the sought counterpart is not necessarily a single object identical with the X-ray source, as was the case for the Crab pulsar. Indeed, about 15 X-ray sources are known to represent just one component of a *binary system*, and the 'optical counterpart' is not the compact object *invisible* in the optical domain of wavelengths, but the *other* component of the system, a more or less 'normal' star still burning its nuclear fuel.

Of course, the first step in any identification of an X-ray source consists of a sufficiently precise *determination of its position*, within a sufficiently small 'error box', i.e. region of uncertainty, on the sky. As in radio astronomy (except when the very long baseline interferometry is used) in X-ray astronomy these error boxes are always much larger than in optical groundbased astronomy. Precision of about a minute of arc would be ncessary to insure an identification by 'positional coincidence' alone. Only HEAO B can yield a precision of this order, or better (2 arc *seconds*). An opposite situation still exists for γ-ray detectors, whose error box is of dimensions of the order of $1°$: in such a box there can lie more than 100 000 stars! One of the main progresses brought by the use of *Uhuru* consisted in a reduction of the error boxes first to an area of $0.02\,\mathrm{deg}^2$ and finally of $0.003\,\mathrm{deg}^2$.

In some exceptional cases, the ingenious method of the radio astronomer C. Hazard, first applied in Dec. 1960 (Hazard, 1961) to the precise measurement of the position of the radio source 3C 212, can be applied to X-ray sources. This method is based on the fact that when a source (radio, X-ray or γ-ray) is occulted by the Moon (whose motion is given with very high precision by optical observations and celestial mechanics) the measurement with radio telescopes or X-ray receivers of the *time* of disappearance and reappearance of the source gives its position with a precision impossible to achieve by orthodox and conventional radio or X-ray techniques.

Thus, for instance (after the identification by this method of the radio source 3C 273, performed by Hazard in 1963, and which will be discussed in Chapter 13), this method has shown, in 1964, that the counterpart of the X-ray source initially called Tau X-1 was lying inside the region occupied by the Crab Nebula. However, only after the discovery of pulsars (see Chapter 7, Section 2) and the identification of the Crab radio pulsar with the neutron star (Baade's star) residue of the supernova explosion which generated the Crab Nebula, it became possible, in 1968, to attempt an identification of Tau X-1 with the Crab pulsar.

Fortunately, many 'observational effects' often allow an identification much more certain than that provided by positional, more or less narrow, proximity. This will appear quite clearly in the following review of these effects, illustrated by adequate examples. Quite generally, the method of identification based on these observational effects can be described as based on '*concurrent variations*' of the X-source and its hypothetical counterpart. And this is possible because most of X-ray sources present more or less regular long-term or short-term variability in brightness, in spectrum, etc. which can be compared (correlated or anticorrelated) with similar variations exhibited in brightness, spectrum, colour, etc. of the counterpart. Such 'timing signatures' from the presumed associates represent a very powerful tool of identification: in the case of the γ-ray sources it represents for the moment the only possible identification criterion.

1.5.1. The 'Pulsar Effect' (X-Ray Pulsars)

Some X-ray sources behave in a way similar to radio pulsars discussed in Chapter 7: their X-brightness is *periodically modulated* (after correction of effects of an eventual orbital motion, which will be discussed below). The corresponding periods $\bar{P}$ of some X-ray pulsars are comparable to those of average radio pulsars, but are usually *much longer*. As for radio pulsars, the most likely origin of this 'pulsar effect' (at least for X-ray pulsars of relatively short period) is the *rotation* (spinning) of a neutron star. However, some of these periods are so long that this interpretation can compete with effects of a rapid *orbital* motion in a binary system.

Thus, for instance, the source SMC X-1, situated in the Small Magellanic Cloud, is an X-ray pulsar with a period $\bar{P}$ of ~ 0.72 s, and the period $\bar{P}$ of the X-ray pulsar Her X-1 is ~ 1.24 s. Both periods are within the range of periods of the radio pulsars. However, already the period of the X-ray pulsar Cen X-3, equal to 4.84 s, is outside this range, and X-rays pulsars with much longer periods have been observed, as shown in Table 9.2. On the 'other side' periods shorter than those of the Crab pulsar (0.033 s) are theoretically possible but not observed.

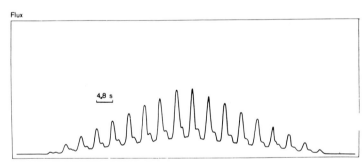

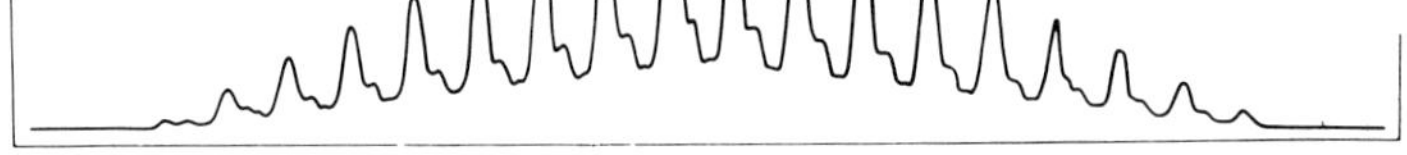

Fig. 9.1. 'Pulsar effect' in the X-brightness of Cen X-3, recorded by *Uhuru*. The curve shows the 4.8 s pulsations as they appear after a Fourier analysis of the raw data. The gradual variation in the height of the peaks from left to right is an effect of the satellite's rotation. [From Gursky and v. d. Heuvel, 1975.]

A very remarkable property of X-ray pulsars is that contrary to radio pulsars their periods $\bar{P}$ generally (regularly and slowly) *decrease*, which means a *spin-up* of the rotation of the neutron star. This phenomenon is generally ascribed to a transfer of angular momentum to the compact component by the accretion process. The corresponding values of $\dot{\bar{P}}/\bar{P}$ range between $\sim (-4) \times 10^{-6}\ \mathrm{yr}^{-1}$ for Her X-1 and $\sim (-8) \times 10^{-4}\ \mathrm{yr}^{-1}$ for SMC X-1.

It is also quite remarkable, that most X-ray pulsars are certainly engaged in binary systems, which is quite exceptional for radio pulsars. Usually no pulsed radio emission from

X-ray pulsars is observed. Apparently X-ray pulsars and radio pulsars represent two different classes of astrophysical objects. According to a suggestion by Illarionov and Sunyaev (1975) the accreting gas could be sufficiently dense to prevent the escape of radio emission from an X-ray pulsar.

TABLE 9.2

Periods of pulses (in seconds) and orbital periods (in days), when known, of some X-ray pulsars

X-ray pulsar (name)	Optical counterpart	Mean period $\overline{P}$ of pulses (s)	Orbital period P_0 (d)
Crab pulsar	Baade's star	0.033	Non binary
SMC X-1	SK 160	0.72	3.89
Her X-1	Hz Her	1.24	1.70
4U 0115 + 63	No name	3.61	24.31
Cen X-3	Krzeminski's star	4.84	2.09
3U 1626 − 67	?	7.68	
No name	SS Cygni	8.9	0.27
A 0535 + 26	HDE 245770	103.8	
GX 1 + 4	?	122	
GX 304 − 1	MMV star	272	
Vela X-1	HD 77581	283	8.95
4U 1145 − 61	HD 102567?	297	~ 12–14
A 1118 − 61	?	405	
{A 1540 − 53 = 4U 1538 − 52}	No name	529	3.73
GX 301 − 2	Hen 787	~ 698	~ 23?
GX 17 + 2	No name	1913	> 2?
4U 1700 − 37	HD 153919	~ 5900 ± 90	3.41

1.5.2. The Doppler Effect Due to the Orbital Motion in Binary Systems

All periods mentioned in the preceding subsection are actually only *mean* values $\overline{P}$ of the *observed* periods P of pulses.

Indeed, because of the orbital motion of those X-ray pulsars which represent one of the components of a binary system, the true rotation periods $\overline{P}$ of the compact component suffers a *Doppler shift*, hence a modulation with a period equal to the orbital period P_0.

The rotating compact object, most often a neutron star, acts as an orbiting clock that enables a measure of the size and of the eccentricity of the orbit by the methods decribed in Chapter 8. This 'orbital information', coupled with other supplementary data, such as the approximate mass of the 'normal' star deduced form its spectral class and its luminosity, enables one to estimate the mass (and sometimes the size) of the compact component.

Table 9.3 gives the masses of some galactic X-ray binaries (and orbital parameters), according to the most recent determinations.

TABLE 9.3

Orbital parameters ($a_x \sin i$, eccentricity) and masses of galactic X-ray sources in binary systems, [After Bahcall, 1978 and *ApJ* (L), **225**, L63 (1978)]

Source	M_x (unit: solar mass $M_\odot$)	Orbital parameters	
		$a_x \sin i$ (10^{-8} pc)	Eccentricity e
Vela X-1	1.0– 3.4	~ 111	0.13 ± 0.04
SMC X-1	0.5– 1.8	53.5	< 0.0007
Cen X-3	0.7– 4.4	39.7	0.0008 ± 2
Her X-1	0.4– 2.2	13.2	< 0.003
4U 1700–37	≳ 0.6		
Cyg X-1	9 –15		
4U 1538 – 52 / A 1540 – 53	1.6– 2.4	~ 55	?
4U 0115 + 63	?	140.1	0.3402 ± 4

If we except the case, discussed later, of Cyg X-1 (which, by the way, is *not* known to be a pulsar!) the mean average value of the mass M_x of the compact component is ~ 1.6 ± 0.3 for all X-ray binaries.

The period of the Doppler modulation is, of course, equal to the orbital period P_0, but its amplitude ΔP depends both on the mean value of P and on the range of linear radial velocities with respect to the observer (see Chapter 8, Section 2). Thus, for instance, the observed pulse period P of Cen X-3 varies between 4.8491 and 4.8357 s (i.e. with an amplitude $\Delta P = 0.0067$ s around the mean value $\overline{P} = 4.8424$ s) with a period equal to $P_0 = 2.09$ days. One can easily verify that these values correspond to the orbital parameters given in Table 9.3.

1.5.3. *Orbital Modulation of the X-Brightness by Total or Partial Eclipses of the Compact Binary Component*

When the observer is situated *in* the orbital plane, or near the orbital plane, of a binary system whose compact component behaves as an X-ray source, total or partial eclipses of this source are observed for adequate configurations of the system with respect to the observer. Usually such X-eclipses are not accompanied by optical eclipses *of the compact* component since this component is optically invisible.

An important criterion of identification can then be provided by the coherence between the phases of the X-eclipses and the phases of the Doppler shifts when the source is an X-pulsar.

Thus, for instance, as can be expected from the general theory discussed in Chapter 8, the mid-eclipse of the X-brightness of Cen X-3 (zero radial velocity) corresponds (see Figure 9.2) to the instant when the observed value of P is equal to its mean value $\bar{P}$. Similarly, P is again equal to $\bar{P}$ in the middle of two successive eclipses of X-brightness.

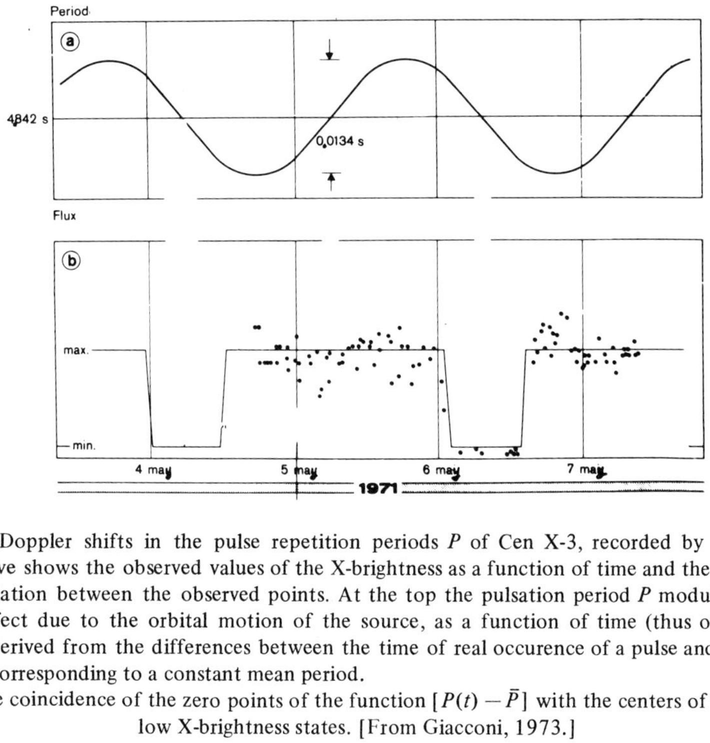

Fig. 9.2. Doppler shifts in the pulse repetition periods P of Cen X-3, recorded by *Uhuru*. The bottom curve shows the observed values of the X-brightness as a function of time and the corresponding interpolation between the observed points. At the top the pulsation period P modulated by the Doppler effect due to the orbital motion of the source, as a function of time (thus of the orbital phase), as derived from the differences between the time of real occurence of a pulse and the time of occurence corresponding to a constant mean period.

Note the coincidence of the zero points of the function $[P(t) - \bar{P}]$ with the centers of the high and low X-brightness states. [From Giacconi, 1973.]

In the most favourable cases the knowledge of orbital parameters deduced from the above mentioned observational data allows the determination of a schematical model of the system. Figure 9.3 gives such a model for Cen X-3.

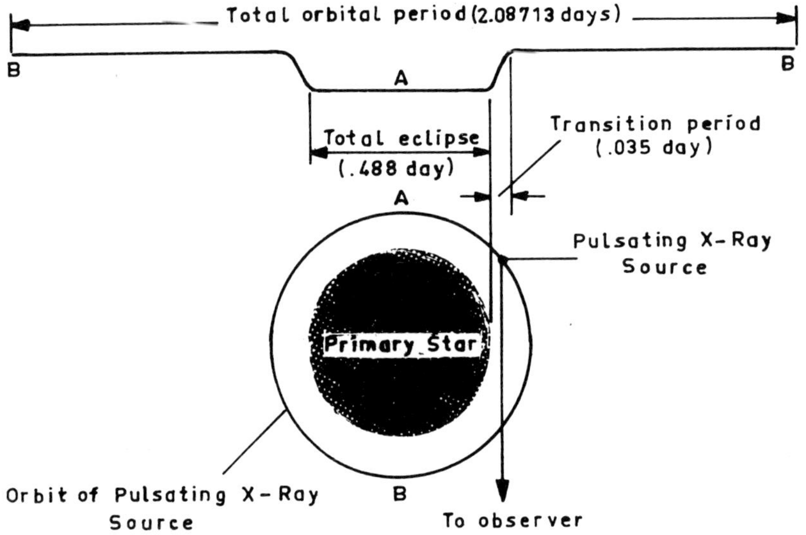

Fig. 9.3. A schematic model of the Cen X-3 binary system. At the top the observed values of the X-brightness of Cen X-3 as a function of time. For the sake of clarity the dimensions of the primary are strongly exagerated with respect to the dimensions of the orbit, hence a distorsion of the ratio of duration of the total eclipse with respect to the orbital period P_0. [From Giacconi, 1973.]

1.5.4. X-ray Heating of the 'Normal' Component: the Illumination Effect

In some particular circumstances accretion can generate another observable effect proper to binary systems whose secondary is an X-ray source: the *'illumination effect'* (or, '*X-ray heating*').

Indeed, when the primary's optical brightness is normally very low, the X-rays emitted in a sufficiently narrow beam by the compact secondary in the direction of the primary, can, by heating a small part of the surface of the primary, produce a *'bright spot'* emitting optical radiation.

Because of its origin this spot will be roughly oriented towards the secondary. This will produce an orbital modulation of the optical brightness of the primary. More generally any asymmetrical heating of the primary by the X-ray compact source will produce a similar modulation.

This illumination effect will manifest itself with a period P_i equal to the orbital period P_0 of the system, since the bright spot on the surface of the primary will be observed under different angles, according to the orbital phase of the system.

Moreover, if the system is an eclipsing one, a supplementary identification criterion will be provided by the coincidence of the mid-eclipse of the X-brightness with the minimum of optical brightness (maximum optical magnitude) of the primary, since at mid-eclipse of the secondary the bright spot will 'turn its back' to the observer.

Of course, by symmetry, the maximum of brightness (minimum of magnitude) of the primary, can correspond to the middle of interval between two successive X-eclipses (if favoured by geometrical proportions of the system).

Thus, for instance the source Her X-1 could be identified in this way as the secondary of a binary system whose primary is the variable star HZ Her. The X-brightness of Her X-1 suffers periodic total eclipses (with a period P_0 equal to 1.70 days), which coincide with the *minima* of optical brightness of HZ Her. Of course, other elements of identification are possible, and they actually exist and will be discussed in Subsection 3.1.2.

When the observer is not rigorously in the orbital plane, but is sufficiently near this plane, partial eclipses may also be accompanied by an effect of X-ray heating. In this case both the X-brightness of the secondary and the optical (or infrared) brightness of the primary will present periodic variations with a period equal to the orbital period P_0 and, typically for the illumination effect, both the minimum of the X-brightness and the minimum of the non-X-brightness will occur almost simultaneously.

This case is illustrated by the source Cyg X-3 (in the constellation *Cygnus*). The X-radiation of Cyg X-3 (3U 2030 + 40) presents minima of X-brightness separated by a period P_0 of about 0.20 days (4.8 h), and each minimum coincides with a minimum of the *infrared brightness* of a primary (unobservable optically because of the effect of interstellar extinction, very strong in the direction of this star lying very near the galactic plane, since its galactic latitude is equal to $\sim 0.7°$). This identification is confirmed by observations in the radio domain of frequencies (Subsection 3.2.2).

When the observer is too far from the orbital plane for occurrence even of partial eclipses, but still sufficiently near the orbital plane to leave the illumination effect perceptible, one can observe the variability of brightness of the primary, without any corresponding variability of the X-brightness.

Such is the case of the source Sco X-1 (3U 1617 − 15) identified with the visual variable star V 818 Sco whose brightness varies with a period P_0 equal to 0.787 318 ± 1 days, whereas no systematic variations in the X-brightness of the X-component are observed (Subsection 3.3.2).

1.5.5. Tidal Distortion of the Primary: the Cigar Effect

The relatively short orbital periods of some binary systems indicate that the orbital semi-major axes are only a few times larger than the primary stars themselves. In such 'close binaries', besides eventual eclipses (when the observer is situated in or near the orbital plane) *tidal distortions* can occur, corresponding in extreme cases to what can be schematically described as a transformation of spherical stars into 'pear-shaped' or 'ellipsoidal' components, elongated along the line between their centers.

Although it obviously represents an exageration of the reality, it can be convenient to describe such ellipsoidal form as a 'cigar', in order to better emphasize the direction of the major axis, and the fact that the cross-section is smaller in a plane perpendicular to this axis than the cross-section containing this axis.

When the axis of this 'cigar' is directed along the line of sight (or near this direction) the visible cross-section of the primary is smaller than when the cigar is perpendicular to the line of sight.

Thus, contrary to what happens for the illumination effect, if the situation of the observer is such that eclipses of the secondary are possible, and result in minima of the X-brightness (phase zero), the orbital period P_0 will contain *two maxima* of the optical brightness (i.e. two *minima* of the visual magnitude) at orbital phases 0.25 and 0.75. Figure 9.4 shows the presence of this 'cigar' (ellipsoidal) effect for the source 4U 1700 − 37.

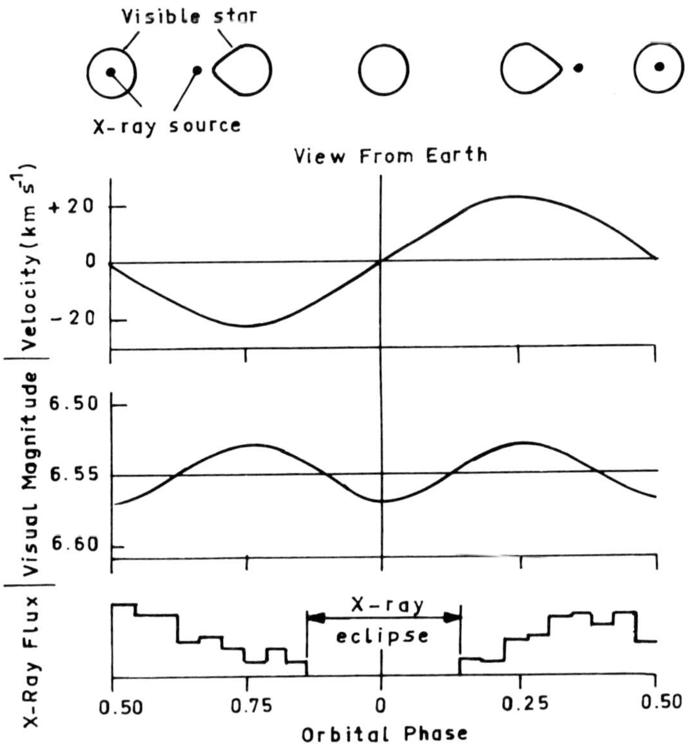

Fig. 9.4. Observations of the source 4U 1700 − 37 and of its visible companion HD 153919. The radial velocity and the visual magnitude of the visible star, compared to the 6–10 keV brightness of the X-component, illustrate the presence of the 'cigar effect' in this system. A model of the point X-source and its tidally distorted optical companion is shown at the top (the 'cigar' is here shaped, more correctly, as a pear), as a function of the orbital phase. [After Wolff and Morrison, 1974; Jones and Liller, 1973; Jones *et al.*, 1973.]

More generally (even in configurations excluding total or partial eclipses), the 'cigar effect' will manifest itself by a period of variations of brightness of the normal star equal to *one half* of the orbital period, with a correspondence of particular phases of the X-radiation of the type illustrated by Figure 9.4.

The source Vela X-1 (4U 0900 − 40) identified with the normal star HD 77581 is also very clearly connected with the 'cigar effect'. Its X-radiation suffers periodic eclipses, with a period P_0 equal to 8.95 ± 2 days, whereas the optical companion presents 'ellipsoidal variations' with a period equal to $\frac{1}{2}P_0$ and one of the minima in coincidence with the mid-eclipse of the X-component. (The X-ray source Vela X-1 must not be confused with the Vela pulsar PSR 0835 − 45!)

1.5.6. *The Burst Effect and the Flickering*

One of the most remarkable discoveries in the field of X-ray astronomy, made in 1975, is that of '*bursts*', i.e. of *sudden enhancements* of X-brightness of some sources, from their much more slowly variable 'persistent level' or from below observational detectability.

(The long-term variability of this persistent level will be discussed in Subsection 1.5.7.)

These bursts reproduce a more or less regular pattern on different time scales.

The burst effect has already been detected in more than 30 X-ray sources (by November 1978).

According to Hoffman *et al.* (1978), one can distinguish two types of bursts: type I and type II.

The type I bursts repeat on time scales of hours, days, weeks, and sometimes longer. Thus, for instance, the source 3U 1820 − 30 have presented bursts with intervals near 3.4 hours in March 1978, which in four days decreased to 2.2 hours. It does not seem possible to ascribe such pseudo-periodic bursts to any spinning or orbital motion.

More generally, some type I bursts are pseudoperiodic, and characterized by 'cycles' of a few hours, a typical type I burst taking ~ 1 s to rise and ~ 5 s to decline, as in the case of the source MX 1742 − 197.

The type II bursts occur at intervals of a few seconds to a few minutes. Thus, for instance, the 'rapid burster' MX 1730 − 335 can produce, beside type I bursts, up to ~ 5000 type II bursts per day. Again, it is not possible to ascribe such bursts to any spinning or orbital motion.

The type II bursts present many meaningful particularities. Thus, each type II burst in MX 1730 − 335 rises to a *nearly the same peak* X-brightness (in less than one second). The duration of these bursts can be very different: they can last from a few to about 100 s. Thus, the energy ΔE involved in different bursts is also different.

The remarkable property is that the more energetic bursts are *followed* by roughly proportionally *longer* intervals, or, to put it otherwise, the larger the burst the longer is the wait until the next one. Even more precisely, the interval of time Δt_n from the nth to the $(n + 1)$th burst is roughly proportional to the energy $(\Delta E)_n$ involved in the nth burst. This relation is often referred to as $\Delta E - \Delta t$ relation.

The average duration of a typical type II burst can correspond to an X-ray luminosity L_x of the order of 10^{38}–10^{39} erg s^{-1}, and can represent about 10 times the average L_x of the source. The energy (ΔE) in the largest bursts may exceed 10^{40} erg.

Sometimes the bursts are accompanied, or replaced, by even more rapid fluctuations or 'flickering', on a time scale *below one second*. The most typical of such fluctuations are exhibited by the binary source Cyg X-1, on a scale between 0.1 and 0.5 s. (However, although there is much variability in Cyg X-1 at about 0.04 s and slower scales, the fluctuations disappear on a millisecond scale!)

Such short-term variability is generally considered as a kind of measure of the *compactness* of the source. Indeed, this correlation between the short time scale of variations of the brightness of an astronomical object and its linear dimensions is generally justified by a somewhat questionable argument according to which the linear dimensions on any variable source cannot exceed the distance travelled by light during the time corresponding to the time scale of the observed variations: "otherwise, the difference in time taken by light to travel from different parts of the source would tend to blur any variations" (Kellerman, 1973). Another argument is that "the output of an object cannot vary significantly in less time that it takes for light to cross it" (Freedman, 1978). Hence, the conclusion that rapid fluctuations imply that their source is very compact. (We come back to this problem below, in connection with the detailed discussion of the properties of Cyg X-1, and, in Chapter 13, in connection with the properties of the quasar 3C 273.)

Usually all types of bursts, and of short time scale fluctuations, are superimposed on a more or less regular, more or less long-term, variations (of more or less large amplitude). The corresponding 'persistent component' of the X-brightness is often improperly called the 'steady state' component, or the 'steady component'.

The burst activity, in all its manifestations, can be very irregular on a long-term scale. Thus, for instance, the 'rapid burster' MX 1730 − 335 turned off in April 1976 and did not turn on again until 13 months later. More generally, this source had been burst active from 1971 to 1977 during one to two months for each 'cycle', with cycles repeated on a 0.5–1-year time scale. In 1978 it became moderately burst active during about 10 days in March, but stopped its activity in the middle of April, after having lasted only roughly three weeks.

Burst activity can be correlated or anticorrelated with the variations of the persistent component. Thus, for instance, on one occasion the persistent component of the X-brightness of the source 3U 1820 − 30 gradually increased (up to about 4/3 of its initial value) but at the same time the burst activity decreased and finally stopped. More generally, it often happens that the burst activity is present when the persistent X-ray emission is weak.

Galactic sources affected by the burst effect have a specific space distribution: some of them are strongly concentrated towards the galactic plane, others, in addition, are concentrated near the center of the galaxy (near the so called 'galactic bulge'). Those which are found in globular clusters are situated mostly in low latitude ones such as NGC 6640 ($b^{II} \sim -1°$).

Most speculations concerning the origin of the bursts fall into two broad categories:

(a) Instabilities in the accretion of matter onto a compact object, such as a neutron star or a black hole.
(b) Thermonuclear flashes in matter accreted onto the surface of a neutron star.

As to very short time scale fluctuations, they may reflect particularly turbulent conditions in an accretion disk.

However, the understanding of the burst effect is, for the time being, far from complete, and the corresponding physical mechanism is still very controversial.

The relaxation phenomenon, obviously implied by the $\Delta E - \Delta t$ relation, has been compared (Clark, 1977) to the one present in a neon-lamp flasher. This relaxation oscillator has a constant trigger level but a variable depth of discharge. After each flash the oscillator must recover for a length of time proportional to the depth of discharge before it can flash again. Apparently, in a similar way, when the gravitational potential energy of infalling matter fills the 'reservoir', at some critical value of the accreted mass the energy reservoir springs a leak and a burst is generated. The energy-conversion process must be self-limiting, perhaps as a result of the effect of the radiation pressure and heating on the motion of the gas falling through the leak in the reservoir.

1.5.7. The Long-Term Variability and the Flare (Nova) Effect

Some X-ray sources, sometimes called the 'transients' or flare sources (but to a lesser degree, almost all sources) present a very considerable long-term variability, this concept being understood as opposed to the short term burst activity considered in the preceding subsection. Sometimes this variability can take the form of an X-ray nova outburst. Of course, none of these long-term variations exclude the presence of short time scale ones. The typical time scale of the variability considered here lies between a few months and a few years.

Li *et al.* (1978) introduced a parameter, they call ξ, expressing quantitatively the degree of this type of variability. The small values of ξ, of the order of five correspond to very steady sources, such as the Crab Nebula ($\xi = 4$), or the radio-source Cas A ($\xi = 5$). The moderate values of ξ, between $\xi = 8$ and $\xi = 18$, correspond to either steady or only slightly variable sources, characteristic of those situated near the galactic bulge, but not necessarily very burst active, such as 3U 1744 − 26 or MX 1709 − 40.

The high values of ξ, between 20 and 155, correspond to highly long-term variable sources, such as the source MX 1746 − 20 associated with the globular cluster NGC 6440, the slow binary pulsar GX 304 − 1 ($\xi = 103$ and $\bar{P} = 272$ s), or the relatively rapid binary pulsar Cen X-3 ($\xi = 153$ and $\bar{P} = 4.8$ s). Thus, we see that the values of ξ of the same order can correspond to very different objects.

The highest long-term variability is the one characterizing the flare or nova-like sources, whose ξ is larger than 200, such as the source Aql X-1 ($\xi = 218$) or the source 3U 1543 − 47 ($\xi = 304$) whose long-term variability is the highest of all sources studied by Li *et al.*, but which is certainly not the most variable of all existing sources from the point of view of long-term variability.

In the case of Aql X-1, its flares have been observed in June 1975, June 1976 and June 1978. On 14 June 1978 its X-brightness was 0.02 'Crab Units' (cu), but it reached 0.5 cu by June 18, and about 1.0 cu by June 22 (in the 3–6 keV energy band). This was the first X-ray outburst of Aql X-1 for two years.

Since an X-flare is usually accompanied by a flare in other frequency domains (optical, radio, etc.), such phenomena (or 'effects') represent a useful identification criterion.

In the case of Aql X-1, in June 1978, the optical counterpart was in outburst at the same time as the X-ray source. Its optical blue brightness rised from its quiescent level of B ~ 20 mag, to B ~ 16.5, which represents about a 25-fold increase in optical luminosity.

Another very interesting recurrent nova-like transient, A 0620 − 00 (formerly called A 0621 − 00), can be mentioned. This source, first detected in August 1975, was 500 times X-brighter eight days later.

These first observations of Ariel V have been continued by the American satellite SAS-3. About two days before its maximum, A 0620 − 00 had the same X-brightness (in the domain 1.5–5 keV) as Sco X-1, but at maximum it reached an X-brightness about four times larger than that of Sco X-1. After that the decline was slow.

During the increase of the X-brightness, the spectral distribution $I_\epsilon = f(\epsilon)$ of this sources changed very considerably: the decrease of $f(\epsilon)$ as a function of ϵ, becoming steeper and steeper.

No 'pulsar effect' is present (between time scales from 1 to 500 seconds).

A 0620 − 00 has been identified with a star whose photographic magnitude was about 17 in 1950, about 12 at the time of discovery of the X-ray source (August 1975), and 10.4 about two weeks later. It seems, according to the photographic archives of Harvard Observatory, that it already suffered a nova outburst in November 1917.

The discovery and the study of still another 'transient', 4U 0115 + 63, which proved to be a *binary*, with a relatively very large separation $a_x \sin i$ of ~ 140×10^{-8} pc between the two components and relatively moderate eccentricity $e = 0.34$ of the orbit (Rappaport *et al.*, 1978) seems to provide a first indication of the binary nature of other transients.

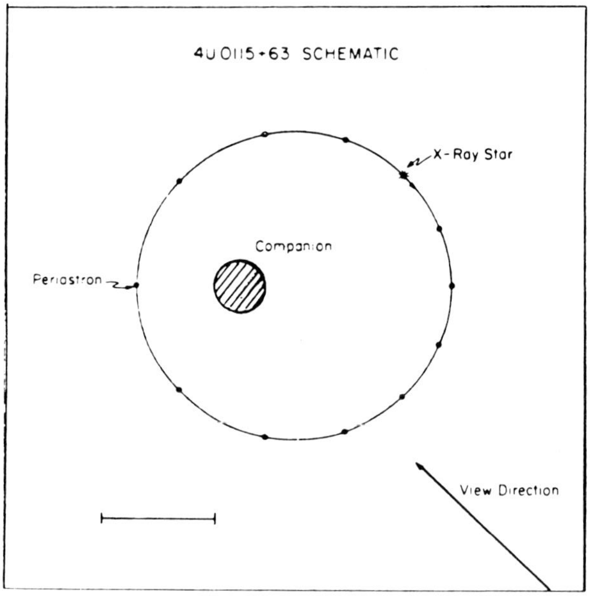

Fig. 9.5. A schematic model of the 4U 0115 + 63 system. The heavy dots represent the positions of the compact X-ray source. The time interval between these points is ~ 2.0 days. The companion star is placed at the center of mass of the system. Its radius is chosen to have an illustrative value of $10R_\odot$ corresponding to its spectral type and luminosity class. The small horizontal segment at the lower left indicates the scale: it represents 100 light-seconds, i.e. about 10^{-6} pc. [From Rappaport *et al.*, 1978]

Apparently, due to the large orbital separation of the components, the mass-transfer rate may be low except during episodes of enhancement of spontaneous mass loss from the primary companion. In this picture, the intervals between outburst are governed by the intervals between the active episodes of the primary and the mass transfer may occur along any portion of the orbit, whereas it was previously assumed that an eventual very high value of eccentricity could be responsible for the intermittent character of the mass-transfer (see Figure 9.5).

1.5.8. Polarization Effects

Strong and variable optical linear and circular polarization, synchronous with X-ray variability of a source, represents a very useful observational effect, which can be used for identification purposes.

Thus, the X-ray source 4U 1813 + 50 was recently identified as the optically invisible companion of the spectroscopic binary AM Her through observation of the synchronism between its polarized optical emission and the variations of its X-brightness, both variable with a 3.1 hr period.

This source is particularly interesting, because the observed polarization is indicative of high magnetic fields of the order of 10^8 gauss, and of the fact that the compact secondary is a 'magnetic white dwarf' rather than a neutron star.

In such a system, the formation of an accretion disk is impossible because the interstar gas is channelled by the magnetic field towards the magnetic *poles*, instead of forming a kind of 'equatorial ring' (accretion disk).

The main dips in optical and X-ray brightnesses are separated by about half of the orbital period. Apart from this period, no coherent oscillations have been found in the optical photometry of AM Her. However, irregular changes on time scales of months and years are observable. AM Her is usually of $\sim$ 12.5th mag, but occasionally is found to be as faint as $\sim$ 15th mag.

It is possible that AM Her represents a 'slow optical pulsar' (Priedhorsky and Krzeminsky, 1978) and that the primary minimum in the visual light curve is the *eclipse* of a region of intense optical emission in the magnetic field near the surface of the white dwarf *by itself*, so that the observed 3.1 hr period would represent the period of rotation of the white dwarf co-rotating as binary.

From the point of view of a probable presence of a magnetic white dwarf, the white dwarf nova SS Cyg, whose optical and X-ray flares are synchronous, could be similar to AM Her.

2. The Classification Problem

Any classification of astronomical objects is always more or less arbitrary and essentially temporary. It depends on criteria chosen for classification and on technical progress in observations. New techniques can suddenly introduce new criteria through discovery of new phenomena.

Moreover, many individual sources exhibit several effects, such as those described in the preceding section, and these effects can be combined in a very different way in each source. So, generally, it is not at all obvious which combination can be taken as classification criterion.

It also happens very often that the same observational effects correspond to objects of different usual *morphological* classes. The bursts of type I are found both in globular clusters and outside globular clusters, both in transient and nontransient sources, etc.

Thus, for instance, the term 'burster' must be understood as indicating a 'burst active' object (of any morphological class) and not a particular class of astronomical objects. For a similar reason, the term 'bulge source' must be understood as indicating that the source is situated in or near the galactic bulge, and not, as is often done, as a particular class of X-ray sources. The same is true concerning the term 'globular sources', etc.

For all these reasons, we preferred to insist upon separate 'observational effects' instead of attempting a questionable classification of galactic X-ray sources.

A few particularly often studied sources, discussed below in a more or less arbitrary order, will illustrate all these statements. The headings of each subsection should not be considered as a classification, but simply as a *summary* of the main effects observed in each group of sources.

3. A Few Particularly Interesting Galactic X-ray Sources*

3.1. ECLIPSING X-RAY PULSARS IN BINARY SYSTEMS

3.1.1. Centaurus X-3

As already briefly mentioned above, this source (3U 1118 − 60) represents one of the components of a binary system. Its companion is a 'normal' star, called 'Krzeminski's star' (Krzeminski is the name of the Polish astronomer who performed this identification in 1973) of spectral type O 6.5 II.

The visual brightness of the optical component varies with a period equal to *one half* of the period P_0 ($\sim$ 2.087 d) of the X-eclipses of Cen X-3, (i.e. of the period of the orbital motion) about a mean value of the visual magnitude $\overline{m}_v$ equal to 13.4, with an amplitude Δm_v of 0.08.

The mean period $\overline{P}$ of the X-ray pulses is equal to 4.8424 s. The observed period P of the X-ray pulses is modulated by the Doppler effect due to the orbital motion: the period of this variability is, of course, equal to P_0 and its amplitude is $\pm$ 0.0067 s around $\overline{P}$.

The X-ray eclipse is total. The 'phase of totality' presents a duration of about 0.488 d, with (on both 'sides') a transition period of about 0.035 d.

As was already seen in Figure 9.2, the zero points of the function $[P(t) - P]$ describing the period $\overline{P}$ of the X-pulses affected by the Doppler effect, correspond to both the middle of the totality phase (the center of the X-component just 'behind' the center of the optical component) and to the middle of the interval between two successive eclipses (the centre of the X-component just 'in front of' the centre of the optical component). In both cases, the radial velocity of the relative motion is zero.

Figure 9.3 schematically represents a model of the binary system, in approximate correspondence with observations represented in Figure 9.2 and with the phases of the X-eclipse reproduced at the top of Figure 9.3. The transition period corresponds to the crossing by the X-radiation of the Roche lobe or of the extended atmosphere, of variable X-ray thickness, surrounding the primary visual star.

The fact that the period of variations of the visual magnitude of the primary is equal to $\frac{1}{2}P_0$ indicates that these variations represent a 'cigar effect' defined and described in Subsection 1.5.5. The different phases of these variations are in excellent agreement with the model of Figure 9.3. A faint 'illumination effect' (Subsection 1.5.4) can also be present.

Application of the general theory of single-line spectroscopic binary stars (Chapter 8, Subsection 2.4) yields the mass-function of the system (corresponding to the assumption that the orbits are circular) through the interpretation of the Doppler effect in the period of the X-pulses. Although the presence of eclipses proves that the inclination i is near 90° (observer in the plane of the orbits), this mass-function only yields the order of magnitude of the *sum* of masses.

The already quoted discussion by Bahcall (1978) seems to indicate that the mass M_x of the compact component is larger than $0.7M_\odot$ and smaller than $4.4M_\odot$.

* All references relative to particular X-ray sources are gathered in (2) of the list of references at the end of this Chapter.

3.1.2. Hercules X-1

As already briefly indicated above, this source (3U 1653 + 35) represents one of the components of a binary system. Its companion is a 'normal' variable star HZ Her. The visual brightness of this optical component varies with the same period $P_0 \sim 1.700$ d as the total eclipses of the X-brightness. The mean value of the visual magnitude of HZ Her is of the order of 14 and the corresponding amplitude is of the order of 1 mag. This indicates the presence of an 'illumination effect' defined and described in Subsection 1.5.4.

In a way similar to the case of Cen X-3, the X-brightness of Her X-1 is periodically modulated with a mean period $\bar{P}$ equal to 1.237 82 s. This mean period is modulated by the Doppler effect due to the orbital motion, so that the period P of pulses varies with a period equal to P_0 around the mean value $\bar{P}$, with phases corresponding to a model similar to the model indicated in Figure 9.3 for Cen X-3.

Analysis of the observational data shows that the mass of HZ Her is of the order of $2M_\odot$, the mass of the X-component being between 0.4 and $2.2M_\odot$. It is thus probable that the X-component is a neutron star.

The distance of the binary system seems to be of the order of 4 ± 2 kpc, hence an X-luminosity of the order of 10^{36} erg s^{-1}. This relatively high X-luminosity, and the absence of the lengthening of the mean period $\bar{P}$ of pulses, indicate that the corresponding source of energy is not (as in radio pulsars) the slowing down of the rotation of the neutron star but, as explained in Subsection 1.5.1, the accretion of matter from the optical primary star onto the neutron star secondary.

In addition to the above mentioned properties, the X-radiation of Her X-1 presents a 35.5 days 'on–off' cycle. The successive phases of this cycle, which corresponds to about 21 orbital periods, are the following:

(1) A sharp increase of the mean X-brightness in *about one day*;
(2) A progressive slow decrease of the mean X-brightness, during *about ten days*;
(3) An X-brightness very small or zero during *about 23 days* ($\sim$ 14 orbital periods).

It is noteworthy that the optical brightness of the visible component HZ Her does not vanish even during the phase (3) of extinction of the X-radiation. However, the corresponding light curve (m_v as a function of time t), during each of the 21 revolutions corresponding to the 35.5 d cycle, is different for each of the three phases of the X-radiation. This also confirms the presence of an 'illumination effect'.

Some theoreticians have proposed to explain the 35.5 cycle of Her X-1 by an absence of coincidence between the rotation axis ('spin axis') and the normal to the orbital plane of the neutron star, generating a motion of *precession* of the accretion disk about the normal to the orbital plane due (as in the case of the 'equatorial bulge' of the Earth) to the gravitational torque acting on the equatorial accretion disk.

It can happen that during a part of the precessional period of 35.5 d the narrow beam responsible for the pulsed X-brightness of the neutron star misses the direction of the Earth during the phase (3), but does not miss the nearby normal component of the system. This could explain the persistence, with a few modifications, of the 'illumination effect' and of the optical brightness of HZ Her even during the phase (3).

3.2. Partial Eclipses and Absence of X-Pulses

3.2.1. The Source 4U 1700 − 37

This X-ray source represents one of the components of an eclipsing binary system, but with no 'pulsar effect' (no periodic short-term variability down to 0.1 s).

Its companion is a rather abnormal star, HD 153919, whose spectrum (of type Of) presents several *emission lines*. The orbital period P_0 of the system is equal to 3.4120 ± 3 d. The duration of X-eclipses is about 1.10 d, which is particularly long since it represents 0.32 of the orbital period.

The mean visual magnitude $\bar{m}_v$ of the primary is (see Figure 9.4) equal to 6.55, about the same as that of the companion of Vela X-1. The period of the variations of m_v is equal to $\frac{1}{2}P_0$ ('cigar effect' plus an eventual very faint 'illumination effect'!). The total amplitude of variations of m_v is of the order of 0.04, i.e. particularly small.

The mass of the primary (HD 153919) seems to exceed $10M_\odot$, that of the X-ray source seems to exceed $0.6M_\odot$. Nothing more precise seems to be known about these masses.

3.2.2. Cygnus X-3

This source (3U 2030 + 40) is considered by some astronomers as a particularly interesting astronomical puzzle (Hjellming, 1973).

It presents partial X-eclipses, with a relatively short period of about 0.20 d, much shorter than the orbital period of most binary X-ray sources (which range from 0.27 d for SS Cygni to 24.31 d for 4U 0115 + 63). This proves that it belongs to a binary system in which the components are very close to each other.

The companion of the X-source cannot be observed visually, or photographically, but only *as a radio or infrared source*. This is due to the proximity of the binary system to the galactic plane, hence a strong extinction by interstellar gas and dust in the visual and photographic domain of the spectrum.

Cyg X-3, like 4U 1700 − 37, deeply differs from sources such as Cen X-3 or Her X-1 by *the absence of X-ray pulses*. Its X-brightness presents fluctuations (short time scale variability) but these fluctuations are perfectly random.

The absence of X-ray pulses is in some way compensated (from the point of view of the presence of regular phenomena) by an excellent phase correspondence between the X and the infrared variations of brightness.

Cyg X-3 is associated with a very peculiar radio source (itself associated with the infrared one), which could increase in radio brightness by a factor of 200 in a few days. For more details the reader is referred to the excellent paper of Hjellming (1973).

3.2.3. The Source OAO 1653 − 40 − [V861 Sco]

This source presents X-eclipses with a period of 7.85 d, and the duration of eclipses is at least 1.75 d. The beginning of the X-eclipse is coincident with the time of occultation of the optically invisible secondary by the variable star V861 Sco, with the same period, thus providing a quite certain identification. The source is not known to be an X-pulsar. The possible mass M_x seems to lie between 5 and $12M_\odot$, thus suggesting that it could be a black hole.

3.3. Binary Sources with No Eclipses and No Pulses

3.3.1. *Cygnus X-1: a 'Black Hole Candidate'*

This source (3U 1956 + 35) represents one of the components of a binary system. Its companion is a normal star HDE 226868, whose mean visual magnitude $\overline{m}_v$ is equal to 8.9. The period $P_0 = 5.5999 \pm 9$ d of the orbital motion is disclosed by the periodical Doppler shift of the spectral lines of HDE 226868. The visual magnitude m_v varies with and amplitude of about 0.06 within a period equal to $\frac{1}{2}P_0$. This shows the presence of a 'cigar effect' (but, of course, a faint 'illumination effect' can also be present).

However, this cannot be confirmed by a correspondence with the phases of eclipses, since no eclipses are observable.

The X-brightness of Cyg X-1 presents quasiperiodic fluctuations (down to 0.001 s), without any detectable pulsar effect. On the other hand, a periodic variation of the mean X-brightness, with a period equal to the orbital period P_0 was detected by Ariel V. This in itself represents a proof that the identification with HDE 226868 is correct.

Moreover, the mean X-brightness varies between a low state, corresponding to 90% of duration of this cycle, and a high state (in which the mean X-brightness is larger by a factor 5–10), during the remaining 10% of the cycle. The transition from the high to the low state in the X-brightness of Cyg X-1 has been twice (in 1971 and 1976) in correspondence with a sudden *increase* of the *radio*-brightness of a radio source known to be associated with HDE 226868 and situated at 0.1 arcsec from the star (according to VLBI very precise interferometric measurements.)

This anticorrelation between the variations of the X-brightness and the radio-brightness has already received several explanations; two interesting suggestions can be found in Braes and Miley (1976).

Another observation also confirms the companionship between Cyg X-1 and HDE 226868: during the cycle of 5.60 d, common to the periodic variations of the radial velocity of the visual star and of the X-brightness, the minimum of the X-brightness corresponds to a situation of the binary system in which the X-component is 'behind' HDE 226868, although a real eclipse does not take place (Cyg X-1 is then simply in superior conjunction with its companion).

Finally, the optical behaviour of HDE 226868 is directly correlated with the variations of the X-brightness. Indeed, in October 1975, with a shift of one orbital period, a sudden increase of the optical brightness was correlated with a passage of Cyg X-1 from its low to its high X-state.

According to recent investigations of Pandharipandhe *et al.* (1976) neutron stars cannot have a mass exceeding $2–2.5M_\odot$: for higher masses they collapse to form a 'black hole'. Above a mass exceeding $\sim 3M_\odot$ any compact object whose equator is shorter than that of the Earth represents a *black hole*.

A typical black hole, with a mass between 3 and $50M_\odot$ for an equator between 60 and 1000 km, represents an 'invisible' object: the escape velocity of any particle (photons included) at the surface of such an object exceeding the speed of light c.

On the other hand, it seems that Cyg X-1 is very compact, since its X-brightness fluctuates with a very short time scale of the order of 0.05 s. Hence, by an argument already explained in Subsection 1.5.6 a maximum linear dimension of about 15 000 km, for the regions responsible for emission of X-radiation. (These regions cannot be a small part of a normal star, since the major part of the surface of this star would emit, in this case, optical radiation, which is not observed.)

The mass of the compact X-component seems to exceed $9M_\odot$ and could even reach $15M_\odot$. In the most pessimistic case it must exceed $3M_\odot$. An object with a mass equal to $3M_\odot$ behaves like a black hole only if its radius is smaller than ~ 9 km. Therefore, from the observed fluctuations of the X-brightness alone, which yield only a radius smaller than ~ 7500 km, one cannot conclude that Cyg X-1 is a black hole.

Nevertheless, one can use the following argument. All objects whose mass exceeds $3M_\odot$ and whose radius is smaller than 7500 km are gravitationally instable and collapse until they reach the state of a black hole.

If future observations confirm that Cyg X-1 is a black hole invisible by itself, but whose presence manifests itself indirectly by the X-rays emitted by the particles falling on it (by accretion from the normal component) this identification will represent one of the greatest discoveries in astronomy.

For more (elementary) informations about the black holes the reader is referred to Thorne (1974) and, at a more advanced level, to Misner *et al.* (1973, Part VII), or Blumenthal and Tucker (1974, p. 41).

3.3.2. Scorpio X-1

In 1974 Soviet astronomers identified Sco X-1 with the variable star V 818 Sco. The mean visual (blue) magnitude of this star, equal to ~ 13, is of the same order as that of companions of Cen X-3, Her X-1, and SMC X-1. These astronomers proposed an orbital period of 3.9 d, but it was shown later by American and Canadian observers that the emission lines of the optical spectrum of V 818 Sco undergo a Doppler shift indicating an orbital motion with a period $P_0 = 0.787\ 313 \pm 1$ d. (No stellar absorption lines are known to be present in the spectrum of V 818 Sco.)

Comparison between the phases of the emission lines with the phases of the visual magnitude, shows that, in spite of a small (~ 0.05 d) time discrepancy between the zero value of the radial velocity and the minimum of the optical brightness (when the normal star is in front of the compact one), we are here in the presence of an 'illumination effect'.

The X-brightness suffers no eclipse, and no pulsar effect has been detected hitherto: only nonperiodic fluctuations are observable down to 1 s.

3.3.3. Circinus X-1

Very rapid fluctuations, similar to those of Cyg X-1, are observed in this source, generally also considered for this reason as a 'black hole candidate'.

Cir X-1, beside these sporadic emissions, presents regular variations of brightness with a period of 16.6 days. This seems to require that this source is binary, while the apparent lack of simultaneous modulation of hard X-rays seems to rule out an eclipse mechanism.

The most repetitive feature of this presumably orbital variation is a sudden decline in the X-brightness (in the 3–6 keV band), taking place at what can be defined as the phase zero. Little or no X-radiation emerges between phases 0 and 0.5, while there is generally a gradual rise between phases 0.75 and 1.00. The time scale of the decline is a few hours.

Cir X-1 has a radio counterpart which, at high frequencies, shows variations with the same periodicity as in X-rays. However, in each cycle the radio enhancement occurs shortly after the X-brightness drop-off. Cir X-1 has also been identified in the optical region with an emission-line object of ~ 22.5 mag in blue. In the infrared region (1.2–4.8 μm) this object is extraordinary bright (a magnitude of ~ 8 near 2.2 μm).

Although Cir X-1 shows signs of a rough 2.5 s pseudoperiodicity, this period is not sufficiently regular to consider the source as an X-pulsar.

3.4. The 'Coronal' Low Luminosity X-Ray Sources

It is well known that the Sun emits X-rays, especially during the periods of intense 'activity' (flares).

Similarly, it seems that extensive flare and 'star spot' activity on the surface of some isolated or binary stars (from the cooler component) have been recently observed with X-ray equipment, and confirmed by simultaneous observations of radio bursts.

Some variable binary stars of the class 'RS Canum Venaticorum' (RS CVn), i.e. the star RS CVn itself, along with α Aur (Capella), HR 1099, UX Ari, have been found to be X-ray sources of very *low luminosity* (L_x in 0.2–2.8 keV energy band of the order of 10^{31} erg s^{-1}), and this emission was explained in terms of coronal emission at temperatures of the order of 10^7K, i.e. of the same order as the temperature of the solar corona (responsible for the steady component of the X-ray emission of the Sun).

Thus, extensive stellar coronae seem to have been detected in several stellar systems with large star-spot complexes. In Capella the observed X-ray emission corresponds to a gas five times hotter and 1000 times more luminous than the corona of the Sun. Detection in Capella of line emission , at 0.85 keV, represents apparently the opening of a new phase of high energy astrophysics: the spectroscopic study, with X-rays, of the coronae of nearby stars.

Of course, the objects mentioned above, because of their low X-luminosity, can be observed with the present X-ray instruments only because of their proximity (between 14 and 145 pc).

3.5. X-Ray Sources in Globular Clusters

Because the burst effect was initially observed predominantly in sources situated in globular clusters (or represented by globular clusters), it was generally assumed that all such sources present the burst effect, and reciprocally that all 'bursters' are connected in some way with globular clusters.

It is now rather well established that such generalizations are incorrect. It is not at all sure that all sources in globular clusters (or represented by globular clusters) present the burst effect: out of the seven or nine X-ray sources associated with globular clusters only five or six seem to present bursts. And many bursters lie outside globular clusters.

However, it is quite remarkable that, right or wrong, the assimilation of bursters with globular clusters possess at least a heuristic value, since a new globulare cluster, heavily hidden by interstellar dust, and previously unknown, was discovered by W. Liller, in 1976, on searching for an optical counterpart of the 'rapid burster' MXB 1730 − 335; this new globular cluster is now called 'Liller I' (Grindlay, 1977).

Appendix. A Dictionary of Abbreviations in the Field of Galactic Sources. Conversion of Names. Tables

TABLE 9.A1

Catalogues, teams, observatories, satellites, instruments, etc.

A	Ariel V satellite
2A	2A Catalogue [1978, *Monthly Notices Roy. Astron. Soc.* **182**, 499]
ANS	Astronomical Netherlands Satellite
AO	Arecibo Observatory (Radio sources)
3C	Third Cambridge Catalogue [1960, *Mem. Roy. Astron. Soc.* **68**, 37]
CG	Galactic coordinates of a γ-ray source found by COS **B**
COS B	European γ-ray satellite
CCD	Charge-coupled detector [1978, *Sky and Telescope* **56**, 12]
CTIO	Cerro Tololo Inter-American Observatory
ESO	European Southern Observatory
GX	(l^{II}, b^{II}) galactic coordinates of an X-ray source
H	Catalogue of X-ray sources observed with HEAO
HEAO	High Energy Astronomy Observatory (satellite)
HD	Henry Draper Catalog of Stellar Spectra
HDE	Extention of the Henry Draper Catalog
ITS	Image Tube Scanner, [1972, *Publ. A.S.P.* **84**, 161]
Kron.	Globular Clusters, [1956, *Publ. A.S.P.* **68**, 125]
LMC	Large Magellanic Cloud
MC	Modulation Collimator [1978, *Sky and Telescope* **56**, 499]
MX	An X-ray source observed by the MIT team
MXB	A burst-active X-ray source observed by the MIT team
NRAO	National Radio Astronomical Observatory, Vest Virginia, U.S.A.
OAO	Orbiting Astronomical Observatory (satellite)
PKS	Parkes (Australia) Catalogue of radio sources
POSS	Palomar Observatory Sky Survey (Photographic Atlas)
PSS	Palomar Sky Survey
RMC	Rotating Modulation Collimator [1978, *Sky and Telescope* **56**, 499]
RPSS	Red Palomer Sky Survey (Photographs with a red filter)
1S	The first SAS-3 list of X-ray sources [1977, *Nature* **269**, 21]
2S	The second SAS-3 list of X-ray sources
SIT	Silicon Intensifier Target
SMC	Small Magellanic Cloud
Terzan.	Globular Clusters [1971, *Astron. Astrophys.* **12**, 477]
3U	Catalog 3U [1974, *Astrophys. J. Suppl. Series* **27**, 37]
4U	Catalog 4U [1978, *Astrophys. J. Suppl. Series* **38**, 357]
V	Variable star
VLA	Very Large Array (of antennae in radio astronomy)
VLBI	Very Long Baseline Interferometry [1977, *ApJ.* **211**, 658]

TABLE 9.A2

Names of galactic sources. The optical counterparts in [. . .] (Constellation, NGC, HD, and GX entries for most studied sources)

Aql X-1 = 4U 1908 + 00 = 2S 1908 + 005	LMC X-3 = 3U 0539 − 64
= 3U 1908 + 00	LMC X-4 = 3U 0532 − 66
[Ari UX] = H 0324 + 28	Mon X-1 = A 0620 − 00
[Cas γ] = MX 0053 + 60 = 1S 0053 + 604	= [Nova Monocerotis 1975]
Cen X-3 = 3U 1118 − 60	[NGC 1851] = MXB 0512 − 40
= [Krzeminski's star]	[NGC 6440] = MXB 1746 − 20
Cir X-1 = 4U 1516 − 56 = 3U 1516 − 56	[NGC 6441] = 3U 1746 − 37
Cyg X-1 = 3U 1956 + 35 = [HDE 226868]	[NGC 6624] = MXB 1820 − 30
Cyg X-2 = 3U 2142 + 38	[NGC 6712] = A 1850 − 08
Cyg X-3 = 3U 2030 + 40	[NGC 7078] = [M 15] = 2A 2127 + 120
[Cyg SS] = No name	= 3U 2131 + 11
GX 1 + 4 = MXB 1728 − 24 (or − 34)	Nor X-1 = 3U 1636 − 53
GX 9 + 1 = 3U 1758 − 20	[Oph Nova 1977] = 4U 1708 − 23
GX 13 + 1 = 3U 1811 − 17	= H 1705 − 25
GX 17 + 2 = 3U 1813 − 14	[Per β] = [Algol]
GX 301 − 2 = 4U 1223 − 62	[Per Nova 1901] = A 0327 + 43
GX 304 − 1 = 4U 1258 − 61 = 2S 1258 − 613	[Per X] = 4U 0352 + 30 = 2A 0352 + 309
= 3U 1258 − 61	= 1S 0352 + 308
GX 339 − 4 = 4U 1658 − 48	Sco X-1 = [V 818 Sco] = 3U 1617 − 15
[HD 153919] = 4U 1700 − 37	[Sco V 861] = OAO 1653 − 40
= 3U 1700 − 37	= [HD 152667]
[Her AM] = 4U 1813 + 50 = 3U 1809 + 50	Ser X-1 = 4U 1837 + 04 = 3U 1837 + 04
Her X-1 = [HZ Her] = 3U 1653 + 35	SMC X-1 = [SK 160] = 3U 0115 − 73
LMC X-1 = 3U 0540 − 69	TrA X-1 = A 1524 − 61
LMC X-2 = 3U 0521 − 72	Vela X-1 = [HD 77581] = 4U 0900 − 40

TABLE 9.A3

Names of galactic sources (α-entries) (For δ and optical counterparts – see Table 9.A2)

0053 – [Cas γ]	1653 − 40 – [Sco V861]
0115 – SMC X-1	1653 + 35 – Her X-1
0324 – [Ari UX]	1658 – GX 339 − 4
0327 – [Nova Per 1901]	1700 – [HD 153919]
0352 – [Per X]	1708 – [Oph Nova 1977]
0512 – [NGC 1851]	1728 – GX 1 + 4
0521 – LMC X-2	1746 − 20 – [NGC 6440]
0532 – LMC X-4	1746 − 37 – [NGC 6441]
0539 – LMC X-3	1758 – GX 9 + 1
0540 – LMC X-1	1809 – [Her AM]] (3U)
0620 – Mon X-1	1811 – GX 13 + 1
0900 – Vela X-1	1813 − 14 – GX 17 + 2
1118 – Cen X-3	1813 + 50 – [Her AM] (4U)
1223 – GX 301 − 2	1837 – Ser X-1
1258 – GX 304 − 1	1908 – Aql X-1
1516 – Cir X-1	1956 – Cyg X-1
1524 – TrA X-1	2030 – Cyg X-3
1617 – Sco X-1	2142 – Cyg X-2
1636 – Nor X-1	

References

The number of papers used in the preparation of this chapter is so considerable that some of them arc not mentioned in the text, but are given below in an abridged form.

1. WORKS CITED IN THE TEXT

(Works that can be used for further study are marked by an asterisk.)

Bahcall, J. N.: 1978, *Ann. Rev. Astron. Astrophys.* **16**, 241.
Blumenthal, G. R. and Tucker, W. H.: 1974, *Ann. Rev. Astron. Astrophys.* **12**, 23.
Braes, L. L. E. and Miley, G. K.: 1976, *Nature* **232**, 246.
Clark, G. W.: 1977, *Sci. Amer.* **237**(4), 42.
Forman, W. *et al.*, 1978, '4U Catalog', *Astrophys. J. Suppl.* **38**, 357.
Freedman, H.: 1978, *Sky and Telescope* **56**, 490.
*Giacconi, R.: 1973, *Physics Today* **26**(5), 38.
Giacconi *et al.*, 1974, '3U Catalog', *Astrophys. J. Suppl.* **27**, 37.
Grindlay, J. E.: 1977, In *Highlights of Astronomy*, Vol. 4, p. 111, Reidel, Dordrecht.
*Gursky, H. and Van den Heuvel, E. P. J.: 1975, *Sci. Amer.* **232**(3), 24.
Hazard, C.: 1961, *Nature* **191**, 58.
Hjellming, R. M.: 1973, *Science* **182**, 1089.
Hoffman *et al.*: 1978, *Nature* **271**, 630.
Illarionov, A. P. and Synyaev, R. A.: 1975, *Astron. Astrophys.* **39**, 185.
Jones, C. *et al.*: 1973, *Astrophys. J.* (*Letters*) **181**, L43.
Jones, C. and Liller, W.: 1973, *Astrophys. J.* (*Letters*) **184**, L65.
Kellerman, K. I.: 1973, *Physics Today* **26**(10), 38.
Li, F. *et al.*: 1978, *Nature* **275**, 723.
*Misner, C. W. *et al.*: 1973, *Gravitation*, Freeman, San Francisco.
Pandharipandhe, V. R. *et al.*: 1976, *Astrophys. J.* **208**, 550.
*Plavec, M. and Kratochvil, P.: 1964, *Bull. Astron. Soc. Techecoslov.* **15**, 165.
Priedhorsky, W. C. and Krzeminski, W.: 1978, *Astrophys. J.* **219**, 597.
Pye, J. P. and Cooke, B. A.: 1976, *Nature* **260**, 410.
Rappaport, S. A. *et al.*: 1978, *Astrophys. J.* (*Letters*) **224**, L1.
*Thorne, K. S.: 1974, *Sci. Amer.* **231**(6), 32.
Trümper, J. *et al.*: 1978, *Astrophys. J.* (*Letters*) **219**, L105.
Villa, G. *et al.*: 1976, *Mon. Not. Roy. Astron. Soc.* **176**, 609.
Wolff, S. C. and Morrison, N. D.: 1974, *Astrophys. J.* **187**, 69.
Woosley, S. E. and Taam, R. E.: 1976, *Nature* **263**, 101.

2. REFERENCES PARTLY USED IN THE TEXT AND WHICH CAN BE USED FOR FURTHER STUDY

(See first the works cited in 1 marked by an asterisk.)

2.1. Particular X-ray source in (α, δ) order

Abbreviations: *ApJ* = *Astroph. J.*; *Nat* = *Nature*; *AA* = *Astron. Astrophys.*; *Sc* = *Science*; *MN* = *Mon. Not. Roy. Astron. Soc.*; *PhT* = *Physics Today*; *AJU* = *Astron. J. USSR.* (Russian Ed.); *ARAA* = *Ann. Rev. A. A.*; *IAUC* = *IAU Circulars*; *AAA* = *Astron. Astrophys. Abst.*; *ST* = *Sky and Telescope*.

4U 0115 + 63: 1978, *ApJ*, **223**, L71; **224**, L1; *Nat*, **273**, 367; *IAUC*, 3163, 3171.
H 0324 + 28; 1978, *ApJ*, **225**, L119.
MX 0513 – 40: 1975, *ApJ*, **199**, L97. 1976, *Nat*, **260**, 410.
3U 0531 + 21: 1974, *AA*, **34**, 305. 1975, *Nat*, **253**, 610. 1976, See *AAA*. 1977, *ApJ*, **216**, 865; **217**, 807; **217**, 809; *Nat*, **266**, 123. 1978, *ApJ*, **225**, 221; *Nat*, **272**, 679.
A 0535 + 26: 1975, *Nat*, **256**, 628; *Nat*, **256**, 630; *Nat*, **256**, 631; *Nat*, **256**, 633; *Nat*, **257**, 203. 1976, *ApJ*, **208**, L119. 1977, See *AAA*. 1978, *ApJ*, **223**, L71; **223**, 530; *IAUC*, 3167, 3208, 3219, 3259.
A 0620 – 00: 1975, *Nat*, **257**, 656; *Nat*, **257**, 657. 1976, 1977. See *AAA*.

4U 0900 − 40: 1974, *ApJ*, **192**, 685; *AA*, **35**, 301; *AA*, **35**, 353. 1976, *ApJ*, **206**, L99; *ApJ*, **206**, L103; *ApJ*, **208**, 550; *Nat*, **259**, 547; *Nat*, **264**, 219. 1977, *ST*, **54**, 26; *AA*(*Suppl.*), **30**, 195; *AA*, **61**, L35; *IAUC*. 3107. 1978, *ApJ*, **221**, 912; *PhT*, **30**(7), 17; *MN*, **183**, 813; *AA*, **69**, 141.

3U 1118 − 60: 1974, *ApJ*, **192**, L135; *MN*, **169**, 63 P; *AA*, **31**, 339; *AA*, **36**, 261. 1975, *MN*, **172**, 473; *MN*, **172**, 483. 1977, *ApJ*, **211**, 552; **212**, 533; **214**, 235. 1978, *ApJ*, **219**, L77.

A 1118 − 61: 1975, *Nat*, **254**, 577; *Nat*, **254**, 578; *Nat*, **256**, 292; *MN*, **172**, 493. 1976, *Nat*, **263**, 34; *ApJ*, **206**, 257. 1977, See *AAA*. 1978, *ApJ*, **223**, L71.

4U 1516 − 56: 1977, *MN*, **181**, 259. 1978, *Nat*, **276**, 44; *MN*, **183**, 335; *ST*, **56**, 490.

4U 1538 − 52: 1977, *ApJ*, **216**, L11; *MN*, **181**, 73 P. 1978, *ApJ*, **225**, L63; *MN*, **184**, 73 P; *Nat*, **275**, 517. *IAUC*, 3201, 3078.

3U 1617 − 15: 1974, *AJU*, **51**, 905. 1975, *ApJ*, **201**, L65. 1976, 1977, See *AAA*. 1978, *ApJ*, **221**, L13; **223**, L75.

3U 1653 + 35: 1973, *ApJ*, **181**, L75; *AJU*, **50**, 3. 1974, *ApJ*, **187**, 575; *ApJ*, **192**, 517; *ApJ*, **192**, L128; *ApJ*, **194**, L147; *AA*, **32**, 7; *AA*, **35** 407; *AJU*, **51**, 1150. 1975, *Nat*, **253**, 249; *Nat*, **253**, 250. 1976, *Nat*, **263**, 484. 1977, See *AAA*. 1978, *ApJ*, **219**, 292; **219**, 605; **219**, L105; **222**, 652; *ApJ*, **222**, L33; **222**, L113; **225**, 988, **225**, 994; *ApJ*, **225**, L53; *Nat*, **271**, 135; **274**, 571; **275**, 195; *Nat*, **275**, 400; *PhT*, **30**(6), 19; *AA*, **63**L19; *IAUC*, 3184.

OAO 1653 − 40: 1978, *Nat*, **275**, 296; *ST*, **56**, 110; *IAUC*, 3234, 2366.

4U 1700 − 37: 1973, *ApJ*, **181**, L43. 1974, *ApJ*, **188**, 341; *ApJ*, **192**, 677; *AA*, **33**, 49; *AA*, **36**, 295; *MN*, **169**, 47P. 1975, 1976, 1977, See *AAA*. 1978, *ApJ*, **224**, L119; *Nat*, **275**, 400; *MN*, **185**, 137; *AA*, **64**, 399; *AA* (*Suppl.*), **31**, 189; *IAUC*, 3193.

MXB 1730 − 335; 1976, *ApJ*, **207**, L95; 1977, See *AAA*. 1978, *ApJ*, **221**, L53; *MN*, **184**, 1 P; *IAUC*, 3204, 3211.

3U 1813 − 14: 1978, *ApJ*, **219**, 613.

4U 1813 + 50: 1977, *ST*, **53**, 351. 1978, *ApJ*, **219**, 597; **222**, 263; **222**, 641; **225**, L113; **226**, 397.

MXB 1820 − 30: 1978, *ApJ*, **224**, 39; **224**, 383.

3U 1908 + 00: 1977, *ApJ*, **212**, 768; *Nat*, **265**, 606. 1978, *ApJ*, **220**, L13; *Nat* **271**, 633; *IAUC*, 3088, 3225, 3243.

3U 1956 + 35: 1973, *AJU*, **50**, 3. 1974, *ApJ*, **189**, L71; *ApJ*, **190**, L59; *ApJ*, **192**, L68; *ApJ*, **192**, L69; *AJU*, **51**, 1150; *AA*, **34**, 161. 1975, *ApJ*, **200**, 269; *MN*, **173**, 63P; *Nat*, **256**, 109. 1976, *Nat*, **261**, 213; *Nat*, **263**, 393; *Nat*, **263**, 752. 1977, See *AAA*. 1978, *ApJ*, **219**, 288; **220**, L123; **221**, 228; **223**, L17; **225**, 599; *Nat*, **271**, 40; **271**, 630; **273**, 338; **275**, 197; **275**, 400; *MN*, **182**, 315; *AA*, **62**, L1; **62**, 265; *ARAA*, **16**, 241.

3U 2030 + 40: 1973, *PhT*, 26 (No. 1), 17. 1974, *ApJ*, **192**, L119. 1977, See *AAA*. 1978, *ApJ*, **220**, 273; **224**, L113; *AA*, **62**, 275; *Nat*, **272**, 679.

3U 2131 + 11: 1975, *ApJ*, **199**, L93, **199**, L97. 1976, *Nat*, **260**, 410.

3U 2142 + 38: 1978, *AA*, **67**, 287; **69**, 391.

2.2. Particular X-ray sources without (α, δ) *name*

Capella(α Aur): 1978, *ApJ*, **223**, L21.

HR 1099: 1978, *Nat*, **274**, 569.

RS CVn: 1978, *Nat*, **274**, 569.

SS Cyg: 1978, *Nat*, **273**, 338; **275**, 721; *MN*, **184**, 79P; *AA*, **63**, L1.

PART IV

COSMOLOGY: ELEMENTARY THEORY AND BASIC OBSERVATIONAL DATA

CHAPTER 10

ELEMENTARY THEORETICAL COSMOLOGY: THE NEWTONIAN APPROACH

1. Introduction

1.1. THE GENERAL PROBLEM

Cosmology attempts to find the physical and the geometrical properties of the universe, i.e. its form, dimensions, mass, composition, 'age', etc., considered *as a whole* and from an *evolutionary point of view* However, some local properties of the universe are also usually considered in elementary cosmology.

1.2. A FEW HISTORICAL AND PEDAGOGICAL REMARKS ON THE NEWTONIAN APPROACH TO COSMOLOGY

All those who tried to study cosmology before 1952 certainly remember the discouraging mathematical barrier of the tensor analysis, blocking the access to this branch of astrophysics.

And yet, as early as 1934, Milne and McCrea discovered that most of the local properties of the expanding universe could be obtained and clearly explained by a proper application of classical dynamics to the Newtonian theory of gravitation.

Heckmann seems to have been the first to systematically apply this '*Newtonian approach*' in a textbook on cosmology; however his work (written in German and published in Berlin in 1942) escaped general attention.

The method became really wide-spread only through the publication, in 1952, of Bondi's *Cosmology*, with just one mention of Heckmann's "excellent, but almost unobtainable book" – as Bondi says!

Bondi was followed by Zel'dovich, Harrison, Peebles, Sciama and many others, so that today it seems almost unthinkable to use another method in an elementary introduction to cosmology, although this pedagogical error is still quite current.

The Newtonian approach enables one to become familiar with a variety of possible 'models of universe' (including the *expanding universe*) before meeting with more abstract concepts of the general relativity.

As noted by several eminent cosmologists, the Newtonian approach is especially valuable for the following reasons:

(1) "It reveals the implicit simplicity of the cosmological equations (. . .) and offers insight into their physical nature" (Harrison, 1965, pp. 437–438).
(2) "It reveals all the essential features of the relativistic cosmology without the mathematical complexity. It also makes the significance of the different terms apparent much more readily" (Bondi, 1952, p. 75).

(3) "[It prepares] the way towards understanding relativistic cosmology (. . .). Not only is the Newtonian theory mathematically simpler, it also leads to many results that are essentially the same as in relativity" (Sciama, 1971, p. 101).
(4) "It is not only of pedagogical value but also of great heuristic value since [some problems] are too difficult to be taken into account within the framework of the general theory of relativity." (Zel'dovich, 1965, p. 242).

It must be emphasized that the Newtonian approach does *not* represent any special variety of cosmology: it is almost meaningless to speak about '*Newtonian cosmology*'. However, as pointed out by Peebles (1969, p. 6), the fact that the expansion rate of the relativistic models is reproduced by a simple Newtonian gravitational model is not a coincidence: "the models agree just because the Newtonian theory of gravity is *the weak-field limit of general relativity*!"

We start by considering in Section 2 some general fundamental principles common to all current elementary cosmologies (even those based on general relativity). The remaining sections of this Chapter deal with those *local* properties of the universe which can be obtained by application of classical dynamics (supplemented by the principle of equivalence between matter and energy) to the Newtonian theory of gravitation.

2. The Fundamental Principles

Instead of the actual universe it is usual to consider an idealized universe obeying the following principles:

2.1. THE COSMOLOGICAL PRINCIPLE

A principle, common to all elementary cosmologies, can be formulated in the following way: '*Except for local irregularities, the universe presents the same aspect, from whatever point it is observed*'.

This statement, called '*the Cosmological Principle*', can be considered as a more or less arbitrary working hypothesis. The modern tendency is however to consider it rather as an extrapolation of the observational results (see, e.g., Subsection 2.4 below).

2.2. THE PRINCIPLE OF PURE EXPANSION

In addition to the Cosmological Principle it is generally assumed, as a first approximation and in conformity with a reasonable extrapolation of astronomical observations, that the universe has *no differential rotation*. This can be called 'the principle of pure expansion (or of pure contraction)'.

2.3. THE PRINCIPLE OF FLUIDITY

The local constituents of the universe (from stars to clusters of galaxies) correspond to *discrete* particles, such as atoms or molecules of statistical physics.

Nevertheless, considering (as in hydrodynamics) *only large-scale motions*, one can ignore the discrete nature of these 'particles' and treat the constituents of the universe as a *continuous perfect fluid*. This can be called 'the principle of fluidity'.

2.4. The Principle of Homogeneity

All elementary cosmologies (whatever the approach: Newtonian or relativistic) assume as a particular application of the cosmological principle that the cosmic fluid is homogeneous (and isotropic). This also can be considered as an extrapolation of the observational results.

Indeed, many modern observations (but not all) seem to indicate that the large-scale distribution of the clusters of galaxies is homogeneous.

As stated by Sandage, 1972:

> "Over the scale of observations which has relevance to the universe itself (rather than to scales which clearly deal with local details) there is no significant departure from homogeneity in effects on dynamics or the average distribution of matter.
>
> The indeniable evidence of local inhomogeneity which gives such delight to the eye – stars, clusters, galaxies, and clusters of galaxies – all these are wrinkles on the surface of an orange compared with the large-scale dominant geometry determined by its overall curvature."

This conclusion, by one of the most competent observers working with giant telescopes, is based on studies of many clusters of galaxies.

2.5. The Concept of 'Typical Cell' of the Universe

Let us now go a little beyond purely Newtonian concepts, and take into account the equivalence between matter and energy introduced by special relativity. Then the mass density ρ – including the mass corresponding to energy – will be represented everywhere, according to the principle of homogeneity, by the same function $\rho(t)$ of time t.

One of the main purposes of elementary cosmology is thus reduced to the determination of the function $\rho(t)$, i.e. to establihsing and to solving the equations satisfied by this function.

For this purpose it is not necessary to consider the whole of the universe: all volume elements having the same density, the study of *one local* (spherical) *'typical cell'* of the universe is sufficeint.

This is particularly obvious in the case of the Newtonian approach, since in classical mechanics the gravitational field at a given point depends only on the local mass density through the Poisson's equation (see Chapter 2, Section 4).

2.5.1. Remark

In the Newtonian approach, the gravitational field at the surface of a typical spherical cell (S) of center O is usually obtained by application of the Newton's theorem (Chapter 2, Subsection 2.1).

However, the validity of this theorem demands that (S) be surrounded by *a mass of spherical shape* stratified in homogeneous concentric layers, which implies that the homogeneous universe surrounding (S) is *spherical* and centered at O.

Thus one must admit that all eventual irregularities at the 'frontier' of the universe (if any) can be 'rejected to infinity'. This is obviously possible in a first approximation if (S) is supposed to be surrounded by a material homogeneous sphere of a sufficiently great radius.

3. The Kinematics of a Model of Cosmic Fluid. Hubble's Law

3.1. THE SCALE FACTOR

Consider a kinematical model of a *spherically symmetric radial expansion* (*or contraction*) of the cosmic fluid with respect to some particular galaxy O. This galaxy will play the role of the center of expansion, and will be chosen as the origin of a system Oxyz of cartesian coordinates. (The problems set by this choice will be discussed later, in Subsection 3.3.)

Let $\mathbf{P} = \mathbf{P}(t)$ denote the radius vector at time t of a galaxy P participating to the radial motion of expansion of the cosmic fluid.

The value of $\mathbf{P}(t)$ will depend only on its value $\mathbf{P}_0 = \mathbf{P}(t_0)$ at some instant t_0. Moreover, $\mathbf{P}$ will be always aligned on $\mathbf{P}_0$ (radial motion!). Finally, the ratio $\mathbf{P}/\mathbf{P}_0$ will keep the same value $a(t)$ – depending only on time t – for all points P of the fluid situated on the surface of the sphere (S) of center O and of radius R equal to $|\mathbf{P}|$ (spherically symmetric motion!).

The ratio $a(t)$, called *the 'scale factor'* (or, less adequately, *the 'expansion parameter'*, in spite of the fact that it can correspond to a contraction) is thus defined by

$$\mathbf{P} = a(t)\mathbf{P}_0. \tag{1}$$

Obviously, the time variations of R are described by the function $R(t)$ given by

$$R(t) = a(t)R(t_0) = a(t)R_0, \tag{2}$$

where R_0 denotes the radius $R(t_0)$ of the sphere (S) at time t_0.

Equations (1) and (2) show that the scale factor $a(t)$ is *dimensionless*, and that

$$\boxed{a_0 = a(t_0) = 1.} \tag{3}$$

On dividing Equation (1) by Equation (2) one introduces the unit vector $\mathbf{u_P}$ associated with the direction $\mathbf{P}$:

$$\mathbf{u_P} = \mathbf{P}/R = \mathbf{P}_0/R_0; \tag{4}$$

hence

$$\boxed{\mathbf{P}(t) = R(t)\mathbf{u_P}.} \tag{5}$$

and

$$\mathbf{P}_0 = R_0\mathbf{u_P}. \tag{5'}$$

On taking into account Equation (2) we obtain:

$$\mathbf{P}(t) = a(t)R_0\mathbf{u_P}. \tag{5''}$$

The expression for $\mathbf{P}(t)$ given by Equation (5″) clearly indicates all of the kinematical characteristics of the motion: for each element associated with P the direction $\mathbf{u_P}$ remains invariable; all points P situated at time t_0 on the same sphere (S_0) of radius R_0 remain at time t on the same sphere (S) of radius $a(t)R_0$; and the law of motion of these spheres is determined by the time variations $a(t)$ of the scale factor.

3.2. HUBBLE'S KINEMATICAL LAW. HUBBLE'S CONSTANT

The same motion can be described in a somewhat different way.

Let us set, as usual,

$$\dot{a}(t) = \frac{da(t)}{dt}; \qquad \dot{R}(t) = \frac{dR(t)}{dt}, \tag{6}$$

i.e. let the dots denote differentiation with respect to time, and let $v(\mathrm{P}, t)$ denote the vector velocity of the point P at time t, in the same reference system Oxyz.

According to Equations (5) and (2), $v(\mathrm{P}, t) = d\mathbf{P}/dt$ will be given by

$$v(\mathrm{P}, t) = \dot{R}(t)\mathbf{u}_\mathrm{P}, \tag{7}$$

or, by

$$v(\mathrm{P}, t) = \dot{a}(t)R_0\mathbf{u}_\mathrm{P} \tag{7'}$$

On dividing Equation (7) by Equation (5), and on taking into account Equation (2), we obtain:

$$v(\mathrm{P}, t)/\mathbf{P} = \dot{R}(t)/R(t) = \dot{a}(t)/a(t). \tag{8}$$

On introducing the function $H(t)$, so denoted in honor of E. Hubble, defined by

$$H(t) = \dot{a}(t)/a(t) = \dot{R}(t)/R(t), \tag{9}$$

hence the notation

$$\boxed{H_0 = H(t_0) = \dot{a}(t_0)/a(t_0) = \dot{a}(t_0) = \dot{a}_0,} \tag{9'}$$

we can replace Equation (8) by

$$\boxed{v(\mathrm{P}, t) = H(t)\mathbf{P},} \tag{10}$$

or replace Equation (7) by

$$v(\mathrm{P}, t) = H(t)R(t)\mathbf{u}_\mathrm{P}, \tag{10'}$$

hence

$$|v(\mathrm{P}, t)| = H(t)R(t), \tag{11}$$

and

$$\boxed{|v(\mathrm{P}, t_0)| = H_0R_0.} \tag{12}$$

Let us now recall *Hubble's law*. It states, in its usual formulation, that – according to observations – *the velocity of recession of a galaxy is proportional to its distance*.

This (empirical) kinematical law, obtained by Hubble in 1929, expresses the linearity of the plot of radial velocities of some individual galaxies *versus* their distance (the radial velocities being obtained by the Doppler–Fizeau interpretation of the red shifts of spectral lines of galaxies).

More precisely, and in view of comparison with Equation (12), let P represent the position of a galaxy. Then the law states that, *at time t_0 of observation*, the ratio $|v(P, t_0)|/R_0$ is the same, and equal to $H(t_0)$, for all of the observed galaxies, *whatever their distance R_0*.

Thus we see that in some way, to be discussed later in Subsection 5.5.1.1, Equation (12) corresponds to Hubble's kinematical law.

In many textbooks the function $H(t)$ is called the *'Hubble constant'*, which is a quite improper term for a function of time, but which is more or less acceptable if only indicating independence with respect to the distance $R(t_0)$ at a given instant t_0.

Henceforth we shall use the term 'Hubble constant' only for $H_0 = H(t_0)$, where t_0 denotes (as everywhere below, except in Section 7) the *present time*.

The value of H_0 is yet not known with precision: it seems to be near 75 km s^{-1} per Megaparsec. Some authors use a value of 100 km s^{-1} Mpc^{-1}, whereas others consider a value of 50 km s^{-1} Mpc^{-1} to be more probable.

In purely theoretical problems, the precise value of H_0 is irrelevant. Nevertheless, it is always important to indicate the value adopted in theoretical calculations or in observational reductions: in this way it is relatively easy to find the consequences of an eventual modification in the choice of the value of H_0.

3.2.1. Remarks

3.2.1.1. Equation (7) describing the kinematical model corresponding to Hubble's law is easily shown to be equivalent to Equation (5) of Subsections 3.1.

Indeed, starting from Equation (7) and taking into account the definition $d\mathbf{P}/dt$ of the vector velocity $v(P, t)$, an integration (with obvious boundary conditions) with respect to t immediately yields Equation (5).

3.2.1.2. According to Equation (9), the function $H(t)$ can be defined either as $\dot{a}(t)/a(t)$ or $\dot{R}(t)/R(t)$. Nevertheless, one must pay attention to the fact (overlooked by many authors) that the scale factor is a dimensionless quantity, whereas R has the *dimensions of a length*.

3.3. THE PROBLEM OF THE CHOICE OF THE ORIGIN OF THE REFERENCE SYSTEM*

Let us consider the purely kinematical problem of an eventual displacement of the origin of coordinates from the *galaxy* O to another *galaxy* O′, with new axes parallel to the initial ones.

(The *dynamical* problem of inertiality of the reference system is usually left out of consideration in the Newtonian approach. Interested readers are referred to the excellent discussion of this important problem by Schücking (1967.)

If M denotes some other galaxy of the fluid (see Figure 10.1), both the motions of M and of the galaxy O′ will be described in the initial reference system Oxyz by the general Equation (1), so that we can write, at time t:

* This subsection can be omitted in a first reading.

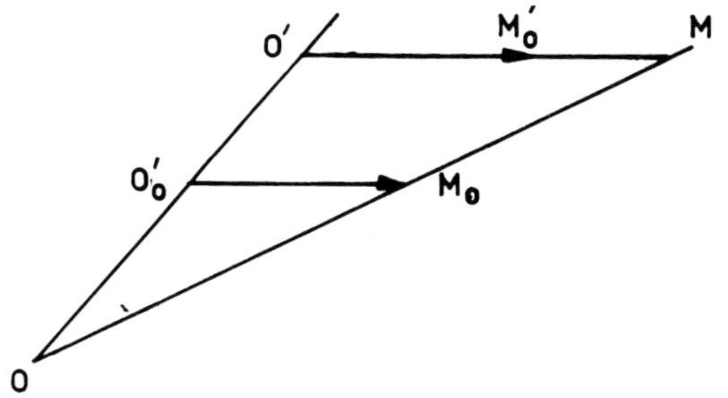

Fig. 10.1. The vectors involved in the change of the origin of the reference system.

$$\mathbf{M} = a(t)\mathbf{M}_0, \tag{13}$$

and

$$\mathbf{O}' = a(t)\mathbf{O}_0'. \tag{14}$$

The subscript (0) corresponds, as above, to time t_0, and *all bold face symbols are vectors taken with origin at* O.

On subtracting Equation (14) from Equation (13) we obtain:

$$\mathbf{M} - \mathbf{O}' = a(t)(\mathbf{M}_0 - \mathbf{O}_0'). \tag{15}$$

In Figure 10.1 we see that $(\mathbf{M} - \mathbf{O}')$ represents the radius vector of the galaxy M with respect to the origin O′ of the new reference system O′x′y′z′, whereas $(\mathbf{M}_0 - \mathbf{O}_0)$ at the right-hand side of Equation (15) *does not* represent any radius vector (neither in the system Oxyz, nor in the system O′x′y′z′).

Let us however consider (see Figure 10.1) the point M_0' defined by the vector $(\mathbf{M}_0' - \mathbf{O}')$ equivalent to $(\mathbf{M}_0 - \mathbf{O}_0')$ but originating at O′. We can then replace Equation (15) by

$$\mathbf{M} - \mathbf{O}' = a(t)(\mathbf{M}_0' - \mathbf{O}'), \tag{16}$$

where again all bold face symbols are vectors taken with origin at O.

Now, $(\mathbf{M} - \mathbf{O}')$ represents the radius vector of the galaxy M in the system O′x′y′z′, whereas $(\mathbf{M}_0' - \mathbf{O}')$ represents the radius vector of the point M_0' in the same reference system O′x′y′z′. It seems, therefore, at first sight, that Equation (16) is entirely equivalent to Equation (1). Some authors stop their analysis here considering the discussion given above as the proof showing that, from the kinematical point of view, the model conserves all of its properties when the 'center' is arbitrarily shifted from its particular position O to some other general position O′.

However, the situation is not as simple as it appears at first sight. Indeed, some attentive readers could experience some trouble because the galaxy M is assigned at time t_0, in this discussion, *two different positions*: M_0' in Equation (16), when this last Equation has to be assimilated with Equation (1), and M_0 in Equation (13).

To clarify the situation (or to skip over this difficulty) instead of a description in terms of positions, like the description given by Equation (1) and in Figure 10.1, one can use a description in terms of vector velocities given by Equation (10) and in Figure 10.2.

Then the passage from the origin O to the origin O′ described in terms of positions by Equation

$$\mathbf{M} - \mathbf{O}' = (\mathbf{M} - \mathbf{O}) - (\mathbf{O}' - \mathbf{O}), \tag{17}$$

is to be replaced by the 'galilean transformation' for velocities:

$$\boldsymbol{v}'(\mathrm{M}, t) = \boldsymbol{v}(\mathrm{M}, t) - \boldsymbol{v}(\mathrm{O}', t). \tag{18}$$

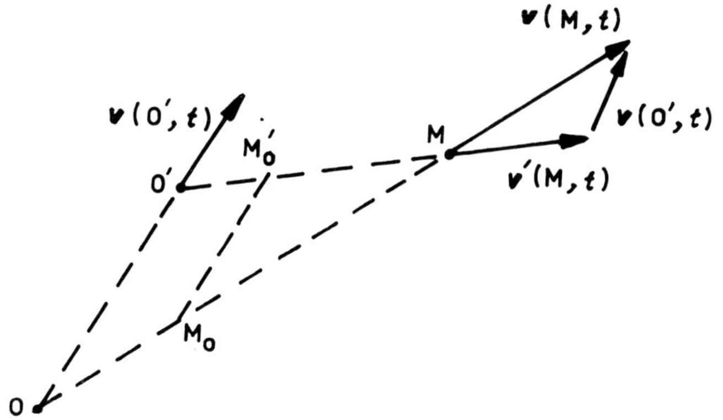

Fig. 10.2. The vectors involved in the application of the 'galilean transformation' for velocities.

In Equation (18) the bold face symbol $\boldsymbol{v}'$ exceptionally represents the velocity vector of M with respect to the reference system O′x′y′z′, whereas $\boldsymbol{v}(\mathrm{M}, t)$ and $\boldsymbol{v}(\mathrm{O}', t)$ are the velocities of M and O′ in the reference system Oxyz respectively, in conformity with the notation defined above.

On applying to galaxies M and O′ the fundamental relation (10) we can write:

$$\boldsymbol{v}(\mathrm{M}, t) = H(t)\mathbf{M}, \tag{19}$$

and

$$\boldsymbol{v}(\mathrm{O}', t) = H(t)\mathbf{O}', \tag{19'}$$

hence, by substitution into Equation (18):

$$\boxed{\boldsymbol{v}'(\mathrm{M}, t) = H(t)(\mathbf{M} - \mathbf{O}').} \tag{20}$$

Equation (20) represents kinematically in the reference system O′x′y′z′ exactly the same relation as Equation (10) in the system Oxyz.

Of course, in Equation (18) *the velocity of the same galaxy* M is again represented by *two different vectors*: $\boldsymbol{v}(\mathrm{M}, t)$ in the reference system Oxyz and $\boldsymbol{v}'(\mathrm{M}, t)$ in the reference system O′x′y′z′. However, now we obviously encounter a quite classical phenomenon: the *relativity of velocities*, i.e. the modification of the direction and of the speed of a motion when it is observed from two reference systems in relative uniform translation. This 'galilean relativity' is already expressed by Equation (18).

Now we are well prepared to understand the paradox encountered in the description corresponding to Figure 10.1. In that case we encountered the phenomenon called in astronomy the '*phenomenon of aberration*'.

Let us consider, to be more concrete, the situation illustrated in Figure 10.3.

In the system Oxyz, represented schematically in two dimensions by the axes Ox and Oy, the galaxy M is supposed to the situated at time t_0 at the point M_0 of the axis Oy.

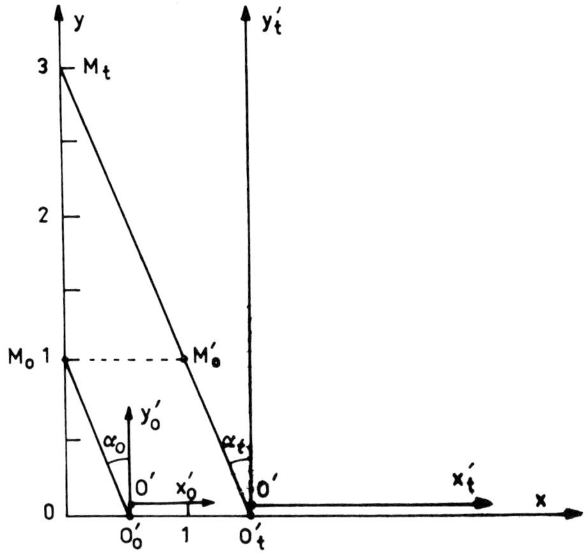

Fig. 10.3. The phenomenon of aberration illustrated by an example.

The scale factor at time t is supposed to be equal to 3. Thus, at time t the galaxy M will be at the point M_t of the axis Oy; with $(OM_t) = 3 \times (OM_0)$.

At time t_0 the galaxy O′ is supposed to be at the point O'_0 of the axis Ox at a distance (OO'_0) from O equal to $(\frac{1}{2})(OM_0)$.

Thus, again at time t_0, the observer O′ situated at O'_0 observes the galaxy M in the direction O'_0M_0 at a distance equal to (O'_0M_0). The angle α_0 between the axis $O'_0y'_0$ parallel to Oy and the direction O'_0M_0 is equal to $\alpha_0 = \tan^{-1}(OO'_0/OM_0) = \tan^{-1}(\frac{1}{2})$.

At time t, the galaxy O′ is at the point O'_t of the axis Ox. Since $(OO'_t) = a(t)(OO'_0) = 3 \times (OO'_0) = (\frac{3}{2})(OM_0)$, the observer O′ sees the galaxy in the direction O'_tM_t. The angle α_t between the axis $O'_ty'_t$ parallel to Oy and the direction O'_tM_t is equal to α_0 since $\alpha_t = \tan^{-1}(OO'_t/OM_t) = \tan^{-1}(\frac{3}{2}/3) = \tan^{-1}(\frac{1}{2})$.

As to the initial position M_0 of the galaxy M obviously it remains fixed in the system O′x′y′z′, but because of its participation to the motion of O′x′y′z′ it comes at time t to the point M'_0 of Oxyz.

Since $(M_0M'_0)$ is equal and parallel to $(O'_0O'_t)$, the figure $O'_0M_0M'_0O'_t$ is a parallelogramm and M'_0 can be considered as defined just as in Figure 10.1.

4. A few Observational Data

According to Oort (1958), for a value of H_0 equal to 75 km s^{-1} Mpc^{-1}, the average ratio of the mass to the luminosity of a galaxy (isolated or situated in a cluster of galaxies) is about 21 times greater than the corresponding ratio $M_\odot/L_\odot$ for the Sun.

Moreover, the average total luminosity L_{gal} of all galaxies contained in a volume V_1 of *one* pc^3 of the universe is of the order of $2.2 \times 10^{-10} L_\odot$. The total mass M_{gal} of all galaxies contained in a volume equal to V_1 will thus be given by

$$M_{\text{gal}} \approx 21(M_\odot/L_\odot)L_{\text{gal}} \approx 21M_\odot(2.2 \times 10^{-10}) \approx 4.6 \times 10^{-9} M_\odot. \qquad (21)$$

Since $M_\odot \approx 2 \times 10^{33}$ g and $V_1 \approx (3.1 \times 10^{18})^3 \text{ cm}^3 \approx 3 \times 10^{55} \text{ cm}^3$, the present average mass density ρ_{mat} of observable 'material particles' such as galaxies will be given, in c.g.s. units, by

$$\rho_{\text{mat}} = M_{\text{gal}}/V_1 \approx \frac{(4.6 \times 10^{-9}) \times (2 \times 10^{33})}{3.0 \times 10^{55}} \approx 3 \times 10^{-31} \text{ g cm}^{-3}. \qquad (22)$$

This value is confirmed by more recent data of Peebles and Partridge (1967) who give $\rho_{\text{mat}} \approx 4.5 \times 10^{-31} \text{ g cm}^{-3}$.

On the other hand, the *mass* density ρ_{rad} corresponding to the present average *energy* density ($u_{\text{rad}} \approx 4 \times 10^{-13} \text{ erg cm}^{-3}$) of the cosmic radiation field (that of a black body at 2.7 K) is of the order of $\rho_{\text{rad}} \approx 4 \times 10^{-34} \text{ g cm}^{-3}$ (see for instance Reinhardt (1969, Subsection 2.1)).

One can therefore neglect, *in the present state of the universe*, i.e. at time t_0, the contribution of the radiation field to the total mass density of the universe.

Let us now consider the pressures. The radiation pressure P_{rad} of the photon gas (equal to one third of u_{rad}) has the same physical dimensions as the energy density $u_{\text{mat}} = c^2\rho_{\text{mat}}$ corresponding to the average mass density ρ_{mat} of material particles at rest.

This explains, to some extent, why in the rigorous treatment of the cosmological problems, the role played by P_{rad} essentially depends on the value of the ratio $u_{\text{rad}}/u_{\text{mat}} = \rho_{\text{rad}}/\rho_{\text{mat}}$. According to the data quoted above, this ratio, in the present state of the universe, is certainly negligible in a first approximation.

For this reason we can neglect P_{rad} with respect to an eventual 'gas pressure' P_{gas} corresponding to the relative motion of galaxies, at time t_0.

However, P_{gas} itself is negligible in the present state of the universe. Indeed, P_{gas} is equal in a first approximation to two thirds of the average kinetic energy density $u_{\text{kin,gal}}$ of the (non-relativistic) relative motion of galaxies. Thus P_{gas} has the physical dimensions of u_{mat} and its importance essentially depends on the ratio $u_{\text{kin,gal}}/u_{\text{mat}}$. Let us show that this ratio is certainly less than 2×10^{-7}:

The average relative velocities $\bar{v}_{\text{gal}}$ of galaxies (and of stars in galaxies) are of the order of 200 km s^{-1}, i.e. of the order of $2 \times 10^7 \text{ cm s}^{-1}$. If $\bar{m}_{\text{gal}}$ denotes the average mass of a galaxy (i.e. of one 'material particle' of the universe) and if n_{gal} denotes the average number density of these particles, $u_{\text{kin,gal}}$ and u_{mat} will be given respectively by

$$u_{\text{kin,gal}} = \tfrac{1}{2}\bar{m}_{\text{gal}}(\bar{v})^2 n_{\text{gal}} = \tfrac{1}{2}(\bar{v})^2 \rho_{\text{mat}}, \qquad (23)$$

and

$$u_{\text{mat}} = \rho_{\text{mat}}/c^2. \qquad (24)$$

Hence the value of the ratio $u_{\text{kin,gal}}/u_{\text{mat}}$ indicated above.

5. The Friedmann Model of Universe

5.1. THE NEGLECT OF THE 'COSMOLOGICAL CONSTANT' Λ

In 1922 (thus before the observational discovery, in 1929, of the recession of galaxies) the Russian mathematician A. Friedmann conceived, in the frame of general relativity, a very simple *non-static* model of universe.

In order to understand the originality of the Friedmann model it is necessary to take into account that, in accordance with the observational data of the time, A. Einstein believed that the universe was *static*.

However, already in the Newtonian approach, it is obvious that the universe cannot be static if the only forces acting upon stars, galaxies, etc. are those of mutual gravitation, and the same is true in the frame of the concepts of general relativity.

Therefore, in order to prevent the gravitational collapse of a static universe, Einstein introduced in 1917, for this sole purpose, in his equations, a term corresponding physically (in the Newtonian approach) to a cosmic *repulsion*. This term was proportional to a quantity denoted by Λ and called the '*cosmological constant*'.

(Misner *et al.* (1970), pp. 409f give a fascinating report of this mistake, which Einstein later called "the biggest blunder of my life" and, p. 751, give Friedmann's biography).

5.2. The Definition of the Friedmann Universe

Friedmann, in his revolutionary papers of 1922 and 1924, assumed that a non-static universe was a quite realistic possibility and constructed a model of universe *neglecting*:

(1) The cosmological constant Λ;
(2) The energy density u_{rad} of the radiation field, and the corresponding pressure P_{rad};
(3) The 'gas pressure' P_{gas} of galaxies.

Thus, in the Friedmann model one considers as the only physical reality the *mass density* ρ_{mat} *of the cosmic fluid* and – in the Newtonian approach – the corresponding mutual gravitational attraction inside each 'typical cell' of the universe.

The observational data, mentioned in Section 4, show that this model, also sometimes called *the 'Friedmann universe'*, is quite compatible (as an elementary approximation) with our nowadays knowledge of the *present* state of the universe and its present evolution.

5.3. The Fundamental Invariant

Let us consider, at time t, a typical spherical cell (S) of the Friedmann universe, of a variable radius R, and of a constant mass M.

It will be shown below that this cell can be either in a state of expansion or in a state of contraction. This phenomenon, described by the function $R(t)$ of time t, leads to a variation of the mass density ρ of the cell (thus of the universe) described by the function $\rho(t)$.

In the Friedmann universe $\rho = \rho_{\text{mat}}$, and the subscript 'mat' becomes unnecessary. Thus, neglecting any eventual transformation of matter into radiation, the constancy of M will be expressed by

$$M = \frac{4\pi}{3} R^3(t)\rho(t) = \text{const.}, \tag{25}$$

i.e. by the invariance of $R^3\rho$ with respect to time:

$$\boxed{R^3(t)\rho(t) = \text{const.}.} \tag{25$'$}$$

On applying this relation both at times t and t_0 we obtain:

$$R^3(t)\rho(t) = R^3(t_0)\rho(t_0) = R_0^3\rho_0. \tag{25$''$}$$

On introducing the scale factor $a(t)$, on taking into account Equations (2) and (3), and on dividing by R_0^3, Equation (25″) takes the form:

$$\rho(t)a^3(t) = \rho(t_0) = \rho_0, \tag{26}$$

hence

$$\rho(t) = \rho(t_0)/a^3(t) = \rho_0 a^{-3}(t). \tag{26'}$$

Thus, we find that, in the Friedmann cosmological model, the mass density of the universe varies proportionally to $a^{-3}(t)$.

5.4. THE FUNDAMENTAL EQUATION. THE PARAMETERS Ω AND q_0

Let us consider (see Figure 4) a spherical cell of the Friedmann universe and a small element (A) of the cosmic fluid, of unit mass, situated at the surface (S) of the cell.

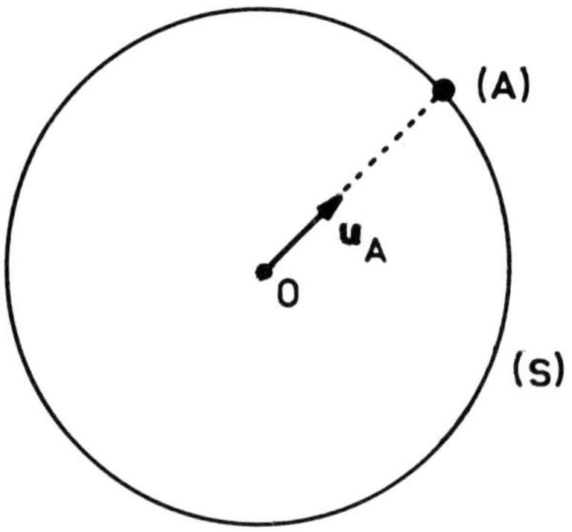

Fig. 10.4. The 'typcial cell' (S) and the element (A) of unit mass, at the frontier of (S).

According to the principle of homogeneity (Subsection 2.4) the mass density ρ, whose dependence on time t is described by the function $\rho(t)$, is the same everywhere inside (S).

As already explained (Subsection 2.5) it is assumed that Newton's theorem (proved in Chapter 2, Subsection 2.1) can be applied. This means that the gravitational field g_A at the point A of (S) does *not* depend on the part of the universe outside (S). Moreover, according to the same theorem, g_A is directed towards the center O of (S), and equivalent to the field produced by the entire mass M of the cell concentrated at O.

Thus, if $\mathbf{u}_A$ denotes the unit vector issued from O and directed towards A, we have

$$g_A = -\frac{GM}{R^2}\mathbf{u}_A, \tag{27}$$

where G is the gravitational constant equal to 6.67×10^{-8} c.g.s. .

Let us now apply the fundamental principle of Newtonian dynamics:

$$m\gamma = F, \tag{28}$$

where F is the force producing the vector acceleration γ of the mass m. Since the mass m of the element (A) is supposed to be equal to unity, and since in the Friedmann model the pressure is supposed to be zero, the force F_A acting upon (A) will be equal to g_A. Thus, the corresponding vector acceleration γ_A of (A) will be given by

$$\gamma_A = g_A. \tag{29}$$

According to Equation (27) this is equivalent to

$$\gamma_A = -\frac{GM}{R^2}\mathbf{u}_A. \tag{30}$$

This relation confirms the physically obvious fact that γ_A is a *purely radial vector*, which reduces to its radial component $(d^2R/dt^2)\mathbf{u}_A$.

Thus we find:

$$\frac{d^2R}{dt^2} = -\frac{GM}{R^2}. \tag{31}$$

Equation (31) represents the differential *equation of motion* of the unit mass of the cosmic fluid situated at the distance $R(t)$ from O.

This equation shows – rather unexpectedly, at first sight, for an *expanding universe* – that the radial acceleration d^2R/dt^2 of (A) is always *negative*.

The paradox vanishes if, as usual, one ascribes the observed expansion to a violent *initial explosion* of a very condensed 'early universe' (the '*big-bang model*' of universe).

If the universe 'began' in a state of rapid expansion from a very nearly condition of near infinite density (and, in the 'hot' standard big-bang model, of near infinite temperature), i.e. in a state of *near infinite radial velocity* $v_{rad} = dR/dt$, then dR/dt can remain *positive* (thus corresponding to an expansion) even if $dv_{rad}/dt = d^2R/dt^2$ is negative. The motion of expansion initiated by the big-bang is constantly *decelerated* by the mutual attraction of different parts of the cosmic fluid!

Note, immediately, that this explanation does not exclude the possibility of an eventual 'exhaustion' of the initial velocity of expansion (at some time t_{max} later than t_0) under the braking action of mutual gravitation. Thus the present expansion could stop at time t_{max} of a maximum of expansion and the universe could then start to contract (with an *increasing* radial velocity of contraction) back to whatever it came from. The conditions for such a reversal will be discussed below.

The equation of motion (31) can be integrated once by multiplying by dR/dt, which gives:

$$\boxed{\frac{1}{2}\left(\frac{dR}{dt}\right)^2 + \left(-\frac{GM}{R}\right) = \text{const.}} \tag{32}$$

This equation corresponds to the law of the conservation of energy: the first term on the left is the kinetic energy of the unit mass (A), the second term on the left is the (negative) potential energy of the system formed by (A) and the cell (S). Thus, the constant on the right is the total energy of the system.

Equation (32) will be called henceforth the '*fundamental equation*', because the relativistic approach shows that (as usual for laws expressed in terms of energy) it is *much more general* than Equation (31), which holds only with Friedmann's assumption of zero pressure. (We discuss more thoroughly this point in Subsection 7.4.)

Returning, for the time being, to the Friedmann universe, we can replace M by $(4\pi/3)R^3\rho$ in the fundamental Equation (32), thus obtaining:

$$\left(\frac{dR}{dt}\right)^2 = \frac{8\pi}{3} G\rho R^2 + \text{const.} \tag{33}$$

With the almost obvious notation

$$\dot{R}_0 = \left(\frac{dR}{dt}\right)_{t=t_0}, \tag{34}$$

we can rewrite Equation (33) as

$$(\dot{R})^2 = \frac{8\pi}{3} G(\rho R^2 - \rho_0 R_0^2) + (\dot{R}_0)^2. \tag{35}$$

Let us now take into account the invariance of $R^3\rho$, characteristic of the Friedmann universe, expressed by Equation (25″), viz.

$$\rho R^3 = \rho_0 R_0^3. \tag{36}$$

This allows us to replace Equation (35) by

$$(\dot{R})^2 = \frac{8\pi G}{3}\left(\frac{\rho_0 R_0^3}{R} - \rho_0 R_0^2\right) + (\dot{R}_0)^2. \tag{37}$$

It is important to note that Equation (37) is satisfied by solutions of the same kinematical form as Equation (2), viz.

$$R(t) = R_0 a(t); \qquad a(t_0) = 1, \tag{38}$$

i.e. by solutions corresponding (according to Subsection 3.2) to Hubble's kinematical law.

Indeed, on substituting $R(t)$ defined by Equation (38) into Equation (37), R_0 cancels out and one is left with the differential equation

$$(\dot{a})^2 = \frac{8\pi G}{3}\rho_0\left(\frac{1}{a} - 1\right) + (\dot{a}_0)^2, \tag{39}$$

which, with the proper choice of initial conditions, determines the time variation of the scale factor a.

According to the definition (9′) of the Hubble constant H_0 this constant is equal to $\dot{a}_0$. This allows us to write Equation (39) as

$$(\dot{a})^2 = \frac{8\pi G}{3}\rho_0\left(\frac{1}{a} - 1\right) + H_0^2, \tag{39′}$$

or

$$(\dot{a})^2 = H_0^2\left[\frac{8\pi G}{3H_0^2}\rho_0\left(\frac{1}{a} - 1\right) + 1\right]. \tag{39″}$$

On introducing, with Zel'dovich (1965) the parameter Ω defined by

$$\boxed{\Omega = \frac{8\pi G}{3H_0^2}\rho_0 = \frac{\rho_0}{(3H_0^2/8\pi G)},} \tag{40}$$

we obtain the differential equation:

$$\left(\frac{da}{dt}\right)^2 = H_0^2\left[\frac{\Omega}{a} - (\Omega - 1)\right], \tag{41}$$

or, in a form more convenient when Ω is smaller than 1:

$$\left(\frac{da}{dt}\right)^2 = H_0^2\left[\frac{\Omega}{a} + (1 - \Omega)\right]. \tag{41'}$$

5.4.1. *Remark*

Many cosmologists instead of the parameter Ω make use of the *'deceleration parameter'* q_0 defined by

$$q_0 = -\ddot{a}_0/(\dot{a}_0)^2 \tag{42}$$

where, of course, $\ddot{a}_0$ represents $(d^2a/dt^2)_{t=t_0}$.

On taking the derivative of both sides of Equation (41) with respect to t, we obtain:

$$2\ddot{a} = 2\frac{d^2a}{dt^2} = -H_0^2\Omega a^{-2} = -(\dot{a}_0)^2\Omega a^{-2}. \tag{43}$$

Since, at time $t_0, a_0 = a(t_0) = 1$, Equation (43) yields, for the Friedmann universe:

$$\boxed{q_0 = \frac{\Omega}{2}.} \tag{44}$$

This relation is *not* general. It will be shown in Subsection 7.4.1 that for a radiation filled model of the universe (the 'standard model of the early universe') q_0 is equal to Ω', where Ω' is a parameter corresponding to Ω.

5.5. THE TIME VARIATION OF THE SCALE FACTOR. THE CRITICAL VALUE OF Ω AND THE CRITICAL DENSITY. THE QUALITATIVE DISCUSSION

Equations (41) and (41′) show that the time variation $a(t)$ of the scale factor, i.e. the evolution of the Friedmann universe, depends on the value of Ω, which by definition is always positive, but can be smaller or greater than one (or equal to one).

When $\Omega \leqslant 1$ the term $(1 - \Omega)$ in Equation (41′) is positive or zero, and nothing prevents the square bracket to remain positive or zero for all positive values of $a(t)$, i.e. forever. Since at the present time da/dt is positive (expanding universe!), and cannot vanish, it remains positive for all positive values of $a(t)$. This means that for these values of Ω the model must *expand forever*.

When $\Omega > 1$ the term $(\Omega - 1)$ in Equation (41) is positive. Therefore, at time t_{max}, defined by the equation

$$a(t_{max}) = a_{max} = \frac{\Omega}{\Omega - 1}, \tag{45}$$

the derivative (da/dt) will vanish, and $a(t)$ will reach its maximum value $a(t_{max}) = a_{max}$.

Indeed, after t_{max} the only possibility for the square bracket to remain positive is that the scale factor $a(t)$ becomes smaller that a_{max}. Thus, after t_{max} the derivative da/dt must become negative, and the model must *collapse back on itself*; this contraction phase being described by the equation

$$\frac{da}{dt} = -H_0\left[\frac{\Omega}{a} - (\Omega - 1)\right]^{1/2} = -H_0(\Omega - 1)^{1/2}\left(\frac{a_{max}}{a} - 1\right)^{1/2} \tag{46}$$

The two possibilities, according to the situation of Ω with respect to its critical value 1, are illustrated schematically by Figures 10.5 and 10.6:

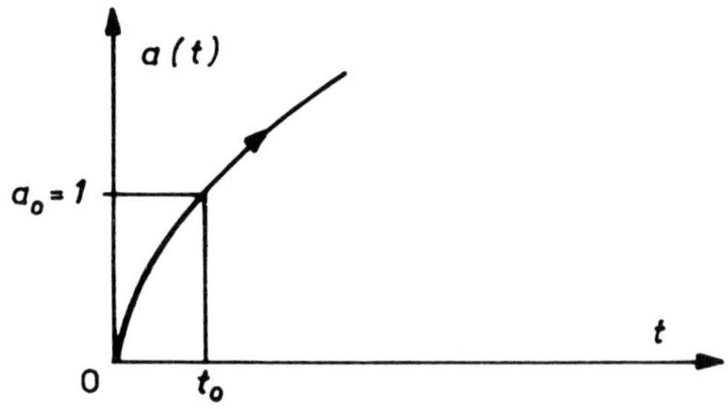

Fig. 10.5. Evolution of the scale factor a in the case $\Omega \leqslant 1$.

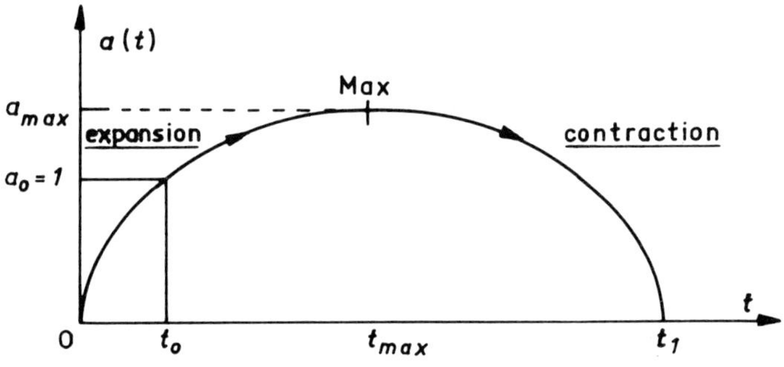

Fig. 10.6. Evolution of the scale factor a in the case $\Omega > 1$, (an 'oscillating universe').

(In this qualitative discussion the precise mathematical form of the function $a(t)$, given in Subsection 5.8, is not yet entirely taken into account!.)

Naturally, as usual, the elegance of this mathematical discussion, hides the corresponding physical conditions discriminating the case (Figure 10.5) of continuous expansion against the case (Figure 10.6) of an expansion followed by a contraction.

Returning to the definition (40) of Ω, we see that the critical value $\Omega = 1$ corresponds physically to an equality between the present (average) mass density ρ_0 of the idealized homogeneous universe, and the quantity ρ_c defined by

$$\rho_c = \frac{3H_0^2}{8\pi G}, \tag{47}$$

homogeneous to a mass density, and called the *'critical density'*.

This critical density depends only on the value of the Hubble constant H_0. On adopting for H_0 the value 75 km s^{-1} Mpc^{-1}, which is equivalent to 2.5×10^{-18} c.g.s., one finds that

$$\boxed{\rho_c \approx 1.1 \times 10^{-29}\,\text{g cm}^{-3}.} \tag{48}$$

Equations (40) and (47) allows us to rewrite the definition of Ω as

$$\Omega = \rho_0/\rho_c. \tag{49}$$

Thus, the mathematical discussion leading qualitatively to Figures 10.5 and 10.6 is physically equivalent to the following statements:

(1) If the present average mass density ρ_0 is less than (or equal to) the critical density ρ_c, then the distance between two given 'elements' of the cosmic fluid will increase forever (continuous expansion!)

This is easily explained, since with a sufficiently small mass density of the universe, the braking action of the mutual gravitational attraction will be insufficient to stop the expansion due to the initial explosion.

(2) If the present average mass density ρ_0 is greater than the critical density ρ_c, then the distance between two given elements of the cosmic fluid will vary as represented schematically in Figure 10.6, i.e. pass through a maximum at time $t_{\max}$ and then undergo a decrease (contraction of the universe!), up to zero at time t_1.

This again is easy to explain: with a sufficiently great mass density of the universe, the braking action of the mutual gravitational attraction will be able first to stop at some time the initial expansion and then produce a 'gravitational collapse'.

Similarly the gravitational attraction of the Earth obliges a cannon bullet fired vertically with a velocty below the 'escape velocity' to fall back on the Earth: this represents the gravitational collapse of the system Earth–bullet.

5.5.1. Remarks

5.5.1.1. Kepler's third empirical (and purely kinematical) law was stating the constancy of the ratio T^2/d^3 for *any* planet (of orbital period T and of semimajor axis d) of the solar system:

$$T^2/d^3 = \text{const.} = C_k^{-1}, \tag{50}$$

where the subscript k means 'keplerian'.

We have already shown, in Chapter 8, how the joined application of classical dynamics and of his gravitational theory enabled Newton to find the expression for Kepler's empirical constant C_k in terms of dynamical factors such as the mass $M_\odot$ of the Sun and the gravitational constant G, viz. (see Subsection A2 of Chapter 8):

$$C_k^{-1} \approx \frac{4\pi^2}{GM_\odot}. \tag{51}$$

One can consider that in some way Hubble's empirical (and purely kinematical) law (12), written as

$$|v(P, t_0)|/R_0 = \text{const.} = H_0, \tag{51'}$$

corresponds to Kepler's third empirical law (50), when it states that the ratio of the radial velocity of *any* galaxy to its distance, observed at time t_0, remains constant, and equal to H_0, for all galaxies of the universe.

As a matter of fact, the solution $a(t)$ of Equation (41), to be given explicitly in Subsection 5.8, obviously depends, through the parameter Ω, on the 'dynamical factor' ρ_0. Thus, to some extent the treatment leading to Equation (41), based on classical dynamics and on the Newton's theory of gravitation, has a physical meaning similar to Newton's treatment leading to the Newtonian form of Kepler's third law.

However, the analogy is not full, since the Newtonian approach to cosmology is not able to give the value of H_0 as a function of ρ_0, because of the dependence on Ω of the expression for H_0 as a function of ρ_0, given by Equation (40), viz.

$$H_0 = (8\pi G\rho_0/3\Omega)^{1/2}. \tag{51''}$$

5.5.1.2. In the case ($\Omega > 1$) of an expanding universe followed by a contraction it is usual to speak about an '*oscillating universe*'; but, of course, the theory can predict for the moment only *one* 'oscillation'; we do not know what happens after the return to the state of infinite condensation.

However, as noted by Peebles (1969, p. 11) there is at least the possibility that the matter can collect itself into a new expanding phase.

5.6. QUANTITATIVE DISCUSSION OF THE PARTICULAR CASE $\Omega \Rightarrow 1$. THE CORRESPONDING AGE OF THE UNIVERSE

The particular case when the present mass density ρ_0 is equal to the critical density ρ_c, i.e. the case $\Omega = 1$, is theoretically very important because it corresponds to a kind of 'standard model', called the 'Einstein–de Sitter (or the EdS) model', in the relativistic treatment.

In this case, the Friedmann universe undergo a continuous expansion. Therefore we must take the square root of the corresponding form of Equation (41) with a positive sign. Hence

$$\frac{da}{dt} = H_0/a^{1/2}, \tag{52}$$

or

$$a^{1/2}da = H_0\,dt. \tag{52'}$$

If, in contradiction with the nowadays current view that the 'early universe', near the time $t = 0$ of the big-bang, was dominated not by matter but by radiation (see Section 7) we assume that the Friedmann model can be applied as early as $t = 0$, with the boundary condition $a(0) = 0$, we can integrate Equation (52′) and obtain:

$$\tfrac{2}{3}a^{3/2} = H_0 t, \tag{53}$$

or

$$a = (\tfrac{3}{2}H_0)^{2/3}t^{2/3}. \tag{53'}$$

Let us consider, as usual, the time elapsed from the big-bang up to the present time t_0 as the *'age of the universe'*. Then, on the very questionable (and purely academic) assumption that the Friedmann model can be applied throughout from the present time back to the time of the big-bang ($t = 0$), the age of the universe would be given by the value of t_0 yielded by Equation (53), viz. (since $a_0 = 1$), by

$$t_0 = \frac{2}{3}\frac{1}{H_0}a_0^{3/2} = \frac{2}{3}\frac{1}{H_0}. \tag{54}$$

With this value of t_0 we can put Equation (53′) into a more elegant form (proper to the EdS model):

$$a(t) = \left(\frac{t}{t_0}\right)^{2/3}. \tag{55}$$

If we take for H_0 our 'standard value' of 75 km s^{-1} Mpc^{-1}, the corresponding value of $(1/H_0)$, called the *'Hubble time'*, will be given (since 1 yr $\approx 30 \times 10^6$ s) by:

$$\boxed{\frac{1}{H_0} \approx 13 \times 10^9 \text{yr}.} \tag{56}$$

Thus, in the present case the age of the universe would be equal to

$$t_0 \approx 8.7 \times 10^9 \text{yr}, \tag{57}$$

different from the 'Hubble time'!

If we take into account the invariance of ρa^3 expressed by Equation (26′), viz.

$$\rho(t) = \rho_0 a^{-3}, \tag{58}$$

we obtain, from Equation (53′):

$$\rho(t) = \rho_0(\tfrac{9}{4}H_0^2t^2)^{-1}. \tag{59}$$

In our particular case $\rho_0 = \rho_c$. Thus, replacing ρ_c by its value given by Equation (47) we find the very simple result:

$$\rho(t) = \frac{1}{6\pi G}t^{-2} \approx (8 \times 10^5)t^{-2} \text{g cm}^{-3}. \tag{60}$$

The apparent independence from ρ_0 of this expression for $\rho(t)$ is somewhat misleading, since it was obtained on assuming a particualr value, ρ_c, for ρ_0.

5.7. THE RELATIVISTIC FORM OF THE FUNDAMENTAL EQUATION. THE ABSOLUTE SCALE FACTOR

Let us now return to the general case of *any* positive value (except $\Omega = 1$) of the parameter Ω.

In the preceding subsections the evolution of the local properties of the Friedmann universe was expressed in terms of the time variations of the scale factor a.

However, instead of eliminating (as in Subsection 5.4) the arbitrary radius R_0 of our 'typical cell' (S) at time t_0, we can take advantage of the very arbitrariness of the value of R_0 and fix its value by some '*normalizing relation*', such as

$$\boxed{R_0^2 H_0^2(\Omega - 1) = kc^2,} \tag{61}$$

where c is the speed of light in vacuum, whereas $k = +1$ if $\Omega > 1$ and $k = -1$ if $\Omega < 1$.

The purpose of this normalization is first of all to privde a link between the Newtonian approach and the relativistic treatment of Chapter 12. Moreover, it also provides a considerable simplification of some important relations in Subsections 5.8 and 5.9.

The normalizing relation (61) is dimensionally homogeneous since Ω and k are dimensionless, and since, according to Hubble's law (12), $R_0 H_0$ has the dimension of a velocity, like c.

The quantity R normalized by Equation (61) will be called the '*absolute scale factor*'.

It will appear in Chapter 12 that in the relativistic treatment the absolute scale factor represents when $\Omega > 1$ the *radius of curvature of the universe* ($k = +1$) of a positive curvature.

Instead of describing the evolution of the Friedmann universe in terms of the function $a(t)$, we can describe it henceforth in terms of the time variation of the absolute scale factor $R(t)$.

Let us now return to Equation (41), viz.

$$\left(\frac{da}{dt}\right)^2 = H_0^2\left[\frac{\Omega}{a} - (\Omega - 1)\right]. \tag{62}$$

On multiplying this equation by R_0^2 and on taking into account Equation (38) we obtain:

$$\left(\frac{dR}{dt}\right)^2 = R_0^2 H_0^2\left[\frac{\Omega R_0}{R} - (\Omega - 1)\right]. \tag{63}$$

On replacing $R_0^2 H_0^2(\Omega - 1)$ by the right-hand side of Equation (61), we find

$$\left(\frac{dR}{dt}\right)^2 = c^2\left[\frac{\Omega R_0 k}{(\Omega - 1)}\frac{1}{R} - k\right]. \tag{64}$$

Let us now introduce the quantity $\tilde{R}$ defined by

$$\tilde{R} = \frac{\Omega R_0 k}{(\Omega \quad 1)}. \tag{65}$$

In the case $k = +1$, i.e. the case $\Omega > 1$, of an 'oscillating universe' the physical meaning of $\tilde{R}$ is particularly simple. Indeed, according to Equations (38), (45) and (65) it represents the value of the maximum of $R(t)$ corresponding to $a_{\max}$, viz.

$$\tilde{R} = R_{\max} = \frac{\Omega R_0}{\Omega - 1} = \frac{c\Omega}{H_0(\Omega - 1)^{3/2}} \quad (\text{if } \Omega > 1). \tag{65'}$$

In the case $k = -1$, i.e. the case $\Omega < 1$, $\tilde{R}$ reduces to the positve quantity:

$$\tilde{R} = R^* = \frac{\Omega R_0}{1-\Omega} = \frac{c\Omega}{H_0(1-\Omega)^{3/2}} \quad (\text{if } \Omega < 1). \tag{65''}$$

Thus Equation (64) finally reduces, according to the sign of k, or of $(\Omega - 1)$, to one of the following very simple differential equations:

$$\left(\frac{dR}{dt}\right)^2 = c^2\left(\frac{R_{\max}}{R} - 1\right) \quad (\text{if } \Omega > 1, \text{ or } k = +1), \tag{66}$$

or

$$\left(\frac{dR}{dt}\right)^2 = c^2\left(\frac{R^*}{R} + 1\right) \quad (\text{if } \Omega < 1, \text{ or } k = -1). \tag{66'}$$

The difference between the case (66) of an 'oscillation' and the case (66′) of a continuous expansion appears quite clearly on these equations, which are even more explicit than Equation (41).

5.8. The Parametric Form of the Time Variation of the Absolute Scale Factor, and the Corresponding Age, of the 'Oscillating' Friedmann Universe

As shown by Friedmann, the solution of the differential Equation (66) of the 'oscillating universe', with the boundary condition $R(0) = 0$, can be represented parametrically by the pair of the following functions of an auxiliary variable η:

$$R[\eta] = \tfrac{1}{2}R_{\max}(1 - \cos\eta); \tag{67}$$

$$t[\eta] = \frac{1}{2c}R_{\max}(\eta - \sin\eta), \tag{68}$$

where we write $R[\eta]$ for R considered as a function of η, in order to reserve the notation $R(t)$ for R considered as a function of t; with the same notation for $t[\eta]$.

Indeed, we immediately find that for $\eta = 0$ the boundary condition $R(0) = 0$ is satisfied. Moreover, on taking the derivative with respect to η we obtain:

$$\frac{dR}{d\eta} = \frac{1}{2}R_{\max}\sin\eta = R_{\max}\sin\frac{\eta}{2}\cos\frac{\eta}{2}; \tag{69}$$

$$\frac{dt}{d\eta} = \frac{1}{2c}R_{\max}(1 - \cos\eta) = \frac{1}{c}R_{\max}\sin^2\frac{\eta}{2}. \tag{70}$$

Hence

$$\frac{dR}{dt} = \frac{dR/d\eta}{dt/d\eta} = \frac{c}{\tan\eta/2}. \tag{71}$$

On the other hand, Equation (67) yields:

$$\frac{R_{\max}}{R[\eta]} - 1 = \frac{1+\cos\eta}{1-\cos\eta} = \frac{\cos^2\dfrac{\eta}{2}}{\sin^2\dfrac{\eta}{2}} = \frac{1}{\tan^2\dfrac{\eta}{2}}. \tag{72}$$

These expressions obviously satisfy the differential equation (66).

Though this does not play any role in the physical interpretation of Equations (67) and (68), one can note that geometrically the curve representing R *versus* t is a *cycloid.*

Table 10.1, based on Equations (67) and (68), gives the variations of $R[\eta]$ and of $t[\eta]$, thus showing how R varies as a function of t.

Obviously, in Table 10.1, η_0 represents the value of η corresponding to t_0. The last line indicates the corresponding variations of the mass density $\rho(t)$, given by Equation (25″), viz.

$$\rho(t) = \frac{R_0^3\rho_0}{R^3(t)}. \tag{73}$$

TABLE 10.1

The variations of t, R and ρ (as a function of η) in the case of $\Omega > 1$ of an 'oscillating universe'

	Primordial explosion	Expansion	Present time	Expansion	Maxim. of Expansion	Contraction	Return to $\rho = \infty$
η	0	$\nearrow$	η_0	$\nearrow$	π	$\nearrow$	2π
t	0	$\nearrow$	t_0	$\nearrow$	$t_{\max}$	$\nearrow$	t_1
R	0	$\nearrow$	R_0	$\nearrow$	$R_{\max}$	$\searrow$	0
ρ	∞	$\searrow$	ρ_0	$\searrow$	$\rho_{\min}$	$\nearrow$	∞

The 'age of the universe', defined as in Subsection 5.6 (and with the same reserves), by the value of t_0, can be obtained as a function of H_0 and Ω(i.e. as a function of H_0 and ρ_0) in the following way:

For $t = t_0$ Equations (67) and (68) give:

$$R_0 = \tfrac{1}{2}R_{\max}(1-\cos\eta_0), \tag{74}$$

or

$$R_0 = R_{\max}\sin^2\frac{\eta_0}{2}; \tag{74'}$$

and

$$t_0 = \frac{1}{2c}R_{\max}(\eta_0 - \sin\eta_0), \tag{75}$$

or

$$t_0 = \frac{1}{c}R_{\max}\left(\frac{\eta_0}{2} - \sin\frac{\eta_0}{2}\cos\frac{\eta_0}{2}\right). \tag{75'}$$

On taking into account the value of $R_0/R_{\max}$ given by Equation (65′), viz.

$$\frac{R_0}{R_{\max}} = \frac{\Omega - 1}{\Omega}, \tag{76}$$

Equation (74′) yields:

$$\sin\frac{\eta_0}{2} = \left(\frac{\Omega - 1}{\Omega}\right)^{1/2}. \tag{77}$$

Hence

$$\frac{\eta_0}{2} = \sin^{-1}\left(\frac{\Omega - 1}{\Omega}\right)^{1/2}; \tag{78}$$

and

$$\cos\frac{\eta_0}{2} = \left(1 - \frac{\Omega - 1}{\Omega}\right)^{1/2} = \frac{1}{\Omega^{1/2}}. \tag{79}$$

When we substitute these values into Equation (75′), and replace $R_{\max}$ by one of its values (65′), viz.

$$R_{\max} = \frac{c\Omega}{H_0(\Omega - 1)^{3/2}}, \tag{80}$$

we find, on introducing the notation

$$\boxed{t_0 = f(\Omega)H_0^{-1},} \tag{81}$$

due to Zel'dovich:

$$f(\Omega) = \frac{\Omega}{(\Omega - 1)^{3/2}}\left[-\frac{(\Omega - 1)^{1/2}}{\Omega} + \sin^{-1}\left(\frac{\Omega - 1}{\Omega}\right)^{1/2}\right]. \tag{82}$$

This gives t_0 as a function of H_0 and ρ_0, since by Equation (40), viz.

$$\Omega = \frac{8\pi G\rho_0}{3H_0^2}, \tag{83}$$

Ω is a function of H_0 and ρ_0.

5.8.1. *Remarks*

5.8.1.1. Equation (82) can be given a slightly different form in the following way: By Equations (74) and (76) we have

$$\cos\eta_0 = 1 - \frac{2R_0}{R_{\max}} = \frac{2 - \Omega}{\Omega}. \tag{84}$$

Hence

$$\sin\eta_0 = (1 - \cos^2\eta_0)^{1/2} = \frac{2(\Omega - 1)^{1/2}}{\Omega}, \tag{85}$$

and

$$\eta_0 = \sin^{-1}\left[\frac{2(\Omega - 1)^{1/2}}{\Omega}\right]. \tag{86}$$

When we substitute these values into Equation (75), and replace R_{max} by its value (80), we find

$$f(\Omega) = \frac{\Omega}{(\Omega-1)^{3/2}}\left\{-\frac{(\Omega-1)^{1/2}}{\Omega} + \tfrac{1}{2}\sin^{-1}\left[\frac{2(\Omega-1)^{1/2}}{\Omega}\right]\right\}. \tag{87}$$

Of course, the identity of Equations (82) and (87) corresponds to the purely trigonometric unusual identity:

$$\sin^{-1}\left(\frac{\Omega-1}{\Omega}\right)^{1/2} = \tfrac{1}{2}\sin^{-1}\left[\frac{2(\Omega-1)^{1/2}}{\Omega}\right]. \tag{88}$$

5.8.1.2. The value of R_{max} given by Equation (80) can be transformed by replacing Ω according to Equation (83), viz.

$$\Omega = \frac{8\pi G\rho_0}{3H_0^2}, \tag{89}$$

and by taking into account Equation (61) $k = 1$, viz.

$$\Omega - 1 = \frac{c^2}{R_0^2 H_0^2}. \tag{90}$$

We thus find:

$$R_{max} = \frac{c\Omega R_0^3 H_0^3}{H_0 c^3} = \frac{\Omega R_0^3 H_0^2}{c^2} = \frac{8\pi G}{3c^2}\rho_0 R_0^3. \tag{91}$$

5.8.1.3. The time t_{max} of the maximum of $R(t)$ is obtained on taking $\eta = \pi$ in Equation (68). We thus obtain:

$$t_{max} = \frac{1}{2}\frac{\pi}{c}R_{max}, \tag{92}$$

or, taking into account Equation (80):

$$t_{max} = \frac{\pi\Omega}{2H_0(\Omega-1)^{3/2}}. \tag{93}$$

On the other hand, on replacing in Equation (92) R_{max} by its expression (91) we find:

$$t_{max} = \frac{8\pi^2 G}{6c^3}\rho_0 R_0^3. \tag{94}$$

5.8.1.4. The return to a state of infinite condensation at time t_1 corresponds to $\eta = 2\pi$. This gives $R = 0$ and $t_1 = (\pi/c)R_{max} = 2t_{max}$.

5.8.1.5. The value of ρ_{min} corresponding to the maximum R_{max} of $R(t)$ will be given by Equation (73), on replacing R_{max} by its value (65′), viz.

$$R_{max} = R_0\frac{\Omega}{\Omega-1}. \tag{95}$$

We thus find:

$$\rho_{\min} = \rho_0\left(1 - \frac{1}{\Omega}\right)^3 \quad (\text{if } \Omega > 1). \tag{96}$$

5.8.1.6. When, according to Equation (44), the parameter Ω is replaced by $2q_0$ defined in 5.4.1 – as done, for instance, by Weinberg (1972, p. 482) – the elimination of $R_{\max}$ by its value (65′) allows one to transform the Friedmann Equations (67) and (68) into (much more complicated) relations:

$$R[\eta] = \frac{q_0 R_0}{2q_0 - 1}(1 - \cos\eta); \tag{67′}$$

$$t[\eta] = \frac{q_0}{H_0}(2q_0 - 1)^{-3/2}(\eta - \sin\eta). \tag{68′}$$

5.9. The Parametric Form of the Time Variation of the Absolute Scale Factor, and the Corresponding Age, of an Ever Expanding Friedmann Universe

In the case $\Omega < 1$, $(k = -1)$, of the Friedmann model in continuous expansion, described by Equation (66′) with the boundary condition $R(0) = 0$, the solution is very similar to the one found in Subsection 5.8, the main difference lying in a suitable replacement of trigonometric functions by the corresponding hyperbolic functions. (N.B. In order to facilitate the comparison between the present subsection and Subsection 5.8, all numbers of the *corresponding* Equations present a constant *difference of 30 units*!)

With the same notation as in Subsection 5.8 the parametric form of the solution is given by

$$R[\eta] = \tfrac{1}{2}R^*(\cosh\eta - 1); \tag{97}$$

$$t[\eta] = \frac{1}{2c}R^*(\sinh\eta - \eta). \tag{98}$$

Indeed, on taking the derivative with respect to η we obtain:

$$\frac{dR}{d\eta} = \frac{1}{2}R^* \sinh\eta; \tag{99}$$

$$\frac{dt}{d\eta} = \frac{1}{2c}R^*(\cosh\eta - 1). \tag{100}$$

Hence

$$\frac{dR}{dt} = \frac{2c\sinh\frac{\eta}{2}\cosh\frac{\eta}{2}}{2\sinh^2\frac{\eta}{2}} = \frac{c}{\tanh\frac{\eta}{2}}. \tag{101}$$

On the other hand, Equation (97) yields:

$$\frac{R^*}{R[\eta]} + 1 = \frac{\cosh \eta + 1}{\cosh \eta - 1} = \frac{\cosh^2 \dfrac{\eta}{2}}{\sinh^2 \dfrac{\eta}{2}} = \frac{1}{\tanh^2 \dfrac{\eta}{2}}. \tag{102}$$

These expressions obviously satisfy the differential equation (66′).

Table 10.2, based on Equations (97) and (98), gives the variations of $R[\eta]$ and of $t[\eta]$, thus showing how R varies as a function of t.

Obviously, in Table 10.2, η_0 represents again the value of η corresponding to t_0. The last line indicates the corresponding variations of the mass density given by Equation (25″), viz.

$$\rho(t) = \frac{R_0^3 \rho_0}{R^3(t)}. \tag{103}$$

TABLE 10.2

The variations of t, R and ρ (as a function of η) in the case $\Omega \leqslant 1$ of an ever expanding universe

	Primordial explosion	Expansion	Present time	Expansion	Infinite time
η	0	↗	η_0	↗	∞
t	0	↗	t_0	↗	∞
R	0	↗	R_0	↗	∞
ρ	∞	↘	ρ_0	↘	0

The value t_0 of the ‘age of the universe’ is easily obtained as a function of H_0 and Ω (i.e. as a function of H_0 and ρ_0) by the following succession of Equations, given without comments because of the similarity with the treatment of Subsection 5.8:

$$R_0 = \tfrac{1}{2}R^*(\cosh \eta_0 - 1) = R^* \sinh^2 \frac{\eta}{2}, \tag{104}$$

$$t_0 = \frac{1}{2c} R^*(\sinh \eta_0 - \eta_0) = \frac{R^*}{c}\left(\sinh \frac{\eta_0}{2} \cosh \frac{\eta_0}{2} - \frac{\eta_0}{2}\right), \tag{105}$$

$$\frac{R_0}{R^*} = \frac{1-\Omega}{\Omega}, \tag{106}$$

$$\sinh \frac{\eta_0}{2} = \left(\frac{1-\Omega}{\Omega}\right)^{1/2}. \tag{107}$$

$$\frac{\eta_0}{2} = \sinh^{-1}\left(\frac{1-\Omega}{\Omega}\right)^{1/2}, \tag{108}$$

$$\cosh \frac{\eta_0}{2} = \Omega^{-1/2}. \tag{109}$$

$$R^* = \frac{c\Omega}{H_0(1-\Omega)^{3/2}}, \tag{110}$$

$$t_0 = f(\Omega)H_0^{-1}, \tag{111}$$

$$f(\Omega) = \frac{\Omega}{(1-\Omega)^{3/2}}\left[\frac{(1-\Omega)^{1/2}}{\Omega} - \sinh^{-1}\left(\frac{1-\Omega}{\Omega}\right)^{1/2}\right]. \tag{112}$$

5.9.1. Remark

The relative probability of different values of $\Omega = \rho_0/\rho_c$. will be thoroughly discussed in Chapter 13. However, the case $\Omega < 1$, considered in the present subsection, is rather probable, since ρ_0, given by Equation (22), viz. $\rho_0 \approx 3 \times 10^{-31}$ c.g.s., is much lower than ρ_c given by Equation (48), viz. (for $H_0 = 75\ \text{km s}^{-1}\ \text{Mpc}^{-1}$) $\rho_c \approx 10^{-29}$ c.g.s. .

Of course, this conclusion could be modified if some 'missing mass' were to be found (in a hitherto hidden form) outside the normal galaxies.

6. The Ratio of the Age of the Friedmann Universe to the 'Hubble Time' as a Function of Ω

Taking into account the results expressed by Equations (54), (82) and (112), the ratio $f(\Omega) = t_0/H_0^{-1}$ of the age t_0 of the Friedmann universe to the 'Hubble time' H_0^{-1} has the values given by Table 10.3.

TABLE 10.3
The numerical values of $f(\Omega) = t_0/H_0^{-1}$ and of t_0 as a function of $\Omega = \rho_0/\rho_c$

Ω	0	0.03	1.00	2.00	6.00
$f(\Omega)$	1	0.96	0.67	0.57	0.40
t_0 yr	13×10^9	12.5×10^9	8.7×10^9	7.5×10^9	5.2×10^9

The last line of Table 10.3 gives the values of t_0 in years, corresponding to $H_0 = 75$ km $\text{s}^{-1}\ \text{Mpc}^{-1}$.

Table 10.3 shows that the value $f(\Omega) = 1$, i.e. $t_0 = 1/H_0$, often given in popular books for the age of the Friedmann universe, is conceptually incorrect (in addition to objections already presented in Subsection 5.6) since $f(\Omega)$ is always different from 1 (except for the quite impossible limiting value $\Omega = 0$).

On the other hand, as noted by Harrison (1973, p. 162), the 'Hubble time' H_0^{-1} plays physically more important role than the age t_0 of the Friedmann universe (simply because H_0 represents an observable quantity, whereas t_0 is a purely theoretical parameter).

7. The Radiation Model of the Early Universe

7.1. Introduction

As explained by Harrison (1973, p. 166) the 'standard model of the early universe' shows that at a temperature of about 4×10^9 K the universe was predominantly composed of photons, non-interacting neutrinos and a small admixture of protons, electrons and slowly decaying neutrons.

It is therefore usually assumed that, during this *'radiation-dominated era'*, the *'radiation model'* of the universe, considered as composed simply of *a gas of photons*, represents an acceptable first approximation, up to the beginning of the *'matter-dominated era'*, reasonably described by the Friedmann model.

This matter-dominated era seems to have started when the temperature dropped below a few thousands (about 3000) K, as a consequence of the expansion (this consequence will be explained quantitatively in Section 7.3).

7.2. The Fundamental Invariant of the Radiation Model

Let $u(t)$ denote, as usual, the (non-directional) energy density of the radiation field (we drop here the subscript 'rad' introduced in Section 4, since no confusion can arise in the present section dealing only with radiation).

If $U(t)$ represents the total (internal) energy of photons contained in a 'typical (spherical) cell' (S) of radius $R(t)$ and of volume $V(t)$ at time t, we obviously have

$$U(t) = u(t)V(t). \tag{113}$$

On the other hand, the radiation pressure $P(t) = P_{\text{rad}}$ created by the radiation field is given (see Chapter 1, Subsection 4.4) by

$$P(t) = P_{\text{rad}}(t) = \tfrac{1}{3}u(t). \tag{114}$$

From the thermodynamic identity:

$$dU = -P\,dV, \tag{115}$$

where dU is the decrease in the internal energy of (S), and $P\,dV$ is the work done in expanding against the pressure P ($dQ = 0$ since the expansion is supposed to be adiabatic).

On dividing Equation (115) by dt, and on taking into account Equations (113) and (114), we obtain:

$$\frac{d(uV)}{dt} = -\frac{u}{3}\frac{dV}{dt}. \tag{116}$$

Since $V = (4\pi/3)R^3$, this is equivalent to

$$\frac{d(uR^3)}{dt} = -\frac{u}{3}\frac{d(R^3)}{dt}. \tag{117}$$

This differential equation is easily shown to be satisfied by the *'fundamental invariant'*:

$$u(t)R^4(t) = u(t_0)R^4(t_0) = u_0R_0^4 = \text{const.} = A. \tag{118}$$

where, *exceptionally* (as *in the rest of the Section* 7) the subscript (0) does *not* corresspond to the present time (since the present time lies outside the range of the radiation model) but to some given time inside the range of the radiation model.

Indeed, if this relation (118) is supposed to hold, we can write Equation (117) in the following form:

$$\frac{d(A/R)}{dt} = -\frac{1}{3}\frac{A}{R^4}\frac{d(R^3)}{dt} = -\frac{A}{R^2}\frac{dR}{dt}, \tag{119}$$

which obviously represents an identity.

On introducing the mass density ρ corresponding, by Einstein's relation $\rho = u/c^2$, to the energy density u, the invariant (118) takes the form:

$$\rho R^4 = \rho_0 R_0^4. \tag{120}$$

Equation (120) can be used to show clearly that the mass M equivalent to the energy U of the photons contained in (S) *is not constant*, but is a function $M(t)$ of time t, given by

$$M(t) = \frac{U(t)}{c^2} = \frac{uV}{c^2} = \frac{u4\pi R^3}{c^2 3} = \frac{4\pi}{3}\frac{\rho R^4}{R} = \frac{4\pi\rho_0 R_0^4}{3R(t)}. \tag{121}$$

Thus, instead of remaining constant, as in the Friedmann model, the mass M of each cell of the radiation universe decreases as $1/R(t)$ when the universe expands.

7.3. Invariance of the Total Number of Photons and Decrease of the Temperature during the Radiation Era

If the 'mass' of an elementary cell (S) of the radiation model is variable, it is easy to show that the total number N_S of all photons inside the cell (S), and consequently *the total number N of photons of the the whole radiative universe*, remains invariable as long as no transformation of photons in other forms of energy takes place.

Indeed, if the radiation field is assimilated to that inside a black body at temperature T, its energy density per unit frequency interval near frequency ν, denoted by $u_\nu^*(T)$ in Chapter 1, will be given by Equation (44) of Chapter 1, viz.

$$u_\nu^*(T) = \frac{8\pi h\nu^3}{c^3}(e^{h\nu/kT} - 1)^{-1}. \tag{122}$$

Since the energy of each photon of frequency $(\nu, d\nu)$ is equal to $h\nu$, the corresponding number density $n_\nu^*(T)$ will be given by

$$n_\nu^*(T) = \frac{u_\nu^*}{h\nu} = \frac{8\pi\nu^2}{c^3}(e^{h\nu/kT} - 1)^{-1}. \tag{122'}$$

The integrated number density n of the field at time t, i.e., according to the terminology defined in Chapter 1, Subsection 1.1.1.2, the number density of photons of all frequencies, will thus be given by

$$n = n^*(T) = \int_0^\infty n_\nu^*(T)\,d\nu = CT^3, \tag{123}$$

an Equation obviously equivalent to

$$n = n(T) = n(T_0)T^3/T_0^3 = n_0 T^3/T_0^3, \tag{123'}$$

where, on introducing $x = h\nu/kT$, the constant C is given by

$$C = \frac{8\pi k^3}{c^3 h^3} \int_0^\infty \frac{x^2}{e^x - 1} dx. \tag{124}$$

On the other hand, the energy density u at time t is given by the classical expression for the integrated (non-directional) energy density $u^*(T)$, (see Chapter 1, Subsection 4.2), viz.

$$u = u^*(T) = aT^4, \tag{125}$$

when T is replaced by the function $T(t)$ which describes the time variation of the temperature.

On taking into account the invariance of uR^4 expressed by Equation (118), we thus find (since a is a constant equal to 7.56×10^{-15} c.g.s.):

$$T(t)R(t) = T(t_0)R(t_0) = T_0 R_0 = \text{const.}, \tag{126}$$

or

$$T(t) = \frac{T_0 R_0}{R(t)} = \frac{\text{const.}}{R(t)}. \tag{126'}$$

This shows that, as announced above, the expansion of the early universe must necessarily produce a decrease of its temperature.

On inserting the expression (126′) for T into Equation (123′) we obtain the value of the number density n at time t:

$$n = n(t) = \frac{n_0 R_0^3}{R^3}. \tag{127}$$

This shows that the total number $N_S(t)$ of all photons present in (S) at any time t is *constant* since it is given by $n(4\pi/3)R^3$, i.e. by

$$N_S(t) = \frac{n_0 R_0^3}{R^3} \frac{4\pi R^3}{3} = \frac{4\pi R_0^3}{3} n_0 = V_0 n_0 = N_S(t_0) \tag{128}$$

Hence, of course, the constance of the total number N of photons of the whole radiative universe, announced above.

7.4. THE FUNDAMENTAL EQUATION

Let us neglect, as in Subsection 5.4, the difficulties concerning the dynamical nature of the reference system and the application of Newton's theorem.

And let us apply once more the 'fundamental equation (32), viz.

$$\frac{1}{2}\left(\frac{dR}{dt}\right)^2 + \left(-\frac{GM}{R}\right) = \text{const.}, \tag{129}$$

though it can no longer be proved by Newtonian approach, as in the case $P = 0$. On doing so we assume, quite exceptionally without proof, that being an *energy law*, the fundamental equation must be more general than the law of motion. The actual proof demands a detailed treatment, based on the general relativity. Moreover, we use this extrapolation of the 'fundamental equation' solely in the present subsection, so that our 'act of faith' does not affect the remainder of our textbook.

Many authors using the Newtonian approach give different pseudo-proofs based simply on elementary but still general relativistic (and even less intuitive) relations. However, such an approach is even less satisfactory than our's, since it introduces the general relativity *implicitly*. See, for instance, McCrea (1951), Zel'dovich (1965, p. 258), Harrison (1965, p. 441) or Peebles (1969, p. 8).

On applying Equation (129) to the case when Λ is still zero, as in the Friedmann model, but the pressure $P = P_{rad}$ is no longer negligible, we must take into account the time variation of the mass M of the cell (S) described by Equation (121), viz.

$$M(t) = \frac{4\pi}{3}\frac{\rho_0 R_0^4}{R(t)}. \tag{130}$$

Hence the differential equation:

$$\left(\frac{dR}{dt}\right)^2 = \frac{8\pi G}{3}\frac{\rho_0 R_0^4}{R^2} + \text{const.}. \tag{131}$$

On introducting the values corresponding to t_0 (some given time inside the interval of time corresponding to the radiation era, and not the present time!) we obtain:

$$(\dot{R})^2 = \frac{8\pi G}{3}\left(\frac{\rho_0 R_0^4}{R^2} - \rho_0 R_0^2\right) + (\dot{R}_0)^2. \tag{132}$$

This equation is satisfied, just as the corresponding equation for the Friedmann model, by solutions of the form (2), viz.

$$R(t) = R_0 a(t), \tag{133}$$

with $a_0 = 1$.

On substituting $R(t)$ defined by Equation (133) into Equation (132), R_0 cancels out and one is left with a differential equation to be satisfied by the scale factor $a(t)$:

$$(\dot{a})^2 = \frac{8\pi G}{3}\rho_0\left(\frac{1}{a^2} - 1\right) + (\dot{a}_0)^2. \tag{134}$$

This equation is very similar, but not identical, to Equation (39) describing the evolution of the scale factor $a(t)$ of the Friedmann universe.

On introducing the following definitions, which must not be confused with similar definitions corresponding to the case when t_0 represents the present time, viz. the definitions given by Equations (9′) and (40):

$$H_0' = \dot{a}_0; \qquad \Omega' = \frac{8\pi G}{3}\rho_0\frac{1}{H_0'^2}, \tag{135}$$

we can transform Equation (134) into

$$\left(\frac{da}{dt}\right)^2 = H_0'^2\left[\frac{\Omega'}{a^2} - (\Omega' - 1)\right]. \tag{136}$$

7.4.1. Remark.

The differentiation of both sides of Equation (136) with respect to time yields

$$\ddot{a} = \frac{d^2a}{dt^2} = -H_0'^2\Omega' a^{-3}. \tag{137}$$

Since $a_0 = 1$, this equation yields the value (for the radiation model) of the deceleration parameter q_0 defined by Equation (42), viz.

$$q_0 = -\ddot{a}_0/(\dot{a}_0)^2 = \Omega'. \tag{137'}$$

7.5. THE TIME VARIATION OF THE SCALE FACTOR AND OF THE DENSITY NEAR THE BIG-BANG

Since we are interested, for the radiative model, only in the early stages of the universe, i.e. in solutions corresponding to very small values of $a(t)$, we can neglect (under certain conditions examined below) the term $(\Omega' - 1)$ in the square brackets of Equation (136).

Thus, near $t = 0$, which [through the boundary condition $a(0) = 0$] corresponds to very great values of $1/a$, Equation (136) reduces to

$$\frac{da}{dt} \approx \frac{H_0'\Omega'^{1/2}}{a}, \tag{138}$$

since only the positive square root suits to an expansion.

On integrating, with the boundary condition recalled above, we obtain:

$$a^2 \approx 2H_0'\Omega'^{1/2}t, \tag{139}$$

or

$$a \approx (2H_0')^{1/2}\Omega'^{1/4}t^{1/2} \approx (\text{const.})t^{1/2}. \tag{139'}$$

It is now easy to find the corresponding law of variation of the density $\rho(t) = u(t)/c^2$. Indeed, the invariance of ρR^4 can be expressed by

$$\rho = \rho_0 a^{-4}, \tag{140}$$

hence, on replacing a^2 by its value (139) and Ω' by its expression (135):

$$\rho(t) \approx \frac{3}{32\pi G}t^{-2} \approx (4.5 \times 10^5)t^{-2}\,\text{c.g.s.}. \tag{141}$$

We see that in this relation all primed quantities cancel out, in spite of the fact that (contrary to the case $\Omega = 1$ of the Friedmann model) no assumption was made concerning the value of Ω'.

Comparison with Equation (60) shows that the function $\rho(t)$ has, for the radiation model and small values of t, the same form as for the $\Omega = 1$ case of the Friedmann model. The only difference lies in the numerical value of the coefficient of t^{-2}.

On taking into account Equation (139′) the (purely academic) limit of validity of our approximation $|(\Omega' - 1)| \ll \Omega'/a^2$ is easily shown to be

$$t \ll \frac{\Omega'^{1/2}}{2H_0'|(\Omega' - 1)|}. \tag{142}$$

References

1. WORKS CITED IN THE TEXT
(Works that can be used for further study are marked by an asterisk.)

Bondi, H.: 1952, *Cosmology*, Univ. Press, Cambridge, G.B. (2nd Edn.: 1960, Univ. Press, Cambridge).
McCrea, W. H. and Milne, E. A.: 1934, 'Newtonian Universes and the Curvature of Space', *Quart, J. Math. (Oxford Ser.)* **5**, 73.
McCrea, W. H.: 1951, *Proc. Roy. Soc.* **206**, 562.
Friedmann, A.: 1922, *Z. Phys.* **10**, 377 and 1924, *Z. Phys.* **21**, 326.
*Harrison, E. R.: 1965, 'Cosmology without General Relativity', *Annals of Physics* **35**, 437.
*Harrison, E. R.: 1973, 'Standard Model of the Early Universe', *Ann. Rev. Astron. Astrophys.* **11**, 155.
Heckmann, O.: 1942, *Theorien der Kosmologie*, Springer, Berlin (Reprinted 1968).
Hubble, E.: 1929, *Proc. Nat. Acad. Sc.* **15**, 168.
Milne, E. A.: 1934, 'A Newtonian Expanding Universe', *Quart. J. Math. (Oxford Ser.)* **5**, 64.
Oort, J.: 1958, In *La Structure et l'Evolution de l'Univers*, Inst. Int. de Phys. Solvay, Stoops, Brussels, Belgium.
Peebles, P. J. E. and Partridge, R. B.: 1967, *Astrophys. J.* **148**, 713.
*Peebles, P. J. E.: 1969, 'Cosmology', *R.A.S.C. J.* **63**, 4.
*Reinhardt, M. von,: 1969, 'Neuere Probleme der Kosmologie', *Naturwiss.* **56** (12), 581.
Sandage, A.: 1972, *Astrophys. J.*, **172**, 253.
*Schücking, E. L.: 1967, *Lectures in Applied Math.* **8**, Publ. of Amer. Math, Soc., pp. 221–224.
Sciama, D. W.: 1971, *Modern Cosmology*, Univ. Press, Cambridge, G.B.
*Weinberg, S.: 1972, *Gravitation and Cosmology*, Wiley, New York.
Zel'dovich, Ya. B.: 1965, 'Survey of Modern Cosmology', In *Advances in Astronomy and Astrophysics*, Vol. 3, pp. 241–379.

2. REFERENCES FOR FURTHER STUDY
(See first the works cited in 1 marked by an asterisk.)

Harrison, E. R.: 1968, 'The Early Universe', *Physics Today*, June issue, p. 31.
Heckmann, O. and Schücking, E.: 1959, 'Newtonsche und Einsteinsche Kosmologie', in *Handbuch der Phys.* **53**, 489.
Kundt, W.: 1971, 'Survey of Cosmology', *Springer Tracts Modern Physics* **58**, 1.
Weinberg, S.: 1977, *The First Three Minutes*, Basic Books, New York.
Zel'dovich, Ya. B.: 1964, 'The Theory of the Expanding Universe as Originated by A. A. Friedmann', *Soviet Physics Uspekhi*, **6**, 475.

CHAPTER 11

BASIC CONCEPTS OF RELATIVISTIC COSMOLOGY

1. Introduction

The Newtonian approach, quite sufficient in a popularized account or in an elementary introduction, still leaves out many important problems and all properties of the universe *as a whole.*

Therefore, we introduce in this chapter some specifically relativistic concepts, all presented in such a way as to be accessible without any preliminary study of general relativity or tensor calculus.

On the other hand, a minimal preliminary knowledge of some aspects of the *special relativity* will be very useful. These can be found in Kourganoff (1964), in Kittel *et al.* (1962); or Smith (1965).

For moderately advanced readers, Section 2, where we simply recall some fundamental relativistic concepts, will replace these references.

Very advanced readers can start with Section 3.

2. Some Elementary Relativistic Concepts

2.1. Geometrization of Kinematics: the Space-Time

2.1.1. The Case of a One-Dimensional Motion

In order to introduce in a very simple way one of the main relativistic concepts, let us consider the motion of a small ball in a narrow tube.

This motion involves only *one space dimension* since the position P of the ball can be indicated by a single parameter: abscissa x_1 of P on an axis Ox_1 along the tube.

Let us consider for the moment only the *kinematical* properties of the motion, i.e. the time variations of the position of the ball, without paying attention to *dynamical parameters* (masses, forces) responsible for the particularities of the motion.

In classical mechanics, these kinematical particularities are essentially characterized by *analytical* properties of some function $x_1 = f(t)$ of the time t.

Of course, the analytical properties of this function can be *illustrated* by a diagram of x_1 *versus* t, similar to those used to control the railway traffic. On such a diagram the *time* t is considered as essentially different, as to its physical dimensions, from the *space parameter* x_1.

In the theory of relativity (either special or general), the time t is associated to a parameter x_0 defined by

$$x_0 = ct, \qquad (1)$$

where c is the speed of light in vacuum. Since the parameter x_0 has physically the dimensions of a *length*, the diagram of x_1 *versus* x_0 can be considered from a *geometrical* point of view, i.e. as a curve (C) in a plane referred to the coordinate system (x_1, x_0).

The curve (C) is a 'geometrized' representation of the history of successive 'events' formed by the passages of the ball at points P at time t. Each such event corresponds to some pair of quantities x_1 and t. It is thus represented by a point $\mathrm{M}(x_1, x_0)$ of (C) in the bi-dimensional mathematical 'space' (x_1, x_0) associated to the genuine *one-dimensional* physical space (the tube) through more or less artificial *addition* of the parameter x_0. The bi-dimensional mathematical space thus introduced is called the *'space-time'*.

In this way all kinematical characteristics of the motion are expressed by the corresponding *geometrical* properties of the curve (C), sometimes called the *'world line'* of the ball, (a rather misleading term, when used in cosmology).

Thus, for instance, when kinematically the motion is *uniform*, i.e. when the speed of the ball is constant, geometrically the curve (C) is a *straight line*. Similarly, when kinematically the motion is more or less *accelerated* (or decelerated), the curve (C) geometrically presents some positive (or negative) *curvature*: (C) is concave and its concavity is oriented in the positive direction of the $\mathrm{O}x_1$ axis (or in the opposite direction).

2.1.1.1. Remark. The correspondence between the kinematical properties of the motion and the geometrical properties of its representation in space-time is not as direct and simple as one could imagine at first sight.

Let us consider, for instance, a motion described kinematically by the function $x_1 = t^2$. The corresponding acceleration is constant and equal to $|\gamma| = +2$. In the space-time (x_1, x_0) this motion is described by a parabola (C), whose equation in analytical geometry is

$$x_1 = \frac{1}{c^2}(x_0)^2. \tag{2}$$

The radius of curvature ρ of (C) is given by the well known relation

$$\rho = \frac{[1 + (dx_1/dx_0)^2]^{3/2}}{|d^2x_1/dx_0^2|} = \frac{[1 + 4(x_0)^2/c^4]^{3/2}}{2c^{-2}}, \tag{3}$$

so that the 'curvature' σ of (C), defined as usual by $(1/\rho)$, is given by

$$\sigma = \frac{2c^{-2}}{[1 + 4(x_0)^2/c^4]^{3/2}}. \tag{4}$$

Equation (4) shows that, as obvious geometrically, the curvature σ of the parabola (C) is not the same at its different points M.

2.1.2. *The Case of a Tri-Dimensional Motion*

More generally, the relativistic treatment of tri-dimensional motions is associated with a similar geometrization of kinematics, obtained through the introduction of a four-dimensional 'space-time'.

This four-dimensional space-time is a mathematical 'space' built by an addition of the parameter x_0, defined again by Equation (1) to the three general space parameters (coordinates) called in relativity x_1, x_2, x_3 (rather than the Cartesian coordinates x, y, z or the spherical coordinates r, θ, φ). This notation is convenient for a general presentation of the main concepts, but other notation will be encountered later.

It is, of course, impossible to draw, even in perspective, four-dimensional curves. Therefore, the *geometrical* properties of the curves (C) describing the kinematical characteristics of tri-dimensional motions, must be investigated *analytically* by a proper generalization of analytical geometry, provided by the mathematical investigation of 'Riemannian spaces', or even more general 'spaces'.

2.2. TIME-LIKE INTERVAL. COPUNCTUALITY. PROPER TIME

The concept of *'interval'* is used extensively both in special relativity and general relativity. Its meaning and its main properties appear very clearly in the case of one-dimensional motion, like the one considered in Subsection 2.1.1.

2.2.1. *The Case of One-Dimensional Motion*

Let us consider two one-dimensional reference systems (or 'laboratories') (S) and (S′) in uniform relative motion: (S) glides over (S′), like a train (reduced to its sole length) gliding with a constant speed over straight rails.

Each laboratory has its own coordinates: x on some Ox axis of (S) and x' on some O′x′ axis of (S′). Moreover, in conformity with the theory of special relativity, each reference system has *its own 'time system'*: t in (S) and t' in (S′).

Thus each single event (E), characterized in (S) by the pair of values (x, t), will be characterized in (S′) by the corresponding pair (x', t').

The pairs (x, t) and (x', t') are connected by the relations of the 'Lorentz transformation', but this is unimportant for our present purpose.

Instead of considering one isolated event, such as (E), let us now consider two events (E_1) and (E_2).

If (x_1, t_1) and (x_1', t_1') characterize (E_1) in laboratories (S) and (S′) respectively; and if, similarly, (x_2, t_2) and (x_2', t_2') characterize (E_2), let $\Delta x = x_2 - x_1$ denote the *distance* in (S) and $\Delta t = t_2 - t_1$ denote the *interval of time* in (S) between (E_1) and (E_2).

(We do not use here the same notation as in Subsection 2.1, but the present notation suits better to description of the concept of 'interval' defined below.)

Both in special and in general relativity the quantity Δs defined, in the case $c^2(\Delta t)^2 \geqslant (\Delta x)^2$, by the positive square root of

$$(\Delta s)^2 = (\Delta t)^2 - c^{-2}(\Delta x)^2, \tag{5}$$

plays a very important role. It is called the *'interval'* between the events (E_1) and (E_2) or, more precisely (for reasons given below) the *'time-like interval'*.

This quantity Δs is *kinematically invariant* with respect to a *replacement of the reference system* (S) by the reference system (S′), or reciprocally. This statement means that Δs keeps its value when both Δx and Δt are replaced by the corresponding quantities relative to the system (S′), i.e. when Δx is replaced by $\Delta x' = x_2' - x_1'$ and Δt is replaced by $\Delta t' = t_2' - t_1'$. This invariance is thus expressed by

$$(\Delta s)^2 = (\Delta t)^2 - c^{-2}(\Delta x)^2 = (\Delta t')^2 - c^{-2}(\Delta x')^2. \tag{6}$$

The condition

$$c^2(\Delta t)^2 \geqslant (\Delta x)^2, \tag{7}$$

formulated above is physically very important: it means that Δs, defined by Equation (5), is a real quantity, with a clear physical significance, viz. that it is possible to find a *third* reference system $(\bar{\mathrm{S}})$, in uniform relative motion with respect to (S) or (S′), where $\Delta\bar{x} = 0$, i.e. where the distance $\Delta\bar{x} = \bar{x}_2 - \bar{x}_1$ between the events (E_1) and (E_2) is zero.

(One can say that the two events are *'copunctual'* with respect to the system $(\bar{\mathrm{S}})$, i.e. take place at *the same point* of the laboratory $(\bar{\mathrm{S}})$. This terminology, which is not quite usual but which is very convenient, will be used everywhere below.)

Indeed, we can express the invariance of Δs with respect to a replacement of (S) by $(\bar{\mathrm{S}})$ on writing

$$(\Delta s)^2 = (\Delta t)^2 - c^{-2}(\Delta x)^2 = (\Delta\bar{t})^2, \tag{8}$$

where

$$\Delta\bar{t} = \bar{t}_2 - \bar{t}_1. \tag{9}$$

This, of course, would not be possible physically if $(\Delta s)^2$ were not positive (or zero) i.e. if the condition (7) were not satisfied.

In the usual terminology of the special relativity, the interval of time $\Delta\bar{t}$ between two events (E_1) and (E_2) which take place *at the same point* of some laboratory $(\bar{\mathrm{S}})$ is called the *'interval of proper time'* between (E_1) and (E_2).

Since according to Equation (8) $\Delta s = \Delta\bar{t}$ we can interpret physically the interval Δs between any pair of events satisfying in some system (S) the condition (7) as the interval of proper time between the considered events. Hence, the term *'time-like interval'* Δs introduced above.

One must pay attention to the difference between the term *'interval'* used to denote Δs and the word interval used in the expression *'interval of time'*. The best way to avoid this type of confusion is to speak simply about Δs, or about the time-like Δs.

The very current term *'line element'* used to denote Δs is extremely unfortunate for different reasons. It suggests, indeed, that Δs represents some kind of 'distance' between the points M_1 and M_2 corresponding, in the space–time of (S) or in the space–time of (S′), to the events (E_1) and (E_2), in the geometrized description of motion.

However, if we return for a while to notation of Subsection 2.1.1, and use x_1 to denote the coordinate x and x_0 to denote ct of some general event (E) in (S), then $(\Delta s)^2$ relative to two events will be expressed in (S) by

$$(\Delta s)^2 = (\Delta t)^2 - c^{-2}(\Delta x)^2 = c^{-2}[(\Delta x_0)^2 - (\Delta x_1)^2], \tag{10}$$

which shows that $(\Delta s)^2$ is quite different from $[(\Delta x_1)^2 + (\Delta x_0)^2]$ which represents the square of the (Euclidean) distance between the two events in the space–time of (S), first of all because of the *minus sign* and next because of the factor $(1/c^2)$. The minus sign is usually taken into account by a (physically rather artificial) introduction of *non-Euclidian* geometry but the factor $(1/c^2)$ remains a very strong objection against the super-geometrized term 'line element'.

2.2.1.1. Remark. Note incidentally that we can suppose that the events (E_1) and (E_2), coponctual in $(\bar{\mathrm{S}})$, take place at the origin $\bar{\mathrm{O}}$ of the coordinates $\bar{x}$ of the reference system $(\bar{\mathrm{S}})$.

Then (Δx) represents the distance travelled in (S) by the point $\bar{O}$ during the time Δt of the laboratory (S), so that $(\Delta x/\Delta t)$ represents the constant speed v of $(\bar{S})$ with respect to (S).

On writing the relation (8) in the form

$$\Delta t = \frac{\Delta \bar{t}}{(1 - v^2/c^2)^{1/2}}, \tag{11}$$

we find the well known relation between the interval of proper time $\Delta \bar{t}$ and the corresponding interval of 'improper time' Δt in the system (S) moving with a constant speed v with respect to the system $(\bar{S})$ of 'copunctuality'.

Equation (11) expresses the so-called *'dilation of times'*.

2.2.2. *The General Case of a Tri-Dimensional Motion*

More generally, in the case of tri-dimensional motions, the reference system (S) of the one-dimensional case is replaced by a tri-dimensional one (S_3) characterized, for instance, by three Cartesian axes Oxyz and its time system t.

Similarly the reference system (S') is replaced by a tri-dimensional one (S'_3) characterized, for instance, by three Cartesian axes O′x′y′z′ and its time system t'.

The great similarity with the one-dimensional case is insured by the fact that (S'_3) is supposed to be in uniform *translation* with respect to (S_3), the axes O′x′, O′y′, O′z′ remaining respectively parallel to the axes Ox, Oy, Oz, and the axis O′x′ gliding with a constant speed v along the axis Ox.

Let us consider two events (E_1) and (E_2), whose respective coordinates and times are (x_1, y_1, z_1, t_1) and (x_2, y_2, z_2, t_2) in (S_3), whereas they are (x'_1, y'_1, z'_1, t'_1) and (x'_2, y'_2, z'_2, t'_2) in (S'_3).

If we use the notation, similar to that used in Subsection 2.2.1, viz.

$$\Delta x = x_2 - x_1; \quad \Delta y = y_2 - y_1; \quad \Delta z = z_2 - z_1; \quad \Delta t = t_2 - t_1 \tag{12}$$

and

$$\Delta x' = x'_2 - x'_1; \quad \Delta y' = y'_2 - y'_1; \quad \Delta z' = z'_2 - z'_1, \quad \Delta t' = t'_2 - t'_1, \tag{13}$$

then the square $(\Delta\lambda)^2$ of the distance $\Delta\lambda$ between (E_1) and (E_2) in (S_3) will be given by

$$(\Delta\lambda)^2 = (\Delta x)^2 + (\Delta y)^2 + (\Delta z)^2, \tag{14}$$

and the corresponding square $(\Delta\lambda')^2$ of the distance $\Delta\lambda'$ between (E_1) and (E_2) in (S'_3) will be given by

$$(\Delta\lambda')^2 = (\Delta x')^2 + (\Delta y')^2 + (\Delta z')^2. \tag{15}$$

The square $(\Delta s)^2$ of the time-like interval Δs will be defined in (S_3) by

$$(\Delta s)^2 = (\Delta t)^2 - c^{-2}(\Delta\lambda)^2, \tag{16}$$

where, as we see by comparison with Equation (5), $\Delta\lambda$ replaces Δx of the one-dimensional case (and represents a genuine 'line element').

The invariance of Δs, proved in the theory of special relativity, will be expressed by

$$(\Delta s)^2 = (\Delta t)^2 - c^{-2}(\Delta\lambda)^2 = (\Delta t')^2 - c^{-2}(\Delta\lambda')^2, \tag{17}$$

which obviously represents a generalization of Equation (6).

2.3. Space-Like Interval. Simultaneity. Proper Distance

2.3.1. *The Case of One-Dimensional Motion*

Let us return to the case of one-dimensional motion, but change once more, for a while, our notation and call for more clarity δx the distance and δt the interval of time between two events (E_a) and (E_b) in the system (S). In the system (S′) the same events will present a distance $\delta x'$ and an interval of time $\delta t'$.

The pair of events $(E_a)(E_b)$ is supposed to be kinematically of a quite different type than the pair $(E_1)(E_2)$ considered in Subsection 2.2.1. This assumption is expressed by the use of notation (δ) instead of (Δ). The difference lies in the fact that now we suppose, contrary to the condition (7) that

$$(\delta x)^2 \geqslant c^2(\delta t)^2. \tag{18}$$

Thus we can now introduce a real 'interval' δs between the pair of events $(E_a)(E_b)$ defined as the positive square root of

$$(\delta s)^2 = (\delta x)^2 - c^2(\delta t)^2, \tag{19}$$

since now, through the assumption (18), $(\delta s)^2$ is a positive (or zero) quantity.

The interval δs can be called, more precisely, a *'space-like interval'*. for reasons given below.

Similarly to the time-like interval defined by Equation (5) the space-like interval δs defined by Equation (19) is *kinematically invariant* with respect to a replacement of the reference system (S) by the reference system (S′), or reciprocally. This invariance is expressed by Equation

$$(\delta s)^2 = (\delta x)^2 - c^2(\delta t)^2 = (\delta x')^2 - c^2(\delta t')^2. \tag{20}$$

The invariance of δs makes it possible, because of the condition (18), to find a system of reference $(\tilde{S})$, in uniform relative motion with respect to (S) or (S′), where $\delta\tilde{t} = 0$, i.e. where the interval of time $\delta\tilde{t} = \tilde{t}_b - \tilde{t}_a$ between the events (E_a) and (E_b) is zero.

It is usual in special relativity to say that *in the reference system* $(\tilde{S})$ the two events (E_a) and (E_b) are *'simultaneous'*.

We can express the invariance of (δs) with respect to a replacement of (S) by $(\tilde{S})$ on writing

$$(\delta s)^2 = (\delta x)^2 - c^2(\delta t)^2 = (\delta\tilde{x})^2, \tag{21}$$

where

$$\delta\tilde{x} = \tilde{x}_b - \tilde{x}_a. \tag{22}$$

This, of course, would not be physically possible if $(\delta s)^2$ were not positive (or zero), i.e. if the condition (18) were not satisfied.

In the terminology proper to some authors in special relativity, the distance $\delta\tilde{x}$ between two events $(E_a)(E_b)$ which are simultaneous in some laboratory $(\tilde{S})$ is called the *'proper distance'* between (E_a) and (E_b).

Since according to Equation (21) $\delta s = \delta\tilde{x}$, we can physically interpret the interval δs between any pair of events satisfying in some system (S) the condition (18) as the proper distance between the considered pair of events. Hence the term 'space-like interval' δs.

2.3.1.1. Remark. Let us take into account the relation (proved in the special theory of relativity):

$$\delta t = c^{-2} v(\delta x), \tag{23}$$

between the interval of time δt in *the system of non-simultaneity* (S) and the *'improper distance'* δx of the events (E_a) and (E_b) in the same system (S) moving with constant velocity v with respect to the *'system of simultaneity'* ($\tilde{S}$).

Then Equation (21) solved for $\delta\tilde{x}$ takes the form

$$\delta x = \frac{\delta\tilde{x}}{(1 - v^2/c^2)^{1/2}}. \tag{24}$$

This relation represents an expression for what may be called, by analogy with the terminology expressing the relation (11), the *'dilation of distances'*.

However, it is usually 'inverted' by the introduction of *lengths* instead of distances.

Indeed, let us consider a rigid rod ab at rest in (S) and parallel to its Ox axis.

The measure of the length of ab in this system is a trivial operation which when performed with a measuring chain gives what can be called the *'proper length'* L_{pr} of the rod.

However the measure of the length of the same rod by the physicists of the system ($\tilde{S}$), in relative motion with respect to (S), cannot be performed in the same usual and trivial way. More precisely, before speaking about the 'length' of the considered rod in ($\tilde{S}$), one must generalize the very *concept of length*, in the way proposed by A. Einstein, or, more clearly, introduce the concept of *'improper length'* (or 'relativistic length') L_{imp} defined as the proper *distance* $\delta\tilde{x}$ in ($\tilde{S}$) of two simultaneous observations (E_a) and (E_b), by the physicists of ($\tilde{S}$), of the two ends a and b of the rod. On the other hand, the corresponding proper length L_{pr} will be given by the corresponding improper distance δx of the same events in (S).

Therefore Equation (24) will take the form

$$L_{imp} = (1 - v^2/c^2)^{1/2} L_{pr}, \tag{25}$$

expressing the well known statement about the so-called *'contraction of lengths'* of rigid rods, when observed in laboratories in motion with respect to the rod.

Thus, the 'dilatation of distances' becomes the 'contraction of lengths': a misleading terminology when the generalization of the concept of length is not sufficiently emphasized.

(Besides, even the word *'dilatation'* is misleading because it suggests a physical effect, whereas in reality we deal here with a purely kinematical *increase* in relative distance!)

2.3.2. The General Case of a Tri-Dimensional Motion

In obvious notation and by obvious similarity with the results of Subsection 2.2.2, we can define in (S_3), without further comments, the square $(\delta s)^2$ of the space-like interval δs by

$$(\delta s)^2 = (\delta\lambda)^2 - c^2(\delta t)^2, \tag{26}$$

and express its invariance by

$$(\delta s)^2 = (\delta\lambda)^2 - c^2(\delta t)^2 = (\delta\lambda')^2 - c^2(\delta t')^2. \tag{27}$$

2.4. The Zero Interval of the Photon Propagation

Let us consider the reference systems (S_3) and (S_3') defined in Subsection 2.2.2 and let us consider as events (E_1) and (E_2) the respective passages of an electromagnetic wave in two different points of (S_3), distant by $\Delta\lambda$ at times differing in (S_3) by Δt, the corresponding quantities in (S_3') being denoted by $\Delta\lambda'$ and $\Delta t'$.

According to the fundamental principle of the special relativity, the speed c of light in vacuum is the same in (S_3) and in (S_3'). This can be expressed by the two relations:

$$\Delta\lambda = c(\Delta t), \tag{28}$$

and

$$\Delta\lambda' = c(\Delta t'). \tag{28'}$$

On substituting into Equation (17) we see that the invariance of Δs takes place, the common values of Δs in (S_3) and (S_3') being

$$\Delta s = 0. \tag{29}$$

A similar result is obtained when one applies to the propagation of light the expression (27) of the invariance of the space-like interval δs. The common value of δs in (S_3) and (S_3') is again

$$\delta s = 0. \tag{30}$$

2.5. The Unified Notation for the Interval

Obviously the use of two different notations for intervals, distances and intervals of time, in the cases of time-like and space-like intervals, involving respectively (Δ) and (δ), very useful for the sake of clarity in *explanations*, represents an unnecessary complication in subsequent *applications*.

Especially for infinitely small intervals, distances and intervals of time the unique usual symbol (d) is sufficient and one can write Equation (16) and (26) respectively as

$$(ds)^2 = (dt)^2 - c^{-2}(d\lambda)^2, \tag{31}$$

and

$$(ds)^2 = (d\lambda)^2 - c^2(dt)^2. \tag{32}$$

However, one must now pay attention to the fact that the right-hand side of Equation (31) defining the time-like interval and the right-hand side of Equation (32) defining the space-like interval, do *not differ only* by the sign but have *different physical dimensions*: the square of a *time* in Equation (31) and the square of a *length* in Equation (32).

For instance, let us consider a pair of events characterized in (S_3) by the differences $(d\lambda, dt)$. If the substitution of these differences into Equation (31) yields a *negative result* this will show that the considered pair of events presents a *space-like interval*. However, the value of the square of this space-like interval will not be given by the absolute value of the right-hand side of Equation (31). This absolute value can, of course, yield the square of the space-like interval through a multiplication by c^2, but it is even more convenient in this case to apply Equation (32).

With a unified notation for the interval, such as ds, it is important to indicate explicitly whether $(ds)^2$ is defined by Equation (31) or by Equation (32).

Misner *et al.* (1973) (red pages) indicate the choice of 37 cosmologists which correspond for 20 of them to Equation (31), for 13 of them to Equation (32), and to four of them is not stated explicitly.

This division into two almost equal parts is not accidental. It is due to the fact that both choices have their own advantages.

The choice of the time-like interval corresponding to a $(ds)^2$ defined by Equation (31) is more useful in the problems dealing with the *evolution* of the universe, whereas the choice of the space-like interval corresponding of a $(ds)^2$ defined by Equation (32) makes it easier to express the geometrical properties of the universe *at a given instant*.

The choice of the definition of $(ds)^2$ corresponding to Equation (31) is called the '*Landau, Lifshitz* (1971) *time-like convention*'; the choice of the definition of $(ds)^2$ corresponding to Equation (32) is called the '*Landau, Lifshitz* (1962) *space-like convention*'.

When the units are chosen in such a way that $c = 1$, the difference between the physical dimensions of the (ds) defined by Equation (31) and the (ds) defined by Equation (32) does not appear explicitly, which is rather unfortunate from the pedagogical point of view.

3. The Characteristic Properties of Non-Euclidean Spaces

3.1. Bi-Dimensional Spaces

3.1.1. Introduction: the Plane Space (E_2)

A plane surface represents a particularly familiar and simple bi-dimensional space devoid of curvature (a 'flat' space).

For reasons given at the end of this subsection, we shall denote the surface of a plane by the symbol (E_2), the subscript indicating its bi-dimensionality.

Any point M of (E_2) can be referred to some coordinate system, the most usual being the Cartesian one (x,y), referred to orthogonal axes Ox, Oy. The notation $M(x,y)$ permits us to explicit the coordinates of M.

Of course, more general coordinate systems (p,q) can be introduced by transformations such as

$$x = x(p,q) = f(p,q); \qquad y = y(p,q) = g(p,q), \tag{33}$$

where $f(p,q)$ and $g(p,q)$ are given functions.

Thus one can introduce the usual polar coordinates (r,φ) (see Figure 11.1) connected with x and y by the transformations

$$x = x(r,\varphi) = r\cos\varphi; \qquad y = y(r,\varphi) = r\sin\varphi. \tag{34}$$

In this case the position of the point M can be indicated by $M(r,\varphi)$.

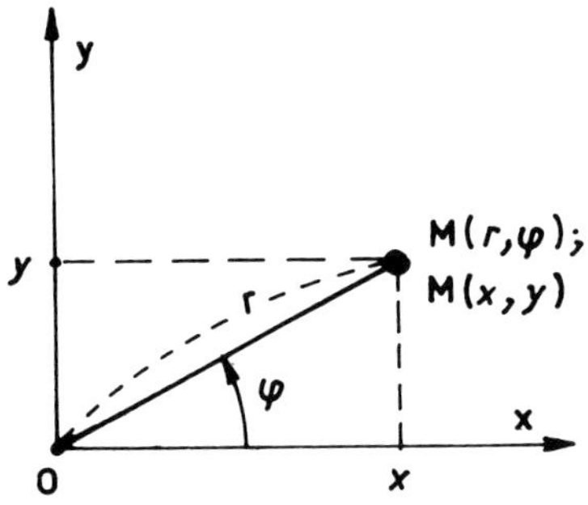

Fig. 11.1. The usual polar coordinates (r, φ).

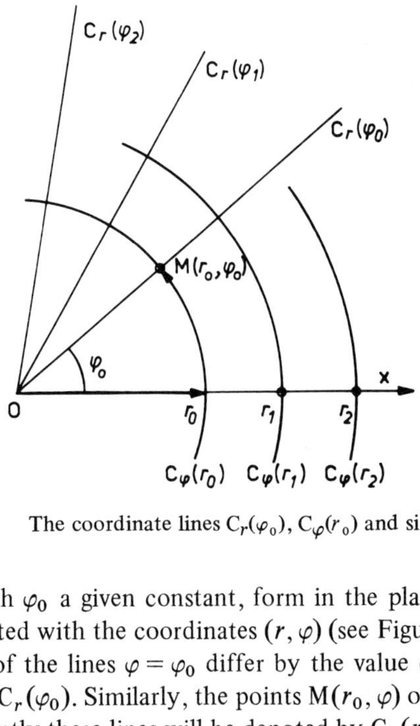

Fig. 11.2. The coordinate lines $C_r(\varphi_0)$, $C_\varphi(r_0)$ and similar ones.

The lines $\varphi = \varphi_0$, with φ_0 a given constant, form in the plane (E_2) a network of the '*coordinate lines*' associated with the coordinates (r, φ) (see Figure 11.2).

The points $M(r, \varphi_0)$ of the lines $\varphi = \varphi_0$ differ by the value of r. For this reason these lines will be denoted by $C_r(\varphi_0)$. Similarly, the points $M(r_0, \varphi)$ of the lines $r = r_0$ differ by the value of φ. Consequently these lines will be denoted by $C_\varphi(r_0)$.

In Cartesian coordinates the lines $C_r(\varphi_0)$ are represented in parametric form (r playing the role of parameter) by

$$x = x(r, \varphi_0) = r \cos \varphi_0; \qquad y = y(r, \varphi_0) = r \sin \varphi_0. \tag{35}$$

whereas the lines $C_\varphi(r_0)$ are represented in parametric form (φ playing the role of parameter) by

$$x = x(r_0, \varphi) = r_0 \cos \varphi; \qquad y = y(r_0, \varphi) = r_0 \sin \varphi. \tag{36}$$

If, starting from the point $M(r_0, \varphi_0)$, we consider (see Figure 11.3) the point $M(r_0 + dr, \varphi_0)$ of $C_r(\varphi_0)$ corresponding to an infinitesimal increment dr of r, the

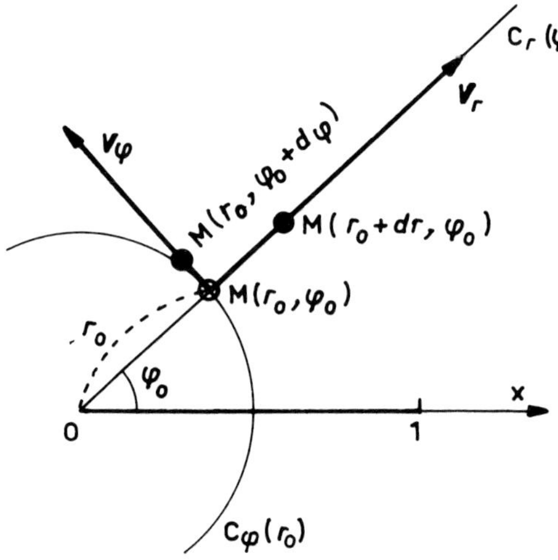

Fig. 11.3. The vector $(dM)_r$ describing the displacement from $M(r_0, \varphi_0)$ to $M(r_0 + dr, \varphi_0)$ and the vector $(dM)_\varphi$ describing the displacement from $M(r_0, \varphi_0)$ to $M(r_0, \varphi_0 + d\varphi)$.

Cartesian components of the vector $(dM)_r$ describing the elementary displacement from $M(r_0, \varphi_0)$ to $M(r_0 + dr, \varphi_0)$ will be given analytically by

$$(dM)_r = \begin{cases} [\partial x(r, \varphi_0)/\partial r]_{r=r_0}\, dr; \\ [\partial y(r, \varphi_0)/\partial r]_{r=r_0}\, dr. \end{cases} \tag{37}$$

This shows that the vector V_r whose components are

$$V_r = \begin{cases} [\partial x(r, \varphi_0)/\partial r]_{r=r_0} = \cos\varphi_0; \\ [\partial y(r, \varphi_0)/\partial r]_{r=r_0} = \sin\varphi_0, \end{cases} \tag{38}$$

and whose modulus is

$$|V_r| = 1, \tag{39}$$

is *tangent* at $M(r_0, \varphi_0)$ to the line $C_r(\varphi_0)$.

Equations (37) are thus equivalent to Equation

$$(dM)_r = (V_r)\,dr. \tag{40}$$

If we denote by $d\lambda_r$ the modulus of $(dM)_r$ we obtain from Equations (39) and (40):

$$d\lambda_r = dr. \tag{41}$$

Note that this expression for $d\lambda_r$, which corresponds to a displacement from $M(r_0, \varphi_0)$ to $M(r_0 + dr, \varphi_0)$, would remain the same if we had started from a more general initial point $M(r, \varphi)$, going from $M(r, \varphi)$ to $M(r + dr, \varphi)$.

Similarly, the vector $(dM)_\varphi$ describing the elementary displacement from $M(r_0, \varphi_0)$ to $M(r_0, \varphi_0 + d\varphi)$ on $C_\varphi(r_0)$ (see Figure 11.3), is easily shown to be given by

$$(dM)_\varphi = (V_\varphi)\,d\varphi, \tag{42}$$

where the Cartesian components of the vector V_φ are

$$V_\varphi = \begin{cases} [\partial x(r_0, \varphi)/\partial\varphi]_{\varphi=\varphi_0} = -r_0 \sin\varphi_0; \\ [\partial y(r_0, \varphi)/\partial\varphi]_{\varphi=\varphi_0} = +r_0 \cos\varphi_0. \end{cases} \tag{43}$$

This vector V_φ has a modulus equal to

$$|V_\varphi| = r_0, \tag{44}$$

and is tangent, at $M(r_0, \varphi_0)$, to the line $C_\varphi(r_0)$.

If $d\lambda_\varphi$ denotes the modulus of $(dM)_\varphi$, we obtain from Equations (42) and (44):

$$d\lambda_\varphi = r_0\, d\varphi. \tag{45}$$

More generally, near the point $M(r, \varphi)$ the modulus $d\lambda_\varphi$ of the displacement from $M(r, \varphi)$ to $M(r, \varphi + d\varphi)$ would be given by

$$d\lambda_\varphi = r\, d\varphi. \tag{46}$$

According to Equations (38) and (43) we have

$$(V_r \cdot V_\varphi) = -r_0 \sin\varphi_0 \cos\varphi_0 + r_0 \sin\varphi_0 \cos\varphi_0 = 0. \tag{47}$$

Thus, at $M(r_0, \varphi_0)$, the tangents V_r and V_φ are orthogonal. This is, of course, a well known property of the polar coordinates.

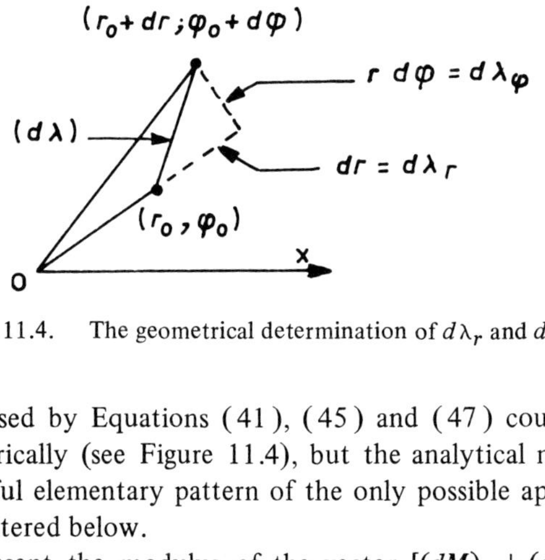

Fig. 11.4. The geometrical determination of $d\lambda_r$ and $d\lambda_\varphi$.

The results expressed by Equations (41), (45) and (47) could be obtained much more rapidly geometrically (see Figure 11.4), but the analytical method applied above represents a very useful elementary pattern of the only possible approach to more complex problems encountered below.

Let $d\lambda$ now represent the modulus of the vector $[(dM)_r + (dM)_\varphi]$ describing the elementary displacement from the point $M(r_0, \varphi_0)$ to the point $M(r_0 + dr, \varphi_0 + d\varphi)$. Since V_r and V_φ are orthogonal, $(d\lambda)^2$ will be given, according to Equations (41) and (45), by

$$(d\lambda)^2 = (dr)^2 + r_0^2 (d\varphi)^2. \tag{48}$$

More generally, the square $(d\lambda)^2$ of the element $d\lambda$ of displacement on (E_2) near the point $M(r, \varphi)$, corresponding to variations dr of r and $d\varphi$ of φ, will be given by

$$\boxed{(d\lambda)^2 = (dr)^2 + r^2(d\varphi)^2.} \tag{48'}$$

Equation (48′) shows that in polar coordinates the modulus $d\lambda$ of the displacement from the point $M(r, \varphi)$ to a point $M(r + dr, \varphi + d\varphi)$ depends not only on the *increments* dr and $d\varphi$ of coordinates but also on the *position* of the 'initial point' $M(r, \varphi)$, more precisely on the initial value of the coordinate r.

It is very important to note that an expression for $(d\lambda)^2$, such as the one given by Equation (48′), is equivalent to *the pair* of relations *giving separately* $d\lambda_\varphi$ and $d\lambda_r$, since Equation (48′) on taking $d\varphi = 0$ reduces to Equation (41) and on taking $dr = 0$ reduces to Equation (46).

Let us now write Equation (48′) in an even more general way (extensively used in general relativity), as the following quadratic form:

$$(d\lambda)^2 = g_{rr}(r, \varphi)(dr)(dr) + g_{r\varphi}(r, \varphi)(dr)(d\varphi) + g_{\varphi\varphi}(r, \varphi)(d\varphi)(d\varphi). \tag{49}$$

Then Equation (48′) will be represented by

$$g_{rr} = g_{rr}(r, \varphi) = 1; \tag{50}$$

$$g_{r\varphi} = g_{r\varphi}(r, \varphi) = 0; \tag{50'}$$

$$g_{\varphi\varphi} = g_{\varphi\varphi}(r, \varphi) = r^2. \tag{50''}$$

In cartesian coordinates the displacement from $M(r_0, \varphi_0)$ to $M(r_0 + dr, \varphi_0 + d\varphi)$ corresponds, by the transformation (34), to a displacement from $M(x_0, y_0)$ to $M(x_0 + dx, y_0 + dy)$.

In terms of (dx, dy), the square of $d\lambda$ is obviously given by

$$(d\lambda)^2 = (dx)^2 + (dy)^2. \tag{51}$$

Note that this expression for $(d\lambda)^2$ in cartesian coordinates *does not depend* on the position of the initial point $M(x, y)$.

If we write Equation (51) in a way analogous to Equation (49), viz.

$$(d\lambda)^2 = g_{xx}(x, y)(dx)(dx) + g_{xy}(x, y)(dx)(dy) + g_{yy}(x, y)(dx)(dy), \tag{52}$$

we can express Equation (51) by the three simple relations:

$$g_{xx} = g_{xx}(x, y) = 1; \tag{53}$$

$$g_{xy} = g_{xy}(x, y) = 0; \tag{53'}$$

$$g_{yy} = g_{yy}(x, y) = 1. \tag{53''}$$

Equations (50′) and (53′) express the trivial property of *orthogonality* of polar and cartesian coordinate lines respectively.

As to Equations (53) and (53″) they express the well known property of the square $(d\lambda)^2$ of the 'line element' $d\lambda$ of the plane (E_2) to be given in cartesian coordinates by the simple quadratic form (51), where by 'simple' we mean henceforth the fact that none of the functions g_{xx}, g_{xy}, g_{yy} depends on x or y.

One could imagine (and such is the wrong opinion of almost all beginners!) that this 'simplicity' of Equation (51) is characteristic of *the absence of curvature* of a surface such as a plane. Or, to put it otherwise, that the curvature of a surface could be indicated by the 'non-simplicity' of the corresponding expression for the square $(d\lambda)^2$ of the line element $d\lambda$.

However, on considering the expression (48′) for the same $(d\lambda)^2$ in polar coordinates or, more precisely, on considering Equation (50″), viz.

$$g_{\varphi\varphi}(r, \varphi) = r^2, \tag{54}$$

we immediately see that *in spite of the absence of curvature* (i.e., in spite of the 'flatness') of the plane (E_2), the expression for the square $(d\lambda)^2$ of the line element is *no longer 'simple'*, as soon as cartesian coordinates are replaced by some other coordinates, e.g. by the polar ones.

Therefore one is led to look for a real criterion of the curvature of a bi-dimensional space, and more generally of a tri- or four-dimensional space. The tensor calculus provides (besides other applications) such a criterion.

However, instead of the tensor calculus, a more accessible method, illustrated by several examples given below, can suffice in an elementary introduction to relativistic cosmology.

For the sake of pedagogical progressivity, we start by two exercises concerning the non-curved space (E_2).

For us, familiar with the tri-dimensional space, the flatness (absence of curvature) of a plane is obvious and represents almost a tautology. However, let us adopt the point of view of '*creeping beings*' dwelling on the surface (E_2), and suppose given, as the only known characteristic of the space (E_2), the $(d\lambda)^2$ in polar coordinates, expressed by Equation (48′), viz.

$$(d\lambda)^2 = (dr)^2 + r^2(d\varphi)^2 . \tag{55}$$

Then, let us suppose that the inhabitants of (E_2) use this expression of $(d\lambda)^2$ for an *analytical* determination of the lengths and areas of simple geometrical figures also defined analytically.

First, let them calculate by an adequate integration the *length* L_e of the closed line $C_\varphi(r_0)$ defined by the constancy of the parameter r. On the line $C_\varphi(r_0)$, $r = r_0$ and $dr = 0$. Thus $d\lambda$ is simply given by

$$d\lambda = r_0\, d\varphi. \tag{56}$$

Hence, by integration from $\varphi = 0$ to $\varphi = 2\pi$, as expected, the 'Euclidean' relation:

$$L_e = 2\pi r_0. \tag{57}$$

Next, let them calculate the area A_e enclosed within $C_\varphi(r_0)$. The element of area dA_e corresponding to $(dr, d\varphi)$ near the point $M(r, \varphi)$ can be assimilated, because of the orthogonality of coordinate lines, with a rectangle of sides $d\lambda_r$ and $d\lambda_\varphi$. Equation (41) gives $d\lambda_r$ but one must pay attention to the fact that dA_e not being necessarily on the very frontier $C_\varphi(r_0)$, one must take for $d\lambda_\varphi$ the general expression (46) and not (45).

Hence

$$dA_e = r\, dr\, d\varphi. \tag{58}$$

On integrating, the inhabitants of (E_2) would obtain, as expected by us, the Euclidean relation:

$$A_e = \int_{\varphi=0}^{2\pi} d\varphi \int_{r=0}^{r_0} r\,dr = 2\pi[r^2/2]_0^{r_0} = \pi r_0^2. \tag{59}$$

Thus, according to Equations (57) and (59), on the flat surface (E_2), the analytical procedure applied above would show that the ratios (L_e/r_0) and (A_e/r_0^2) possess the Euclidean properties (i.e. the properties by *Euclidean geometry*) of being respectively equal to 2π and to π.

The plane surface was denoted by (E_2) precisely in order to recall the Euclidean properties of this bi-dimensional flat space.

The discussion, given in the next subsection, of the properties of a spherical surface (S_2), the simplest *curved* bi-dimensional space, will show that none of such Euclidean properties belong to the figures defined analytically on (S_2).

The *presence* of a few Euclidean properties in a given space does not *prove*, of course, that this space *is* Euclidean (as long as all possible tests were not performed). They are considered simply as an exercise concerning a geometry already known to be Euclidean.

However, *the absence* of a single Euclidean property in a given space, can be conversely considered as a *rigorous proof* of its curvature. This statement will be clarified in Subsection 3.1.2.

3.1.2. The Spherical Surface (S_2)

3.1.2.1. The geometry of the creeping inhabitants of (S_2). Let us consider a point M on the surface (S_2) of a sphere and let us adopt one of its points P as the origin (the 'north pole') of the following coordinate system (see Figure 11.5).

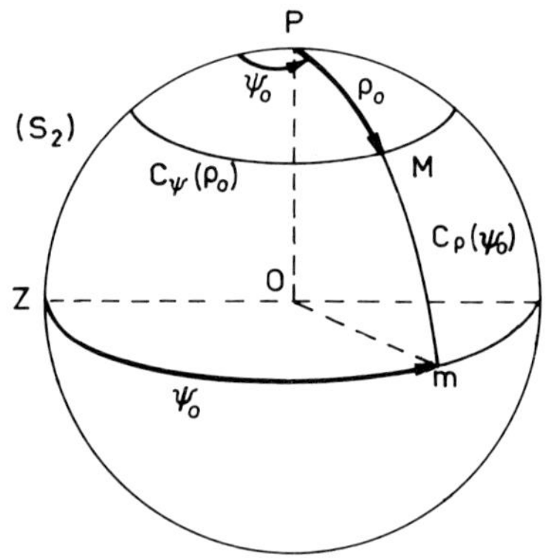

Fig. 11.5. The coordinate system ρ, ψ on (S_2).

The network of coordinate lines will be formed by 'meridians' and 'parallels' as in geography. One of the coordinates of M will be its 'longitude' ψ, reckoned positively in some specified direction from some 'zero meridian' PZ.

However, the second coordinate of M will not be provided by its 'latitude' but by its *'linear polar distance'* ρ defined as *the length of the arc of meridian* PM.

By analogy with the notation used in Subsection 3.1.1, the meridian and the parallel passing through the point $M(\rho_0, \psi_0)$ will be denoted by $C_\rho(\psi_0)$ and $C_\psi(\rho_0)$ respectively.

For us, familiar with the tri-dimensional space, the *curvature* of the spherical surface (S_2) is *intuitive* and obvious.

However, we must study the properties of (S_2) from the point of view of the two-dimensional 'creeping beings' dwelling on (S_2). Our problem is to show how these inhabitants of (S_2) could discover the curvature of their space without any use of the tensor calculus.

Note, first of all, that qualified land-surveyors or physicists dwelling on (S_2) could easily find that the modulus $d\lambda_\rho$ of the elementary displacement $(dM)_\rho$ from $M(\rho_0, \psi_0)$ to $M(\rho_0 + d\rho, \psi_0)$ along the meridian $C_\rho(\psi_0)$ is given by

$$d\lambda_\rho = d\rho, \tag{60}$$

and is thus independent of ψ_0.

On the other hand, after some trials, they would find that the modulus $d\lambda_\psi$ of the elementary displacement $(dM)_\psi$ from $M(\rho_0, \psi_0)$ to $M(\rho_0, \psi_0 + d\psi)$ along the parallel $C_\psi(\rho_0)$ *is not equal to* $d\psi$ but is given by the 'empirical relation' (empirical *for them*!):

$$d\lambda_\psi = \frac{1}{C}(d\psi)\sin(C\rho_0), \tag{61}$$

where C is a certain constant.

More generally, near a point M at a 'distance' ρ from the pole P the elementary displacement corresponding to $d\psi$ would be given by

$$d\lambda_\psi = \frac{1}{C}(d\psi)\sin(C\rho), \tag{61$'$}$$

whereas the expression for the elementary displacement corresponding (for a given value of ψ) to $d\rho$ would still be given by Equation (60), viz.

$$d\lambda_\rho = d\rho. \tag{61$''$}$$

Thus the square $(d\lambda)^2$ of the element $d\lambda$ of displacement on (S_2) near the point $M(\rho_0, \psi_0)$ corresponding to the variations $d\rho$ of ρ and $d\psi$ of ψ would be given by

$$(d\lambda)^2 = (d\rho)^2 + C^{-2}(d\psi)^2\sin^2(C\rho_0); \tag{62}$$

and, more generally, near a point $M(\rho, \psi)$, by

$$(d\lambda)^2 = (d\rho)^2 + C^{-2}(d\psi)^2\sin^2(C\rho). \tag{62$'$}$$

This expression for $(d\lambda)^2$ is equivalent to the pair of relations giving separately $d\lambda_\rho$ and $d\lambda_\psi$ since Equation (62′) on taking $d\psi = 0$ reduces to Equation (61″) and on taking $d\rho = 0$ reduces to Equation (61′).

If we write the expression (62′) for $(d\lambda)^2$ in the general form:

$$(d\lambda)^2 = g_{\rho\rho}(d\rho)^2 + g_{\rho\psi}(d\rho)(d\psi) + g_{\psi\psi}(d\psi)^2, \tag{63}$$

then Equation (62′) will be expressed by

$$g_{\rho\rho} = 1; \tag{64}$$

$$g_{\rho\psi} = 0; \tag{64'}$$

$$g_{\psi\psi} = C^{-2} \sin^2 (C\rho). \tag{64''}$$

Of course, we can easily confirm, by application of our tri-dimensional geometry, all of these empirical results, and discover the geometrical meaning of the constant C.

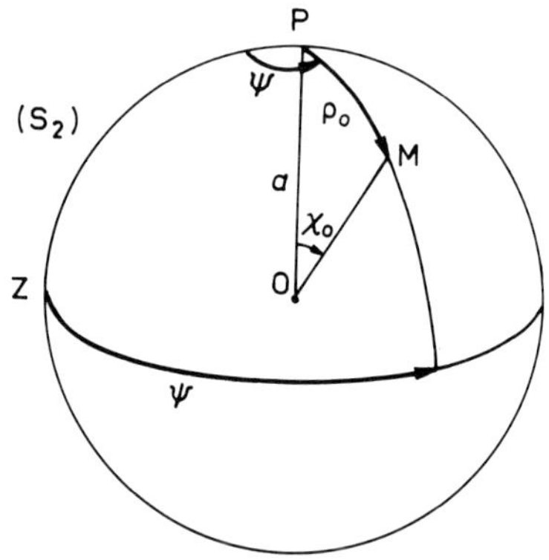

Fig. 11.6. The radial angular (dimensionless) coordinate χ corresponding to the linear polar distance ρ.

Indeed (see Figure 11.6), if we denote by χ_0 the angle $\widehat{POM}$ between the points P and M(ρ_0, ψ_0) seen from the center O of (S_2), and by a the radius of the sphere (S_2) (unknown to inhabitants of the surface of the sphere!) then *we* shall find:

$$\chi_0 = \rho_0/a; \tag{65}$$

or more generally

$$\chi = \rho/a; \tag{65'}$$

and, r_0 being the distance of M from OP:

$$r_0 = a \sin \chi_0 = a \sin (\rho_0/a); \tag{66}$$

$$d\lambda_\psi = r_0 \, d\psi = a(d\psi) \sin (\rho_0/a). \tag{67}$$

Thus, from our point of view the constant C, introduced empirically by the creeping observers is simply *the reciprocal* ($1/a$) of the radius a of the sphere (S_2):

$$C = 1/a. \tag{68}$$

From our point of view C is simply a parameter characteristic of *the curvature* of (S_2), which allows us to write Equation (62') in the form:

$$\boxed{(d\lambda)^2 = (d\rho)^2 + a^2(d\psi)^2 \sin^2 (\rho/a).} \tag{68'}$$

If we introduce the parameter χ defined by Equation (65'), this takes the form

$$(d\lambda)^2 = (ad\chi)^2 + (a \sin \chi d\psi)^2. \tag{68''}$$

Let us now return to the geometry of inhabitants of (S_2). At first sight, the line element given by Equation (62′) is not much more complicated than the line element given by Equation (48′). And yet the geometry of the creeping observers of (S_2) associated with Equation (62′) is completely different from the Euclidean geometry of the observers creeping over the plane (E_2).

Indeed, let them investigate the relation between ρ_0 and the length L_s of the closed line $C_\psi(\rho_0)$ corresponding to a constant value ρ_0 of ρ, i.e. a line they will consider as a 'circle' of radius ρ_0.

Either empirically or by integration of $d\lambda_\psi$ from $\psi = 0$ to $\psi = 2\pi$, they will find

$$L_s = \int_{\psi=0}^{2\pi} C^{-1} \sin(C\rho_0) d\psi = \frac{2\pi}{C} \sin(C\rho_0). \tag{69}$$

Thus on (S_2) the length of a circle of radius ρ_0 cannot be obtained on multiplying ρ_0 by 2π as in an Euclidean space like (E_2).

For us this 'deviation' from Euclidean properties of a circle is a first manifestation of *the curvature* of the space (S_2).

It is very interesting to note in this connection that *for a given value of* ρ_0 a sufficiently small value of the curvature C allows one to replace $\sin(C\rho_0)$ by $C\rho_0$, so that in this limiting case (of a sphere of very great radius a) we find that Equations (66) and (68) yield

$$(r_0)_{C\to 0} = \rho_0, \tag{70}$$

and Equation (69) takes the Euclidean form:

$$(L_s)_{C\to 0} \to 2\pi\rho_0 \to 2\pi r_0 \to L_e. \tag{71}$$

Thus the plane surface (E_2) can be considered as the limit of the curved surface (S_2) when the curvature C of (S_2) tends to zero.

Besides, the curvature of the space (S_2) manifests itself to the creeping geometers in many other ways. Thus, on studying analytically or empirically the variations of the function $L_s(\rho_0)$ they will find something quite different (and very strange from our point of view) from the result found by the creeping geometers of the flat space (E_2).

We can make the result of such investigations quite immediate and easy to visualize by considering geometrically the evolution of L_s in the tri-dimensional space.

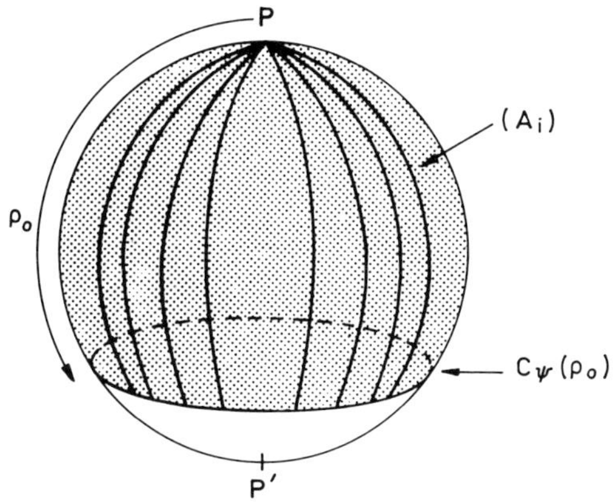

Fig. 11.7. Geometrical investigation of the evolution of L_s and of A_i in the tri-dimensional space.

Thus we 'see' that as ρ_0 increases from $\rho_0 = 0$ to $\rho_0 = \pi a$, the length $L_s(\rho_0)$ of the circle $C_\psi(\rho_0)$ *increases* up to a maximum value $(L_s)_{max} = 2\pi a = 2\pi/C$ reached for $\rho_0 = (\pi/2)a = (\pi/2)/C$ when the parallel becomes the 'equator'; but when the 'radius' ρ_0 of the circle increases beyond this value the circle *begins to shrink* and finally *reduces to a point* (the 'south pole') when $\rho_0 = \pi a = \pi/C$, i.e. when χ_0 takes the value π, (see Figure 11.7).

This result is quite different from *the indefinite* increase of the length $L_e(r_0)$ of a circle $C_\varphi(r_0)$ in the space (E_2) when the radius r_0 of the circle increases.

If, instead of considering the lengths, the creeping geometers consider the areas they will also find many oddities (from our point of view).

Note first that the parallel $C_\psi(\rho_0)$ marks the boundaries not of one but of two finite areas on (S_2): an *'interior area'* (A_i) enclosing the pole P and an *'exterior area'* (A_{ext}) situated between $C_\psi(\rho_0)$ and the south pole P′.

Let us study only the variations of A_i as a function of ρ_0. On computing the area of an element (dA_i) *within* $C_\psi(\rho_0)$ one must pay attention to the fact that this 'general' element is not situated at the linear polar distance ρ_0 from the pole P, but at the linear polar distance ρ which varies from 0 to ρ_0. Therefore, in the computation of dA_i, one must use the general expression (61′) for $d\lambda_\psi$ instead of the particular one (61). As to $d\lambda_\rho$ we have already noted that its expression (61″) does not depend on the position of the point M. Hence

$$dA_i = (d\lambda_\rho)(d\lambda_\psi) = \frac{(d\rho)(d\psi)}{C}\sin(C\rho). \tag{72}$$

On integrating with respect to ρ from $\rho = 0$ to $\rho = \rho_0$ and with respect to ψ from $\psi = 0$ to $\psi = 2\pi$, the creeping geometers would obtain

$$\begin{aligned} A_i = A_i(\rho_0) &= \int_0^{2\pi} d\psi \int_0^{\rho_0} C^{-1}\sin(C\rho)\,d\rho \\ &= 2\pi C^{-2}[1 - \cos(C\rho_0)]. \end{aligned} \tag{73}$$

The inhabitants of (S_2) could study this function analytically or measure empirically the evolution of A_i for different values of ρ_0.

However, *we* can foresee and visualize their results very easily by considering geometrically in the tri-dimensional space the strange (from our point of view) evolution of the area A_i.

Indeed, we 'see' in Figure 11.7 that as ρ_0 increases from $\rho_0 = 0$ to $\rho_0 = \pi a$ the area $A_i(\rho_0)$ increases monotonically from zero to a maximum value equal to the total area of the space (S_2), viz. the area of a sphere of radius a [given also by Equation (73) with $\cos(C\rho_0) = -1$]:

$$[A_i(\rho_0)]_{max} = \text{The total area of the space } (S_2) = 4\pi a^2 = 4\pi C^{-2}. \tag{74}$$

Hence, incidentally:

$$A_{ext}(\rho_0) = [A_i(\rho_0)]_{max} - A_i(\rho_0) = 2\pi C^{-2}[1 + \cos(C\rho_0)]. \tag{75}$$

Most interesting is that, contrary to what happens in the flat space (E_2), the variation of the area $A_i(\rho_0)$ enclosed within the circle $C_\psi(\rho_0)$ does not always correspond to the variation of the length $L_s(\rho_0)$ of the frontier $C_\psi(\rho_0)$ of (A_i), since for values of ρ_0 increasing beyond $(\pi/2)a$, the length $L_s(\rho_0)$ *decreases*, whereas the area A_i continues to *increase*.

The area A_i reaches its maximum value, given by Equation (74), when the length $L_s(\rho_0)$ reduces to zero (i.e. when the circle of radius $\rho_0 = \pi a$ reduces to the point P′). Thus, not only the variations of L_s and of A_i as a function of the 'radius' ρ_0 of the circle $C_\psi(\rho_0)$ do not follow the Euclidean pattern, but both L_s and A_i *cannot exceed some limiting values*, whereas in the Euclidean space (E_2) both L_e and A_e of a circle can become infinite.

Moreover, the fact that the total area of the space (S_2) is *finite* shows that this space (of 'positive curvature') is not only curved but also *'closed'*. Such would not be the case, e.g., for the curved bi-dimensional space (of 'negative curvature') represented by a hyperboloid.

3.2. TRI-DIMENSIONAL EUCLIDEAN GEOMETRY AND THE METHOD OF EMBEDDING

3.2.1. *Introduction*

The usual tri-dimensional *geometry* is by definition 'Euclidean', since we use the deviations from its results as an expression for the non-Euclidean character of other geometries, as those of the geometers creeping on (S_2).

However, it is not at all sure that *our physical* space can be adequately described at large scale by Euclidean geometry.

Therefore, in the next subsections, we shall investigate the properties of curved tri-dimensional spaces defined by the corresponding expression for $(d\lambda)^2$.

Nevertheless, in order to become familiar with the analytical methods by which the non-Euclidean character of these spaces are revealed, it can be useful to show how very trivial results of Euclidean tri-dimensional geometry can be obtained by similar analytical methods from Euclidean expressions for $(d\lambda)^2$.

Of course, such an exercise can be profitable only if the corresponding expressions for $(d\lambda)^2$ are sufficiently 'complicated' to leave a doubt concerning the Euclidean properties *implied* by the given $(d\lambda)^2$. This was already the case of the $(d\lambda)^2$ corresponding to the plane Euclidean geometry of the space (E_2) expressed in polar coordinates.

That is the reason why, before starting the study of curved tri-dimensional spaces, we treat by similar analytical methods the $(d\lambda)^2$ of the usual tri-dimensional Euclidean geometry in spherical coordinates.

3.2.2. *Some Trivial Properties of Euclidean Geometry Deduced Analytically from the Corresponding Line Element in Spherical Coordinates of the Space* (E_3).

Let us show how some very trivial properties of the tri-dimensional Euclidean geometry can be deduced *analytically* from the well known expression of $(d\lambda)^2$ in usual spherical coordinates (r, θ, φ), viz.

$$(d\lambda)^2 = (dr)^2 + (r\,d\theta)^2 + (r\sin\theta\,d\varphi)^2, \tag{76}$$

where r is supposed to be unlimited, whereas $0 \leqslant \theta \leqslant \pi$ and $0 \leqslant \varphi \leqslant 2\pi$.

According to Equation (76), the elementary displacements $d\lambda_r$, $d\lambda_\theta$, $d\lambda_\varphi$ corresponding to the increments $dr, d\theta, d\varphi$ of the coordinates r, θ and φ are

$$d\lambda_r = dr; \tag{77}$$

$$d\lambda_\theta = r\,d\theta; \tag{78}$$

$$d\lambda_\varphi = r\sin\theta\,d\varphi, \tag{79}$$

respectively, and the system of coordinates (r, θ, φ) is orthogonal since

$$g_{r\theta} = g_{\theta\varphi} = g_{\varphi r} = 0. \tag{80}$$

As a first exercise let us determine analytically the area $A(r_0)$ of the surface $S(r_0)$ defined by a constant value r_0 of the coordinate r, i.e. the area of the sphere of radius r_0.

On such a surface $dr = 0$, hence, according to Equations (78), (79) and (80):

$$d\lambda_r = 0; \tag{81}$$

$$d\lambda_\theta = r_0\,d\theta; \tag{82}$$

$$d\lambda_\varphi = r_0\sin\theta\,d\varphi. \tag{83}$$

Therefore the element dA of area of $S(r_0)$ will be given (because of the orthogonality of the coordinate system) by

$$dA = r_0^2\sin\theta\,d\theta\,d\varphi, \tag{84}$$

and the area $A(r_0)$ of $S(r_0)$ will be given by

$$A(r_0) = \int_0^{2\pi} d\varphi \int_0^{\pi} r_0^2\sin\theta\,d\theta = 4\pi r_0^2. \tag{85}$$

Thus we obtain the trivial well known expression for the area of a sphere of radius r_0.

As a second exercise let us analytically determine the volume $V(r_0)$ of a sphere of radius r_0.

By analogy with a similar procedure used to determine, in Subsection 3.1.2.1, the area $A_i(\rho_0)$, we take as element of volume dV the product

$$dV = (d\lambda_r)(d\lambda_\theta)(d\lambda_\varphi), \tag{86}$$

which, according to Equations (78), (79) and (80) is equal to

$$dV = r^2\sin\theta\,d\theta\,d\varphi. \tag{87}$$

Hence the value of $V(r_0)$ given by

$$V(r_0) = \int_0^{2\pi} d\varphi \int_0^{\pi} \sin\theta\,d\theta \int_0^{r_0} r^2\,dr = \frac{4\pi}{3}r_0^3. \tag{88}$$

Once more we find the trivial Euclidean expression for the volume of a sphere of radius r_0.

As a third, and last, exercise let us determine analytically the length of a circle (C), which in a tri-dimensional space (E_3), corresponds to the intersection of the sphere $r = r_0$ with the plane $\theta = \theta_0$. This circle (C) is thus described analytically by $r = r_0, \theta = \theta_0, dr = 0$ and $d\theta = 0$, so that the corresponding element of length $d\lambda$ will be given, according to Equation (80), by

$$d\lambda_\varphi = r_0 \sin \theta_0 \, d\varphi. \tag{89}$$

If we call r_1 the quantity

$$r_1 = r_0 \sin \theta_0, \tag{90}$$

which obviously represents the radius of the circle (C), we obtain:

$$d\lambda_\varphi = r_1 \, d\varphi, \tag{91}$$

and, by integration with respect to φ, the length $L(r_1)$:

$$L(r_1) = 2\pi r_1. \tag{92}$$

Once more we find the Euclidean expression for the length of a circle of radius r_1.

Of course, all these results do not prove that the space (E_3) is Euclidean, but make the absence of curvature of (E_3) very plausible. Thus we can venture to consider it as an Euclidean tri-dimensional space, and denote it, as we did, by (E_3).

We repeat that the actual interest of the tests performed above does not lie in the three trivial results deduced from Equation (76) by the analytical method, but in the non-trivial, i.e. non-Euclidean, results given by the same method for some tri-dimensional spaces considered below, thus proving that they are curved.

3.2.3. *The Legitimacy of the Study of the Curved Space* (S_2) *in the Tri-Dimensional Euclidean Space* (E_3). *The Method of Embedding*

In Subsection 3.1.2.1 we have studied the properties of the curved bi-dimensional space (S_2) by a deliberately *hybrid method* alternating the purely analytical (or empirical) approach accessible to the creeping inhabitants of (S_2) with our own tri-dimensional geometrical intuition. The purpose of this 'mixture' was first of all *to visualize* the properties generated by the curvature of the space (S_2), and incidentally to avoid too many elementary calculations implied in the analytical study of the functions $L_s(\rho_0)$ and $A_i(\rho_0)$.

This enabled us to show how much these properties had to appear as non-Euclidean, compared with those known to Euclidean geometers of the flat space (E_2).

However, in view of applications to more general problems, concerning the curved tri-dimensional spaces, it can be worth presenting the study of the properties of the space (S_2) by another method called the 'method of embedding'.

In this method the n-dimensional curved space is studied by purely analytical procedures as embedded in a $(n + 1)$-dimensional space supposed to possess the properties of Euclidean geometry.

In the case of the two-dimensional space (S_2) this amounts precisely to consider it as a part of the tri-dimensional space obeying the Euclidean geometry, i.e. as a part of the space (E_3).

Therefore, as a last exercise before attacking more complicated problems, we shall justify our previous introduction of the tri-dimensional Euclidean space by showing that even the inhabitants of (S_2) had the possibility of considering (S_2) as a *'slice'* (section) of a (for them theoretical) tri-dimensional Euclidean space (E_3), and use systematically the properties of (E_3) instead of doing this occasionally as we did.

A parametric representation of the spherical surface (S_2) in the tri-dimensional Euclidean geometry is obtained when each of the three spherical coordinates (r, θ, φ) is expressed as a function of only two parameters.

Taking into account the choice (ρ, ψ) of coordinates made by the creeping geometers of (S_2), and in order to facilitate the comparison with the results obtained above we can take as parameters precisely ρ and ψ defined in Subsection 3.1.2.1 and indicated in Figure 11.5.

We see in Figure 11.5 that the three functions describing in (E_3) the surface (S_2) in parametric form (when usual definitions of the spherical coordinates r, θ, φ are used) are:

$$r = r(\rho, \psi) = a; \tag{93}$$

$$\theta = \theta(\rho, \psi) = \rho/a; \tag{94}$$

$$\varphi = \varphi(\rho, \psi) = \psi - \frac{\pi}{2}, \tag{95}$$

where a is the constant equal to the radius of (S_2).

Hence also:

$$dr = 0; \tag{96}$$

$$d\theta = \frac{1}{a} d\rho; \tag{97}$$

$$d\varphi = d\psi. \tag{98}$$

Let us show that if (S_2) were defined by its *two-dimensional* $(d\lambda)^2$, given for instance by Equation (62′), viz.

$$(d\lambda)^2 = (d\rho)^2 + C^{-2}(d\psi)^2 \sin^2(C\rho), \tag{99}$$

one would be able to consider it as a surface *'embedded'* in the tri-dimensional Euclidean space (E_3). Indeed, starting from the $(d\lambda)^2$ of the tri-dimensional Euclidean space (E_3), expressed in spherical coordinates by Equation (76), viz.

$$(d\lambda)^2 = (dr)^2 + (r\, d\theta)^2 + (r \sin\theta\, d\varphi)^2, \tag{100}$$

we see that on a surface defined in (E_3) by Equations (93), (94), (95), the $(d\lambda)^2$ given by Equation (100) would yield Equation (99) with $a = 1/C$.

On the other hand, this result would show that (S_2) can be considered as a bi-dimensional slice (or section) of the tri-dimensional space (E_3), a slice obtained through the fixation of *one* of the three coordinates of (E_3) by Equation (93), viz.

$$r = a = \text{const.}. \tag{101}$$

As a by-product of these results the representation of (S_2) in Cartesian coordinates of (E_3) would be immediately given either in implicit form on replacing r^2 by $(x^2 + y^2 + z^2)$, hence:

$$x^2 + y^2 + z^2 = a^2 = \text{const.}, \tag{102}$$

or in parametric form, on replacing r, θ and φ by their values (93), (94), (95) in the usual relations

$$x = r \sin\theta \cos\varphi; \tag{103}$$

$$y = r \sin\theta \sin\varphi; \tag{104}$$

$$z = r \cos\theta. \tag{105}$$

3.2.4. *The Slices* (S_{n-1}) *of Spaces* (S_n)

Let us consider, as an introduction to the study of curved spaces the 'slices' (subspaces S_{n-1}) of n-dimensional spaces (S_n), introduced by a particularization of a coordinate of (S_n).

Thus, for instance, in the flat space (E_2), studied by means of the standard polar coordinates (r, φ), a circle $C_\varphi(r_0)$ can be introduced, by a particular choice of the coordinate r, viz. $r = r_0$. This circle represents a one-dimensional 'space' (S_1) which is non-Euclidean since it is curved. The position of a point in the space (S_1) depends on a single parameter φ.

Reciprocally, one can consider the plane (E_2) as 'generated' by the circles (S_1) of variable radius r_0.

This ultra-elementary example shows that *a curved subspace* (S_{n-1}), such as (S_1), *can generate a perfectly Euclidean (non-curved) space* (S_n) such as (E_2).

Similarly, let us return for a while to the curved bi-dimensional space (S_2), already studied, in Subsection 3.1.2, by means of the coordinates (ρ, ψ), where ρ is the linear polar distance and ψ is the longitude of a point $M(\rho, \psi)$.

If we choose a particular value ρ_0 of ρ, we introduce a circle $C_\psi(\rho_0)$, which is again a one-dimensional curved line, i.e. a one-dimensional non-Euclidean space (S_1). The position of a point in (S_1) depends on the single parameter ψ.

Thus (S_2) can be considered as generated by the circles $C_\psi(\rho_0)$ of variable ρ_0, i.e. by the *curved* one-dimensional subspaces (S_1).

However, as shown by the first example, this does not imply that (S_2) is itself a curved bi-dimensional space. Actually the curvature of (S_2) was proved by the 'tests' showing that on (S_2) the length $L_s(\rho_0)$ of a circle of 'radius' ρ_0 was *not* given by the Euclidean expression $2\pi\rho_0$, and that the area $A_i(\rho_0)$ interior to this circle was *not* given by the Euclidean expression $\pi\rho_0^2$. Moreover, the fact that the space (S_2) was not only curved, but was also *closed* was revealed by the 'test' showing that its total area was finite and could not exceed $4\pi a^2 = 4\pi C^{-2}$.

All these preliminary remarks adequately generalized will greatly facilitate the exploration of the geometrical properties of curved tri-dimensional spaces.

3.3. GENERAL TRI-DIMENSIONAL SPACES

3.3.1. *Introduction*

The general relativistic approach to cosmology to be presented in Chapter 12 makes use of 'Riemannian spaces' defined by the following square of line element $d\lambda$:

$$(d\lambda)^2 = (R\,d\chi)^2 + (R\,\sigma_k\,d\theta)^2 + (R\,\sigma_k \sin\theta\,d\varphi)^2, \tag{106}$$

where, for the moment, we shall consider R as a given constant, and where σ_k represents, according to the value of the subscript k, one of the following elementary functions of a dimensionless (angular) parameter (coordinate) χ:

$$\sigma_0 = \chi; \tag{107}$$

$$\sigma_{+1} = \sin\chi; \tag{108}$$

$$\sigma_{-1} = \sinh\chi. \tag{109}$$

The coordinates θ and φ are supposed to correspond to the standard spherical coordinates θ and φ, by which we simply mean the θ varies between 0 and π, whereas φ varies between 0 and 2π.

The spaces thus defined by Equation (106) are tri-dimensional since they depend on three parameters χ, θ, φ. Therefore we shall denote them by (Σ_3^0), (Σ_3^+) or (Σ_3^-) respectively, according to the value 0, + 1 or − 1 of k.

3.3.2. *The Euclidean Space* $(\Sigma_3^0) = (E_3)$

The square of the line element $d\lambda$ of the space (Σ_3^0) corresponding to $k = 0$ is given by Equations (106) and (107), viz. by

$$(d\lambda)^2 = (R\,d\chi)^2 + (R\,\chi\,d\theta)^2 + (R\,\chi \sin\theta\,d\varphi)^2, \tag{110}$$

where, as stated above, R is considered for the moment as a given constant.

If we define a parameter r by

$$r = R\chi, \tag{111}$$

Equation (110) takes the form

$$(d\lambda)^2 = (dr)^2 + (r\,d\theta)^2 + (r \sin\theta\,d\varphi)^2. \tag{112}$$

This form of $(d\lambda)^2$ is identical with Equation (76) which gives (in spherical coordinates) the square of the line element of the space (E_3) already considered in Subsection 3.2.2. Thus the space (Σ_3^0) is identical with the space (E_3) known to be Euclidean (i.e. without any curvature) and *unbound*.

3.3.3. *The Non-Euclidean (Curved) Closed Space* (Σ_3^+)

The square of the line element $d\lambda$ of the space (Σ_3^+) corresponding to $k = +1$ is given by Equations (106) and (108), viz. by

$$(d\lambda)^2 = (R\,d\chi)^2 + (R \sin\chi\,d\theta)^2 + (R \sin\chi \sin\theta\,d\varphi)^2, \tag{113}$$

where as above, R is considered as a given constant.

Let us, by analogy with the discussion of Subsection 3.2.4, introduce a bi-dimensional slice (Σ_2^+) of the tri-dimensional space (Σ_3^+), by giving a constant value χ_0 to the coordinate χ. Thus, with

$$\chi = \chi_0, \tag{114}$$

we have $d\chi = 0$, and according to Equations (113) and (114) the square of the line element of the space (Σ_2^+) will be given by

$$(d\lambda)^2 = (R \sin \chi_0 \, d\theta)^2 + (R \sin \chi_0 \sin \theta \, d\varphi)^2. \tag{115}$$

On introducing the parameters a, ρ and ψ defined by

$$a = R \sin \chi_0; \tag{116}$$

$$\rho = (R \sin \chi_0)\theta = a\theta; \tag{117}$$

and

$$\psi = \varphi. \tag{118}$$

Equation (115) takes a form identical with the expression given for $(d\lambda)^2$ of the bi-dimensional curved space (S_2) by Equation (68′), viz.

$$(d\lambda)^2 = (d\rho)^2 + a^2(d\psi)^2 \sin^2(\rho/a). \tag{119}$$

Thus the slice (Σ_2^+) of (Σ_3^+) corresponding to $\chi = \chi_0$ is simply the *surface* of a sphere $S_2(\chi_0)$ of radius $a = R \sin \chi_0$.

This allows us to consider the space (Σ_3^+) as 'generated' by the spheres $S_2(\chi_0)$ of variable radius $a = R \sin \chi_0$, but *this radius is not unbound*.

Indeed, when χ_0 reaches the value π, the radius a of the spheres $S_2(\chi_0)$, given by Equation (116), *reduces to zero*, which means that the corresponding *area* of (S_2) is itself zero. The situation is quite similar to the non-Euclidean phenomenon encountered on the curved space (S_2) itself, where the *length* of the one-dimensional slice $(S_1) = C_\psi(\rho_0)$ was reduced to zero for $\rho_0 = \pi a$, i.e. for $\chi_0 = \pi$, thus limiting the area $A_i(\rho_0)$ enclosed by the one-dimensional space (S_1) to a maximum value $4\pi a^2$.

Let us now, considering that $S_2(\chi_0)$ plays in (Σ_3^+) (with one more dimension) the role of the 'frontier' played by (S_1) in (S_2), determine the *volume* $V(\chi_0)$ of the space enclosed in (Σ_3^+) by this two-dimensional frontier.

According to Equation (113) the expression for the volume element dV corresponding to $(d\chi, d\theta, d\varphi)$ is given by

$$dV = (R\, d\chi)(R \sin \chi \, d\theta)(R \sin \chi \sin \theta \, d\varphi). \tag{120}$$

Hence the value of $V(\chi_0)$:

$$\begin{aligned} V(\chi_0) &= R^3 \int_0^{2\pi} d\varphi \int_{-1}^{+1} d(\cos\theta) \int_0^{\chi_0} \sin^2 \chi \, d\chi \\ &= 2\pi R^3(\chi_0 - \tfrac{1}{2} \sin 2\chi_0). \end{aligned} \tag{121}$$

Since χ_0 cannot exceed the value π, the volume of the space (Σ_3^+) cannot exceed the total volume $V_{\text{tot}}(\pi)$ given by

$$\boxed{V_{\text{tot}} = V_{\text{tot}}(\Sigma_3^+) = 2\pi^2 R^3.} \tag{122}$$

This proves that the space (Σ_3^+) is *non-Euclidean* (i.e. curved), since the relation (121) is non-Euclidean, and *closed* (bound), since it has a finite volume.

3.3.3.1. Remark. One could, of course, study the properties of the space (Σ_3^+) by the method of embedding (see Subsection 3.2.3). This would imply an embedding of the tri-dimensional curved space (Σ_3^+) in a four-dimensional Euclidean space (E_4).

However, this would not clarify the situation (contrary to what happened for the embedding of S_2 in E_3) since we do not have an intuitive representation of such a purely conceptual space (E_4), which would have to be studied, for instance, in '*hyper-spherical coordinates*' $(r, \theta, \varphi, \chi)$, quite different from the coordinates of the four-dimensional space-time $(t, \chi, \theta, \varphi)$ to be considered below.

The interested reader is referred to Misner *et al.* (1973, p. 723) where it is shown that (Σ_3^+) represents a tri-dimensional slice of (E_4), corresponding to r (of the hyper-spherical coordinates) equal to R. Thus (Σ_3^+) is a tri-dimensional hyper-sphere (a volume!) in the four dimensional Euclidean space.

3.3.4. *The Non-Euclidean (Curved) Open Space* (Σ_3^-)

The square of the line element $d\lambda$ of the space (Σ_3^-) corresponding to $k = -1$ is given by Equations (106) and (109), viz. by

$$(d\lambda)^2 = (R\, d\chi)^2 + (R \sinh \chi\, d\theta)^2 + (R \sinh \chi \sin \theta\, d\varphi)^2, \tag{123}$$

where, as above, R is considered as a given constant.

In a way similar to the one used in the discussion of the properties of the space (Σ_3^+), let us introduce a bi-dimensional slice (Σ_2^-) by giving a constant value χ_0 to the coordinate χ. Thus with

$$\chi = \chi_0, \tag{124}$$

we have $d\chi = 0$, and according to Equations (123) and (124) the square of the line element of the space (Σ_2^-) will be given by

$$(d\lambda)^2 = (R \sinh \chi_0\, d\theta)^2 + (R \sinh \chi_0 \sin \theta\, d\varphi)^2. \tag{125}$$

On introducing the parameters a, ρ and ψ defined by

$$a = R \sinh \chi_0; \tag{126}$$

$$\rho = (R \sinh \chi_0)\theta = a\theta; \tag{127}$$

and

$$\psi = \varphi, \tag{128}$$

Equation (125) takes the form identical with the expression given for $(d\lambda)^2$ of the bi-dimensional curved space (S_2) by Equation (68′), viz.

$$(d\lambda)^2 = (d\rho)^2 + a^2(d\psi)^2 \sin^2 (\rho/a). \tag{129}$$

Thus the slice (Σ_2^-) of (Σ_3^-) corresponding to $\chi = \chi_0$ is again the surface of a sphere $S_2(\chi_0)$. However, its radius is no longer $R \sin \chi_0$ but $a = R \sinh \chi_0$.

The space (Σ_3^-) can thus be considered as generated by the spheres $S_2(\chi_0)$ of variable radius $a = R \sinh \chi_0$. However, here this radius is *unbound*.

Indeed, the radius a of the spheres $S_2(\chi_0)$ never reduces, in the present case, to zero, since $\sinh \chi_0 \to \infty$ when $\chi_0 \to \infty$.

Let us now determine the volume $V(\chi_0)$ of the space enclosed in (Σ_3^-) by the bi-dimensional surface $S_2(\chi_0)$.

According to Equation (123) the expression for the volume element dV corresponding to $(d\chi, d\theta, d\varphi)$ is given by

$$dV = R^3 \sinh^2 \chi \sin \theta \, d\chi \, d\theta \, d\varphi. \tag{130}$$

Hence the value of $V(\chi_0)$:

$$\begin{aligned} V(\chi_0) &= R^3 \int_0^{2\pi} d\varphi \int_{-1}^{+1} d(\cos\theta) \int_0^{\chi_0} \sinh^2 \chi \, d\chi \\ &= 2\pi R^3(-\chi_0 + \tfrac{1}{2}\sinh 2\chi_0). \end{aligned} \tag{131}$$

Contrary to the case of (Σ_3^+), in the present case χ_0 can become infinite. Since, asymptotically, for $\chi_0 \to \infty$, $\sinh 2\chi_0 \to \frac{1}{2}e^{2\chi_0}$, the volume $V(\chi_0)$ tends to infinity.

Thus the space (Σ_3^-) is both *non-Euclidean* (i.e. curved) and *open* (unbound).

4. The Geodesic Principle

4.1. THE GEOMETRIZATION OF DYNAMICS AND THE GEODESIC PRINCIPLE

According to Section 2, in the geometrized conception of *kinematics*, the history of the motion of a particle M is described by a curve (C) of the four-dimensional space-time (a 'world line'). Such a curve (C) can be represented in a parametric form by four functions, e.g. (if the space coordinates are χ, θ, φ) by four functions $\chi = f_\chi(\mu)$; $\theta = f_\theta(\mu)$; $\varphi = f_\varphi(\mu)$; $t = f_t(\mu)$ of some parameter μ.

According to an additional principle, the *dynamical* description of the motion of a mass point (or a photon) M is achieved in two steps.

As a first step one determines the expression for the 'interval' ds describing *the gravitational effects of the distribution of masses upon the structure of the four-dimensional space-time* (S).

As a second step (called the *'geodesic principle'*), one assumes that the motion of the mass point (or a photon) M in the given gravitational field is described by a particular kind of curves (C) of the space-time (S) called 'geodesics' of (S), defined in the following way.

A *geodesic* (G) joining two points P_0 and P_1 of (S) is defined to be *a curve along which the integral of the interval ds has a stationary value* compared with a similar integral measured along any other neighbouring curve joining P_0 and P_1.

When the orthogonal spherical coordinates (χ, θ, φ) are used, as in Section 3, in the description of the spatial part $(d\lambda)^2$ of the space-like square of the interval ds, viz.

$$(ds)^2 = -c^2(dt)^2 + (d\lambda)^2, \tag{132}$$

$(d\lambda)^2$ takes, in the most general of the cases considered below, the form given by Equation (106), viz.

$$(d\lambda)^2 = R^2(t)[(d\chi)^2 + \sigma_k^2(\chi)(d\theta)^2 + \sigma_k^2(\chi)\sin^2\theta(d\varphi)^2]. \tag{133}$$

In the last Equation σ_k represents a function of χ defined, according to the value of k, by Equations (107), (108), (109), viz.

$$\sigma_0 = \chi; \qquad \sigma_{+1} = \sin\chi; \qquad \sigma_{-1} = \sinh\chi \tag{134}$$

In a more general way, this $(ds)^2$ can be written:

$$(ds)^2 = g_{tt}(dt)^2 + g_{\chi\chi}(d\chi)^2 + g_{\theta\theta}(d\theta)^2 + g_{\varphi\varphi}(d\varphi)^2, \tag{135}$$

where

$$\begin{cases} g_{tt} = -c^2; \quad g_{\chi\chi} = +R^2(t); \quad g_{\theta\theta} = +R^2(t)\,\sigma_k^2(\chi); \\ g_{\varphi\varphi} = +R^2(t)\,\sigma_k^2(\chi)\sin^2\theta. \end{cases} \tag{136}$$

We suppose here that the first step indicated above has already been achieved and that the functions $g_{tt}, g_{\chi\chi}, g_{\theta\theta}, g_{\varphi\varphi}$ are already so chosen as to describe the effect of the distribution of masses on the space-time (S). The problem is then *to express analytically* the definition of a geodesic given above and to determine the four functions $f_\chi, f_\theta, f_\varphi$ and f_t of μ describing a geodesic (G) of the four-dimensional space-time (S).

When these four functions will be found the law of motion of the particle M in the tri-dimensional space will be easily obtained in the following way.

An inversion of the relation $t = f_t(\mu)$ would then give $\mu = F_\mu(t)$, and on replacing μ by $F_\mu(t)$ in $f_\chi(\mu), f_\theta(\mu)$ and $f_\varphi(\mu)$ one would obtain the functions $\chi = F_\chi(t), \theta = F_\theta(t)$ and $\varphi = F_\varphi(t)$ parametrically describing the motion of M in the usual way.

Unfortunately, the functions $f_\chi, f_\theta, f_\varphi, f_t$ of μ representing the geodesic (G) are not given by any algebraic transformation of the functions $g_{tt}, g_{\chi\chi}, g_{\theta\theta}, g_{\varphi\varphi}$, but by a system of four differential equations of second order, where the four unknown functions are precisely $f_\chi(\mu), f_\theta(\mu), f_\varphi(\mu)$ and $f_t(\mu)$.

4.2. THE GENERAL VARIATIONAL METHOD GIVING THE DIFFERENTIAL EQUATIONS OF GEODESICS

According to the definition of a geodesic given above, these curves must be obtained by *a variational method* similar to the one used in Newtonian description of mechanical systems by means of *Lagrange's equations*. This analogy is far from being fortuitous!

The system of differential equations describing the geodesics takes a more systematic form when one uses the notation:

$$x_0 = ct; \quad x_1 = \chi; \quad x_2 = \theta; \quad x_3 = \varphi. \tag{137}$$

With this notation Equation (135) takes the form:

$$(ds)^2 = \sum_{i=0}^{3} g_{ii}(x_0, x_1, x_2, x_3)(dx_i)^2. \tag{138}$$

The variational principle defining the geodesics leads to consider besides the curves (C), represented parametrically by the functions $x_i = x_i(\mu)$, with $(i = 0, 1; 2, 3)$, the curves $(C_{\epsilon,\omega})$ represented parametrically by $x_i = \bar{x}_i(\mu)$, where the functions $\bar{x}_i(\mu)$ are defined by

$$\bar{x}_i(\mu) = x_i(\mu) + \epsilon\omega_i(\mu) = x_i + \epsilon\omega_i \quad (i = 0, 1, 2, 3). \tag{139}$$

In Equations (139) ϵ is supposed to represent a *small* quantity whose powers *above* ϵ^2 are negligible, and where $\omega_i(\mu)$ are four *arbitrary* functions supposed only to vanish for $\mu = \mu_0$ and $\mu = \mu_1$. Thus all curves $(C_{\epsilon,\omega})$ are situated very near the curve (C), and all pass through two fixed points P_0 and P_1 of (C) corresponding to $\mu = \mu_0$ and $\mu = \mu_1$.

In the neighbourhood of a point of $(C_{\epsilon,\omega})$ corresponding to a given value of μ, the variation of $(ds/d\mu)^2$ due to a variation $d\mu$ of μ will be given by

$$\left(\frac{ds}{d\mu}\right)^2 = \sum_{i=0}^{3} g_{ii}(\bar{x}_0, \bar{x}_1, \bar{x}_2, \bar{x}_3)(\bar{x}_i')^2, \tag{140}$$

where we indicate by *primes* the derivation with respect to μ (the same notation is used everywhere below).

For a given value of ϵ and a given choice of the functions $\omega_i(\mu)$, the left-hand side of Equation (140) is a certain function $\Lambda(\mu, \epsilon)$ and therefore $(ds/d\mu)$ is also a certain function $H(\mu, \epsilon)$ of μ and ϵ.

Let $I(\epsilon)$ denote the function defined by:

$$I(\epsilon) = \int_{\mu_0}^{\mu_1} H(\mu, \epsilon)\, d\mu. \tag{141}$$

It is almost obvious that the extremum of $I(\epsilon)$, when ϵ varies in the neighbourhood of $\epsilon = 0$, corresponds to the extremum of the function $J(\epsilon)$ defined by

$$J(\epsilon) = \int_{\mu_0}^{\mu_1} \Lambda(\mu, \epsilon)\, d\mu = \int_{\mu_0}^{\mu_1} H^2(\mu, \epsilon)\, d\mu. \tag{142}$$

(The rigorous proof of this statement can be found, for instance, either in Landau and Lifshitz (1960, Ch. I) or in Misner *et al.* (1973, p. 322)).

The use of $J(\epsilon)$ instead of $I(\epsilon)$ avoids the complications connected with the presence of a square root of $(ds/d\mu)^2$ in Equation (141).

Consider now the following limited expansion of $\Lambda(\mu, \epsilon)$ in powers of ϵ:

$$\Lambda(\mu, \epsilon) = \Lambda_0(\mu) + \epsilon\Lambda_1(\mu) + \epsilon^2\Lambda_2(\mu). \tag{143}$$

This expansion corresponds to an expansion of $J(\epsilon)$:

$$J(\epsilon) = J_0 + \epsilon J_1 + \epsilon^2 J_2, \tag{144}$$

where

$$J_1 = \int_{\mu_0}^{\mu_1} \Lambda_1(\mu)\, d\mu, \tag{145}$$

must vanish if $J(\epsilon)$ has to become extremal with respect to the variations of ϵ in the neighbourhood of $\epsilon = 0$.

Let us, thus, determine the coefficient $\Lambda_1(\mu)$ of ϵ in the expansion of $\Lambda(\mu, \epsilon)$ limited to the terms in ϵ^0 and ϵ^1 alone.

In the computation of $\Lambda_1(\mu)$ we can limit our expansions in the following way, where, for the sake of brevity, we introduce the notation

$$G_{ii,j} = \frac{\partial g_{ii}}{\partial x_j} \quad (i = 0, 1, 2, 3)(j = 0, 1, 2, 3); \tag{146}$$

and take into account the definition (139):

$$\bar{x}_i'^2 = (x_i' + \epsilon\omega_i')^2 \approx x_i'^2 + \epsilon(2x_i'\omega_i'), \tag{147}$$

and the first term of the Taylor expansion of g_{ii}:

$$\left\{\begin{aligned} g_{ii}(\bar{x}_0, \bar{x}_1, \bar{x}_2, \bar{x}_3) &= g_{ii}(x_0 + \epsilon\omega_0; x_1 + \epsilon\omega_1; \ldots) \\ &\approx g_{ii} + \epsilon \sum_{j=0}^{3} G_{ii,j}\omega_j. \end{aligned}\right. \tag{148}$$

On inserting these limited expansions into Equation (140) we obtain:

$$\left(\frac{ds}{d\mu}\right)^2 \approx \sum_{i=0}^{3}\left[\left(g_{ii} + \epsilon \sum_{j=0}^{3} G_{ii,j}\omega_j\right)(x_i'^2 + \epsilon 2x_i'\omega_i')\right]. \tag{149}$$

Hence the coefficient $\Lambda_1(\mu)$ of ϵ:

$$\Lambda_1(\mu) = \sum_{i=0}^{3} 2g_{ii}x_i'\omega_i' + \sum_{i=0}^{3}\sum_{j=0}^{3} G_{ii,j}x_i'^2\omega_j. \tag{150}$$

On developing the sums and on exchanging the names of indices we can write Λ_1 as

$$\Lambda_1(\mu) = \sum_{j=0}^{3}\left[2g_{jj}x_j'\omega_j' + \omega_j \sum_{i=0}^{3} G_{ii,j}x_i'^2\right]. \tag{150'}$$

On taking into account, at the right-hand side of Equation (150′), the identity:

$$(g_{jj}x_j')\,\omega_j' = \frac{d}{d\mu}[(g_{jj}x_j')\,\omega_j] - \omega_j\frac{d}{d\mu}(g_{jj}x_j'), \tag{151}$$

we introduce the term $d(g_{jj}x_j'\omega_j)/d\mu$, which, by integration, becomes $(g_{jj}x_j')\,\omega_j$ and gives, because of $\omega_j(\mu_0) = \omega_j(\mu_1) = 0$, a zero contribution to J_1.

Thus, in Equation (145) $\Lambda_1(\mu)$ can be limited to $\tilde{\Lambda}_1(\mu)$ defined by

$$\tilde{\Lambda}_1(\mu) = \sum_{j=0}^{3} \omega_j\left[-2\frac{d}{d\mu}(g_{jj}x_j') + \sum_{i=0}^{3} G_{ii,j}x_i'^2\right]. \tag{152}$$

The integral J_1 defined by

$$J_1 = \int_{\mu_0}^{\mu_1} \tilde{\Lambda}_1(\mu)\,d\mu, \tag{153}$$

must be zero for any choice of the functions $\omega_j(\mu)$. This is possible only if the coefficient of each ω_j in the integrand, i.e. in Equation (152), is separately zero, and therefore the system of differential equations of a geodesic is represented by the four equations

$$\begin{cases} 2\dfrac{d}{d\mu}(g_{00}x_0') = \displaystyle\sum_{i=0}^{3} G_{ii,0}x_i'^2; & 2\dfrac{d}{d\mu}(g_{11}x_1') = \displaystyle\sum_{i=0}^{3} G_{ii,1}x_i'^2; \\ 2\dfrac{d}{d\mu}(g_{22}x_2') = \displaystyle\sum_{i=0}^{3} G_{ii,2}x_i'^2; & 2\dfrac{d}{d\mu}(g_{33}x_3') = \displaystyle\sum_{i=0}^{3} G_{ii,3}x_i'^2. \end{cases} \tag{154}$$

(A more elegant, but more abstract, proof can be found, for instance, in McVittie (1965, Sect. 2.3)).

It is important to remember that in the left-hand side of Equations (154) each $g_{ii}(x_0, x_1, x_2, x_3)$ must be considered as a function of μ, since each $x_i = x_i(\mu)$.

When the $(ds)^2$ of the space-time is described by Equation (135), some of the coefficients g_{ii} do not depend on certain of the variables x_0, x_1, x_2, x_3. Thus, according to Equation (136) we have simply:

$$\begin{cases} g_{00} = -1; \quad g_{11} = g_{11}(x_0); \quad g_{22} = g_{22}(x_0, x_1); \\ g_{33} = g_{33}(x_0, x_1, x_2). \end{cases} \tag{155}$$

Hence, according to the definition (146):

$$\begin{cases} G_{00,0} = 0; \quad G_{00,1} = G_{11,1} = 0; \quad G_{00,2} = G_{11,2} = G_{22,2} = 0; \\ G_{00,3} = G_{11,3} = G_{22,3} = G_{33,3} = 0, \end{cases} \tag{156}$$

and the system of differential Equations (154) of a geodesic reduces to:

$$\begin{cases} 2\dfrac{d}{d\mu}(g_{00}x_0') = G_{11,0}x_1'^2 + G_{22,0}x_2'^2 + G_{33,0}x_3'^2; \\ 2\dfrac{d}{d\mu}(g_{11}x_1') = G_{22,1}x_2'^2 + G_{33,1}x_3'^2; \\ 2\dfrac{d}{d\mu}(g_{22}x_2') = G_{33,2}x_3'^2; \\ 2\dfrac{d}{d\mu}(g_{33}x_3') = 0. \end{cases} \tag{157}$$

4.3. The Geodesics Describing the Motion of Photons

According to the geodesic principle formulated in Subsection 4.1, when the expression of $(ds)^2$ of a space-time (S) is given, the motion of photons, like the motion of any other particle, is represented by a geodesic of the space-time (S).

Thus, for a space-time whose $(ds)^2$ is given by Equations (135) and (136), the motion of photons must be represented parametrically, according to Subsection 4.2 by the system of four functions $x_0 = ct = x_0(\mu)$; $x_1 = \chi = x_1(\mu)$; $x_2 = \theta = x_2(\mu)$; $x_3 = \varphi = x_3(\mu)$, *solutions of the system of Equations* (157).

Now, it is relatively easy to show that the system:

$$\begin{cases} x_0 = x_0(\mu) = ct(\mu); \quad x_1 = x_1(\mu) = \chi(\mu); \\ x_2 = \theta_0 = \text{const.}, \quad x_3 = \varphi_0 = \text{const.}, \end{cases} \tag{158}$$

which corresponds to a '*radial propagation*' of the photons observed at the origin of coordinates, does actually describe such a solution.

Indeed, when x_0, x_1, x_2, x_3 are given by Equations (158) we have

$$x_2' = \frac{dx_2}{d\mu} = 0; \qquad x_3' = \frac{dx_3}{d\mu} = 0. \tag{159}$$

Thus, the last two equations of the system (157) are automatically satisfied and the system (157) reduces to a system of only two equations:

$$\begin{cases} 2\dfrac{d}{d\mu}(g_{00}x_0') = G_{11,0}x_1'^2 = \dfrac{\partial g_{11}}{\partial x_0}x_1'^2; \\ 2\dfrac{d}{d\mu}(g_{11}x_1') = 0, \end{cases} \tag{160}$$

equivalent, according to Equations (146) and (158) to the system:

$$\begin{cases} 2\dfrac{d}{d\mu}\left(-c^2\dfrac{dt}{d\mu}\right) = \dfrac{dg_{11}}{dt}\left(\dfrac{d\chi}{d\mu}\right)^2; \\ \dfrac{d}{d\mu}\left(g_{11}\dfrac{d\chi}{d\mu}\right) = 0, \end{cases} \tag{161}$$

or, in a more condensed form:

$$\begin{cases} -2c^2\dfrac{d}{d\mu}(t') = \dfrac{dg}{dt}\chi'^2; & \text{(162a)} \\ \dfrac{d}{d\mu}(g\chi') = 0, & \text{(162b)} \end{cases}$$

where g, according to Equation (136), is a given function of t defined by:

$$g = g(t) = g_{11} = R^2(t). \tag{163}$$

Since the system (161) contains only two unknown functions $t(\mu)$ and $\chi(\mu)$ it *always has a solution*:

$$t = f_t(\mu); \qquad \chi = f_\chi(\mu), \tag{164}$$

which through elimination of μ between $f_t(\mu)$ and $f_\chi(\mu)$ will give the time variation $\chi = \chi(t)$ of the radial coordinate χ.

However, this solution will be acceptable only if we succeed in showing that it is compatible with the condition $ds = 0$ which characterizes the propagation of light in vacuum (see Subsection 2.4).

Let us solve the system (162) and show that the condition $ds = 0$ can be satisfied. On multiplying Equation (162a) by $t' = dt/d\mu$ we obtain:

$$-c^2\frac{d(t'^2)}{d\mu} - \frac{dg}{d\mu}\chi'^2 = 0. \tag{165}$$

On derivating once $(g\chi')$ in Equation (162b) and on multiplying by $2\chi'$, we obtain:

$$g\frac{d(\chi'^2)}{d\mu} + 2\frac{dg}{d\mu}\chi'^2 = 0. \tag{166}$$

By addition of the left-hand side of Equations (165) and (166) we obtain:

$$-c^2\frac{d(t'^2)}{d\mu} + g\frac{d(\chi'^2)}{d\mu} + \frac{dg}{d\mu}\chi'^2 = 0, \tag{167}$$

equivalent to

$$-c^2\frac{d(t'^2)}{d\mu} + \frac{d}{d\mu}(g\chi'^2) = 0, \tag{167$'$}$$

or, since $g = R^2$, equivalent to

$$\frac{d}{d\mu}(\chi'^2R^2 - c^2t'^2) = 0. \tag{168}$$

Thus the system (162) of differential equations is equivalent to the system formed by Equation (168) and *one* of Equations (162), for instance the Equation (162b). This new system can be integrated *once* immediately, and gives *two* '*first integrals*':

$$R^2(t)(d\chi)^2 - c^2(dt)^2 = \alpha = \text{const.}; \tag{169}$$

$$R^2(t)\frac{d\chi}{d\mu} = \beta = \text{const..} \tag{170}$$

The first of these two equations shows that by taking $\alpha = 0$ this solution can be made compatible with the condition $ds = 0$, since we have $d\theta = d\varphi = 0$ so that $(ds)^2$ defined by Equation (135) reduces precisely to

$$(ds)^2 = R^2(t)(d\chi)^2 - c^2(dt)^2. \tag{171}$$

The second equation, i.e. Equation (170), can be satisfied by an arbitrary choice of the constant β, and it simply gives the relation between the parameter μ and the time t, since, on taking into account the relation $ds = 0$, i.e.

$$R(t)\,d\chi = \pm c\,dt, \tag{172}$$

it becomes

$$d\mu = \pm\frac{cR(t)}{\beta}dt, \tag{173}$$

hence

$$\mu - \mu_0 = \pm\int\frac{cR(t)}{\beta}dt. \tag{174}$$

Thus, finally, we have proved that in a space-time described by a $(ds)^2$ given by Equation (135), the photons observed at the origin move along the '*radial lines*' $\theta = \theta_0$, $\varphi = \varphi_0$, according to a law of motion defined by

$$\boxed{d\chi = -\frac{c}{R(t)}dt,} \tag{175}$$

where the sign (—) was chosen in order to express that for the photons observed at the origin the radial coordinate χ *decreases* when the time t increases. This relation applies whatever the value $(-1, 0, +1)$ of k, since σ_k is eliminated by the radial nature of the path.

Very many authors skip over the proof given above and consider as 'obvious' that the lines $\theta = \theta_0$, $\varphi = \varphi_0$ are necessarily geodesic, so that they indulge in deriving the law of motion directly from the condition $(ds) = 0$.

References

WORKS CITED IN THE TEXT

(Works that can be used for further study are marked by an asterisk.)

Kittell, C., Knight, W. D., and Ruderman, M. A.: 1962, *Mechanics* (Berkeley Physics Course, Vol. I), McGraw-Hill, N.Y.

Kourganoff, V.: 1964, *Initiation à la Theorie de la Relativité*, Presses Univ. de France, Paris. (This elementary textbook written in French, exists in translation into Spanish, Russian and Polish. A

mimeographed English translation can be bought, on request, by writing to the author: 20 Av. Paul Appell, 75014, Paris, France.)
Landau, L. D. and Lifshitz, E. M.: 1960, *Mechanics*, Addison-Wesley, Reading, Mass.
* McVittie, G. C.: 1965, *General Relativity and Cosmology*, 2nd edn., Chapman and Hall, London.
* Misner, C. W., Thorne, K. S., and Wheeler, J. A.: 1973, *Gravitation*, Freeman, San Francisco.
Smith, J. H.: 1965, *Introduction to Special Relativity*, Benjamin, N.Y.

CHAPTER 12

RELATIVISTIC EFFECTS IN OBSERVATIONAL COSMOLOGY. THE COSMOLOGICAL REDSHIFT IN EXPANDING UNIVERSE

1. A Bi-Dimensional Model of an Expanding Universe. Fixity in Mobility: The Comoving Coordinates

Let us consider, as an introduction to the study of relativistic effects in observational cosmology, a *bi-dimensional model* of a curved expanding universe.

The corresponding 'concretization' is provided by an *expanding spherical plastic balloon* upon which several points are painted. These points are supposed to represent both the observers and the galaxies (or clusters of galaxies) of that bi-dimensional universe.

Kinematically, this model corresponds to a sphere (S_2) of variable radius $a(t)$. Its relativistic representation in (tri-dimensional) space-time can be given by a space-like $(ds)^2$:

$$(ds)^2 = -c^2(dt)^2 + (d\lambda)^2, \tag{1}$$

where $(d\lambda)^2$ is given by Equation (68″) of Chapter 11, viz.

$$(d\lambda)^2 = a^2(t)[(d\chi)^2 + (d\psi)^2 \sin^2 \chi]. \tag{2}$$

Consider a 'creeping observer' at the origin P (the 'north pole') of the coordinate system (χ, ψ) introduced in Subsection 3.1.2.1 of Chapter 11. Let (χ_0, ψ_0) denote the *constant* angular coordinates of a galaxy M painted on (S_2). These coordinates of the galaxy M would not be modified by the expansion of the balloon. The constancy of (χ_0, ψ_0) would express a kind of *fixity*, or *insertion*, of the galaxy M in the *moving* space (S_2).

Of course, because of the variability of the radius a of (S_2), described by the function $a(t)$, the *linear* polar distance ρ_0 of M becomes a function $\rho_0(t)$ of time t (and of the angular polar distance χ_0 of M):

$$\rho_0 = \rho_0(t) = a(t)\chi_0. \tag{3}$$

Thus, in spite of the *fixity* of M in the space (S_2), expressed by the constancy of χ_0 and ψ_0, the creeping astronomers placed at the origin P will observe a *recession* (increase of the distance ρ_0) of the galaxy M and of all other galaxies, if the variation of $a(t)$ corresponds to a phase of expansion of (S_2).

In cosmology, it is rather usual to express this phenomenon by saying that the dimensionless (angular) coordinates (i.e. the coordinates χ and ψ in the case of a bi-dimensional expanding space) represent for each galaxy the '*comoving coordinates*'.

This terminology (also used in the tri-dimensional expanding space considered below) is not very satisfactory, since it risks to introduce the idea of *mobility* in a concept where the most important is the idea of *fixity* (with respect to the corresponding space). However, no better terminology being available, we shall use it below.

Of course, the variability of $\rho_0(t)$ will manifest itself by a kind of Doppler shift (a redshift if a increases), but the only purpose of this subsection being the introduction of the concept of the fixity of the comoving coordinates in an expanding space, we shall restrict our discussion of this 'cosmological redshift' and of the other relativistic effects to the tri-dimensional spaces considered below.

2. Tri-Dimensional Friedmann Relativistic Models of Expanding Universe

2.1. THE ROBERTSON–WALKER SPACE-TIME AND ITS METRIC

Robertson, in 1935–36, and independently Walker, in 1936, introduced the space-time metric defined by an equation formally identical with Equation (1), but where $(d\lambda)^2$ is given by

$$(d\lambda)^2 = (R\,d\chi)^2 + (R\sigma_k\,d\theta)^2 + (R\sigma_k \sin\theta\,d\varphi)^2. \tag{4}$$

In Equation (4) R is a function $R(t)$ of time t, whereas σ_k is a function of χ defined, according to the value $(-1, 0, +1)$ of k, by

$$\sigma_0 = \chi; \qquad \sigma_{+1} = \sin\chi; \qquad \sigma_{-1} = \sinh\chi, \tag{5}$$

as in Chapter 11, Subsection 3.3.1.

Both Robertson and Walker gave the first proof that this '*metric*' defines the most general Riemannian geometry compatible with the cosmological principle, i.e. with the homogeneity and isotropy of the idealized universe.

These very complex and abstract proofs are rather clearly exposed by Robertson–Noonan (1968) and by Misner *et al.* (1973). However, we strongly recommend to study these proofs only after a complete assimilation of the present chapter of our textbook, in which we limit ourselves to the presentation of the main *consequences* of the space-time metric defined by Equations (1), (4) and (5).

It was already shown, in Chapter 11, Subsection 3.3, that *at a given moment* (this concept is discussed, on a more advanced level, by Misner *et al.* (1973, p. 713f), corresponding in Equation (1) to $dt = 0$, the tri-dimensional slice of the four-dimensional space-time of the Robertson–Walker metric is, for $k = 0$, the open Euclidean space (Σ_3^0); for $k = +1$, the closed non-Euclidean space (Σ_3^+); and for $k = -1$, the open non-Euclidean space (Σ_3^-). It was also shown that in the cases $k = \pm 1$ the value of $1/R(t)$ characterizes the degree of (positive or negative) curvature of the corresponding ('spherical' or 'hyperbolic') space.

2.2. THE EINSTEINIAN EQUATIONS AND COMPARISON WITH THE EQUATIONS GIVEN BY THE NEWTONIAN APPROACH

The Newtonian gravitational theory can be considered as summarized by Poisson's Equation (see Chapter 2, Section 4), viz.

$$\operatorname{div} \boldsymbol{g} = -4\pi G\rho, \tag{6}$$

where $\boldsymbol{g}$ is the gravitational field at a point of local mass density ρ and G is the gravitational constant.

In the relativistic, 'Einsteinian' treatment, based on the concept of space-time, described for instance by the Robertson–Walker metric, the term div $\mathbf{g}$ of Equation (6) is replaced by an expression for the '*curvature of the space-time*' (which must not be confused with the curvature of the tri-dimensional spaces $\Sigma_3^{\pm}$), and which, in the Friedmann model, is attributed to the presence of a certain mass density ρ of the cosmic fluid alone.

After very long and tedious computations (see, for instance, McVittie (1965) or Weinberg (1972)) one finds that one can conserve in Equation (6) the term $(-4\pi G\rho)$, provided that one replaces div $\mathbf{g}$ by the following function of $R(t)$:

$$\operatorname{div} \mathbf{g} \to -\tfrac{3}{2}R^{-2}[(dR/dt)^2 + kc^2], \tag{7}$$

where k can take only one of the three values $(-1); 0; (+1)$.

Hence, the Einsteinian differential equation for $R(t)$, obtained by Friedmann:

$$\tfrac{3}{2}R^{-2}[(dR/dt)^2 + kc^2] = 4\pi G\rho = (4\pi G\rho_0)\left(\frac{\rho}{\rho_0}\right). \tag{8}$$

On introducing the notations of Chapter 10, Subsections 3.1, 3.2, and 5.4, viz.

$$H_0 = \left(\frac{dR/dt}{R}\right)_{t=t_0}; \qquad \Omega = \frac{8\pi G\rho_0}{3H_0^2}; \qquad a = \frac{R}{R_0}, \tag{9}$$

and the invariant of the Friedmann universe, viz.

$$\rho R^3 = \rho_0 R_0^3, \tag{10}$$

Equation (8) takes the form:

$$\left(\frac{da}{dt}\right)^2 = H_0^2\left(\frac{\Omega}{a} - \frac{kc^2}{R_0^2 H_0^2}\right). \tag{11}$$

As already explained in Chapter 10, Subsection 5.7, this Equation can be identified with the Newtonian Equation (41) of Ch. 10, provided that R_0 is fixed in the Newtonian theory by the 'normalizing relation'

$$(\Omega - 1)R_0^2 H_0^2 = kc^2, \tag{12}$$

which can also be considered as the relativistic expression for the curvature of the universe $(1/R_0)$ at time t_0 in terms of H_0 and Ω.

Equation (12) shows that $k = +1$ must correspond to $\Omega > 1$ and $k = -1$ must correspond to $\Omega < 1$.

It also shows that $k = 0$ corresponds to $\Omega = 1$. In this particular case, no normalization is actually necessary since Equation (11) reduces for $k = 0$ and $\Omega = 1$ to Equation

$$\frac{da}{dt} = \dot{a} = H_0 a^{-1/2}, \tag{13}$$

identical with Equation (52) of Ch. 10, which no longer depends on R_0, and which determines directly the (relative) scale factor a as a function of time t.

2.3. THE EXPANDING OPEN EUCLIDEAN FRIEDMANN UNIVERSE (Σ_3^0)

The space (Σ_3^0) was already studied in Chapter 11, Subsection 3.3.2 from a purely geometrical point of view, and was shown to be an open Euclidean space.

Here we are in presence of the so-called 'Einstein–de Sitter' (or EdS) model of Friedmann universe, and the comparison with the Newtonian treatment shows that it corresponds to $\Omega = 1$. Therefore, this model is ever-expanding, and the details of this expansion have already been discussed in Chapter 10, Subsection 5.6.

Since now R is supposed to vary with time t, we are (with one more dimension) in a situation similar to that of an expanding balloon discussed in Section 1.

The linear radial distance r defined in the ordinary tri-dimensional space by Equation (111) of Chapter 11, viz.

$$r = R\chi, \tag{14}$$

becomes here

$$r(\chi, t) = R(t)\chi. \tag{15}$$

This shows that a galaxy M whose angular ('comoving') coordinates (χ, θ, φ) have a given fixed value (χ_0, θ_0, φ_0) can be observed, if the space (Σ_3^0) itself is in a phase of expansion, as an object receeding from the observer placed at the origin of the coordinates, since $r(t)$ increases with $R(t)$.

2.4. THE OSCILLATING CLOSED NON-EUCLIDEAN FRIEDMANN UNIVERSE (Σ_3^+)

The space (Σ_3^+) was already studied in Chapter 11, Subsection 3.3.3 from a purely geometrical point of view and was shown to be a closed non-Euclidean space.

Since, as shown in Subsection 2.2, $k = +1$ corresponds to $\Omega > 1$, it follows that we are here in presence of an *'oscillating universe'*. All details of its oscillation have already been studied in Chapter 10, Subsection 5.8, in terms of the time variation $R(t)$ of R.

2.5. THE EXPANDING OPEN NON-EUCLIDEAN FRIEDMANN UNIVERSE (Σ_3^-)

The space (Σ_3^-) was already studied in Chapter 11, Subsection 3.3.4 from a purely geometrical point of view and was shown to be an open non-Euclidean space.

Since, as shown in Subsection 2.2, $k = -1$ corresponds to $\Omega < 1$, it follows that we are here in presence of *an ever-expanding universe*. All details of this expansion have already been studied in Chapter 10, Subsection 5.9.

3. The Cosmological Redshift

3.1. FUNDAMENTAL RELATIONS

3.1.1. The Relation between the Radial Coordinate χ of a Source and the Scale Factor

Consider the emission of an electromagnetic signal at time t_e by a 'source' (a galaxy, a quasar, etc.) (S). The source is supposed to be *fixed* in terms of its 'comoving' coordinates (χ, θ, φ), (see Sections 1 and 2). For the sake of brevity the subscripts (0) are here omitted from $\chi_0, \theta_0, \varphi_0$.

This emission represents an event (E_e), whose coordinates in the four-dimensional space-time are $(\chi, \theta, \varphi, t_e)$.

The receipt of this signal, at present time t_0, by an observer (or by a 'receiver') placed at the origin O of coordinates, represents an event (E_0), whose coordinates in the space-time are $(0, 0, 0, t_0)$.

The history of the 'radial' propagation (i.e. a propagation such that $d\theta = 0$ and $d\varphi = 0$, see Chapter 11, Subsection 4.3) of this signal is described (whatever the value $-1, 0, +1$ of k) by the fundamental Equation (175) of Ch. 11, viz.

$$d\chi = -\frac{c}{R(t, \Omega)}dt, \tag{16}$$

where we have written $R(t, \Omega)$ instead of $R(t)$ in order to emphasize the dependence (proved in Chapter 10, Subsection 5) of $R(t)$ on Ω (or on q_0), and, more generally, on the model of universe. Of course, R also depends on H_0, but this will prove less important in the present section.

Let t' denote some time *intermediate* between the departure (emission) time t_e and the arrival (receipt) time t_0. At time t' the radial coordinate of the signal will be χ' such that $0 \leqslant \chi' \leqslant \chi$.

With this notation Equation (16) takes a more general form:

$$d\chi' = -\frac{c}{R(t', \Omega)}dt'. \tag{17}$$

On integrating from t_e to t_0 we obtain (after a double reversal of sign):

$$\chi = \int_{t_e}^{t_0} \frac{c}{R(t', \Omega)}dt'. \tag{18}$$

On introducing the (relative) scale factor $a(t', \Omega)$ defined by Equation (2) of Ch. 10 but with a notation emphasizing the dependence on Ω, this can also be written as

$$\boxed{\chi = \frac{c}{R(t_0, \Omega)} \int_{t_e}^{t_0} \frac{1}{a(t', \Omega)}dt'.} \tag{18'}$$

Note that it would be more consistent, in our notation, to call t'_e the departure time and t'_0 the arrival time, but it is traditional to replace t'_e by t_e and t'_0 by t_0.

The most important, however, is not to lose sight of the fact that χ is *independent of time*.

We assume everywhere below that even for an oscillating universe ($\Omega > 1$) the time t_0 still corresponds to *a phase of expansion*.

One can then represent schematically the variations of $R(t', \Omega)$ as a function of time t' by the upper part (1a) of Figure 12.1, and the time variations of $c/R(t', \Omega)$ by the lower part (1b) of Figure 12.1, both figures corresponding to the boundary condition $R(0, \Omega) = 0$ characteristic of the 'big-bang model' of universe.

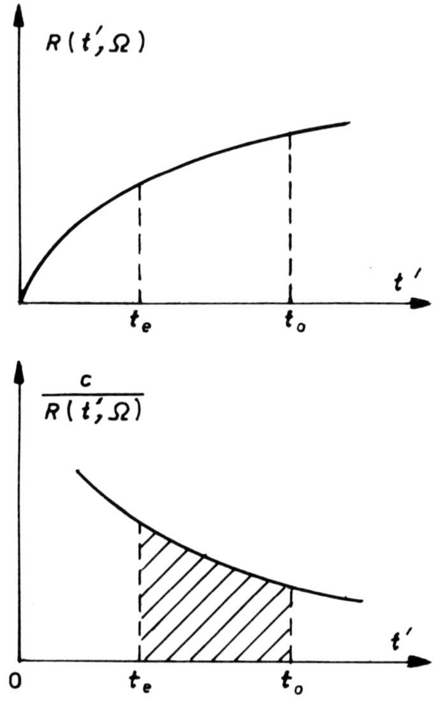

Fig. 12.1. The upper part (a) of the figure represents the variations of the absolute scale factor R as a function of time (in a phase of expansion). The lower part (b) of the figure represents the time variations of c/R. The cross-hatched area represents the value of the radial coordinate χ of the source.

According to Equation (18), the cross-hatched area in Figure 12.1b represents the value of χ.

3.1.2. *The Relation between the Scale Factor and the Cosmological Redshift of a Source*

Consider now two successive maxima M_n and M_{n+1} of an electromagnetic wave, and let T_e denote the period of this wave at departure from the source and T_0 denote the period of the same wave observed at its arrival at O. This means that if the maximum M_n is emitted at time t_e the next maximum M_{n+1} is emitted at time $(t_e + T_e)$; and similarly that if M_n arrives at O at time t_0, M_{n+1} arrives at time $(t_0 + T_0)$.

All these events can be summarized in the following way (see Table 12.1):

TABLE 12.1
The coordinates of different events in the four-dimensional space-time

	Phase M_n	Phase M_{n+1}
Emission	$E_e(\chi, \theta, \varphi, t_e)$	$E_e'(\chi, \theta, \varphi, t_e + T_e)$
Receipt	$E_0(0, 0, 0, t_0)$	$E_0'(0, 0, 0, t_0 + T_0)$

In Figure 12.2, similar to Figure 12.1b, which represents schematically $c/R(t', \Omega)$ as a function of t' (but on a different scale) the successive times t_e, $(t_e + T_e)$, t_0, $(t_0 + T_0)$ correspond to the points A, A′, D, D′ of the Ot′ axis respectively.

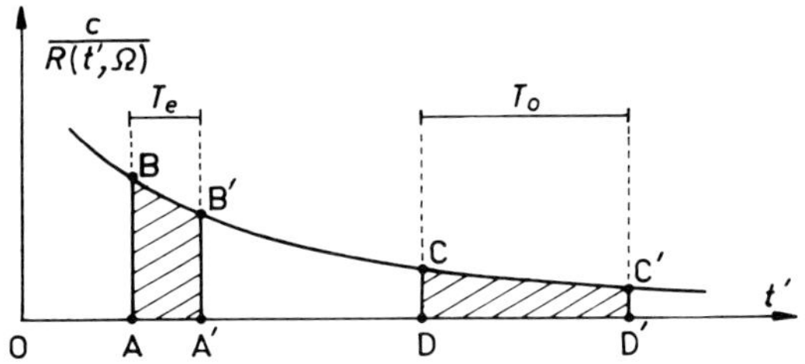

Fig. 12.2. The time variation of c/R. The cross-hatched areas must be equal.

Of course, since the comoving coordinates of the source and those of the receiver remain constant, the value of χ given by the area ABCD corresponding to the propagation of M_n, between the times t_e and t_0, must be equal to the value of χ given by the area A′B′C′D′ corresponding to the propagation of M_{n+1}. In Figure 12.2 the ratio T_e/t_e and the difference between T_e and T_0 are deliberately exagerated for the sake of clarity.

On subtracting the common part A′B′CD we are left with two equal areas ABB′A′ and DCC′D′ cross-hatched in Figure 12.2 (these areas must not be confused with the cross-hatched area in Figure 12.1).

Since during the expansion phase, when $R(t', \Omega)$ increases, $c/R(t', \Omega)$ decreases, we have: $\mathrm{DC}(t_0) < \mathrm{AB}(t_e)$. Therefore the equality between the areas ABB′A′ and DCC′D′ demands that $\mathrm{DD}' > \mathrm{AA}'$.

Thus it is physically obvious that

$$T_0 > T_e. \tag{19}$$

If we consider the corresponding wave-lengths λ in vacuum we have

$$\lambda_0 = cT_0; \qquad \lambda_e = cT_e, \tag{20}$$

and

$$\lambda_0 > \lambda_e. \tag{21}$$

The wave-length λ_e of the emitted wave is, by definition, equal to the wave-length of *the comparison wave in the laboratory at* O (hence a possibility of confusion, since in usual spectroscopy this wave length λ_e is called λ_0 and not λ_e!).

Thus we see that, by a kind of Doppler effect, the observed cosmic radiation will be of a greater wave-length λ_0 than the laboratory wave-length λ_e. Actually, what is observed is the '*cosmological redshift*' generated, for a source *whose comoving coordinates* χ, θ, φ *are constant*, by *the expansion of the space* itself. This cosmological redshift is *a purely relativistic effect*, an effect of *general* relativity, since even in special relativity the Doppler effect corresponds to the relative motion *in* space and not to the motion *of* the space!

More precisely, on assimilating the areas ABB′A′ and DCC′D′ with narrow rectangles, which is permitted because T_0 and T_e are very small, we have

$$\text{Area (ABB'A')} = \frac{c}{R(t_e, \Omega)} T_e, \tag{22}$$

and

$$\text{Area (DCC'D')} = \frac{c}{R(t_0, \Omega)} T_0. \tag{23}$$

Therefore the equality of the areas ABB′A′ and DCC′D′ demands that

$$\frac{T_e}{R(t_e, \Omega)} = \frac{T_0}{R(t_0, \Omega)}, \tag{24}$$

hence

$$\frac{\lambda_0}{\lambda_e} = \frac{T_0}{T_e} = \frac{R(t_0, \Omega)}{R(t_e, \Omega)}. \tag{25}$$

The ratio λ_0/λ_e will be denoted henceforth by ζ, so that we can write:

$$\zeta = \frac{\lambda_0}{\lambda_e} = \frac{R(t_0, \Omega)}{R(t_e, \Omega)}. \tag{26}$$

Some authors (for instance Andrillat (1970) or Heidmann (1973)) call this function ζ the '*spectral ratio*', but as many other important quantities in cosmology, it has not any standard name.

A more complicated quantity z defined by

$$z = \frac{\lambda_0 - \lambda_e}{\lambda_e} = \zeta - 1, \tag{27}$$

is more usual (and less convenient) and is called the '*redshift*'.

Thus we have, according to Equations (26) and (27):

$$\zeta = 1 + z = \frac{R(t_0, \Omega)}{R(t_e, \Omega)}. \tag{28}$$

On introducing the (relative) scale factor $a(t', \Omega)$ defined by

$$a(t') = a(t', \Omega) = \frac{R(t', \Omega)}{R(t_0, \Omega)}, \tag{29}$$

we find that $\zeta = 1 + z$ is also given by

$$\zeta = 1 + z = \frac{1}{a(t_e, \Omega)}. \tag{30}$$

During an expansion phase of the universe one necessarily has $\zeta \geqslant 1$ and $z \geqslant 0$.

3.2. The Redshift of a Source as a Function of its Travel-Time

3.2.1. *The Definition of the Travel-Time of a Source. The Scale Factor as a Function of the Travel-Time*

The difference between the arrival-time t_0 and the departure-time t_e, i.e. the quantity τ defined by

$$\tau = t_0 - t_e, \tag{31}$$

is called the '*travel time*' of the considered source.

More generally, every value t' of the passage of the signal at some intermediate point P of radial coordinate χ' can be associated with the '*residual travel-time*' τ' defined by

$$\tau' = t_0 - t', \tag{32}$$

so that t' can be expressed by

$$t' = t_0 - \tau'. \tag{32'}$$

If the propagation of the signal is described, for a given value of t_0, by the function $\chi'(\tau')$, it becomes convenient to consider, besides the function $a(t', \Omega)$, the function $\tilde{a}(\tau', \Omega)$, which represents the same physical quantity but is defined mathematically by

$$\tilde{a}(\tau', \Omega) = a(t_0 - \tau', \Omega) = a(t', \Omega). \tag{33}$$

In terms of the variable τ' the departure-time is expressed by $\tau' = \tau$ whereas the arrival-time is expressed by $\tau' = 0$, so that the corresponding scale factors will be given by $\tilde{a}(\tau, \Omega)$ and $\tilde{a}(0, \Omega)$ respectively.

We see that τ' can be considered as the time '*reckoned backwards*' from an origin of times corresponding to the receipt of the signal by the observer.

In terms of the function $\tilde{a}(\tau', \Omega)$ the fundamental relation (30) becomes:

$$\boxed{\zeta = 1 + z = 1/\tilde{a}(\tau, \Omega).} \tag{34}$$

3.2.2. *The Approximate Expression for the Redshift z of a Source as a Function of its Travel-Time τ for Near Sources*

When the source is sufficiently near the origin, i.e. when z and τ are sufficiently small, one can use limited power expansions to easily obtain an approximate expression for z as a function of τ and reciprocally.

Indeed, we can expand $\tilde{a}(\tau', \Omega)$, equal according to Equation (33) to $a(t_0 - \tau', \Omega)$, in powers of τ' in the neighbourhood of t_0. This gives:

$$\tilde{a}(\tau', \Omega) = a(t_0 - \tau', \Omega) \approx a(t_0, \Omega) - \dot{a}(t_0, \Omega)\tau' + \tfrac{1}{2}\ddot{a}(t_0, \Omega)\tau'^2. \tag{35}$$

This expansion can also be written on taking into account the definitions of a_0, H_0 and

q_0 given by Eqs. (3), (9') and (42) of Ch. 10 (and the fact that for the Friedmann model $q_0 = \Omega/2$):

$$\tilde{a}(\tau', \Omega) \approx 1 - H_0\tau' - \tfrac{1}{2}q_0 H_0^2 \tau'^2 \approx 1 - H_0\tau' - \frac{\Omega}{4} H_0^2 \tau'^2, \tag{35'}$$

or

$$\tilde{a}(\tau, \Omega) \approx 1 - H_0\tau - \frac{\Omega}{4} H_0^2 \tau^2. \tag{35''}$$

On combining Equations (34) and (35″) one obtains:

$$\zeta = 1 + z \approx 1 + H_0\tau + H_0^2\left(1 + \frac{\Omega}{4}\right)\tau^2, \tag{36}$$

or

$$z \approx H_0\tau + \left(1 + \frac{\Omega}{4}\right)(H_0\tau)^2, \tag{36'}$$

or, more generally

$$z' \approx H_0\tau'\left[1 + \left(1 + \frac{\Omega}{4}\right)H_0\tau'\right]. \tag{36''}$$

On solving Equation (36′) for $H_0\tau$, we obtain in a first approximation $H_0\tau \approx z$ and on replacing $(H_0\tau)^2$ by z^2:

$$H_0\tau \approx z - \left(1 + \frac{\Omega}{4}\right)(H_0\tau)^2 \approx z - \left(1 + \frac{\Omega}{4}\right)z^2. \tag{37}$$

3.3. THE CASE OF SEVERAL SOURCES OBSERVED AT THE SAME TIME

If, instead of considering as above just *one* source (S), characterized by the parameters t_e, t_0, z, ζ, τ, etc., we consider n sources (S_i) $(i = 1, 2, 3, \ldots n)$ all observed at the same arrival-time t_0, each (S_α) of these sources (S_i) can be considered as overtaken (bypassed) at time t' by the radiation of a more remote source (S) (observed at the same arrival time t_0) just at the moment of emission of a signal by (S_α).

Thus, with the notation defined in Subsection 3.2.1, the radial coordinate of (S_α) can be denoted by χ' and its (total) travel-time can be denoted by the same symbol τ' as the residual travel-time of the radiation of (S) overtaking (S_α). If, moreover, we denote by z' and ζ' the respective values of the redshift and of the spectral ratio of the source (S_α) we obtain from Equation (30):

$$\zeta' = 1 + z' = \frac{1}{a(t', \Omega)}, \tag{38}$$

and from Equation (34)

$$\zeta' = 1 + z' = 1/\tilde{a}(\tau', \Omega). \tag{38'}$$

3.4. THE RADIAL COORDINATE χ OF A SOURCE AS A FUNCTION OF ITS REDSHIFT

3.4.1. *The Approximate Relations Between χ and z for Near Sources*

When χ and z are sufficiently small, one can use limited power expansions to easily obtain an approximate expression for χ as a function of z and reciprocally.

Indeed, on taking into account the fundamental Equation (18′), the definition (32) of τ', the definition (33) of $\tilde{a}(\tau', \Omega)$ and the first two terms of the expansion (35′), we first obtain the following expansion of $\chi R(t_0, \Omega)/c$ as a function of τ:

$$\frac{\chi R(t_0, \Omega)}{c} = \int_0^\tau \frac{d\tau'}{\tilde{a}(\tau', \Omega)} \approx \int_0^\tau (1 + H_0\tau')\,d\tau'$$

$$\approx \tau + \tfrac{1}{2}H_0\tau^2 \approx \frac{1}{H_0}[(H_0\tau) + \tfrac{1}{2}(H_0\tau)^2]. \tag{39}$$

Then, on replacing $H_0\tau$ by its expansion (37) we finally find:

$$\chi \approx \frac{c}{H_0 R(t_0, \Omega)}\left[z - \left(\frac{1}{2} + \frac{\Omega}{4}\right)z^2\right], \tag{40}$$

which, on taking into account the normalizing Eq. (61) of Ch. 10, can also be written, for $k = \pm 1$:

$$\chi \approx \left(\frac{\Omega - 1}{k}\right)^{1/2} z\left[1 - \left(\frac{1}{2} + \frac{\Omega}{4}\right)z\right]. \tag{40′}$$

In a first approximation Equation (40) can be reduced to

$$z \approx \frac{\chi H_0 R(t_0, \Omega)}{c}, \tag{40″}$$

which shows that z and χ are of the same order. On replacing in Equation (40) z^2 by the square of its approximate value (40″) and on solving for z we obtain the relation, reciprocal of Equation (40):

$$z \approx \frac{1}{c}\chi H_0 R(t_0, \Omega) + \left(\frac{1}{2} + \frac{\Omega}{4}\right)\left[\frac{\chi H_0 R(t_0, \Omega)}{c}\right]^2. \tag{41}$$

3.4.2. *The Radial Coordinate χ of a Source as a General Function of its Redshift*

Of course, it is not necessary to take into account the physical interpretation of ζ' given in Subsection 3.3 and one can simply consider ζ', from a purely mathematical point of view, as an auxiliary variable defined as a function of t' by Equation (38).

We thus obtain, by derivation with respect to t'

$$\frac{d\zeta'}{dt'} = -\frac{1}{a^2(t', \Omega)}\frac{da(t', \Omega)}{dt'} = -\zeta'^2\frac{da(t', \Omega)}{dt'}. \tag{42}$$

On the other hand, according to Eq. (41) of Ch. 10 applied at time t', we have (for a Friedmann universe):

$$\frac{da(t',\Omega)}{dt'} = H_0\left[\frac{\Omega}{a(t',\Omega)}-(\Omega-1)\right]^{1/2} = H_0[\Omega\zeta'-(\Omega-1)]^{1/2}. \tag{43}$$

Hence,

$$dt' = \frac{-d\zeta'}{\zeta'^2 H_0[\Omega\zeta'-(\Omega-1)]^{1/2}}. \tag{44}$$

Let us now return to Equation (18′). If, by writing χ as $\chi(\zeta,\Omega)$ we indicate that χ is a function not only of $\zeta = 1 + z$ but also of Ω, we obtain:

$$\chi(\zeta,\Omega)R(t_0,\Omega) = \int_{t_e}^{t_0}\frac{c\,dt'}{a(t',\Omega)} = c\int_{\zeta'=\zeta}^{\zeta'=1}\zeta'\,dt', \tag{45}$$

and on replacing $\zeta'\,dt'$ by its expression (44) we finally find:

$$\chi(\zeta,\Omega) = \frac{c}{R(t_0,\Omega)H_0}\int_1^{\zeta}\frac{d\zeta'}{\zeta'[\Omega\zeta'-(\Omega-1)]^{1/2}}. \tag{46}$$

In the particular case $\Omega = 1$ (i.e. $k = 0$) of the *EdS model*, Equation (46) yields:

$$\begin{aligned}\chi(\zeta,1) = \chi_{\mathrm{EdS}}(\zeta) &= \frac{c}{R(t_0,1)H_0}\int_1^{\zeta}\zeta'^{-3/2}d\zeta' \\ &= \frac{2c}{R(t_0,1)H_0}(1-\zeta^{-1/2}),\end{aligned} \tag{47}$$

which, for small values of z gives the same expansion as Equation (40), viz.

$$\chi(z,1) \approx \frac{cz}{R(t_0,1)H_0}(1-\tfrac{3}{4}z). \tag{48}$$

In the more general case $\Omega \neq 1$ (i.e. $k = \pm 1$) we can make use of the normalizing Equation (61) of Ch. 10 to transform Equation (46) into:

$$\chi(\zeta,\Omega) = \int_1^{\zeta}\frac{d\zeta'}{\zeta'}\left[\frac{\Omega k\zeta'}{(\Omega-1)}-k\right]^{-1/2}. \tag{49}$$

A computation quite elementary but extremely long and tedious, in which one uses the change of variables

$$\zeta' = \frac{\Omega-1}{\Omega\sin^2(\eta/2)}, \quad \text{or} \quad \zeta' = \frac{1-\Omega}{\Omega\sinh^2(\eta/2)}, \tag{50}$$

gives the value of the integral at the right-hand side of Equation (49).

However, the same result can be obtained much more easily if, starting again from Equation (18), we make use of the parametric representation of the function $R(t,\Omega)$ given in Chapter 10, Section 5.

Indeed, in the case $k = +1$ of an oscillating Friedmann universe, we have by Equations (65′), (67) and (68) of Ch. 10, when we consider the function $R(t',\Omega)$, denoted sometimes below, as in Chapter 10, by $R(t')$, for the sake of brevity:

$$R[\eta] = \tfrac{1}{2}R_{\max}(1-\cos\eta); \tag{51}$$

$$t'[\eta] = \frac{1}{2c} R_{\max}(\eta - \sin \eta), \tag{52}$$

where

$$R_{\max} = \frac{\Omega R_0}{\Omega - 1} = \frac{c\,\Omega}{H_0(\Omega - 1)^{3/2}}. \tag{53}$$

Hence, on taking the derivative of Equation (52), and taking into account Equation (51):

$$c\,dt' = \tfrac{1}{2} R_{\max}(1 - \cos \eta)\,d\eta = R[\eta]\,d\eta. \tag{54}$$

Thus Equation (18), viz.

$$\chi = \int_{t_e}^{t_0} \frac{c\,dt'}{R(t')}, \tag{55}$$

takes for form

$$\chi = \int_{\eta_e}^{\eta_0} \frac{c\,dt'}{R[\eta]} = \int_{\eta_e}^{\eta_0} d\eta = \eta_0 - \eta_e, \tag{56}$$

where η_0 and η_e correspond to t_0 and t_e respectively.

The value of η_0 as a function of Ω was already calculated in Chapter 10, and is given by Equations (78) and (85) of that Chapter, viz.

$$\tfrac{1}{2}\eta_0 = \sin^{-1}\left(\frac{\Omega - 1}{\Omega}\right)^{1/2};\ \tfrac{1}{2}\sin \eta_0 = \frac{(\Omega - 1)^{1/2}}{\Omega}, \tag{57}$$

Let us now determine η_e as a function of ζ and Ω. According to Equation (51) and (53) we have:

$$R(\eta_e, \Omega) = \tfrac{1}{2} R_0 \frac{\Omega}{(\Omega - 1)}(1 - \cos \eta_e). \tag{58}$$

On eliminating $R(\eta_e, \Omega)$ by Equation (28), viz.

$$R(\eta_e, \Omega) = \frac{1}{\zeta} R(t_0, \Omega) = \frac{1}{\zeta} R_0, \tag{59}$$

we find:

$$\tfrac{1}{2}(1 - \cos \eta_e) = \frac{\Omega - 1}{\Omega\,\zeta}, \tag{60}$$

or

$$\cos \eta_e = 1 - \frac{2(\Omega - 1)}{\Omega\,\zeta}. \tag{60'}$$

Hence,

$$\sin(\eta_e/2) = \left(\frac{1 - \cos \eta_e}{2}\right)^{1/2} = \left(\frac{\Omega - 1}{\Omega\,\zeta}\right)^{1/2}, \tag{61}$$

$$\sin \eta_e = (1 - \cos^2 \eta_e)^{1/2} = \frac{2(\Omega - 1)^{1/2}}{\Omega} \frac{(1 + z\Omega)^{1/2}}{\zeta} \tag{61'}$$

and

$$\eta_e/2 = \sin^{-1}\left(\frac{\Omega - 1}{\Omega\,\zeta}\right)^{1/2}. \tag{61''}$$

Thus, according to Equation (56), (57) and (61″), the value of $\chi_+(\zeta, \Omega)$, is given by given by

$$\chi_+(\zeta, \Omega) = 2[(\eta_0/2) - (\eta_e/2)]$$

$$= 2\left[\sin^{-1}\left(\frac{\Omega-1}{\Omega}\right)^{1/2} - \sin^{-1}\left(\frac{\Omega-1}{\Omega\,\zeta}\right)^{1/2}\right], \tag{62}$$

where the subscript (+) indicates that this relation corresponds to $k = +1$.

We note that $\chi_+(\zeta, \Omega)$ depends only on Ω and ζ: it does not depend explicitly on H_0.

Since $\zeta = 1 + z$, we can consider that Equation (62) gives us also χ_+ as a function of the redshift z.

We shall not discuss in detail the case $k = -1$, since we have already seen in Chapter 10, that in Equations such as (62) one can obtain the result corresponding to $k = -1$ on replacing $(\Omega - 1)$ by $(1 - \Omega)$ and sin by sinh. Hence the value of χ_-:

$$\chi_-(\zeta, \Omega) = 2\left[\sinh^{-1}\left(\frac{1-\Omega}{\Omega}\right)^{1/2} - \sinh^{-1}\left(\frac{1-\Omega}{\Omega\zeta}\right)^{1/2}\right]. \tag{62'}$$

3.5. THE TRAVEL-TIME OF A SOURCE AS A FUNCTION OF ITS REDSHIFT

3.5.1. *The Approximate Relations between τ and z for Near Sources*

When a source is sufficiently near the origin, τ can be easily expressed as a function of z by a limited power expansion.

Indeed, according to Equation (37), and on introducing the quantity $\tilde{\omega}$ defined by

$$\boxed{\tilde{\omega} = \frac{\Omega}{4} - \frac{1}{2},} \tag{63}$$

we find

$$\tau \approx \frac{z}{H_0} - \frac{1}{H_0}\left(\tfrac{3}{2} + \tilde{\omega}\right)z^2. \tag{64}$$

The introduction of $\tilde{\omega}$ will simplify the calculations to be done later.

3.5.2. *The Travel-Time τ of a Source as a General Function of its Redshift*

For $k = +1$ Equations (52) and (53) yield:

$$t_e = \frac{\Omega}{H_0(\Omega - 1)^{3/2}}\left(\tfrac{1}{2}\eta_e - \tfrac{1}{2}\sin\eta_e\right), \tag{65}$$

and

$$t_0 = \frac{\Omega}{H_0(\Omega - 1)^{3/2}}\left(\tfrac{1}{2}\eta_0 - \tfrac{1}{2}\sin\eta_0\right). \tag{65'}$$

We thus find for t_0 the Equation already obtained in Chapter 10, Subsection 5.8, viz.

$$t_0 = \frac{\Omega}{H_0(\Omega - 1)^{3/2}}\left[-\frac{(\Omega - 1)^{1/2}}{\Omega} + \sin^{-1}\left(\frac{\Omega - 1}{\Omega}\right)^{1/2}\right], \tag{66}$$

which gives (with the reserves already exprsssed in Chapter 10) the *'age of the universe'*. Of course, t_0 does not depend on the redshift of any source.

On replacing $(\eta_e/2)$ and $\sin \eta_e$ by the values given by Equations (61″) and (61′), we obtain (with $z = \zeta - 1$):

$$t_e = \frac{\Omega}{H_0(\Omega - 1)^{3/2}} \left[-\frac{(\Omega - 1)^{1/2}(1 + z\Omega)^{1/2}}{\Omega\ \zeta} + \sin^{-1} \left(\frac{\Omega - 1}{\Omega\ \zeta} \right)^{1/2} \right]. \quad (67)$$

Hence the travel-time (for $k = +1$) of a signal from a source of redshift $z = \zeta - 1$, defined, as in Subsection 3.21, by

$$\tau = t_0 - t_e. \quad (68)$$

where one has just to replace t_0 and t_e by their expressions given by Equations (66) and (67).

This result can be considered as an explicit result of integration of Equation (44), viz.

$$d\tau' = -dt' = H_0^{-1}\zeta'^{-2}[\Omega\zeta' - (\Omega - 1)]^{-1/2} d\zeta', \quad (69)$$

from $\tau' = 0$ to $\tau' = \tau$.

4. The Metric (Mathematical) Linear Distance of a Source

In many investigations concerning the relativistic effects to be expected in observational cosmology, it is very convenient to associate the 'radial' angular (dimensionless) coordinate χ of a source and the time t, through a function $u_k(\chi, t, \Omega)$ defined, for any value of Ω (thus also of k) by the general relation

$$u_k(\chi, t, \Omega) = R(t, \Omega)\sigma_k(\chi), \quad (70)$$

where $\sigma_k(\chi)$ is given, as above, by

$$\sigma_0 = \chi; \qquad \sigma_{+1} = \sin \chi; \qquad \sigma_{-1} = \sinh \chi. \quad (71)$$

McVittie (1965, p. 161) calls $u_k(\chi, t, \Omega)$ the *'mathematical distance'* from the origin to a galaxy $S(\chi, \theta, \varphi)$ at time t, in order to emphasize that, in spite of its importance, u_k is not directly connected with most of operational methods by which the distance is measured (to be discussed below).

Some other writers (e.g. Andrillat (1970)) call it the *'metric distance'*, but oddly enough this very important function has *no standard name*. The term metric distance will be used below.

More generally, when several sources are considered or when one wishes to consider the radial coordinate χ' of a signal overtaking at time t' a source whose radial coordinate is also χ' (see Subsection 3.3) we can rewrite the definition (70) as

$$u_k(\chi', t', \Omega) = R(t', \Omega)\sigma_k(\chi'). \quad (72)$$

It is very important to realize that in the definition (70) the variables χ and t are (for a given value of Ω) fully independent: a given source (S), of constant radial coordinate χ, can be characterized by different metric distances $u_k(\chi, t, \Omega)$. Indeed, when different stages of the expanding universe are considered, i.e. at different times t, to each value of t corresponds a particular value of the absolute scale factor $R(t, \Omega)$.

Therefore, in principle, one should never speak about *the* metric distance of a source $S(\chi, \theta, \varphi)$, but should speak about *the metric distance of* (S) *at time t*. However, the term metric distance *without any specification of time* can be used when it is implicitly understood that it corresponds to the present time t_0.

Of course, the metric distance $u_k(\chi', t', \Omega)$ of any source $S(\chi')$ at time t' can be expressed as a function of the metric distance $u_k(\chi', t_0, \Omega)$ of the same source at time t_0. This procedure permits elimination of the function $\sigma_k(\chi')$ and successively yields:

$$u_k(\chi', t', \Omega) = \frac{R(t', \Omega)}{R(t_0, \Omega)} u_k(\chi', t_0, \Omega) = a(t', \Omega) u_k(\chi', t_0, \Omega), \tag{73}$$

or, taking into account Equation (38):

$$u_k(\chi', t', \Omega) = \frac{1}{\zeta'} u_k(\chi', t_0, \Omega). \tag{74}$$

4.1. THE METRIC DISTANCE AT TIME t_0 FOR THE EdS MODEL AS A FUNCTION OF THE REDSHIFT

For $\Omega = 1$ (thus for $k = 0$), the function $u_k(\chi, t_0, \Omega)$ reduces to $u_0(\chi, t_0, 1)$, which can also be denoted by $u_{\text{EdS}}(\chi, t_0,)$, defined according to Equation (71) by

$$u_{\text{EdS}}(\chi, t_0) = u_0(\chi, t_0, 1) = R(t_0, 1)\chi. \tag{75}$$

This is identical with the function $r(\chi, t_0)$ introduced by Equation (15), written in a more general way.

On taking into account Equation (47), the right-hand side of Equation (75) can be expressed as a function of the redshift z or as a function of $\zeta = 1 + z$:

$$u_{\text{EdS}}(\chi, t_0) = u_0(\chi, t_0, 1) = \frac{2c}{H_0}(1 - \zeta^{-1/2}), \tag{76}$$

or

$$u_{\text{EdS}}(\chi, t_0) = \frac{2c}{H_0}[1 - (1 + z)^{-1/2}]. \tag{76'}$$

For small values of z we can replace χ in Equation (75) by its limited expansion (48), or expand $(1 + z)^{-1/2}$ in Equation (76'). We thus obtain:

$$u_{\text{EdS}}(\chi, t_0) \approx \frac{cz}{H_0}(1 - \tfrac{3}{4}z). \tag{77}$$

4.2. THE METRIC DISTANCE AT TIME t_0 IN THE CASE $\Omega \neq 1$ FOR SMALL VALUES OF THE REDSHIFT

For $\Omega \neq 1$ (thus for $k \neq 0$) the function $u_k(\chi, t_0, \Omega)$ reduces, according to Equations (70) and (71) to

$$u_{\pm 1}(\chi, t_0, \Omega) = \begin{cases} R(t_0, \Omega) \sin \chi & \text{if } k = +1; \\ R(t_0, \Omega) \sinh \chi & \text{if } k = -1. \end{cases} \tag{78}$$

For small values of χ, both $\sin \chi$ and $\sinh \chi$ can be approximated (on neglecting only the terms in χ^3) by χ.

Hence, in both cases $k = +1$ and $k = -1$:

$$u_{\pm 1}(\chi, t_0, \Omega) \approx R(t_0, \Omega)\chi. \tag{79}$$

On replacing χ by its limited expansion (40) we find that for the Friedmann model, with $\tilde{\omega}$ defined by Equation (63):

$$u_k(\chi, t_0, \Omega) \approx \frac{cz}{H_0}\left[1 - \left(\frac{1}{2} + \frac{\Omega}{4}\right)z\right] \approx \frac{cz}{H_0}[1 - (1 + \tilde{\omega})z]. \tag{80}$$

We note that this approximate relation holds even for $k = 0$ (thus for $\Omega = 1$), since for $\Omega = 1$ it yields the same result as Equation (77). For this reason we have written the left-hand side of Equation (80) as $u_k(\chi, t_0, \Omega)$ and not as $u_{\pm}(\chi, t_0, \Omega)$.

On replacing z by its expansion in terms of τ, given by Equation (36′), we can put Equation (80) into the form:

$$u_k(\chi, t_0, \Omega) \approx c\tau(1 + \tfrac{1}{2}H_0\tau), \tag{80'}$$

which is remarkable by the absence of the parameter Ω.

4.3. THE METRIC DISTANCE AT TIME t_0 AS A GENERAL FUNCTION OF THE REDSHIFT

When z is not necessarily small, we can, on taking into account Equation (56), put $u_{+1}(\chi, t_0, \Omega)$ given by Equation (78), into the form

$$\begin{aligned} u_{+1}(\chi, t_0, \Omega) &= R(t_0, \Omega) \sin(\eta_0 - \eta_e) \\ &= R(t_0, \Omega)(\sin \eta_0 \cos \eta_e - \sin \eta_e \cos \eta_0). \end{aligned} \tag{81}$$

A similar equation holds (with hyperbolic functions) for the case $k = -1$, i.e. for $u_{-1}(\chi, t_0, \Omega)$.

The necessary values of $\sin \eta_0$, $\cos \eta_0$, $\sin \eta_e$, $\cos \eta_e$ are given by Equations (84), and (85) of Ch. 10 and by Equations (60′) and (61′) of this Chapter.

On introducing for the sake of brevity the notation

$$\omega = (\Omega - 1)^{1/2}; \quad x = 1 - \frac{\Omega}{2}; \quad p = (1 + z\Omega)^{1/2}; \quad \zeta = 1 + z, \tag{82}$$

we can write:

$$\sin \eta_0 = \frac{2\omega}{\Omega}; \quad \cos \eta_e = 1 - \frac{2(\Omega - 1)}{\Omega\zeta}; \quad \sin \eta_e = \frac{2\omega p}{\Omega\zeta}; \quad \cos \eta_0 = \frac{2x}{\Omega}. \tag{83}$$

In the case $k = -1$, when $(\Omega - 1) < 0$, similar expressions for sinh and cosh hold with $(1 - \Omega)$ replacing $(\Omega - 1)$ in the definition of ω_0.

Hence, in both cases $k = +1$ and $k = -1$, the same expression for $u_k(\chi, t_0, \Omega)$:

$$u_k(\chi, t_0, \Omega) = \frac{4\omega R(t_0, \Omega)}{\Omega^2 \zeta} [(\tfrac{1}{2}\Omega\zeta + 1 - \Omega) - xp]. \tag{84}$$

The first two terms in the square bracket are easily transformed into $(\frac{1}{2}\Omega z + x)$, so that finally $u_k(\chi, t_0, \Omega)$ is given by

$$u_k(\chi, t_0, \Omega) = 4\omega R(t_0, \Omega)\Omega^{-2}\zeta^{-1}[(\tfrac{1}{2}\Omega z + x) - xp]. \tag{85}$$

On taking into account the normalizing Equation (61) of Ch. 10, viz.

$$R_0\omega = \frac{c}{H_0}, \tag{86}$$

Equation (85) take the form

$$u_k(\chi, t_0, \Omega) = 4cH_0^{-1}(1+z)^{-1}\Omega^{-2}\left\{\frac{z\Omega}{2} + \left(\frac{\Omega}{2} - 1\right)[(1+\Omega z)^{1/2} - 1]\right\}, \tag{87}$$

sometimes called the 'Mattig formula'.

For $\Omega = 1$ this relation becomes identical to Equation (76′), thus justifying the general index k in Equation (87).

5. The Classical Distance of a Source as a Function of its Redshift

The concept of '*classical*' (non-relativistic) distance of a cosmic source, used by many authors without any clear explanation, can trouble some beginners.

Indeed, this non-relativistic distance, which we shall denote by d_{clas} is often expressed in terms of the redshift z, whereas we have introduced the redshift as an essentially relativistic concept, related to sources fixed in comoving coordinates.

The situation is easily clarified by adopting an historical perspective. At the time of the first observations of the redshift of spectral lines of galaxies this phenomenon was attributed to a *real motion* of recession of galaxies in a *static space*, instead of the present view of *static sources* (in comoving coordinates) inserted in an *expanding space* (the 'particular' relative motion of galaxies, in both conceptions, is here, of course, left out of consideration).

The first order Doppler relation between the redshift and the radial velocity V_{rad} of 'real' recession, was classically written as

$$z = \frac{\lambda_{\text{obs}} - \lambda_{\text{lab}}}{\lambda_{\text{lab}}} = \frac{V_{\text{rad}}}{c}, \tag{88}$$

where λ_{lab} is the wave-length measured in a terrestrial laboratory and λ_{obs} is the wave-length observed astronomically.

This was combined with Hubble's Kinematical law

$$V_{\text{rad}} = H_0\, d_{\text{clas}}, \tag{89}$$

where H_0 was the Hubble constant (see Chapter 10, Section 3), and d_{clas} the distance of the considered galaxy in a static (Euclidean) space.

The result was the 'classical' relation expressing the distance d_{clas} as a function of the observed redshift z:

$$\boxed{d_{\text{clas}} = \frac{cz}{H_0}.} \tag{90}$$

It is very remarkable that this classical distance becomes identical with the relativistic 'metric distance' (defined in Section 4) $u_k(\chi, t_0, \Omega)$ at the present time t_0 for *very* near sources, as those observed by Hubble, whose z was smaller than 0.004.

Indeed, if we reduce the expansion of $u_k(\chi, t_0, \Omega)$ given by Equation (80) to its first term, we find (for any value of Ω) precisely:

$$u_k(\chi, t_0, \Omega) \approx \frac{cz}{H_0}. \tag{91}$$

However, for a little more distant (but still near sources) we cannot reduce the expansion (80) to its *first* term, and, of course, for very distant sources of large z, such as quasars in the cosmological interpretation of their redshift, discussed in Chapter 13, the rigorous relations (76′) or (87) must be used.

6. The Relativistic Variation of the Angular Diameter, for Sources of a Given Linear Diameter, as a Function of their Redshift

6.1. INTRODUCTION: THE CLASSICAL VARIATION OF THE ANGULAR DIAMETER

In classical astronomy the observed angular diameter $d\theta_0$ of a non-punctual given source was thought to be inversely proportional to its (classical) distance d_{clas}. Thus, for a source of a linear diameter dD, sufficiently far from the observer to appear under a small angle $d\theta_0$, this *'angular diameter'* was expected to be given by a relation of the form:

$$\boxed{d\theta_0 = \frac{dD}{d_{\text{clas}}},} \tag{92}$$

with adequate units.

This relation, combined with the classical expression (90) for d_{clas} as a function of z, gives

$$(d\theta_0)_{\text{clas}} = \frac{dD}{(cz/H_0)} = \frac{1}{c} H_0(dD) \frac{1}{(\zeta - 1)}, \tag{93}$$

i.e. a law of variation as a function of z in which, for a given value of dD, the angular diameter $(d\theta)_{\text{clas}}$ varies in a way inversely proportional to z.

It will be shown now, that a quite different law of variation is to be expected (except for very small values of z) in relativistic cosmology.

6.2. The Relativistic Law of Variation of the Angular Diameter of Sources of a Given Linear Diameter as a Function of their Redshift

6.2.1. *General Relations (for any z and any Ω)*

In the relativistic interpretation of observations the linear diameter dD of a source represents its 'proper diameter' measured locally, whereas $d\theta_0$ represents its angular diameter measured at time t_0 by an observer at the origin of the coordinates.

A relativistic relation between dD and $d\theta_0$ can be obtained in a way rather similar to the one used in Subsection 3.1.2.

Let us return to the notation used in that subsection and consider two events (E_e) and (E_e'') formed by a simultaneous emission, at time t_e, of two signals, from the two ends A and B of a diameter of the source perpendicular to the line of sight. The points A and B are supposed to correspond to the same value of the coordinate φ (they lie on the same 'meridian' of the coordinate system), but their θ coordinates are supposed to be respectively θ and $(\theta + d\theta_s)$, whereas the 'radial' coordinate is χ for A and B.

We use in the four-dimensional space-time an expression of $(ds)^2$ given by Equations (1) and (4), viz.

$$(ds)^2 = R^2(t, \Omega)\{(d\chi)^2 + \sigma_k^2(\chi)[(d\theta)^2 + (d\varphi)^2 \sin^2 \theta]\} - c^2(dt)^2. \tag{94}$$

Thus, the square of the linear distance between the two ends of the considered diameter AB will be given by the value of the invariant $(ds)^2$ in the 'system of simultaneity' (see Chapter 11, Subsection 2.3) of the events (E_e) and (E_e''), in which, by definition, $dt_e = 0$. In this four-dimensional reference systen the coordinates of (E_e) will be ($\chi, \theta, \varphi, t_e$) and the coordinates of (E_e'') will be ($\chi, \theta + d\theta_s, \varphi, t_e$).

Therefore $(dD)^2$ is represented by the value taken by $(ds)^2$ for $d\chi = 0$; $d\theta = d\theta_s$; $d\varphi = 0$; $dt = 0$ in the neighbourhood of $\chi, \theta, \varphi, t_e$.

Hence, according to Equation (94):

$$(dD)^2 = R^2(t_e, \Omega)\sigma_k^2(\chi)(d\theta_s)^2, \tag{95}$$

or, since all quantities involved here are essentially positive:

$$dD = R(t_e, \Omega)\sigma_k(\chi)\, d\theta_s. \tag{95'}$$

On introducing the redshift by the fundamental Equation (28), viz.

$$R(t_e, \Omega) = \frac{1}{\zeta}R(t_0, \Omega) = \frac{1}{(1+z)}R(t_0, \Omega), \tag{96}$$

we obtain:

$$dD = \frac{1}{\zeta}R(t_0, \Omega)\sigma_k(\chi)\, d\theta_s. \tag{97}$$

According to the general definition of the metric distance given by Equation (70) we have at time t_0:

$$u_k(\chi, t_0, \Omega) = R(t_0, \Omega)\sigma_k(\chi), \tag{98}$$

so that Equation (97) take the form:

$$dD = \frac{1}{\zeta} u_k(\chi, t_0, \Omega)\, d\theta_s. \tag{99}$$

Since each signal, from A and B, follows a 'radial' path (i.e. a path of invariable θ and φ), the angular diameter $d\theta_0$ observed at time t_0 is the same as the difference $d\theta_s$ between θ_A and θ_B. Hence

$$d\theta_0 = \frac{dD}{(1/\zeta) u_k(\chi, t_0, \Omega)}. \tag{100}$$

If we compare Equation (100) with the classical Equation (92), we see that the quantity

$$\boxed{d_A = \frac{1}{\zeta} u_k(\chi, t_0, \Omega)} \tag{101}$$

plays for the source (S) the same role as the classical distance d_{clas}.

As for many other important quantities in cosmology the quantity d_A defined by Equation (100) has no standard name.

It is called the '*distance by apparent size*' by McVittie (1965, p. 162); the '*angular size distance*' by Robertson and Noonan (1968, p. 358); the '*proper distance*' by Peebles (1971, p. 171); the '*angular diameter distance*' by Weinberg (1972, p. 421); and the '*angle effective distance*' by Misner *et al.* (1973, p. 795) etc.

If in Equation (101) we replace $u_k(\chi, t_0, \Omega)$ by its general explicit expression (87) as a function of z and Ω we can rewrite Equations (100) and (101) as

$$d\theta_0 = \frac{dD}{d_A}, \tag{102}$$

and

$$d_A = d_A(z, \Omega) = 4cH_0^{-1}(1+z)^{-2}\Omega^{-2}\left\{\tfrac{1}{2}z\,\Omega + \left(\frac{\Omega}{2} - 1\right)[(1+\Omega z)^{1/2} - 1]\right\}. \tag{103}$$

In the particular case $\Omega = 1$ (the EdS model), we obtain on replacing $u_0(\chi, t_0, 1) = u_{\text{EdS}}(\chi, t_0)$ by its expression (76′):

$$(d_A)_{\text{EdS}} = d_A(z, 1) = \frac{2c}{H_0(1+z)}[1 - (1+z)^{-1/2}], \tag{104}$$

hence, according to Equation (102), in terms of $\zeta = 1 + z$:

$$(d\theta_0)_{\text{EdS}} = \frac{1}{2c} H_0(dD) \frac{\zeta}{(1 - \zeta^{-1/2})}. \tag{105}$$

6.2.2. *Particular Relations Between $d\theta_0$ and dD for Moderately Small Values of the Redshift z (and Arbitrary Ω)*

For moderately small values of z, i.e. the values of z for which one can limit the power expansions to terms proportional to z^2, (but not to terms proportional to z) we can use Equation (80), viz.

$$u_k(\chi, t_0, \Omega) \approx \frac{cz}{H_0}[1 - (1 + \tilde{\omega})z], \tag{106}$$

and replace Equation (100) by

$$dD \approx \frac{cz}{H_0(1+z)}[1 - (1 + \tilde{\omega})z](d\theta_0) \approx \frac{cz}{H_0}[1 - (2 + \tilde{\omega})z](d\theta_0), \tag{107}$$

which, solved for $d\theta_0$ yields:

$$d\theta_0 \approx \frac{dD}{(cz/H_0)}[1 + (2 + \tilde{\omega})z]. \tag{108}$$

Now we see that once more the classical relation (93) (independent of Ω) corresponds only to *the first term* of the relativistic relation (108).

It is easy to verify that for $\Omega = 1$ (thus for $k = 0$ and $\tilde{\omega} = -\frac{1}{4}$) the rigorous relativistic relation (105) expanded in powers of z gives the same limited expansion as (108) viz.

$$(d\theta_0)_{\text{Eds}} \approx \frac{dD}{(cz/H_0)}(1 + \tfrac{7}{4}z). \tag{109}$$

6.2.3. Comparison Between the Classical and the Relativistic Laws of Variation of the Angular Diameter (for a Given Linear Diameter) as a Function of the Redshift

Let us consider the function $y(\zeta, \Omega)$ defined by

$$y(\zeta, \Omega) = \frac{d\theta_0}{(1/c)H_0(dD)}. \tag{110}$$

According to Equation (93), this function is given in the classical theory (and for very small values of z in the relativistic theory) by

$$y_{\text{clas}}(\zeta) = \frac{1}{\zeta - 1}. \tag{111}$$

According to Equations (102) and (103), the relativistic general expression for $y(\zeta, \Omega)$ is given by a much more complicated relation:

$$y_{\text{rel}}(\zeta, \Omega) = \frac{\zeta^2\Omega^2}{2(\zeta - 1)\Omega + 2(\Omega - 2)[(1 - \Omega + \Omega\zeta)^{1/2} - 1]}. \tag{112}$$

The very considerable difference between $y_{\text{clas}}(\zeta)$ and $y_{\text{rel}}(\zeta, \Omega)$ can be illustrated in the particular case $\Omega = 1$ (EdS model, $k = 0$).

Indeed, according to Equation (105), we have in this case:

$$y_{\text{Eds}}(\zeta) = y_{\text{rel}}(\zeta, 1) = \frac{\zeta}{2(1 - \zeta^{-1/2})}. \tag{113}$$

The variations of the function $y_{\text{Eds}}(\zeta)$ are easily obtained by taking the derivative of Equation (113). This derivative is a fraction whose sign is the same as that of its numerator $N(\zeta)$ equal to

$$N(\zeta) = 2 - 3\zeta^{-1/2}. \tag{114}$$

Thus $dy_{\text{EdS}}(\zeta)/d\zeta$ is negative for $\zeta < \frac{9}{4}$, is zero for $\zeta = \frac{9}{4}$ and is positive for $\zeta > \frac{9}{4}$, so that $\zeta = \frac{9}{4}$ corresponds to a *minimum* of $y_{\text{EdS}}(\zeta)$, equal to $y_{\text{EdS}}(\frac{9}{4}) = \frac{27}{8}$.

The law of variation of $y_{\text{clas}}(\zeta)$ is quite different. Its derivative $dy_{\text{clas}}(\zeta)/d\zeta$, equal to $[-(\zeta-1)^{-2}]$, is always negative. Thus $y_{\text{clas}}(\zeta)$ *decreases* in a monotonic way from ∞ (for $\zeta = 1$) to zero (for $\zeta \to \infty$). For $\zeta = \frac{9}{4}$ its value is $y_{\text{clas}}(\frac{9}{4}) = \frac{4}{5}$.

This means that when one considers different galaxies of *the same linear diameter dD* observed with *increasing redshifts*, the corresponding observed angular diameters $d\theta_0$ must, according to the relativistic theory, first *decrease* (as in the classical theory, but at a different rate), then pass by a *minimum*, and finally *increase* (contrary to the classical theory).

In the Friedmann model corresponding to $\Omega = 1$ (i.e. to $\rho_0 = \rho_c$, see Chapter 10, Subsection 5.5) the difference in the rate of decrease *before the minimum* is already very considerable, since at the minimum reached for $\zeta = \frac{9}{4}$ (i.e. for $z = 1.25$) the relativistic $d\theta_0$ is $(\frac{27}{8})/(\frac{4}{5}) \approx 4.2$ times larger than the classical one. And, of course, after the minimum, for $z > 1.25$, this ratio increases.

All this is illustrated in Figure 12.3 (overleaf), where the dotted line represents the variations of $y_{\text{clas}}(\zeta)$, proportional to $(d\theta_0)_{\text{clas}}$, whereas the full line represents the variations of $y_{\text{EdS}}(\zeta) = y_{\text{rel}}(\zeta, 1)$ proportional to $(d\theta_0)$ in the relativistic theory – with the same coefficient of proportionality, viz. $(1/c)H_0(dD)$.

The study of the general case, based on Equation (112), made by Sandage (1961), shows that the existence of a minimum, analogous to that existing for $\Omega = 1$, represents a general feature of $y_{\text{rel}}(\zeta, \Omega)$ for all usual values of Ω.

Misner *et al.* (1973, p. 795) express this relativistic effect by saying that '*curved space* should act as a lens of great focal length'. However, as illustrated by the case $\Omega = 1$, which corresponds, according to Subsection 2.3, to a *non-curved* (*Euclidean*) *space*, this nice statement is somewhat misleading, as most comments to be found in different works on this strange phenomenon.

As a matter of fact, the existence of a variation of $y_{\text{rel}}(\zeta, \Omega)$ opposite, after the minimum, to the classical one, results from an interplay of *two* different factors. Let us show it in the case $\Omega = 1$ (i.e. $k = 0$).

Indeed, physically, according to Equations (28), (71) and (97) $(d\theta_0)_{\text{rel}} = d\theta_s$ is equal to

$$(d\theta_0)_{\text{rel}} = \frac{dD}{R(t_0, \Omega)}\zeta\frac{1}{\chi} = \frac{dD}{R(t_0,\Omega)}\frac{R(t_0,\Omega)}{R(t_e,\Omega)}\frac{1}{\chi(\zeta)}, \tag{115}$$

where we have written $\chi(\zeta)$ for χ in order to emphasize the dependence of χ on ζ.

When we consider different galaxies of increasing ζ the factor $1/\chi(\zeta)$ *decreases* monotonically. This is easily proved by the use of Equation (47), which shows that the derivative of $\chi^{-1}(\zeta)$ with respect to ζ is always negative. Thus the factor $1/\chi$ produces the same type of variation (though with a different rate) as the classical factor $1/d_{\text{clas}}$.

On the other hand, in an expanding (even Euclidean) universe, since $t_e < t_0$ the factor $R(t_0, \Omega)/R(t_e, \Omega)$, equal to ζ, *increases linearly* when galaxies of increasing ζ are considered. This factor acts in a way opposite to the action of the factor $1/\chi$, and dominates for great values of ζ.

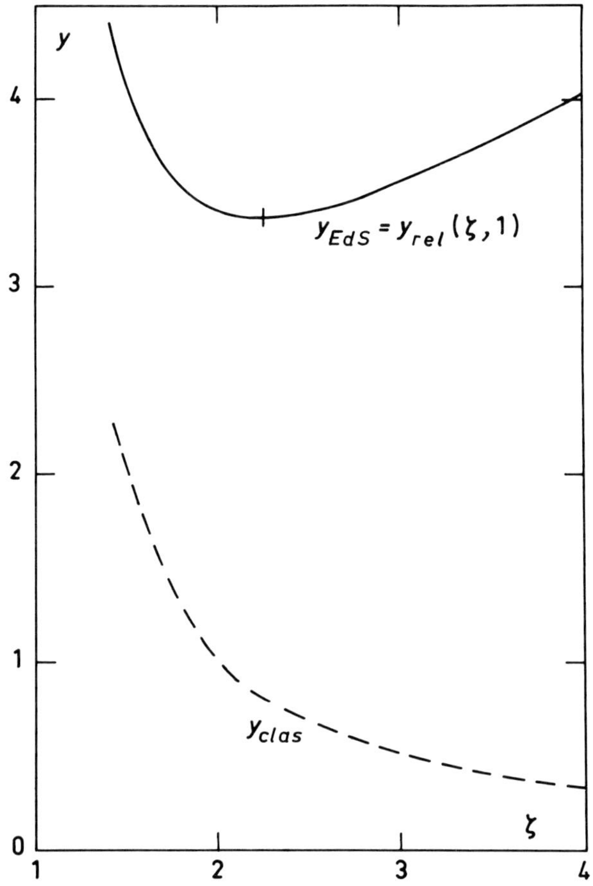

Fig. 12.3. The variation of $y(\zeta, \Omega)$, proportional to the observed angular diameter for a given linear diameter, as a function of ζ. The dotted line corresponds to the classical theory, the full line corresponds to the case EdS ($\Omega = 1$) in the relative theory.

7. A Physical Interpretation of the Metric Distance

The metric distance $u_k(\chi, t_0, \Omega)$ at time t_0, defined in Section 4, and considered as purely 'mathematical' by McVittie (1965, p. 161), can be given (as noted by Weinberg (1972, p. 423)) a physical meaning in the following way.

Consider a linear distance defined by *a transverse displacement dD*, perpendicular to the line of sight, of a *point source*, during an interval of proper time dt_e. If V denotes the corresponding linear velocity of the source, dD will be given by

$$dD = V\,dt_e. \tag{116}$$

This linear displacement dD will be observed astronomically *as an angular displacement* $d\theta_0$ during the time dt_0 near t_0. In astronomical terminology, this will be expressed by saying that the source presents a '*proper motion*' equal to $d\theta_0/dt_0$.

Classically, $d\theta_0$ is obviously given as a function of the classical distance d_{clas} by

$$(d\theta_0)_{\text{clas}} = \frac{V\,dt_0}{d_{\text{clas}}}. \tag{117}$$

In the relativistic theory, the distance dD travelled by the source plays, with respect to the observed $d\theta_0$ the same role as when dD represents the linear diameter of a static extended source, so that the relation between $d\theta_0$ and dD is given by Equation (12.100), viz.

$$d\theta_0 = \frac{\zeta\,dD}{u_k(\chi, t_0, \Omega)}. \tag{118}$$

On replacing dD by its value (116) we obtain

$$(d\theta_0)_{\text{rel}} = \frac{\zeta V\,dt_e}{u_k(\chi, t_0, \Omega)} = \frac{V\,dt_0\zeta(dt_e/dt_0)}{u_k(\chi, t_0, \Omega)}, \tag{119}$$

but here dt_e and dt_0 play the same role as T_e and T_0 in Subsection 3.1.2, so that, according to Equations (25) and (26),

$$dt_0/dt_e = \zeta. \tag{120}$$

Finally we find that

$$(d\theta_0)_{\text{rel}} = \frac{V\,dt_0}{u_k(\chi, t_0, \Omega)}. \tag{121}$$

On comparing this relativistic expression for $d\theta_0$ with the corresponding classical expression given by Equation (117), we see that, for a given value of $V\,dt_0$, the function $u_k(\chi, t_0, \Omega)$ plays in the relativistic cosmology exactly the same role as the distance d_{clas} in the classical theory.

Thus, the metric distance of a source at time t_0 could be called, by analogy with the names given to a distance d_A defined by Equation (101), the '*proper motion distance*'.

8. The Relativistic Variation of the Integrated (Bolometric) Brightness, for Point Sources of a Given Luminosity, as a Function of Their Redshift

8.1. Introduction: The Classical Variation of the Integrated Brightness

In classical astrophysics, the observed integrated (bolometric) brightness B of a point source, i.e. the energy (integrated in frequency) arriving per second on each cm^2 of a receiver perpendicular to the line of sight, was thought to be inversely proportional to the square d_{clas}^2 of its classical distance d_{clas}.

Thus, for a source of luminosity L, the bolometric brightness was expected to be given (for a propagation in vacuum) by a relation of the form:

$$B_{\text{clas}} = \frac{L}{4\pi d_{\text{clas}}^2}, \tag{122}$$

with adequate units.

This, combined with the classical expression (90) for d_{clas} gives

$$B_{\text{clas}} = \frac{LH_0^2}{4\pi c^2}\frac{1}{z^2} = \frac{LH_0^2}{4\pi c^2}\frac{1}{(\zeta - 1)^2}, \tag{123}$$

i.e. a law of variation of B as a function of z, in which, for sources of given L, B is *inversely proportional to the square of the redshift* z of the source.

8.2. THE RELATIVISTIC VARIATION OF THE BRIGHTNESS $B(t_0)$, FOR SOURCES OF GIVEN $L(t_e)$, AS A FUNCTION OF THEIR REDSHIFT

Let us consider, in a Friedmann universe, a source of given constant comoving coordinates (χ, θ, φ). The source is supposed to emit isotropically, in the neighbourhood of time t_e, photons of different frequencies.

Let $N_e(t_e, dt_e, \nu_e, d\nu_e)$ denote *the number* of photons of frequencies between ν_e and $(\nu_e + d\nu_e)$, i.e. $(\nu_e, d\nu_e)$ photons, *emitted* by the source between times t_e and $(t_e + dt_e)$, i.e. in time (t_e, dt_e), in *the totality of all directions of space*. The use of a capital N is meant here and below to specify this *last* particularity!

Let (Σ) denote the pseudo-spherical surface of area A_0 passing through the origin of coordinates (where is placed the observer), over which are distributed, at time t_0 of observation, the photons considered in $N_e(t_e, dt_e, \nu_e, d\nu_e)$. Because of the cosmological redshift, these photons are observed on frequencies $(\nu_0, d\nu_0)$.

Let then $N_\Sigma(t_0, dt_0, \nu_0, d\nu_0)$ denote the number of photons $(\nu_0, d\nu_0)$ crossing (Σ) between times (t_0, dt_0). The photons considered in $N_\Sigma(t_0, dt_0, \nu_0, d\nu_0)$ are the same as the photons considered in $N_e(t_e, dt_e, \nu_e, d\nu_e)$. Hence the fundamental relation:

$$N_\Sigma(t_0, dt_0, \nu_0, d\nu_0) = N_e(t_e, dt_e, \nu_e, d\nu_e). \tag{124}$$

Per unit interval of time (e.g. per second, since at a cosmic scale a second is a very small quantity) we can write, with the same notation as above:

$$N_e(t_e, 1, \nu_e, d\nu_e) = \frac{1}{dt_e} N_e(t_e, dt_e, \nu_e, d\nu_e), \tag{125}$$

and

$$N_\Sigma(t_0, 1, \nu_0, d\nu_0) = \frac{1}{dt_0} N_\Sigma(t_0, dt_0, \nu_0, d\nu_0). \tag{126}$$

On dividing Equation (126) by Equation (125), and on taking into account Equation (124), we obtain:

$$\frac{N_\Sigma(t_0, 1, \nu_0, d\nu_0)}{N_e(t_e, 1, \nu_e, d\nu_e)} = \frac{dt_e}{dt_0}. \tag{127}$$

If now we use the same type of argument as in Subsection 3.1.2 and let dt_e and dt_0 play the same role as T_e and T_0, we shall find, according to Equations (25) and (26), the same Equation as (120), viz.

$$\frac{dt_0}{dt_e} = \zeta. \tag{128}$$

Thus the *Hubble's number-effect*, in McVittie's (1965, p. 164) terminology, will be expressed by

$$\frac{N_\Sigma(t_0, 1, \nu_0, d\nu_0)}{N_e(t_e, 1, \nu_e, d\nu_e)} = \frac{1}{\zeta}. \tag{129}$$

If now we denote by $n_0(t_0, 1, \nu_0, d\nu_0)$ the number of photons $(\nu_0, d\nu_0)$ which are not distributed isotropically over the totality of directions of space, but only cross, per unit time, in the neighbourhood of time t_0, the unit area of (Σ), this n_0 will be related to the corresponding N_Σ by

$$\frac{n_0(t_0, 1, \nu_0, d\nu_0)}{N_\Sigma(t_0, 1, \nu_0, d\nu_0)} = \frac{1}{A_0}. \tag{130}$$

Hence, on multiplying Equation (129) by Equation (130):

$$\frac{n_0(t_0, 1, \nu_0, d\nu_0)}{N_e(t_e, 1, \nu_e, d\nu_e)} = \frac{1}{\zeta A_0}. \tag{131}$$

It is remarkable that the right-hand side of Equation (131) *does not depend on* ν_e *or* ν_0. This explains why the authors (e.g. McVittie, 1965, p. 164; Heidmann, 1973, p. 166; Mavridès, 1973, p. 254; and many others) who incorrectly discuss this problem *as if all photons had the same frequency* finally find a correct result.

The value of A_0 can be easily obtained if one considers that (Σ) has the same area A_0 as the pseudo-sphere (Σ_s), passing through the real source $S(\chi, \theta, \varphi)$, which would be reached at time t_0 by the photons emitted at time t_e by a 'theoretical source' situated at the origin.

Indeed, according to Equations (4) and (5), which give the Robertson–Walker $(d\lambda)^2$ at time t_0, viz.

$$(d\lambda)^2 = [R(t_0, \Omega)\, d\chi]^2 + [R(t_0, \Omega)\sigma_k(\chi)\, d\theta]^2 + \\ + [R(t_0, \Omega)\sigma_k(\chi) \sin\theta\, d\varphi]^2, \tag{132}$$

the same type of argument as the one used in Subsection 3.2.2 of Chapter 11 shows that the element dA_0 of area of (Σ_s) characterized by the values (χ, θ, φ) of the source (S) and by the values $d\chi = 0, d\theta, d\varphi$ characteristic of a surface such as (Σ_s), is given by

$$dA_0 = R^2(t_0, \Omega)\sigma_k^2(\chi)|d \cos\theta|\, d\varphi. \tag{133}$$

This, integrated from 0 to 2π with respect to φ and from (-1) to $(+1)$ with respect to $\cos\theta$ gives

$$A_0 = 4\pi R^2(t_0, \Omega)\sigma_k^2(\chi). \tag{134}$$

On introducing the metric distance at time t_0 defined by Equation (70), we finally obtain:

$$A_0 = 4\pi u_k^2(\chi, t_0, \Omega). \tag{135}$$

Let now $L(t_e)$ denote the luminosity of the source (S) in the neighbourhood of time t_e, i.e. the integrated energy (all frequencies!) emitted in the *totality* of all directions of space per unit interval of time (e.g. one second). (Because luminosity is always defined as an energy emitted per *unit time*, it is no longer necessary to explicit $dt_e = 1$ in the arguments of L.)

If $L(t_e, \nu_e, d\nu_e)$ denotes the part of $L(t_e)$ emitted on frequencies $(\nu_e, d\nu_e)$ the '*monochromatic luminosity*' $L_{\nu_e}(t_e)$, i.e. the luminosity per unit frequency interval near the frequency ν_e and near the time t_e, will be defined, by analogy with the notation used in Chapter 1, as

$$L_{\nu_e}(t_e) = \frac{1}{d\nu_e} L(t_e, \nu_e, d\nu_e). \tag{136}$$

Let, on the other hand, $B(t_0)$ denote the '*integrated brightness*' (or the '*bolometric brightness*') of the source (S), i.e. the energy received per unit area, per unit time near t_0, by the observer, after a propagation in vacuum. If $B(t_0, \nu_0, d\nu_0)$ denotes the part of $B(t_0)$ received on frequencies $(\nu_0, d\nu_0)$, the '*monochromatic brightness*' $B_{\nu_0}(t_0)$, i.e. the brightness per unit frequency interval near the frequency ν_0 and near the time t_0 will be defined by

$$B_{\nu_0}(t_0) = \frac{1}{d\nu_0} B(t_0, \nu_0, d\nu_0). \tag{137}$$

(Some authors denote $B_{\nu_0}(t_0)$ by s_0 and call it the '*received flux density*'.)

Since the energy of each photon of frequency ν is equal to $h\nu$ we have, according to our definitions:

$$L(t_e, \nu_e, d\nu_e) = h\nu_e N_e(t_e, 1, \nu_e, d\nu_e), \tag{138}$$

and

$$B(t_0, \nu_0, d\nu_0) = h\nu_0 n_0(t_0, 1, \nu_0, d\nu_0). \tag{139}$$

Now, since $\nu = c/\lambda$, the definition (26) of $\zeta = 1 + z$ yields:

$$\nu_0 = \frac{1}{\zeta} \nu_e, \tag{140}$$

and

$$d\nu_0 = \frac{1}{\zeta} d\nu_e. \tag{140'}$$

Hence, according to Equations (131), (138), (139) and (140):

$$\frac{B(t_0, \nu_0, d\nu_0)}{L(t_e, \nu_e, d\nu_e)} = \frac{\nu_0 n_0(t_0, 1, \nu_0, d\nu_0)}{\nu_e N_e(t_e, 1, \nu_e, d\nu_e)} = \frac{1}{\zeta^2 A_0}. \tag{141}$$

Similarly, on taking into account Equations (136), (137), (140′) and (141) we find:

$$B_{\nu_0}(t_0)/L_{\nu_e}(t_e) = \frac{(d\nu_e) B(t_0, \nu_0, d\nu_0)}{(d\nu_0) L(t_e, \nu_e, d\nu_e)} = \frac{1}{\zeta A_0}. \tag{142}$$

On taking into account the expression for A_0 given by Equation (135) we obtain:

$$B(t_0, \nu_0, d\nu_0) = \frac{1}{4\pi\zeta^2 u_k^2(\chi, t_0, \Omega)} L(t_e, \nu_e, d\nu_e), \tag{143}$$

and

$$B_{\nu_0}(t_0) = \frac{1}{4\pi\zeta u_k^2(\chi, t_0, \Omega)} L_{\nu_e}(t_e). \tag{144}$$

According to our definitions

$$B(t_0) = \int_{\nu_0=0}^{\infty} B_{\nu_0}(t_0)\, d\nu_0, \tag{145}$$

and

$$L(t_e) = \int_{\nu_e=0}^{\infty} L_{\nu_e}(t_e)\, d\nu_e. \tag{146}$$

Hence, on taking into account Equations (140′) and (144):

$$\boxed{B(t_0) = \frac{L(t_e)}{4\pi[\zeta u_k(\chi, t_0, \Omega)]^2} = \frac{L(t_e)}{4\pi d_L^2}.} \tag{147}$$

Note that the power of ζ is not the same in Equations (144) and (147).

On comparing with Equation (122) we see that in the relativistic theory the quantity d_L defined by

$$\boxed{d_L = \zeta u_k(\chi, t_0, \Omega),} \tag{148}$$

or, according to Equation (70), by

$$d_L = \zeta R(t_0, \Omega)\sigma_k(\chi), \tag{148′}$$

plays in the relation (147) between $B(t_0)$ and $L(t_e)$ the same role as the classical distance d_{clas} in the classical theory.

On comparing Equation (101) defining the distance d_A, viz.

$$d_A = \frac{1}{\zeta} u_k(\chi, t_0, \Omega), \tag{149}$$

with Equation (148) defining the distance d_L we see once more that in the relativistic theory the concept of distance has no absolute meaning, as in Newtonian theory.

On substituting into Equation (147) the general expression for $u_k(\chi, t_0, \Omega)$ given by Equation (87) we obtain:

$$\frac{B(t_0)}{L(t_e)} = \frac{H_0^2\Omega^2}{64\pi c^2}\left\{\frac{z\Omega}{2} + \left(\frac{\Omega}{2} - 1\right)[(1 + \Omega z)^{1/2} - 1]\right\}^{-2}. \tag{150}$$

8.2.1.1. Remark. As many other important quantities in relativistic cosmology, neither $B(t_0)$ nor d_L have any standard name. None is denoted by a standard symbol.

Thus, for instance, $B(t_0)$ is denoted by l_2 in Bondi, (1960, p. 108); by l in Robertson and Noonan (1968, p. 353), in Sciama (1971, p. 120) and in Weinberg (1972, p. 421); whereas Misner *et al.* (1973, p. 783) call it the '*flux measured at the earth*' and denote it by S. As to Peebles (1971, p. 177), he calls it the '*observed total energy flux*' and denotes it by f.

Similarly, the symbol d_L is used by Weinberg (1972, p. 421) but it is denoted by r_b in Robertson and Noonan (1968, p. 353) and by D in McVittie (1965, p. 165), in Zel'dovich *et al.*, (1967, Section 17.3), in Sciama (1971, p. 120) and in Mavridès (1973, p. 286).

As to its name, it is called the *'luminosity distance'* by Bondi (1960, p. 108), McVittie (1965, p. 165), Sciama (1971, p. 120), Weinberg, (1972, p. 421); whereas Robertson and Noonan (1968, p. 353) and Zel'dovich *et al.*, (1967, Section 17.3) call it the *'bolometric distance'*. As to Misner *et al.* (1973, p. 783), they do not give to it any name.

8.2.2. *Comparison between the Classical and the Relativistic Laws of Variation of the Integrated Brightness (for a Given Luminosity) as a Function of the Redshift*

In the particular case of the EdS model ($\Omega = 1$), we have, according to Equations (76) and (148)

$$(d_L)_{\text{Eds}} = \frac{2c}{H_0}(\zeta - \zeta^{1/2}), \tag{151}$$

hence, by Equation (147)

$$B(t_0)_{\text{Eds}} = \frac{L(t_e)H_0^2}{16\pi c^2}\frac{1}{(\zeta - \zeta^{1/2})^2}. \tag{152}$$

Consider now the function $F(\zeta, \Omega)$ defined by

$$F(\zeta, \Omega) = \frac{B(t_0)4\pi c^2}{L(t_e)H_0^2}. \tag{153}$$

This function $F(\zeta, \Omega)$ generally describes the variations of the observed integrated brightness $B(t_0)$ as a function of $\zeta = 1 + z$, hence of z, for different sources of the same luminosity $L(t_e)$.

According to Equation (123), in the classical theory $F(\zeta, \Omega)$ is given by

$$F_{\text{clas}}(\zeta) = \frac{1}{(\zeta - 1)^2} = \frac{1}{z^2}. \tag{154}$$

The function $F_{\text{clas}}(\zeta)$ corresponds to a monotonic decrease of $B(t_0)$, for a given value of $L(t_e)$, when the redshift z of the considered sources increases.

In the relativistic theory, for the particular case EdS ($\Omega = 1$), we have, according to Equation (152):

$$F_{\text{EdS}}(\zeta) = F_{\text{rel}}(\zeta, 1) = \tfrac{1}{4}(\zeta - \zeta^{1/2})^{-2}. \tag{155}$$

In Section 6 we have shown that the function $y(\zeta, \Omega)$, which describes the variations of the observed angular diameter $d\theta_0$ as a function of ζ for different sources of the same linear diameter dD, had not only different rates but also different trends in the classical theory and in the relativistic theory (see Figure 12.3).

Let us now investigate whether we encounter a similar difference between the variations of $F_{\text{clas}}(\zeta)$ and $F_{\text{EdS}}(\zeta)$.

The derivative $dF_{\text{EdS}}(\zeta)/d\zeta$ has the sign of $(\frac{1}{2}\zeta^{-1/2} - 1)$ which is always negative, as $dF_{\text{clas}}(\zeta)/d\zeta$.

Thus, contrary to the case of $y(\zeta, \Omega)$, the function $F_{\text{EdS}}(\zeta)$ *decreases monotonically*, for increasing ζ or z, as the function $F_{\text{clas}}(\zeta)$, but of course, with a different rate of variation.

For $z = 0$, i.e. for $\zeta = 1$, both functions are infinite. But already for $z = 3$, i.e. for $\zeta = 4$, we have

$$F_{\text{clas}}(4) = \tfrac{1}{9}; \qquad F_{\text{EdS}}(4) = \tfrac{1}{16}. \tag{156}$$

Thus we see that, contrary to the case of $y(\zeta, \Omega)$ before the minimum, the decrease of $F(\zeta, \Omega)$ is a little more rapid in the relativistic case EdS than in the classical theory (see Figure 12.4, where the dotted line represents the variations of F_{clas} and the full line represents the variations of F_{EdS}).

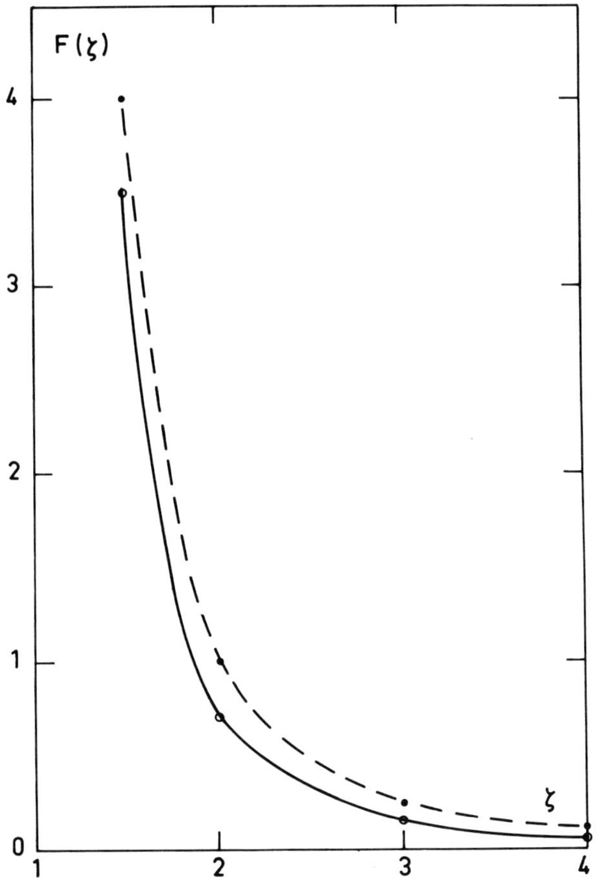

Fig. 12.4. The variation of $F(\zeta, \Omega)$, proportional to the observed integrated (bolometric) brightness for a given luminosity, as a function of ζ. The dotted line corresponds to the classical theory, the full line corresponds to the case EdS ($\Omega = 1$) in the relativistic theory.

The study of the general case ($\Omega \neq 1$, i.e. $k \neq 0$) based on Equation (150) made by Sandage (1961) shows that the monotonic decrease of $F_{\text{rel}}(\zeta, \Omega)$, found here for $F_{\text{EdS}}(\zeta)$, holds for all usual values of Ω.

8.2.3. The Relation Between $B(t_0)$ and $L(t_e)$ for Small Values of z. The Limited Expansion of d_L in Powers of z

For small values of z we can replace $u_k(\chi, t_0, \Omega)$ by its limited expansion in powers of z given by Equation (80), viz.

$$u_k(\chi, t_0, \Omega) \approx \frac{cz}{H_0}[1-(1+\tilde{\omega})z]. \tag{157}$$

Then the limited expansion of d_L will be given, according to Equation (148), by

$$d_L \approx \frac{cz}{H_0}(1-\tilde{\omega}z). \tag{158}$$

On replacing d_L by this limited expansion in Equation (147) we find:

$$B(t_0) \approx \frac{L(t_e)H_0^2}{4\pi c^2}\frac{1}{z^2}(1+2\tilde{\omega}z). \tag{159}$$

This relation becomes identical with the classical relation (123) if we reduce the expansion to its first term.

9. The Brightness of an Extended Source Per Unit Solid Angle

9.1. INTRODUCTION: THE CLASSICAL INVARIANT

In the preceding section we have considered the integrated brightness of *a point source*. This quantity was defined as the energy (integrated in frequency) arriving per second on each cm^2 of a receiver perpendicular to the line of sight.

In the case of an *extended* source, one can consider again the integrated energy arriving per second on each cm^2 of a receiver (perpendicular to the line of sight) but limited to the part of the source (supposed of uniform brightness) *seen within a unit solid angle*. This will be denoted by $b(t_0)$.

In French B is called *'éclat'*, whereas b is called *'brilliance'*, but in English none of the two concepts has a standard name.

The solid angle $d\Omega'$, subtended by a spherical source of angular diameter $d\theta_0$, is given for small values of $d\theta_0$ by

$$d\Omega' = \int_0^{2\pi} d\varphi \int_0^{(d\theta_0/2)} \sin\theta\, d\theta = 2\pi[1-\cos(d\theta_0/2)] \approx \tfrac{1}{4}\pi(d\theta_0)^2. \tag{160}$$

According to the classical expression (92) for $d\theta_0$ as a function of the linear diameter dD of an extended source, viz.

$$(d\theta_0)_{\text{clas}} = \frac{dD}{d_{\text{clas}}}, \tag{161}$$

the solid angle $d\Omega'$ will be given, in the classical theory, as a function of dD by

$$d\Omega' = \frac{\pi(dD)^2}{4d_{\text{clas}}^2}. \tag{162}$$

On the other hand, the integrated brightness of a source of luminosity L situated at a distance d_{clas} is given in the classical theory by Equation (122), (when the light propagates in vacuum), viz.

$$B_{\text{clas}} = \frac{L}{4\pi d_{\text{clas}}^2}. \tag{163}$$

Hence, according to the definition of b:

$$b_{\text{clas}} = \frac{B_{\text{clas}}}{d\Omega'} = \frac{L}{\pi^2 (dD)^2} = \text{const.}. \tag{164}$$

Thus b is, in the classical theory, a quantity which is constant with respect to d_{clas}, and according to Equation (90) is also invariant with respect to z.

Equations (162) and (163) clearly show the reason of this invariance, since in the classical theory both B and $d\Omega'$ are inversely proportional to the square of the distance.

9.2. THE RELAVISTIC VARIATION OF THE BRIGHTNESS PER UNIT SOLID ANGLE (FOR SOURCES OF GIVEN LUMINOSITY AND GIVEN LINEAR DIMENSION) AS A FUNCTION OF THE REDSHIFT

9.2.1. *The Variation of the Integrated Brightness per Unit Solid Angle*

In the relativistic theory of the brightness per unit solid angle of an extended source, the *total* brightness $B(t_0)$ is given by Equation (147), viz.

$$B(t_0) = \frac{L(t_e)}{4\pi\zeta^2 u_k^2(\chi, t_0, \Omega)}. \tag{165}$$

The corresponding solid angle $d\Omega'$ is given, as a function of the angular diameter $d\theta_0$ by Equation (160), but the relativistic expression for $d\theta_0$ is given by Equation (100), viz.

$$d\theta_0 = \frac{\zeta(dD)}{u_k(\chi, t_0, \Omega)}. \tag{166}$$

Hence

$$d\Omega' = \frac{\pi\zeta^2(dD)^2}{4u_k^2(\chi, t_0, \Omega)}. \tag{167}$$

Therefore, according to the definition of $b(t_0)$, this quantity will be given, in relativistic cosmology, by

$$\boxed{b_{\text{rel}}(t_0) = \frac{B(t_0)}{d\Omega'} = \frac{L(t_e)}{\pi^2(dD)^2}\frac{1}{\zeta^4}.} \tag{168}$$

Thus we see that, for given values of $L(t_e)$ and dD, we have

$$b_{\text{rel}}(t_0) = \frac{b_{\text{clas}}}{\zeta^4}. \tag{169}$$

The relativistic brightness per unit solid angle $b(t_0)$ is no longer an invariant with respect to z, but decreases very rapidly when the sources (of given L and dD) of increasing z are considered.

9.2.2. *The Variation of the Monochromatic Brightness per Unit Solid Angle*

If, instead of the integrated (bolometric) total brightness, we consider the monochromatic total brightness $B_{\nu_0}(t_0)$ of an extended source, the corresponding monochromatic brightness per unit solid angle $b_{\nu_0}(t_0)$ will be given in relativistic cosmology by

$$b_{\nu_0}(t_0) = B_{\nu_0}(t_0)/(d\Omega'). \tag{170}$$

On taking into account Equation (167) and the expression (144) for $B_{\nu_0}(t_0)$, viz.

$$B_{\nu_0}(t_0) = \frac{L_{\nu_e}(t_e)}{4\pi\zeta u_k^2(\chi, t_0, \Omega)}, \tag{171}$$

we find that $b_{\nu_0}(t_0)$ is given by

$$b_{\nu_0}(t_0) = \frac{L_{\nu_e}(t_e)}{\pi^2(dD)^2}\frac{1}{\zeta^3}. \tag{172}$$

10. The Source Counts

10.1. Introduction

We consider in this section some of the numerous problems connected with the *source counts*, i.e. with statistical problems concerning the observed *number* of sources corresponding to different *observational* criteria such as the redshift, the integrated brightness, etc.

These problems are seldom expounded quite clearly, for many reasons. Some of them make use of concepts which are seldom encountered in other branches of physics or astrophysics. Each theoretical model combine a great variety of assumptions concerning the geometry of the space (Euclidean or non-Euclidean), the distribution of the sources in space (homogeneous or non-homogeneous), the kinematics of the universe (static or expanding universe), the distribution of luminosities (all sources of identical or different luminosity), the distance of the sources (sources of small or of large redshift), the evolution of sources (stationary sources or sources of variable luminosity; the sources all 'born' at the same rate; the sources born at the same rate as they 'die', etc.), sources whose spectrum obeys some simple law or a more general law, etc.

Moreover, different parameters can be used to characterize a given situation. Thus, for instance, the distance of a source can be described by its redshift, its travel-time, its radial coordinate χ, its metric distance u_k, its 'luminosity distance' d_L, etc.

These unavoidable and intrinsic difficulties are often considerably aggravated by a very unfortunate *ambiguity in notation* (due generally, as already mentioned in our Preface, to an abusive care for 'elegance'): the same symbols being used with different assumptions, for different quantities, or, conversely, the same quantity being given different names, etc.

However, as also emphasized in our Preface, the worst in the current presentation of these problems is the abusive *skipping over* too many intermediate stages of argumentation or calculations, which, elementary as they can be, are usually indispensable for logical continuity and comfortable reading.

We try to spare our reader as many as possible of these difficulties, by insisting even more heavily, than in the rest of our textbook, upon some concepts which can appear as 'obvious' but which become such only through concrete and elementary examples.

We shall also use a unified system of notation and a minimum of parameters to represent the distance (in a loose sense) of a source. And, above all, we shall start with very 'rough' and simple models and introduce more refined ones only very progressively. For those not considered below the reader is referred to the more advanced textbook of Weinberg (1972).

10.2. THE CLASSICAL MODELS OF SOURCES UNIFORMLY DISTRIBUTED IN A STATIC EUCLIDEAN SPACE

In classical cosmology the (Euclidean) space is supposed to be *static* and the distribution of the sources is assumed to be homogeneous, i.e. of the same constant number density everywhere.

10.2.1. Sources Selected by their Redshift

The first and the simplest theoretical problem concerning the source counts uses as statistical criterion the *redshift* z of the observed sources.

The theoretical problem is then to predict the number of sources which will be observed with a redshift z' smaller than z, with suitable assumptions concerning the properties of the space (model of universe) and the distribution of the sources in space.

The redshift z is here considered (see Section 5) as expressing the Doppler effect of a real recession of galaxies, and is connected with the classical distance d_{clas} of a source through Hubble's law, hence Equation (90), viz.

$$d_{\text{clas}} = \frac{cz}{H_0}. \tag{173}$$

All sources S of redshift z must be situated at the same distance and lie on the very surface of the sphere $\Sigma(z)$ of radius d_{clas} defined by Equation (173); all sources S′ of redshift z' *smaller than* z must be situated *within* the sphere $\Sigma(z)$. The volume $V(z)$ of $\Sigma(z)$ is equal to

$$V(z) = \frac{4\pi}{3} d_{\text{clas}}^3 = \frac{4\pi}{3} \frac{c^3}{H_0^3} z^3. \tag{174}$$

Therefore, if the number density of the sources is denoted by n, the total number $N[z' < z]$ of sources S′ observable per unit solid angle of one steradian (equal to the number of sources observable in all directions divided by 4π), will be given by

$$N[z' < z] = \frac{1}{3} \frac{c^3}{H_0^3} n z^3. \tag{175}$$

With a given value of H_0, Equation (175) theoretically permits to deduce n from observed values of $N[z' < z]$ and to verify by its eventual *constancy* the correspondence between the model and reality. However, one must take into account the low *brightness* (considered in Subsection 8.2.2) of very remote sources: some of the sources inside $\Sigma(z)$ could escape observation if their brightness were too low for a given observational equipment.

10.2.2. Sources Selected by their Integrated Brightness

This problem is more complex than that of the counts of $N[z' < z]$ because the integrated brightness B of a source depends not only on the distance d_{clas} of the source, but also on its luminosity L, whereas in classical cosmology, the redshift z is determined by the distance d_{clas} alone.

In searching the theoretical expression for the number $N[B' > B]$ of sources observed, per steradian, with a brightness B' *exceeding* some limiting brightness B, one is thus obliged to make more or less arbitrary assumptions (and to consider different models of increasing generality) concerning the distribution and the evolution of luminosities, as done below.

10.2.2.1. Sources of Identical Stationary Luminosities. In the most elementary classical models selecting the sources by their brightness B, all sources are supposed to have the same standard stationary luminosity L_s erg s^{-1}. Each of these standard sources emits per second and *per unit solid angle* an energy P_s given, in erg s^{-1} sterad^{-1}, by

$$P_s = \frac{1}{4\pi} L_s. \tag{176}$$

We shall call P_s the '*luminosity parameter*'.

According to the trivial classical relation, already considered in Subsection 8.1, the brightness B of each source is given, as a function of its classical distance d_{clas}, by

$$B = \frac{L_s}{4\pi d_{\text{clas}}^2} = \frac{P_s}{d_{\text{clas}}^2}. \tag{177}$$

Since, in our model, P_s is identical for all sources, Equation (177) defines d_{clas} as a function of B, depending on the parameter P_s:

$$d_{\text{clas}} = \varphi(B, P_s) = B^{-1/2} P_s^{1/2}. \tag{177'}$$

Thus, all sources S of brightness B must be situated at the same distance and lie *on the very surface* of the sphere $\Sigma(d_{\text{clas}})$ of radius d_{clas} defined by Equation (177').

Consider now only those S' of the standard sources whose observed integrated brightness B' *exceeds* B. Their distance d'_{clas} will be given, according to Equation (177'), by $\varphi(B', P_s) = B'^{-1/2} P_s^{1/2}$.

It is physically and mathematically obvious that d'_{clas} is smaller than d_{clas}, so that all sources S' will be situated *within* the sphere $\Sigma(d_{\text{clas}})$.

Since, in our simple model, the space distribution of sources is assumed to be homogeneous, their number density is a constant n independent of assumptions concerning the luminosity of the sources. Nevertheless, in view of extension to more complex models, it can be useful to explicit the fact that we consider the sources of luminosity $4\pi P_s$ and denote their number density by $n(P_s)$.

Since all sources S′ lie within the sphere $\Sigma(d_{\text{clas}})$ of volume $V(d_{\text{clas}})$, their *total* number $N_{\text{tot}}(\mathrm{S}')$ will be given, as a function of $n(P_s)$, by

$$N_{\text{tot}}(\mathrm{S}') = n(P_s)V(d_{\text{clas}}), \tag{178}$$

or, on taking into account Equation (177′) by

$$N_{\text{tot}}(\mathrm{S}') = \frac{4\pi}{3}B^{-3/2}P_s^{3/2}n(P_s). \tag{179}$$

Let us now consider the part $N_s[B' > B]$ of $N_{\text{tot}}(\mathrm{S}')$ observed within a solid angle of one steradian, the subscript (s) corresponding to the particular assumption (to be dropped later) that all sources are of a standard luminosity L_s:

$$N_s[B' > B] = \frac{1}{4\pi}N_{\text{tot}}(\mathrm{S}'). \tag{180}$$

Finally we see that $N_s[B' > B]$ is a relatively simple function $f(B, P_s)$ of B depending on the parameter P_s:

$$N_s[B' > B] = f(B, P_s) = \tfrac{1}{3}B^{-3/2}C(P_s), \tag{181}$$

where $C(P_s)$ is defined by

$$C(P_s) = P_s^{3/2}n(P_s). \tag{182}$$

Hence,

$$\log_{10} N_s[B' > B] = \log_{10} f(B, P_s) = -\tfrac{3}{2}\log_{10} B + \log_{10}[\tfrac{1}{3}C(P_s)]. \tag{183}$$

If the considered model were corresponding to reality, a representation of observed $\log_{10} f(B, P_s)$ *versus* $\log_{10} B$ would give a *straight line* of a slope equal to $(-3/2)$. This representation would give the value of $\log_{10} C(P_s)$, hence of $P_s^{3/2}n(P_s)$, but not of P_s and $n(P_s)$ separately!

10.2.2.1.1. The function $N_s[m'_{\text{bol}} < m_{\text{bol}}]$. The integrated brightness can be expressed in terms of (apparent) '*bolometric magnitudes*' m_{bol} by the usual relation

$$m_{\text{bol}} - m^*_{\text{bol}} = -2.5\log_{10}\frac{B}{B^*}, \tag{184}$$

where B^* is the brightness of a reference source S* to which is attributed a more or less arbitrary magnitude m^*_{bol} characteristic of the considered '*system of magnitudes*'.

Equation (184) can also be written as

$$m_{\text{bol}} = -2.5(\log_{10} B) + C^*, \tag{185}$$

where C^* is a given constant for a given system of magnitudes.

Thus $\log_{10} B$ is a function of m_{bol} given by

$$\log_{10} B = (0.4)(C^* - m_{\text{bol}}). \tag{186}$$

On substituting this expression for $\log_{10} B$ into Equation (183) we obtain:

$$\log_{10} N_s[B' > B] = (0.6)m_{\text{bol}} + D^*(P_s), \tag{187}$$

where $D^*(P_s)$, for a given P_s, is a constant defined by

$$D^*(P_s) = -(0.6)C^* + \log_{10}[\tfrac{1}{3}C(P_s)]. \tag{188}$$

Since $B' > B$ corresponds by Equation (185) to $m'_{\text{bol}} < m_{\text{bol}}$, we can also denote $N_s[B' > B]$ by $N_s[m'_{\text{bol}} < m_{\text{bol}}]$ and rewrite Equation (187) as

$$\boxed{\log_{10} N_s[m'_{\text{bol}} < m_{\text{bol}}] = (0.6)m_{\text{bol}} + D^*(P_s).} \tag{189}$$

In this Equation $N_s[m'_{\text{bol}} < m_{\text{bol}}]$ represents the number of sources of bolometric magnitude *smaller* than m_{bol}, observed per unit solid angle in a classical model corresponding to a homogeneous distribution of stationary sources of identical luminosity $L_s = 4\pi P_s$ in a static Euclidean space.

10.2.2.1.2. The function $N_s(B)$. Let us now fix our attention on the values of the brightness within the interval dB in the neighbourhood of B. This interval between B and $(B + dB)$ will be denoted as usual by (B, dB).

If we denote by $N_s(B, dB)$ the number of sources S″, per steradian, of luminosity L_s whose brightness lies in the interval (B, dB), this quantity will be given (see Figure 12.5) by

$$\begin{aligned} N_s(B, dB) &= N_s[B' > B] - N_s[B' > (B + dB)] \\ &= f(B, P_s) - f(B + dB, P_s), \end{aligned} \tag{190}$$

since the sources corresponding to $B' > (B + dB)$ lie within a sphere *smaller* than the sphere containing the sources corresponding to $B' > B$.

Thus, the value of $N_s(B, dB)$ per unit interval ($dB = 1$) of brightness, a quantity which can be denoted by $N_s(B)$, will be given by

$$N_s(B) = -\left[\frac{f(B + dB, P_s) - f(B, P_s)}{dB}\right]. \tag{190'}$$

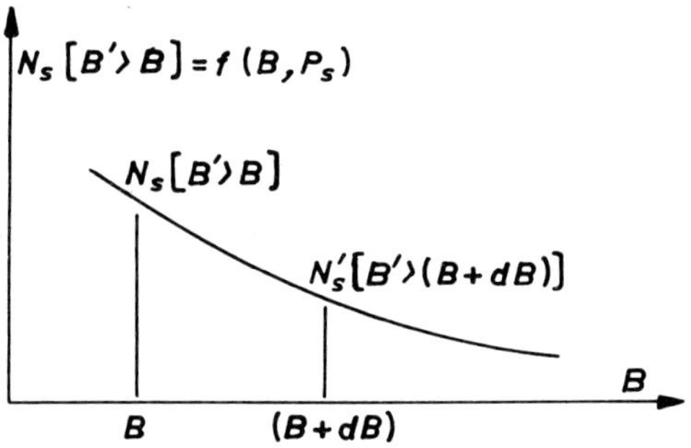

Fig. 12.5. The function $f(B, P_s)$ and the quantity $N_s(B, dB)$.

For a sufficiently small unit of interval of brightness we can consider, in a first approximation, that the square bracket on the right-hand side of Equation (190′) is equal to the derivative of $f(B, P_s)$ with respect to B, and write

$$N_s(B) \approx -\frac{\partial f(B, P_s)}{\partial B}. \tag{191}$$

Thus, according to Equations (181) and (191), we find:

$$N_s(B) = \tfrac{1}{2} C(P_s) B^{-5/2}, \tag{192}$$

or

$$\log_{10} N_s(B) = -\tfrac{5}{2} \log_{10} B + \log_{10} [\tfrac{1}{2} C(P_s)]. \tag{192'}$$

If the model were sufficiently realistic, a plot of $\log_{10} N_s(B)$ *versus* $\log_{10} B$ would give a straight line of slope $(-\frac{5}{2})$, and would yield the same value of $C(P_s)$ as the one corresponding to Equations (183) and (189).

10.2.2.2. Sources of Different Stationary Luminosities

10.2.2.2.1. The case of a discrete distribution of luminosities. Let us consider now again a homogeneous distribution of sources in space, but instead of sources of a standard unique luminosity L_s, let us assume that we deal with m 'families' of sources F_1, F_2, $F_3, \ldots, F_k, \ldots, F_m$, each source S_k of the family F_k being characterized by a specific luminosity L_k, the luminosities $L_1, L_2, L_3, \ldots, L_k, \ldots, L_m$ forming *a discrete sequence*.

As in Subsection 10.2.2.1 we introduce a *'luminosity parameter'* P_k defined by

$$P_k = \frac{1}{4\pi} L_k, \tag{193}$$

describing the energy emitted by each source S_k of the family F_k per second and per steradian.

Of course, all results obtained in Subsection 10.2.2.1 for the sources of standard luminosity $L_s = 4\pi P_s$ hold for the sources S_k of a given family F_k, provided that we change the subscript (s) into (k).

Thus we obtain, according to Equations (181) and (182):

$$N_k[B' > B] = f(B, P_k) = \tfrac{1}{3} B^{-3/2} C(P_k), \tag{194}$$

where

$$C(P_k) = P_k^{3/2} n(P_k). \tag{195}$$

In Equation (195) $n(P_k)$ obviously represents the number density of the sources S_k of the family F_k.

Let us now denote by $N[B' > B]$ the total number of all sources of all m families F_1, $F_2, F_3, \ldots, F_m$ observable per steradian with a brightness B' exceeding B. According to Equations (194) and (195) this quantity will be given by

$$N[B' > B] = \sum_{k=1}^{m} N_k[B' > B] = \tfrac{1}{3} B^{-3/2} \sum_{k=1}^{m} C(P_k). \tag{196}$$

A discussion similar to that made in Subsection 10.2.2.1.2 would give the total number $N(B)$ of all sources of all m families $F_1, F_2, F_3, \ldots, F_m$ observable per steradian with a brightness within the unit of interval of brightness in the neighbourhood of B:

$$N(B) = \tfrac{1}{2} B^{-5/2} \sum_{k=1}^{m} C(P_k). \tag{197}$$

We thus find that the proportionality of $N_s[B' > B]$ to $B^{-3/2}$ and the proportionality of $N_s(B)$ to $B^{-5/2}$ hold in spite of the division of sources in several families F_k of different luminosity $L_k = 4\pi P_k$.

10.2.2.2.2. The case of a continuous distribution of luminosities. Suppose now that the number m of families of sources S_k considered above is very considerable and that their luminosities L form statistically a *continuous* distribution.

Let us again associate to each luminosity L a *'luminosity parameter'* P defined by

$$P = \frac{L}{4\pi}. \tag{198}$$

Obviously P will play in this model the role played by P_k in Subsection 10.2.2.2.1. However, the concept of number density $n(P_k)$ of sources S_k must now be replaced by the concept of density $n(P, \Delta P)$ of sources whose luminosity parameter P lies between P and $(P + \Delta P)$.

Since $n(P, \Delta P)$ is a quantity of first order with respect to ΔP, it is natural to consider (see Chapter 1, Subsection 1.1)

$$n(P) \underset{\text{p.v.}}{=\!=} \frac{n(P, \Delta P)}{\Delta P}, \tag{199}$$

i.e. the *'plateau value'* of the ratio $n(P, \Delta P)/(\Delta P)$, which represents the number of sources per unit interval ($\Delta P = 1$) of the luminosity parameter in the neighbourhood of P.

On assimilating the family F_k of Subsection 10.2.2.2.1 with the family of sources whose luminosity parameter lies between P and $(P + dP)$, we can replace $n(P_k)$ by $n(P, dP)$, which for an adequate order of magnitude of dP will be given by

$$n(P, dP) = n(P)\, dP. \tag{200}$$

Then the sum $\Sigma_{k=1}^{m} P_k^{3/2} n(P_k)$ will be replaced by the integral I defined by:

$$I = \int_0^\infty P^{3/2} n(P)\, dP = \text{const.}, \tag{201}$$

and Equations (196) and (197) will take the form:

$$N[B' > B] = \tfrac{1}{3} I B^{-3/2}, \tag{202}$$

and

$$N(B) = \tfrac{1}{2} I B^{-5/2}. \tag{203}$$

We thus find again that $N[B' > B]$ varies as $B^{-3/2}$ and that $N(B)$ varies as $B^{-5/2}$.

Let us now consider $n(P)$ as a fraction $W(P)$ of the *total* number density n_{tot} of all sources of different luminosities, i.e. introduce $W(P)$ defined by

$$W(P) = \frac{n(P)}{n_{\text{tot}}}, \tag{204}$$

so that

$$\int_0^\infty W(P)\, dP = 1. \tag{205}$$

Now the definition (201) of I can be written as

$$I = n_{\text{tot}} \int_0^\infty P^{3/2} W(P)\, dP. \tag{206}$$

We can now introduce the average value $(\overline{P^{3/2}})$ of P by

$$(\overline{P^{3/2}}) \int_0^\infty W(P)\, dP = \int_0^\infty P^{3/2} W(P)\, dP, \tag{207}$$

and because of the normalizing Equation (205) obtain:

$$\int_0^\infty P^{3/2} W(P)\, dP = (\overline{P^{3/2}}), \tag{208}$$

so that Equation (206) becomes:

$$I = n_{\text{tot}}(\overline{P^{3/2}}). \tag{209}$$

Thus Equations (202) and (203) can be given a condensed and physically more clear form:

$$N[B' > B] = \tfrac{1}{3} n_{\text{tot}}(\overline{P^{3/2}}) B^{-3/2}, \tag{210}$$

and

$$N(B) = \tfrac{1}{2} n_{\text{tot}}(\overline{P^{3/2}}) B^{-5/2}, \tag{211}$$

where $N[B' > B]$ and $N(\mathrm{B})$ both correspond to the unit of solid angle (the steradian), and where $N[B' > B]$ denotes the number of sources brighter than B, whereas $N(B)$ denotes the number of sources per unit of brightness interval near the integrated brightness B.

10.3. THE RELATIVISTIC MODELS OF SOURCES UNIFORMLY DISTRIBUTED, AT DIFFERENT TIMES, IN AN EXPANDING, NON-NECESSARILY EUCLIDEAN, SPACE

10.3.1. The Bi-Dimensional Model of Sources Selected by their Redshift

In order to introduce some of the main particularities of tri-dimensional relativistic models it can be useful to start by considering once more the bi-dimensional model of sources *painted* on an expanding balloon representing the bi-dimensional spherical space (Σ_2).

Let us denote by R the radius of the sphere (Σ_2) and by $R(t)$ the function describing its expansion as a function of the time t. On the other hand, let N_0 denote the *constant* number of sources painted and *uniformly distributed* over (Σ_2). The number density, i.e. the number n of sources per unit area of (Σ_2) will be a function $n(t)$ of time, since it will be given by

$$n(t) = \frac{N_0}{4\pi R^2(t)}. \tag{212}$$

Thus, t_e being the 'emission-time' of the signals observed at time t_0, we shall have:

$$n(t_e)R^2(t_e) = n(t_0)R^2(t_0) = \frac{1}{4\pi} N_0 = \text{const.} \tag{213}$$

On introducing the (relative) scale factor $a(t)$ defined by

$$a(t_e) = R(t_e)/R(t_0) = R(t_e)/R_0, \tag{214}$$

we can express $n(t)$ by

$$n(t_e) = n(t_0)/a^2(t_e). \tag{215}$$

Suppose now the existence in the space (Σ_2) of an observable physical effect of expansion, analogous to the cosmological redshift (measured by $\zeta = 1 + z$) in tri-dimensional space. And suppose that the amount ζ of this effect is connected to the value of the scale factor a at time t_e by a relation of the same form as Equation (30), viz.

$$\zeta = 1 + z = \frac{1}{a(t_e)}, \tag{216}$$

where we have omitted the parameter Ω not relevant to the present problem.

Thus t_e is a known function of ζ:

$$t_e = F(\zeta). \tag{216'}$$

Hence, by Equation (215):

$$\boxed{n(t_e) = \zeta^2 n(t_0).} \tag{217}$$

Let now χ denote, as in Chapter 11, Subsection 3.1.2.1, the constant 'radial' coordinate of a source painted over (Σ_2), i.e. its *angular polar distance* from the origin (situated at the 'north pole').

The radial propagation of the signal emitted at time t_e on the sphere $\Sigma_2(t_e)$ from a source of radial coordinate χ takes place (because of the expansion) on spheres (Σ_2) whose radius $R(t)$ changes continuously.

However, if c denotes the 'velocity of light' on (Σ_2), the distance $c\,dt$ travelled by the signal between times t and $(t + dt)$ will be equal as in the tri-dimensional space to $R(t)|d\chi|$, because the amount of displacement $R(t + dt)d\chi$ on the sphere of radius $R(t + dt)$ differs from $R(t)d\chi$ only by a quantity of second order.

Thus χ will be given, as in the tri-dimensional space, by Equation (18) viz;

$$\chi = \int_{t_e}^{t_0} \frac{c\,dt'}{R(t')} = \frac{c}{R_0}\int_{t_e}^{t_0} \frac{dt'}{a(t')}. \tag{218}$$

According to Equation (216), for a given law of expansion, we have, for the time t' intermediate between t_e and t_0, a well defined value of the radial coordiate ζ' of the signal:

$$\zeta' = \frac{1}{a(t')} = f(t'). \tag{219}$$

Of course, ζ' is equal to ζ for $t' = t_e$ and ζ' is equal to 1 for $t' = t_0$.

Conversely t' can be defined by Equation (219). Therefore, t' and $dt'/d\zeta'$ can be considered as known functions of ζ':

$$t' = F(\zeta'). \tag{219'}$$

Hence,

$$\chi = \frac{c}{R_0}\int_{\zeta'=\zeta}^{\zeta'=1} \zeta'\left(\frac{dt'}{d\zeta'}\right) d\zeta' = \frac{c}{R_0}\int_{\zeta}^{1} \zeta'\left(\frac{dF}{d\zeta'}\right) d\zeta' = X(\zeta), \tag{220}$$

where the function X, for a given value of $R_0 = R(t_0)$, depends only on ζ.

Thus, a given value of the observed redshift $z = \zeta - 1$, determines both, by Equation (216′), the time t_e of emission of the observed signal, i.e. the sphere $\Sigma_2(t_e)$ on which the signal is emitted, and, by Equation (220), the particular angular polar distance χ of the point at which the source was situated on $\Sigma_2(t_e)$.

More generally, each of the sources observed with a redshift $(\zeta', d\zeta')$, i.e. with a redshift between ζ' and $(\zeta' + d\zeta')$, was situated on the sphere $\Sigma_2(t')$, in a zone $\Sigma_2(t', \chi', d\chi')$ defined by the values of $\chi' = X(\zeta')$ and of $d\chi' = (dX/d\zeta')d\zeta'$.

The area $A(t', \chi', d\chi')$ of this zone is determined by the geometry of the space (Σ_2), i.e. by the expression for its $(d\lambda)^2$. In the particular case of a spherical bi-dimensional space under consideration, $(d\lambda)^2$ is given by Equation (68″) of Chapter 11 with a replaced by R, hence the element dA of area corresponding to $(d\chi, d\psi)$ equal to $R^2 \sin\chi\, d\chi\, d\psi$, which integrated with respect to ψ, from 0 to 2π, gives:

$$\begin{aligned} A(t', \chi', d\chi') &= 2\pi R^2(t') \sin\chi'\, d\chi' \\ &= 2\pi R_0^2 \zeta'^{-2} \sin[X(\zeta')] \frac{dX(\zeta')}{d\zeta'} d\zeta'. \end{aligned} \tag{221}$$

The number of sources present on the zone $\Sigma_2(t', \chi', d\chi')$ of (Σ_2) at time t' was equal to $A(t', \chi', d\chi')n(t')$, i.e., according to Equation (217), equal to $n(t_0)\zeta'^2 A(t', \chi', d\chi')$.

Thus, the total number $4\pi N[\zeta' < \zeta]$ of all sources observed with a redshift ζ' smaller than ζ will be given by

$$4\pi N[z' < z] = n(t_0) \int_{\zeta'=1}^{\zeta'=\zeta} \zeta'^2 A(t', \chi', d\chi'), \tag{222}$$

where χ' must be replaced by its expression as a function of ζ', i.e. by the function given by Equation (220).

10.3.2. *The Tri-Dimensional Model of Sources Selected by their Redshift*

A generalization of the results obtained in Subsection 10.3.1 to the tri-dimensional space (Σ_3) is immediate.

Indeed, the fundamental Equations (219) and (220), viz.

$$\zeta' = \frac{1}{a(t')}, \tag{223}$$

and

$$\chi = \frac{c}{R_0} \int_{\zeta'=\zeta}^{\zeta'=1} \zeta' \left(\frac{dt'}{d\zeta'}\right) d\zeta' = X(\zeta), \tag{224}$$

remain of the same form, with of course other functions $a(t')$ and $X(\zeta)$, whereas Equation (217) is replaced by

$$n(t_e) = \zeta^3 n(t_0). \tag{225}$$

Equation (221) is replaced by the expression for the volume $V(t', \chi', d\chi')$ of the pseudo-spherical layer $\Sigma_3(t', \chi', d\chi')$ corresponding to the $(d\lambda)^2$ of the space (Σ_3) given by Eq. (106) of Ch. 11. The element of volume dV corresponding to $(d\chi', d\theta, d\varphi)$ is equal to $R^3 \sigma_k^2 \sin\theta\, d\theta\, d\varphi\, d\chi'$, which integrated with respect to θ from 0 to π and with respect to φ from 0 to 2π yields

$$V(t',\chi',d\chi') = 4\pi R^3(t')\sigma_k^2(\chi')\,d\chi'. \tag{226}$$

After the replacement of $R(t')$ by $R_0a(t')$ and the replacement of $a(t')$ by ζ'^{-1} Equation (226) becomes:

$$\zeta'^3 V(t',\chi',d\chi') = 4\pi R_0^3\sigma_k^2(\chi')\,d\chi'. \tag{226'}$$

Finally the number $N[z'<z]$ of sources observed per steradian with a redshift z' smaller than z will be given by an Equation similar to Equation (222):

$$4\pi N[z'<z] = n(t_0)\int_{\zeta'=1}^{\zeta'=\zeta} \zeta'^3 V(t',\chi',d\chi'). \tag{227}$$

On taking into account Equations (226′) and (224) we finally obtain:

$$\begin{aligned} N[z'<z] &= n(t_0)R_0^3\int_{\zeta'=1}^{\zeta'=\zeta} \sigma_k^2[X(\zeta')]\,\frac{dX(\zeta')}{d\zeta'}\,d\zeta' \\ &= n(t_0)R_0^3\int_{z'=0}^{z'=z} \sigma_k^2[\chi'(z')]\,\frac{d\chi'(z')}{dz'}\,dz'. \end{aligned} \tag{227'}$$

10.3.2.1. The Particular Case of Small Values of the Limiting Redshift. For small values of z we have

$$\sigma_k(\chi') \approx \chi', \tag{228}$$

and the function $\chi'(z')$ is represented by its limited expansion given by Equation (40), which, on replacing $(\frac{1}{2}+\Omega/4)$ by $(1+\tilde{\omega})$ according to the definition (63), can be written:

$$\chi' \approx \frac{cz'}{R_0H_0}[1-(1+\tilde{\omega})z']. \tag{229}$$

Hence,

$$d\chi'/dz' \approx \frac{c}{R_0H_0}[1-2(1+\tilde{\omega})z'], \tag{230}$$

and

$$\sigma_k^2[\chi'(z')]\,\frac{d\chi'(z')}{dz'} \approx \frac{c^3}{R_0^3H_0^3}[z'^2-4(1+\tilde{\omega})z'^3]. \tag{231}$$

On substituting into Equation (227′) one finds after integration:

$$N[z'<z] \approx \frac{1}{3}\frac{c^3}{H_0^3}n(t_0)z^3[1-3(1+\tilde{\omega})z]. \tag{232}$$

Thus, for very small values of z, when the second term in the square bracket can be neglected, we find the ‘classical’ expression (175) for $N[z'<z]$.

Equation (232) shows, on the other hand, that already for moderately small values of z, the second term in square brackets will generate *a slower increase* of $N[z'<z]$ as a function of z than in the classical theory, and so with a rate depending on $\tilde{\omega}=\Omega/4-\frac{1}{2}$.

10.3.2.2. The Effect of the Decrease in Brightness with Increasing Distance of the Sources. Theoretically the value of the redshift of any source is an observable quantity. However, to observe the redshift it is necessary *to disperse* the electromagnetic waves arriving from the source, in order to obtain a measurable spectrum. This is possible only if the telescope is sufficiently 'powerful' (large aperture) and the source sufficiently bright.

As noted above (see Subsection 8.2.2), the observed integrated brightness $B(t_0)$ decreases monotonically, for a given luminosity $L(t_e)$, when the redshift z of the corresponding source increases.

Thus, if the luminosity L of the real existing sources lies between two extreme values $L_{\min}$ and $L_{\max}$, it will be possible to represent schematically the corresponding variations of the brightness B as a function of z in the way shown in Figure 12.6, the two 'external' lines corresponding to $L_{\max}$ and $L_{\min}$ respectively.

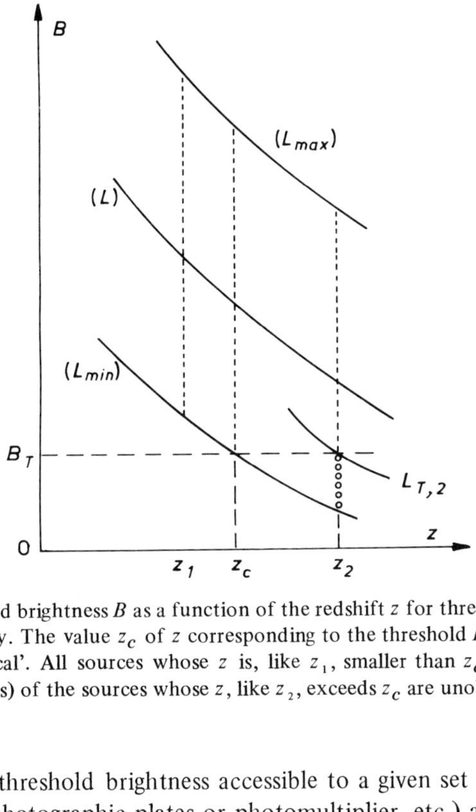

Fig. 12.6. The integrated brightness B as a function of the redshift z for three different values $L_{\min}$, L and $L_{\max}$ of luminosity. The value z_c of z corresponding to the threshold B_T and to the minimum luminosity $L_{\min}$ is 'critical'. All sources whose z is, like z_1, smaller than z_c are observable. A part (open circles) of the sources whose z, like z_2, exceeds z_c are unobservable.

If B_T denotes the threshold brightness accessible to a given set of instruments (telescope, spectrograph, photographic plates or photomultiplier, etc.) and if $z = f(B, L, \Omega)$ represents, for given values of L and Ω, the dependence of z on B, the value z_c of z defined by

$$z_c = f(B_T, L_{\min}, \Omega), \tag{233}$$

will represent the *critical value* z_c of z such that for $z > z_c$ the sources of luminosity $L_{\min}$ will be *unobservable*.

For a value of z smaller than z_c, such as z_1 in Figure 12.6, all sources will have an 'observable brightness' (above B_T), but for a value of z, exceeding z_c, such as z_2 in Figure 12.6, the sources of a luminosity L below the threshold luminosity $L_{T,2}$ defined by

$$z_2 = f(B_T, L_{T,2}, \Omega), \tag{234}$$

will be unobservable (open circles in Figure 12.6), and the observed source counts will not correspond exactly to the theory.

10.3.3. The tri-Dimensional Models of Sources Selected by their Integrated Brightness

Instead of counting the number $N[z' < z]$ of sources observed per steradian with a red-shift z' smaller than some limiting value z, one can count the number $N[B' > B]$ of sources observed per steradian with a brightness B' *exceeding* some limiting value B.

An obvious disadvantage of the latter procedure, in comparison with the former one, is that the brightness of a source depends not only on the 'luminosity distance' (or the red-shift) of a source but also on its luminosity. This obliges to make more or less arbitrary assumptions concerning the distribution and the evolution of luminosities, as for the classical models.

10.3.3.1. Sources of Identical Stationary Luminosities. Let us suppose, as for the corresponding classical models, that all sources are characterized by the same standard luminosity L_s, or, equivalently, by the same 'luminosity parameter' P_s, defined by Equation (176), viz.

$$P_s = \frac{1}{4\pi} L_s. \tag{235}$$

Consider now the ratio β defined by

$$\beta = \frac{H_0}{c}\left(\frac{P_s}{B}\right)^{1/2} = \left[\frac{H_0^2 L_s(t_e)}{4\pi c^2 B(t_0)}\right]^{1/2}. \tag{236}$$

Equation (150) shows that β^2 is a known function Φ of Ω and z:

$$\beta^2 = \Phi(\Omega, z). \tag{237}$$

On considering this relation as an equation defining z as a function of Ω and β, we can solve it for z and obtain:

$$z = f(\beta, \Omega), \tag{238}$$

where f is a known function of β and of Ω, or

$$z = F(\beta, \tilde{\omega}), \tag{239}$$

where F is a known function of β and of $\tilde{\omega}$. We recall that $\tilde{\omega}$ is defined by Equation (63), viz.

$$\tilde{\omega} = \frac{\Omega}{4} - \frac{1}{2}. \tag{240}$$

In order to illustrate these general considerations by more explicit relations let us consider the particular case of small values of z. Then, according to Equation (159), $B(t_0)$ will be given by

$$B \approx \frac{P_s H_0^2}{c^2} \frac{1}{z^2}(1 + 2\tilde{\omega} z). \tag{241}$$

On introducing the quantity β defined by Equation (236) this takes the form

$$1/\beta^2 \approx \frac{1}{z^2}(1 + 2\tilde{\omega}z), \tag{241'}$$

hence

$$\frac{z}{\beta} \approx (1 + 2\tilde{\omega}z)^{1/2} \approx 1 + \tilde{\omega}z. \tag{242}$$

Thus, in a first approximation we have

$$z \approx \beta, \tag{243}$$

and on replacing z by its expression (243) in the small term $\tilde{\omega}z$ at the right-hand side of Equation (242) we obtain:

$$z \approx \beta(1 + \tilde{\omega}\beta), \tag{244}$$

or, taking into account the definition (236) of β:

$$z_B \approx B^{-1/2}P_s^{1/2}\frac{H_0}{c}\left(1 + \tilde{\omega}B^{-1/2}P_s^{1/2}\frac{H_0}{c}\right), \tag{245}$$

where we have written z_B at the left-hand side in order to emphasize the correspondence between z and B.

Now, as already shown in Subsection 8.2.2, B increases monotonically, for a given value of L_s, when z decreases. Thus the number $4\pi N[B' > B]$ of sources whose brightness B' exceeds B is equal to the number of sources $4\pi N[z' < z_B]$ whose z' is smaller that the value z_B of z corresponding to B. The latter is given for small values of z_B by Equation (245), and more generally by Equation (239).

Thus, nothing has to be changed in the discussion of Subsection 10.3.2, leading to the expression for $N[z' < z]$, except that now n represents the number of sources of a particular luminosity $L_s = 4\pi P_s$, which can be made explicit by writing $n(t, P_s)$ instead of $n(t)$, and $N_s[B' > B]$ instead of $N[B' > B]$.

According to Equation (227′) we shall now have

$$N_s[B' > B] = N_s[z' < z] = n(t_0, P_s)R_0^3 \int_{z'=0}^{z'=z_B} \sigma_k^2[\chi'(z')]\frac{d\chi'}{dz'}\,dz', \tag{246}$$

and, for small values of z_B, i.e. for large values of B corresponding to sufficiently near sources, according to Equation (232):

$$N_s[B' > B] = N_s[z' < z_B] \approx \tfrac{1}{3}n(t_0, P_s)\frac{c^3}{H_0^3}z_B^3[1 - 3(1 + \tilde{\omega})z_B]. \tag{247}$$

This last Equation can be made more explicit by replacing z_B by the limited expansion (244). We find successively

$$z_B^3 \approx \beta^3(1 + 3\tilde{\omega}\beta); \tag{248}$$

$$[1 - 3(1 + \tilde{\omega})z_B] \approx 1 - 3(1 + \tilde{\omega})\beta; \tag{249}$$

where z_B was limited to β, because Equation (249) is used in

$$z_B^3[1 - 3(1 + \tilde{\omega})z_B] \approx \beta^3(1 - 3\beta) \approx \beta^3 - 3\beta^4. \tag{250}$$

Hence, finally

$$N_s[B'>B] \approx \tfrac{1}{3}P_s^{3/2}n(t_0,P_s)B^{-3/2}\left[1-\frac{3}{c}H_0P_s^{1/2}B^{-1/2}\right], \tag{251}$$

or

$$N_s[B'>B] \approx \tfrac{1}{3}[P_s^{3/2}n(t_0,P_s)]B^{-3/2}-[P_s^2n(t_0,P_s)]\frac{1}{c}H_0B^{-2}. \tag{252}$$

Thus, in a first approximation, for very small values of z_B, i.e. for very great values of B (very near sources), we find the same 'classical' decrease of $N_s[B'>B]$, proportional to $B^{-3/2}$, as in Equation (181). However, for moderately small values of z_B, we have a 'relativistic correction', given by the square bracket in Equation (251), corresponding to the expansion of the space.

10.3.3.2. Sources of Different Stationary Luminosities. By analogy with the discussion of Subsection 10.2.2.2, one easily understands that in this case one must introduce the number density $n(t_0, P)$ per unit of interval of (dP), replacing in Equation (252) the first square bracket by an integral analogous to the integral I defined by Equation (201), i.e. by

$$I=\int_0^\infty P^{3/2}n(t_0,P)\,dP, \tag{253}$$

and the second square bracket by the integral J defined by

$$J=\int_0^\infty P^2n(t_0,P)\,dP. \tag{254}$$

Thus Equation (252) becomes

$$N[B'>B] \approx \tfrac{1}{3}IB^{-3/2}(1-\mu B^{-1/2}), \tag{255}$$

where $N[B'>B]$ corresponding to all sources replaces $N_s[B'>B]$ and

$$\mu=\frac{3}{c}H_0\frac{J}{I}. \tag{256}$$

The first term in the expansion (255) of $N[B'>B]$ yields again the classical law. However, as for $N[z'<z]$, the relativistic correction due to the expansion produces for increasing B (whatever the value of Ω, in a first approximation) a slower decrease of $N[B'>B]$ than the classical decrease.

Appendix. Relations between the Magnitudes, the Luminosities and the Cosmological Redshift

In observational cosmology the integrated ('bolometric') brightness, and the 'partly monochromatic' ones, relative to the radio, optical, infrared, X-ray or Gamma-ray range of frequencies, respectively, are usually expressed in (apparent) magnitudes. The bolometric brightness B is expressed in bolometric magnitudes m_{bol}, the visual ('optical') brightness B_v is expressed in visual magnitudes V (or m_v), and so on, as in other branches of astronomy.

By definition the '*absolute magnitude*' M of an object of (apparent) magnitude m represents the apparent magnitude of an object of the same luminosity but placed (in vacuum) at a distance of 10 parsecs, and by definition a ratio of brightnesses equal to 10 corresponds (on a logarithmic scale) to a difference in magnitudes equal to 2.5 (with decreasing magnitudes corresponding to increasing brightnesses).

Therefore, according to the classical relation between the luminosity of an object and its brightnesss, if the object is situated at a distance of d parsecs, the corresponding absolute magnitude M will be related to the (apparent) magnitude m and to the distance (in pc) of the same object, by Equation:

$$M = m + 5 - 5 \log_{10} d \quad (d \text{ expressed in pc}), \tag{A1}$$

where adequate subscripts must be added both to M and m.

The classical relation between the distance d and the redshift z (see Chapter 12, Section 5) is:

$$d_{\text{clas}} = \frac{cz}{H_0}. \tag{A2}$$

For $H_0 = 50\,\text{km s}^{-1}\,\text{Mpc}^{-1}$ and $c = 300\,000\,\text{km s}^{-1}$ the combination of Equations (A1) and (A2) yields:

$$M_{\text{clas}} = m - 43.89 - 5 \log_{10} z \quad (\text{For } H_0 = 50\,\text{km s}^{-1}\,\text{Mpc}^{-1}), \tag{A3}$$

valid for small values of z whatever the model of universe.

For a value of H_0 different from $50\,\text{km s}^{-1}\,\text{Mpc}^{-1}$ an additional term equal to $+ 5 \log_{10}(H_0/50)$ will appear at the right hand side of Equation (A3). Thus, for instance for $H_0 = 100\,\text{km s}^{-1}\,\text{Mpc}^{-1}$ the constant (-43.89) will be replaced by (-42.39), which means that for a given m_v a less negative value of M_v will be obtained, corresponding to a smaller value of the luminosity. This is physically obvious since, for a given z, a larger value of H_0 yields a smaller value of the distance; and, for a given value of the brightness, a smaller distance yields a smaller luminosity. Hence:

$$M_{\text{clas}} = m - 43.89 - 5 \log_{10} z + 5 \log_{10} (H_0/50). \tag{A3'}$$

When the value of z exceeds about 0.1 one must replace the classical distance d_{clas}, given as a function of z by Equation (A2), by one of its relativistic generalizations derived in Chapter 12 (Section 8), and take into account the value of $q_0 = \Omega/2$ of the considered model of universe.

Two assumptions are particularly usual:

(1) The parameter q_0 is equal to zero.

Then the limited expansion of d_L given by Eqs. (87) and (148) in powers of q_0 yields, after some tedious but elementary calculations:

$$d_{\text{bol}}(q_0 = 0) = \left(1 + \frac{z}{2}\right) d_{\text{clas}}, \tag{A4}$$

hence

$$M(q_0 = 0) = m - 43.89 - 5 \log_{10} z - 5 \log_{10} \left(1 + \frac{z}{2}\right) + 5 \log_{10} (H_0/50). \tag{A5}$$

(2) The parameter $q_0 = \frac{1}{2}$ (the EdS case).
Then Equations (76′), (148) and (A2)

$$d_L(q_0 = \tfrac{1}{2}) = d_{\mathrm{EdS}} = \frac{2}{z}\,[(1+z) - (1+z)^{1/2}]\,d_{\mathrm{clas}}, \tag{A6}$$

hence

$$M_{\mathrm{EdS}} = m - 45.39 - 5\log_{10}\left[(1+z) - \left(1+\frac{z}{2}\right)^{1/2}\right] + 5\log_{10}(H_0/50). \tag{A7}$$

In principle, Equations (A4) to (A7) can be applied only to *integrated* (*bolometric*) brightness and luminosities, since it was shown in Chapter 12, Section 8 that the factor $(1+z)^2$ in the relation between integrated brightnesses and luminosities is replaced by the factor $(1+z)^1$ in the relation between the *monochromatic* brightnesses and luminosities.

However (see Chapter 13, Subsection 2.2.2) in observational cosmology, as long as only very rough approximations are involved, it is usual to apply Equations (A4) to (A7) even to '*partly monochromatic*' quantities.

For large values of z, say $z = 3.5$, the choice of H_0 and of q_0 can modify rather considerably the value of M corresponding to a given (observed) value of m. Thus, one easily finds that on replacing the choice ($H_0 = 100, q_0 = \frac{1}{2}$) by the choice ($H_0 = 50, q_0 = 0$) the value of $(-M)$ is increased for this z by 3, which means an increase in luminosity, for a given brightness, by a factor of about 17. (Those observers who claim to have observed 'the most luminous object in the universe' usually make the choice which gives the maximum luminosity, i.e. $H_0 = 50$ and $q_0 = 0$.)

Often the Sun is used to fix the 'zero point' of the bolometric magnitudes. Then, on assuming that the bolometric brightness of the Sun is equal to $1.360 \times 10^6\,\mathrm{erg\,cm^{-2}\,s^{-1}}$ and its bolometric magnitude equal to (−26.82), one finds, after some tedious but elementary calculations, that the bolometric luminosity L_{bol} of an object of bolometric absolute magnitude M_{bol} is given, in c.g.s. units ($\mathrm{erg\,s^{-1}}$), by

$$L_{\mathrm{bol}} \approx (3.05 \times 10^{35})\,10^{(-M_{\mathrm{bol}})/(2.5)}\,\mathrm{erg\,s^{-1}}. \tag{A8}$$

When the luminosities are expressed in units equal to a conventional (approximate) value of the optical luminosity L_G of our Galaxy, taken to be equal to $4 \times 10^{43}\,\mathrm{erg\,s^{-1}}$, Equation (A8) takes the form

$$L_{\mathrm{bol}} \approx 10^{(-M_{\mathrm{bol}} - 20.3)/(2.5)}\,L_G. \tag{A9}$$

Since the difference between the visual luminosity and the bolometric luminosity of the Sun differ by less thant 4% Equations (A8) and (A9) can be applied to visual absolute magnitudes and to visual luminosities on replacing M_{bol} and L_{bol} by M_{vis} and L_{vis}, in a first approximation.

A rather tedious but elementary calculation shows that L_x/L_G is given as a function of B_x (the X-ray brightness, or the X-ray 'flux') and of the redshift z (for small values of z and for H_0 in $\mathrm{km\,s^{-1}\,Mpc^{-1}}$) by

$$L_x/L_G \approx 1.1 \times 10^{14} z^2 B_x (50/H_0)^2; \tag{A10}$$

and by

$$L_x/L_G \approx 1.1 \times 10^{14} z^2 \left(1+\frac{z}{2}\right)^2 B_x (50/H_0)^2, \tag{A10′}$$

for any value of z, when $q_0 = 0$, and when H_0 is again expressed in $\mathrm{km\,s^{-1}\,Mpc^{-1}}$.

References

1. WORKS CITED IN THE TEXT
(Works that can be used for further study are marked by an asterisk.)

Andrillat, H.: 1970 *L'introduction à l'Etude des Cosmologies*, Colin, Paris.
Bondi, H.: 1960, *Cosmology* (2nd edn.), Cambridge Univ. Press, G.B.
Heidmann, J.: 1973, *Introduction à la Cosmologie*, PUF, Paris.
*McVittie, G. C.: 1965, *General Relativity and Cosmology*, (2nd edn.) Chapman and Hall, London.
Mavridès, S.: 1973, *L'Univers Relativiste*, Masson, Paris.
*Misner, C. W., Thorne, K. S., and Wheeler, J. A.: 1973, *Gravitation*, Freeman, San Francisco.
Robertson, H. P.: 1935–1936, 'Kinematics and World Structure', *Astrophys. J.* **82**, 248; **83**, 187 and 257.
*Robertson, H. P. and Noonan, T. W.; 1968, *Relativity and Cosmology*, Saunders, Philadelphia.
Sandage, A.: 1961, *Astrophys. J.* **133**, 380.
Sciama, D. W.: 1971, *Modern Cosmology*, Cambridge Univ. Press, G.B.
Walker, A. G.: 1936, 'On Milne's Theory of World-Structure', *Proc. London Math. Soc.* **42**, 90.
*Weinberg, S.: 1972, *Gravitation and Cosmology*, Wiley, N.Y.
*Zel'dovich, Ya, B. and Novikov, I. D.: 1967, *Relativistic Astrophysics* (in Russian), Moscow, U.S.S.R.
Zel'dovich, Ya. B.: 1974, In *Proceedings of IAU Symposium* **63**, IX–XI, Reidel, Dordrecht.

2. REFERENCES FOR FURTHER STUDY
(See first the works cited in 1 marked by an asterisk.)

Novikov, I. D. and Zel'dovich, Ya. B.: 1973, 'Physical Processes Near Cosmological Singularities', In *Ann. Rev. Astron. Astrophys.* **11**, 387.

CHAPTER 13

BASIC DATA IN OBSERVATIONAL COSMOLOGY: ACTIVE GALACTIC NUCLEI AND CLUSTERS OF GALAXIES

1. Introduction

The basic data of the *classical* observational cosmology (Hubble's Law) were provided by the spectroscopic and photometric study of relatively nearby galaxies; those of the modern observational cosmology are provided by the study of two major types of extragalactic objects: the *active galactic nuclei* and the *clusters of galaxies*.

The concept of 'active galactic nuclei' (AGNs) has recently emerged from discoveries showing a rather clear cut *continuity* in properties of Seyfert galaxies, N galaxies, quasars, BL Lacertae objects, and explosively active nuclei of some peculiar galaxies. All these objects will be presented and discussed below. One of their common properties is a very high luminosity. This high luminosity allows their observation at very large distances and justifies several attempts of their utilization in 'cosmological tests' based on relativistic effects described in Chapter 12.

On the other hand, the clusters of galaxies attract more and more general attention because of the growing evidence of the considerable mass of the '*intercluster gas*' and perhaps of the gas filling the space *between* the clusters of galaxies grouped in (more or less hypothetical) '*superclusters*'. This mass, if real, might provide a major part of the mass missing for the 'closure' of the universe.

2. Active Galactic Nuclei (Quasars and Similar Objects)

2.1. PRELIMINARY REMARKS

In the field of classification there always exists an evolution from 'classes' defined by a combination of *apparent* ('immediate') and *intrinsic* characteristics to definitions almost entirely based on intrinsic astrophysical properties of some group of objects, and sometimes finally on purely physical properties when these become recognized.

Thus, for instance, the class of 'nebulae' introduced by the Messier and the NGC catalogues, based on the common *apparent morphology* of globular clusters, galaxies, and bright diffuse nebulae observed with a limited resolution, is now replaced by three new 'classes' better defined astrophysically and physically, as stellar systems (of different structure and 'richness') or fluorescent gas masses, through a better resolution and the use of spectroscopy.

It also often happens that different subclasses of a wide natural general class of objects are not immediately recognized as such (most often because they are defined partly by such apparent criteria as the 'nebulosity' of nebulae). Indeed, when the partly apparent characteristics of some newly discovered objects cannot be made to fit any existing class (or subclass) of existing classifications, the temptation is strong to introduce a 'new category' of objects.

Thus, for instance, the pulsars were considered (and still are sometimes considered) as a 'new category' of astronomical objects, instead of being classified as a spinning subclass of the general category of neutron stars.

The fascinating story of the progressive emergence of the 'unified concept' of active galactic nuclei (AGNs) provides a typical example of both types of evolution in classification and terminology.

We start by a presentation of different kinds of AGNs as 'new categories' of extragalactic astronomical objects (as they appeared historically), defined partly by apparent properties, and show afterwards how the discovery of various new properties, more intrinsic, disclosed the deep similarity of all these objects.

2.2. Discovery and the First Definitions of Different Active Galactic Nuclei Considered as 'New Categories'

2.2.1. Seyfert Galaxies and N Galaxies

C. Seyfert discovered, in 1943, that some *spiral* galaxies possess a small *very bright nucleus* characterized by *very strong and broad emission lines* never observed in galaxies not possessing such nuclei. Thus, a new category of astronomical objects, that of 'Seyfert galaxies' (or simply 'Seyferts') was introduced.

When *elliptical* galaxies possessing a small very bright nucleus (but not necessarily a Seyfert-like spectrum) were observed, a second new category, that of N galaxies, was created.

Seyferts are usually subdivided on the basis of the *relative width* of their emission lines in two subclasses: the Seyferts 1 and the Seyferts 2 (Weedman, 1977).

The Seyferts 1 (often denoted by Sy 1) have broad hydrogen lines (e.g. H_β) but narrower forbidden lines (e.g. the [O III] line). A typical Sy 1 galaxy is the galaxy NGC 5548.

The Seyferts 2 (Sy 2) have hydrogen and forbidden lines of about the same width. A typical Sy 2 galaxy is NGC 1068.

In some Seyfert and N galaxies the galactic envelope surrounding the bright nucleus is barely perceptible and appears rather as a 'faint wisp'. Such galaxies are often classified morphologically as *'compact galaxies'*. Many compact galaxies, classified as such in Zwicky's catalogues (1961–1968), have a Seyfert-like spectrum.

Some N galaxies have the same type of spectrum as Seyferts 1 (e.g. the one associated with the radio source 3C 120), some other N galaxies have, on the contrary, very weak and narrow spectral lines in the spectrum of their nucleus (e.g. the one associated with the radio source 3C 371).

Astronomers who attribute more importance to spectroscopic characteristics than to morphology classify a galaxy such as the one associated with 3C 120 in the category of Seyferts 1.

It will appear later that only the nuclei of Seyfert 1 galaxies and only some of the nuclei of N galaxies belong to the wider class of *active* galactic nuclei (AGNs). The nucleus of NGC 1068 generally is not considered as an AGN, whereas that of 3C 371 will be shown to represent an AGN.

(N.B. In the 'names' used above, 3C represents an abbreviation of the *'Third Cambridge Catalogue of Radio Sources'* (Edge *et al.*, 1960).

In what follows some objects associated with a radio source will be given the same name as that of the radio source. Thus, instead of saying *the galaxy associated with* 3C 120, we shall say *the galaxy* 3C 120.

Moreover, one usually says that the nucleus of a Seyfert or of an N galaxy is 'surrounded by a galaxy'. This is obviously incorrect, but is very convenient, and never leads to confusion.)

2.2.2. *Nuclei of Some Radio Galaxies*

The *'radio luminosity'* L_{rad}, i.e. the total energy emitted per second at radio frequencies (defined as extending from 10 MHz to 10^5 MHz) of the so-called normal galaxies (such as, for instance, the 'Andromeda Nebula', M 31) is generally much smaller than their *'optical luminosity'* L_{opt}. (In what follows we shall call such luminosities as L_{rad} or L_{opt}, corresponding to a limited range of frequencies, the *'partly monochromatic luminosities'*, as opposed to *'bolometric'* or *'integrated'* luminosities.)

Both L_{rad} and L_{opt} can be expressed in units equal to a conventional (approximate) value of the *optical* luminosity L_G of our Galaxy, taken to be equal to 4×10^{43} erg s^{-1}.

For the normal galaxy M 31, L_{rad} is only of the order of $2.5 \times 10^{-6} L_G$, whereas L_{opt} is of the order of $2.5 L_G$.

For other normal galaxies the ratio $L_{\mathrm{rad}}/L_{\mathrm{opt}}$ seldom exceeds 10^{-2}.

On the other hand, many optical counterparts of intense extragalactic radio sources are giant elliptical galaxies (generally situated in clusters of galaxies), called the *'radio galaxies'* because of a very high ratio $L_{\mathrm{rad}}/L_{\mathrm{opt}}$ exceeding 1 and which can become as high as 100.

The radio galaxy NGC 5128 (called Cen A by the radio astronomers) is optically an elliptical galaxy, famous for its obscuring dust lane girdling its equator. It possesses a very active nucleus, which emits radiation over the entire electromagnetic spectrum. (Other particularities of this nucleus will be discussed later.)

Other radio galaxies also possess AGNs.

2.2.3. *The Explosive Galaxy M 87 and its Nucleus*

The giant very peculiar 'explosive' elliptical galaxy Messier 87 (M 87 is also called NGC 4486, or, as radio source, 3C 274 and Virgo A) represents the brightest member of the great Virgo cluster of galaxies.

Though its L_{rad} is of the order of $0.05 L_G$ and its L_{opt} is of the order of $5 L_G$ (hence a ratio L_{rad}/L_G of the order of only 0.01) it is often considered (contrary to the definition of the preceding subsection) as a 'radio galaxy'.

Jutting out of its nucleus is a 'jet', about 2 kpc long, consisting of a series of bright knots that suggest a sequence of violent explosions at the galaxy's core. The radio source associated with the nucleus of M 87 is extremely compact (about 0.09 pc across), hence perhaps the temptation of considering it as a 'radio galaxy'.

Recently this peculiar galaxy (quite 'normal' down to about 0.1 kpc from its centre) has attracted much attention from astronomers because two separate sets of observations (Sargent *et al.*, 1978; Young *et al.*, 1978) have suggested the existence in the core of M 87 of a supermassive object, possibly a black hole, as a probable explanation for the excess of mass-to-light ratio at the center of the galaxy. To explain their results the observers propose a central dark mass of about $5 \times 10^9 M_\odot$ within a radius of no more than 110 pc.

From other evidences, to be discussed later, it is probable that the most central region of M 87 represents a peculiar kind of AGN.

2.2.4. Quasars

Seldom a scientific discovery included so much suspense as the discovery of quasars, and seldom a 'new category' of objects was given a name as improper and as misleading as 'quasar'.

In 1960 some of the radio sources of the 3C catalogue were already identified with many quite trivial optical counterparts such as galaxies.

This permitted a determination of the (classical) distance of these radio sources, through the measure of the redshift z (see Chapter 12, Section 3) of spectral lines of the associated galaxies, and the application of the Hubble's Law, (see Chapter 10, Section 3). Only the classical distance (defined in Chapter 12, Section 5) was involved for these relatively nearby objects with a very small redshift.

However, some other radio sources, of *small angular diameter* were not yet known to be associated with any galaxy. On assuming that all radio sources had the same linear diameter (a wrong assumption, but, as often happens, a fruitful one) one could expect that such radio sources were particularly remote, thus very useful as possible data for cosmological tests. Indeed, it was shown in Chapters 10 and 12 that, in principle, different models of universe are best discriminated by observation of remote objects.

On assuming further the existence of a correlation between the radio luminosity L_{rad} of a radio source and the optical luminosity L_{opt} of the associated galaxy, one could expect to perform a new identification for a radio source such as 3C 48 both of *small angular diameter* and of *large radio brightness*.

However, as for all identifications, it was necessary that the position given by radio techniques be known with a sufficient precision and that a sufficiently powerful optical telescopes be able to yield a photographic image of a faint remote galaxy.

Unfortunately, before the introduction of the VLBI (Very Long Baseline Interferometry), radio techniques were unable to provide the position of radio sources with great accuracy: the uncertainties were, in 1960, of *several arc minutes*.

However, T. A. Matthews and his colleagues succeeded at Owens Valley Radio Observatory (California, U.S.A.), to localize 3C 48 with a precision six times better (through an adequate use of conventional radio interferometry) than that given in the Cambridge Catalogue.

This made it worthwhile to attempt optical identification with the 200-inch Palomar telescope, and such an attempt was undertaken by A. Sandage, who soon announced (Sandage, 1961) that he has found something within the new error box of the radio position of 3C 48.

However, instead of the expected faint *galaxy* this 'something' was a 16th magnitude *star-like* object (of a strange blue colour) classified by Sandage as a 'relatively nearby *star* with *most peculiar properties*'.

Shortly after Sandage made a similar identification of two other small (angular) diameter sources 3C 196 and 3C 286 with star-like objects.

Of course, it still remained a remote possibility that these 'star-like' objects represented very distant galaxies, too distant to appear photographically as extended objects.

However, photometric records soon showed that the optical brightness of 3C 48 was not constant, as usual for normal galaxies, but rather rapidly variable. Thus the 'star-like object' could not be a normal galaxy.

The spectrum of the 'star' associated with 3C 48 obtained by Sandage was particularly strange. It was dominated by *broad emission lines*, *none of which could be identified* with any plausible combination of lines of a known neutral or ionized element.

We know today the reason of this difficulty: the relatively *large redshift* $z = 0.37$ of these emission lines. Though this redshift was about 10 times smaller than that of some quasars discovered later, it was sufficient to bring into the photographic domain of the spectrum some ultraviolet lines (such as the line L_α of H) which have never been observed previously in the spectrum of any star (because of the strong atmospheric extinction in this part of the spectrum) though they have already been observed in laboratory and, from missiles, in the solar spectrum.

The second act of the drama took place when Hazard *et al.* published, in March 1963, the details concerning the identification of another radio source 3C 273 with a '13th magnitude *star* . . . a peculiar object with a jet'. This second identification was made possible by observation of the lunar occultation of 3C 273 with the giant (70 m) steerable radio telescope at Parkes (Australia), a method already described in Chapter 9, Subsection 1.5.

In the case of 3C 273, the optical identification was easily performed after application of Hazard's method and a thorough examination of the photographic plates of the corresponding region of the sky, already obtained earlier with the 200-inch Palomar telescope. Indeed, the 'star' was the only optical object in the very small error box left by Hazard's ingenious technique. Moreover, this 'star' was 3.5 magnitudes *brighter* than the 'star' associated with 3C 48.

Once more, the spectrum was quite unusual. However, this time, by chance (which, as stated by L. Pasteur, favours only 'prepared minds') the redshift of 3C 273 was sufficiently *large* ($z = 0.16$) to display an unusual spectrum, but *much smaller* than the redshift ($z = 0.38$) of the 'star' associated with 3C 48.

This allowed the major breakthrough achieved by M. Schmidt in 1963: an identification of a few features of the observed spectrum as four redshifted emission lines (H_ϵ, H_δ, H_γ, H_β) of the Balmer series of H and a redshifted emission line of ionized Mg ($\lambda = 2798$ Å), *all corresponding to the same value of* z. The line H_α ($\lambda = 6563$ Å) was measured later in the infrared by Oke (1963) corresponding again to the redshift of 0.16.

After this major discovery, Greenstein and Matthews (1963) were able to identify the 'ununderstandable' spectroscopic features in the optical associate of 3C 48 and obtain a value of z equal to 0.367 already mentioned above. The broad resonance line of Mg II already observed in the redshifted spectrum of the 'star' associated with 3C 273, the line $\lambda = 1549$ Å of C IV and even the line L_α ($\lambda = 1216$ Å) of H, situated in the far ultraviolet, were brought into the photographic part of the spectrum by this large value of z.

These two discoveries were soon followed by identification of many other radio sources with optical objects characterized by a *star-like* (apparent) morphology, a *variable brightness* (on time scales never observed in galaxies), and a spectrum with *highly redshifted* very *broad emission* lines.

These, first discovered, properties have been taken as a definition of a 'new category' of astronomical objects, called the '*quasars*' because of their star-like appearance on the Palomar Sky Survey plates and because of the role played by the radio sources in their detection (*Quasi Stellar Radio Sources* being abbreviated as Quasar).

2.2.5. *BL Lacertae Objects* ('*Blasars*')

In 1969 a radio source with a particularly 'flat' radio spectrum was identified with a variable 'star' (called as such BL Lacerta) with very peculiar optical properties. Indeed, it soon appeared that the object was *extragalactic* and could not be classified as previously in the category of ordinary variable stars.

About 30 similar extragalactic objects were soon discovered, all characterized (in the beginning): (1) by a flat or even '*inverted*' (when compared to other radio sources) *radio spectrum* (i.e. a spectrum in which the radio brightness increased with frequency); (2) a *featureless optical* spectrum (no emission or absorption lines, a purely continuous spectrum); (3) a *star-like* appearance on photographic plates; (4) an extreme *variability* in optical brightness (on a time scale from days to months) and in optical polarization (10% to 20%, variable at a time scale of a few hours).

This unusual combination of strange properties suggested the creation of an extra 'new category' of astronomical objects, called, on taking BL Lacerta as the 'prototype' of the category, the 'BL *Lacertae objects*'.

This rather clumsy name is obviously unacceptable and misleading since the 'BL Lacertae Objects' have generally no connection with the particualr constellation Lacerta. A much better name – the '*blasars*' – was proposed once, more or less seriously, by E. Spiegel, in order to emphasize the now well established similarity between quasars and BL Lacertae objects (also called the 'BL Lacs'). The name '*blasar*' is so convenient that, in spite of the fact that it is not yet generally accepted, we shall use it henceforth everywhere. (Somebody must dare to start!)

2.3. THE FIRST DIFFICULTIES IN THE DEFINITION OF QUASARS

2.3.1. *Radio Quiet Quasars*

A few years after the discovery of the first quasar, Sandage discovered several star-like objects *non-associated* with any detectable radio source but distinctly possessing the optical properties of a strong 'ultraviolet excess' and a large redshift.

When a star is observed photoelectrically with adequate filters the 'partly monochromatic' brightnesses are observed and expressed in V (visual), B (blue, photographic), and U (ultraviolet) magnitudes. For normal stars the colour indexes $(U-B)$ and $(B-V)$ are statistically correlated, so that in a plot of $(U-B)$ against $(B-V)$ for each star, one obtains a relatively narrow band of points near an average line corresponding to the standard 'main series' line of the HR-diagram, if only the main sequence stars are represented. (See Figure 13.1 overleaf).

On such a diagram white dwarfs, novae (during their decline after the flare) and irregular variable stars take place *outside* the main sequence band at a place indicating an excess of the 'ultraviolet brightness' for a given $(B-V)$ colour. Such is also the case for the objects discovered by Sandage.

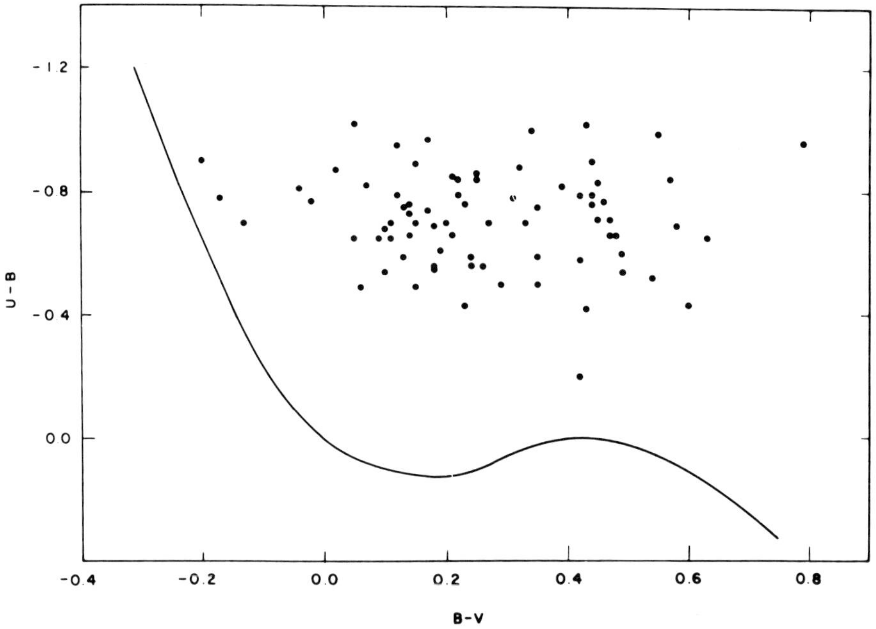

Fig. 13.1. Two-colour plot. The curve is the locus of normal (main sequence) unreddened galactic stars (highest temperature at upper left). The points correspond to objects which present an 'ultraviolet excess', such as white dwarfs, novae (after the flare), some irregular variable stars, and some AGNs.

These 'radio quiet' quasars very rapidly outnumbered those associated with sources of a measurable radio brightness.

After this discovery the term 'quasar' was replaced by the term 'QSO' (*Quasi Stellar Object*) and the minority of QSOs associated with radio sources was called QSS (Quasi Stellar Sources, the word Source being taken as meaning a *Radio* Source).

However, the most recent tendency is to neglect the original meaning of the word 'quasar', i.e. to neglect the presence of the letter R (radio) in this term, and to use the word 'quasar' as a synonym of QSO (which remains in current use).

Thus, the former QSSs become the '*radio quasars*' (or, in a more popular style, the '*radio-loud*' quasars) and those observed only optically become the '*optical quasars*', or the '*radio quiet quasars*'.

Those discovered by radio techniques are then called the 'radio selected quasars' and those discovered by optical techniques are called the 'optically selected quasars' (OSQ).

This complicated terminology (also quite inadequate as will be explained below) becomes even more confusing when, as is often done, the optically *selected* quasars (in the sense defined above) are called 'optical quasars' even when they possess a non negligible radio brightness.

2.3.2. First Indications of a Stellar Envelope

As soon as 1973, Kristian demonstrated (Kristian, 1973) that many of the quasars near enough and intrinsically faint enough for an underlying 'host galaxy' to be seen (without being masked by the brilliant nucleus represented by the quasar) actually had an associated *nebulous wisp* or, at least, the suggestion of a *nonstellar* image. This was a first blow to the S in the term QSO and in the term quasar. More recent discoveries reported in Subsection 2.4 will show how misleading was the initial criterion of the 'star-like' photographic image, both for quasars and for blasars.

2.4. Progressive Emergence of the AGN Concept through New Discoveries

2.4.1. Discovery of the 'Host Galaxies' of Quasars and Blasars

Already in 1975, Wampler *et al.* have observed that 3C 48 had a blue *halo*. But it is only quite recently that the *galactic envelopes* could be put into evidence, in an undubitable way, around a few quasars and blasars.

The failure up to 1977–1978 to photograph the 'host galaxy' of a quasar or of a blasar is mainly due to the high luminosity of the 'active nucleus' (quasar or blasar) with respect to the luminosity of the galactic envelope. Since the luminosity L_{nuc} of the nucleus is highly variable, whereas the luminosity L_{env} of the envelope is essentially constant, the ratio $L_{\text{nuc}}/L_{\text{env}}$ is variable, but at some phases and for some objects it can be as large as 100 or 1000. Only when the nucleus is in a 'low state', and for a few objects for which $L_{\text{nuc}}/L_{\text{env}}$ is then of the order of unity, and by the use of very sophisticated and powerful techniques, can the 'host galaxy' be put into evidence.

This probably explains why no 'nebula' or even 'fuzzy borders' have been detected around some blasars such as OJ 287, 3C 66 A, or Q 1418 + 54 even in the most deeply exposed photographic plates.

However, if on photographic plates with short exposures, a blasar like BL Lac appears completely starlike, on very long exposures a *fuzzy halo* already appears surrounding the bright core. Of course, the structural information embodied in a photographic 'nebulosity' is very poor, and its true physical nature is not immediately obvious.

The main progress came, between 1975 and 1978.

First, when spectrographic observations of the nebulosity surrounding some blasars permitted detection of *absorption* lines, and established that they were typically *stellar* 'galactic' lines (such as the lines H and K of ionized calcium) with a redshift between $z = 0.03$ and $z = 0.05$, thus showing that the blasars AP Lib, Mrk 421 and Mrk 501 actually represented active galactic *nuclei of giant elliptical galaxies* with redshifts comparable to those of the least redshifted quasars.

Second, when Miller and Hawley, succeeded, in 1977 (through the use of image tube scanners capable of recording very faint lines and more special ingenious techniques protecting the image of the galactic 'halo' of BL Lac against contamination by the light of the bright blasar nucleus) to observe both some emission lines of the formerly 'featureless' spectrum of this blasar, and some absorption lines of the surrounding nebulosity. Both series of lines proved to be shifted by *the same amount* (z_e of emission lines and z_a of absorption lines being equal to 0.0688 ± 0.0002). It was not yet possible to state with absolute confidence that the 'nebulosity' was a giant elliptical galaxy, as in the case of AP Lib, Mrk 421 and Mrk 501. It was not even possible to be quite sure that there existed a galaxy of stars associated with the blasar BL Lac. But the proof was found that the 'featurelessness' of the blasar spectra could no longer be considered as a characteristic property of these objects. Moreover, the relatively large redshift of BL Lac introduced even more deeply the blasars in the z-region of quasars.

The third hard blow to the old categories came from the study of the object Fairall-9, discovered by the South African astronomer Fairall, and described by him in Fairall, (1977) as 'an extreme Seyfert – almost a quasar'.

This object soon was proved, by West *et al.*, (1978), to possess not only a 'Seyfert morphology' detected by Fairall (a well developed spiral structure) but also a typically quasar nucleius, both the nucleus and the surrounding galaxy with a Seyfert-quasar spectrum with emission and absorption lines redshifted by $z = 0.046$. The ratio $L_{\text{nuc}}/L_{\text{env}}$ was unusually high for a galaxy: probably around 3–10.

The visual (optical) luminosity L_v of the nucleus of F-9 (on assuming a value of 50 km s^{-1} Mpc^{-1} for H_0) is of the order of $30L_G$ (absolute magnitude M_v near -24), far above the visual luminosity of the least luminous of the objects already classified as a quasar: Mrk 501, whose luminosity (with the same value of H_0) is only of the order of $7L_G$ (M_v near -23). On the other hand, the visual luminosity of F-9 is of the same order as that of the most luminous nucleus of Seyfert 1 galaxies (I Zw 1 with L_v also near $30L_G$).

2.4.2. *Confirmation of the 'Cosmological' Character of the Quasar Redshifts and the Luminosity Problem*

In the preceding subsection we implicitly assumed that the (visual) luminosity L_v of a quasar (expressed in visual absolute magnitudes M_v) can be deduced from its apparent magnitude m_v or V by the relations of the Appendix to Chapter 12, in which the redshift z plays the role of the 'distance parameter' (with the values of H_0 and q_0 supposed known from other evidences).

Indeed, it was shown in Chapter 12 that, in relativistic cosmology, the redshifts of the galaxies are considered as a 'cosmological effect', i.e. are not attributed to a 'genuine' mutual recession, producing a Doppler effect, but are ascribed to the expansion of the space. Astronomical objects are considered as inserted (like ornamental accessories) in an expanding space and fixed with respect to the space itself (as would be the case for the points of a bi-dimensional space *painted* on an expanding balloon).

Of course, the relations expressing the luminosity of an object of given brightness as a function of its 'cosmological redshift' z cannot depend on the *physical nature* of the object, and should apply to quasars in the same way as to galaxies.

However, as more and more quasars were discovered (the Catalogue of QSOs of G. R. Burbidge *et al.* (1977) already contains a description of 639 quasars) a double evidence progressively emerged.

The first was provided by the discovery of several quasars for which the combination of their observed visual (and, for very few, of their bolometric) brightness with their observed redshift z (considered as a 'cosmological distance indicator') yielded visual and bolometric luminosities of unprecedent magnitude.

Table 13.1, in which the bolometric luminosities are expressed in L_G units, and in which different objects are listed in order of increasing bolometric luminosities, immediately shows the huge values of luminosities reached by a few quasars at their peak (variable) brightness, compared with the most luminous galaxy.

TABLE 13.1

Bolometric luminosities of three quasars and the bolometric luminosity of the most luminous normal galaxy, for $H_0 = 50 \text{ km s}^{-1} \text{ Mpc}^{-1}$ and $q_0 = 0.1$ (after 1978, *Nature*, **275**, 298)

The Name	The type	L_{bol}/L_G	z	Comments
Markarian 231	–	~ 400		The most luminous normal galaxy
3C 273	quasar	~ 3,000	0.158	The most studied quasar
3C 232	quasar	~ 20,000	0.529	
Q 0420 – 388	quasar	~ 50,000	3.12	The most luminous, for small q_0

The approximate visual luminosities of some 30 AGNs (most luminous nuclei of Seyfert 1 and N galaxies, blasars, quasars) listed in order of increasing luminosity (compiled from about 20 different sources) in Table 13.2 show to which extent the most luminous quasars outshine all other objects and exceed the visual luminosity of a galaxy like our Galaxy (of visual luminosity of the order of L_G).

The exotic luminosities of the most luminous quasars have caused some astronomers, first of all G. R. Burbidge, to question whether the quasars actually are at the great 'cosmological' distances implied by their redshifts. And many even more 'exotic' hypotheses have been advanced by several unorthodox astronomers as alternative to the 'cosmological hypothesis'. However, the same Table 13.2 shows a second line of evidence, due to most recent discoveries, which progressively have proved that the 'cosmological hypothesis' represents the most reasonable *extrapolation* of our present knowledge.

Of course, the controversy over whether the measured redshifts of emission lines of quasars could be 'cosmological' or not is still not completely over and we shall rediscuss this problem in a later subsection (2.6) in connection with other unsolved problems concerning more generally all AGNs.

On the other hand, the *overlapping* of luminosities of Seyfert and N nuclei, of blasars and of many quasars of small or moderate luminosity, clearly shown by Table 13.2, proves that when one accepts the cosmological interpretation of the redshift of Seyferts and of blasars (as even G. R. Burbidge does) there is no more reason to remain sceptical about quasars, which appear to exhibit the same phenomenon as the other AGNs, simply with more intensity, at least from the point of view of their luminosities. Even in the X-ray domain, the same overlapping of X-luminosities exists as shown by the corresponding column in Table 13.2.

It results from the data of Table 13.2 that it would be just as unreasonable to reject the cosmological interpretation of the redshifts of quasars because of their great luminosities as it would be unreasonable to reject the laws of photometry because of the existence of superluminous stars (improperly called 'supergiants'). Unfortunately, the great luminosity of quasars is not the only difficulty encountered in their study and, more generally, in the study of all AGNs. However, before discussing these difficulties we shall briefly review the various properties of AGNs from an unified point of view emerging from the preceding subsections.

TABLE 13.2

Values of L_v/L_G and of L_x/L_G for some of the most studied AGNs. S means Seyfert 1; N means N galaxy; B means blasar (BL Lac Object); Q means quasar. The values of z_a of blasars are given in parentheses. All L_x are relative to the 2–10 keV or to the 2–11 keV range of energies. All data from the quoted references have been reduced to the same system: $H_0 = 50\ \mathrm{km\ s^{-1}\ Mpc^{-1}}$, $q_0 = 0$. Special details are indicated in notes (. . .), where ARAA = *Ann. Rev. Astron. Astrophys.*; *Nat.* = *Nature*; *ApJ* = *Astrophys. J.*; *AA* = *Astron. Astrophys.*

AGN type	$\frac{L_v}{L_G}$	z_e or (z_a)	Usual name	X-ray name	$\frac{L_x}{L_G}$	Abridged references
(a)		~ 0.001	NGC 5128	3U 1322 − 42	(a)	1977, *ARAA* **15**, 541.
(b)		~ 0.003	M 87	3U 1228 + 12	(b)	*Idem*
S		~ 0.003	NGC 4151	4U 1206 + 39	(c)	*Idem*; 1977, *Nat*, **270**, 319.
S	2	0.014	ESO 140 − G43	–	–	1978, *AA*, **65**, 151.
S, N. Q	3.6	0.076	PHL 1070	–	–	1978, *ApJ*, **219**, 381.
B	7.5	(0.034)	Mrk 501	4U 1651 + 39	6.0	1978, *ApJ*, **226**, L65.
B	15	(0.031)	Mrk 421	2A 1102 + 382	(d)	*Idem*
B	~ 25(e)	(0.069)	BL Lac	H 0548 − 322	15	1978, *ApJ*, **224**, 812.
S	~ 25	0.061	I Zw 1	–	–	1978, *AA*, **62**, L13.
S, Q	~ 35	0.046	Fairall-9	4U 0106 − 59	~ 5	1978, *Nat*, **273**, 450.
Q	~ 35	0.064	MR 2251 − 178	2A 2251 − 179	40(f)	1978, *Nat*, **271**, 35 + *Idem*.
B	~ 70(g) (m)	(0.2)	0J 287	–	–	1978, *ApJ*, **224**, 812.
Q	~ 100(h)	0.044	Q 0241 + 622	4U 0241 + 61	7.5	1978, *ApJ*, **224**, L43.
Q	~ 630(i)	0.158	3C 273 (k)	4U 1226 + 02	150(j)	1978, *AA*, **62**, L13.
Q	~ 1 440(m)	0.594	3C 345	–	–	1978, *ApJ*, **224**, 812.
B	~ 7 600(m) (n)	(0.851)	A0 0235 + 164	–	(p)	1976, *Nat*, **260**, 754.
Q	~ 13 000	2.22	PKS 0237 − 23	–	–	1978, *ApJ*, **222**, L1.
Q	~ 17 000	3.53	OQ 172 (q)	–	–	*Idem*
Q	~ 25 000	2.2	B2 1225 + 31	–	–	*Idem*
Q	~ 25 000	3.27	PKS 2126 − 15	–	–	1978, *ApJ*, **223**, L1.
Q	~ 25 000	2.69	PHL 957	–	–	1978, *ApJ*, **222**, L1.
Q	~ 48 000(m)	0.536	3C 279	–	–	*Idem*
Q	~ 57 500(m)	1.930	PHL 61 (r)	–	–	*Idem*

(Notes to Table 13.2)

Notes:

(a) Giant radio source Cen A (the nearest radio galaxy); L_x/L_G variable from 0.01 to 0.06 (from radio lobes).
(b) Peculiar 'explosive' galaxy; $L_x/L_G = \sim 0.1$.
(c) According to *1977 ARAA*, **15**, 541, $L_x/L_G = \sim 0.1$; According to 1978, *ApJ*, **225**, L115, L_x/L_G varies from 0.04 to 0.08.
(d) L_x/L_G is variable from ~ 1 to ~ 80.
(e) A peak value; According to 1978, *ApJ*, **226**, L65, the average value is ~ 7.5.
(f) A peak value; L_x/L_G varies from ~ 5 to ~ 35.
(g) The value of z_a is not known with precision.
(h) According to 1978, *Nat*, **275**, 298, this object is probably identical to the gamma-ray source CG 135 + 1; According to 1978, *Nat*, **273**, 450, L_v/L_G can lie between 30 and 830 because the correction for interstellar absorption is still uncertain (the most probable value is ~ 160).
(i) Average value: L_v/L_G varies from ~ 570 to ~ 1100 according to 1978, *Nat*, **273**, 450.
(j) Average value: according to 1978, *Nat*, **273**, 450, L_x/L_G varies from ~ 50 to ~ 250.
(k) Also the gamma-ray source CG 291 + 65; according to 1978, *Nat*, **275**, 298, L_γ (50–500 MeV)$/L_G$ is of the order of 500.
(m) A peak value.
(n) This object possesses also a system of absorption redshifts with $z_a = (0.524)$; 1978, *Sky & Telescope*, **56**, 490, reports a peak value, in 1975, of 25 000 without references.
(p) According to 1978, *Nat*, **276**, 375, $L_x/L_G \lesssim 3\,000$.
(q) Also PKS 1442 + 101.
(r) Also PKS 2134 + 004.

2.5. UNIFIED VIEW OF VARIOUS AGN PROPERTIES

2.5.1. *Colour Characteristics*

The 'ultraviolet excess' used as a characteristic property of quasars in their 'old' definition becomes somewhat ambiguous in the presence of a considerable redshift.

Indeed, the standard 'main series' diagram for normal stars (already mentioned in Subsection 2.3.1) in a plot of $[-(U-B)]$ colour index *versus* $(B-V)$ colour index represents a relatively narrow band of points near the average line represented in Figure 13.1.

Since B represents the 'blue' *magnitude* and since V represents the 'visual' (photoelectric) *magnitude* of an object, the presence of the minus sign in the relation between the brightnesses and the corresponding magnitudes, viz.

$$B = -2.5 \log_{10} b_B + \text{const.}, \tag{1}$$

$$V = -2.5 \log_{10} b_V + \text{const.}, \tag{2}$$

and

$$U = -2.5 \log_{10} b_U + \text{const.}. \tag{3}$$

(where the brightnesses are denoted by b_B, b_V, b_U, respectively, in order to avoid confusion between b_B and B), means that an increase in $(B-V)$ corresponds to a 'more red' colour (the visual brightness b_V exceeding the blue brightness b_B).

Similarly, an increase in $[-(U-B)] = B-U$ corresponds to a 'more ultraviolet' colour (ultraviolet brightness b_U exceeding the blue one).

On a diagram like the one in Figure 13.1, white dwarfs, novae (after explosion), irregular variable stars take place 'above' the main sequence, since the $(U-B)$ axis is positively oriented *downwards*. This is indicated schematically by the isolated points in Figure 13.1, which correspond to the 'classical objects' mentioned above, presenting an UV-excess (abnormally large ultraviolet brightness).

As to AGNs, such as quasars, (also represented schematically by isolated points in Figure 13.1), they also take place 'above' the main sequence standard line. However, without taking into account the redshift one cannot immediately know whether they are 'displaced' *upwards* by a real UV-excess or displaced to the *right* by their redshift. Moreover, as noted by Jauncey *et al.* (1978) as a result of the redshifting of the deep and complex *absorption* spectrum shortward of Lα line into the blue filter and the redshifting of strong C IV emission into the red filter, high-redshift quasars (such as e.g. PKS 2126 − 15) acquire a 'neutral colour' or a red colour.

A 'neutral colour' is not always produced by the mechanism just mentioned above. Blasars are usually more red than an average (low or moderate redshift) quasars, even when allowance is made for interstellar reddening. For most blasars, the neutral colours (which technically means the same 'intensity' on red and blue plates of the Palomar Sky Survey) are usually caused by a *steep optical continuum* spectrum.

Finally, one must take into account an eventual influence of the 'host galaxy' on the colour of the observed nucleus. The colours of AGNs, such as Mrk 501 and 421, apparently embedded in a host galaxy, are (as reported by Maza *et al.* (1978)) redder than those of AGNs with no apparent envelope, such as 0J 287. Moreover, the reddening of an AGN produced by the host galaxy is more pronounced when the galaxy is elliptical (like in Mrk 501 or Mrk 421) instead of being a spiral (as for most Seyferts).

2.5.2. *Spectral Characteristics*

2.5.2.1. Optical Spectra (Continuous and Line Spectra). As already briefly mentioned above, no difference seems to exist between Seyfert 1 nuclei and quasars from the point of view of *emission lines*: both are characterized by broad and strong 'thermal' emission lines. Permitted lines are particularly broad ($\sim$ 10 000 km s^{-1}) on the average, forbidden lines ([O III] for instance) are usually more narrow ($<$ 1000 km s^{-1}), but still broad. The width of Hα emission line can take extraordinary values. Thus, Margon and Kwitter (1978) report a full width zero intensity of 24 000 km s^{-1} in the rest frame, for Q 0241 + 622.

On the other hand, many *systems* of absorption lines with different redshifts occur in some quasars (but this property will be discussed later, in Subsection 2.6).

As to blasars, when observed with conventional techniques, they are characterized by *featureless* purely continous spectra. However, with more refined techniques, applied near the minimum of the variable brightness of the AGN, one can observe very weak emission lines (from the nucleus) and weak absorption 'stellar' lines from the 'host galaxy'. As to the continuum of the nucleus, it is often very similar to that of most quasars. Such is the case, for instance, for the blasar 0J 287.

This *similarity of continuous spectra* of all AGNs, characterized by a 'nonthermal' *power law* energy distribution, already discussed in Chapter 9, Section 1, proper to sources radiating by the '*synchrotron process*', provides one of the most important links between different AGNs.

The nonthermal continuum is the dominant component even in the spectra of such 'enveloped' AGNs as the blasars Mrk 501 and Mrk 421 (Maza *et al.*, 1978).

On the other hand, the relative weakness of emission lines in blasars is certainly an *intrinsic* phenomenon, which does not fit very easily in the unified concept of AGNs. It seems however, that the unified view can be preserved if one assumes that in blasars the synchrotron emission in the optical region is boosted to such a level that it swamps the emission lines.

Emission lines are supposed to be produced when a cloud of *gas* is heated by an intense source of UV (or even higher energy) radiation. These high energy photons can be produced either *directly* by the synchrotron process, characterized by a very broad continuous emission (which can extend from the radio waves to hard gamma rays) or indirectly through the '*inverse Compton process*' in which low energy photons colliding with high energy electrons can gain a considerable energy: radio frequency 'photons' (an incorrect but easy to understand expression) can be converted in this way into X-ray photons.

High energy photons heating the surrounding gas (if such gas is present) produce several effects: *thermal excitation*, *photoexcitation*, and a (probably turbulent) *expansion* of the gas. Expansion of the gas explains why in quasars and Seyferts the emission lines are so broad: some parts of the expansion cloud receed, other parts approach, hence a 'Doppler broadening'.

It is generally assumed, by analogy with the theory of planetary nebulae, that permitted lines are formed in a region of high density of the gas, and one explains the weakness of forbidden lines in Seyferts and quasars by collisional de-excitation due to this high density. Forbidden lines are supposed to be formed in a region farther from the nucleus, a region of smaller density. However, as shown by Shull and McCray (1978) when the high energy continuum has an energy density as high as in AGNs photoexcitation and photoionization from metastable levels can generate the weakness and even the absence of forbidden lines even at low density of the gas.

Thus, the weakness of emission lines in blasars can be due not only to excessive strength of the continuum but also to a 'gas deficiency' around the core of the blasar.

2.5.2.2. X-ray and Gamma-Ray Spectrum. It has been recently discovered that Seyfert nuclei, quasars, blasars and other AGNs (e.g. M 87) are able to emit X-rays or gamma-rays. About 20 Seyferts, a few quasars, a few blasars, etc. represent more or less powerful X-ray sources. Moreover, as already indicated in Table 13.2, the quasar 3C 273 is a gamma-ray source (CG 291 + 65) and the quasar Q 0241 + 622 is also probably a gamma-ray source (CG 135 + 1).

X-luminosities L_x (in the 2 to ~ 10 keV band) of some of these AGNs are indicated in Table 13.2.

According to Mushotzky *et al.* (1978) even in the band 2–60 keV, the blasars Mrk 501 and Mrk 421 possess power law spectra with a relatively small 'energy index' between (−0.4) and (+0.4). Thus, these and other blasar X-ray spectra are remarkable for their hardness (and, incidentally, for their lack of low energy cut off, which confirms a probable deficiency in gas). In the 2–30 keV band, the best fit for Mrk 501 gives a power law with 'photon number' index of about 1.2.

The non negligible X-luminosity of all these AGNs is very significant in establishing a physical connection between all these objects, and in confirming the width of the energy domain of their continuous spectrum, typical of the synchrotron process. Interested readers are referred to Swanenburg *et al.* (1978) which shows the spectrum of 3C 273 from radio to gamma-ray energies: even at 10^5 keV its intensity is considerable.

2.5.2.3. Radio Spectrum. The radio spectrum of most quasars and of some blasars is nearly flat. In most blasars it is (as already mentioned above) 'inverted', i.e. the radio brightness increases with the frequency, contrary to most 'normal' radio sources.

According to Apparao *et al.* (1978) the radio spectrum of the quasar Q 0241 + 622 is typical of multi-component synchrotron spectra of AGNs, in which recent synchrotron events add optically thick components to optically thin nonthermal spectra due to older events.

However, one must remember that only about 10% of known quasars appear to be 'radio' active with the present radio techniques, which does not exclude, of course, a *weak* radio emission in all AGNs.

On the other hand, the fact that all known blasars *are* radio sources obviously represents an effect of observational selection since most of them have been discovered in a search for optical counterparts of compact radio sources particularly bright at high radio frequencies.

2.5.3. The Variability and the Sizes

One of the most important and characteristic properties common (with some specific differences) to all AGNs is their *variability*. All present variations in brightness (thus in luminosity, if one neglects an eventual influence of the intervening medium). All present a more or less considerable variable optical polarization (though, according to Stockman and Angel (1978) for most quasars this polarization may be nonintrinsic and arise from interstellar polarization alone).

However, although common to all AGNs, the variability is the characteristic which, along with the luminosity, displays the greatest differences, for AGNs of different 'subclasses', such as quasars and blasars. The *amplitude* scale of variations in brightness, can be very different for different individual objects. And, most important, the *time scale* of variations in brightness and in polarization can be different for individual objects and more generally for different subclasses of AGNs.

Quasars and blasars are variable both at radio, optical and X-ray frequencies. And several noteworthy correlations exist in this connection. Thus, for instance, it seems that radio quasars are more likely to be optically variable than radio quiet quasars. In some blasars (e.g. BL Lac), the optical and radio variations seem entirely independent but a correlation has been observed in others (e.g. 0J 287). In the very peculiar AGN Cen A radio and X-ray brightnesses show some concurrent variations but do not track one another throughout the observations (1978, *ApJ*, **219**, 836). On the other hand, for the same object, other observers (1978, *ApJ*, **220**, 790) find that the radio variability is in phase with X-ray variability. Blasars Mrk 501 and Mrk 421 (and some other AGNs) have been identified as X-ray sources, by the correlated variability in the X-ray, optical and radio band.

A very remarkable lack of simultaneity of flare-ups viewed in different colours (U, B, V) has been observed in several blasars and quasars. Thus, for instance, for the quasar 3C 345 and for the blasars 0J 287 and BL Lac the optical flare-up occurs first almost simultaneously in B and V and only later (about eight days for the two blasars and about 30 days for 3C 345) in U, according to Bekenstein and Rosenkrantz (1978).

The amplitude scale of variations of some AGNs can be enormous. Thus, the blasar A0 0235 + 164 increased in brightness, in 1975, more than 100 fold in a few months, both at radio, infrared and optical wavelengths (and probably also, but not observed, in the X-ray and gamma ray region of the spectrum).

The total range of amplitude variations of two AGNs (Q 1418 + 54 and B2 1101 + 38) are given by Miller (1978) to be of the order of five magnitudes, which represents a factor of 100 for the brightness.

Even more important is the *shortness of the time scale* of variations in brightness (and polarization for blasars) of some AGNs. Indeed, according to the conventional light travel argument, already presented in Chapter 9, no signal can travel faster than light and so the variable region cannot have dimensions larger than the distance light travels during the shortest time scale of variation. Therefore, usually the shortest period in which an AGN can change its brightness or its polarization is taken to represent a measure of the maximum size of the object or, at least, of the size of the emitting region.

Table 13.3 is very convenient when one wants to 'visualize' different time scales in terms of linear sizes according to the conventional assumption.

It shows, in particular, that a time scale of a few *days*, typical for some blasars (whose variations may be 10 times faster than those of average quasars) correspond to sizes of *a few milliparsecs*, which represents dimensions of the order of only $10^{-7} D_G$ (where D_G represents the conventional value of the linear diameter of our Galaxy: 30 000 pc).

TABLE 13.3

Correspondence between the time scale of brightness (or polarization) variations and the maximum size of the emitting region (based on the conventional 'light travel argument'). The approximate linear diameter of our Galaxy, D_G, is supposed to be equal to 30 000 pc

Time scale of variations	Maximum linear size (in pc)	Maximum linear size (in D_G units)
1 min	6×10^{-7}	2×10^{-11}
1 hr	3.6×10^{-5}	1.2×10^{-9}
1 day	8.6×10^{-4}	2.9×10^{-8}
1 week	6.1×10^{-3}	2.0×10^{-7}
1 month	2.6×10^{-2}	8.7×10^{-7}
1 yr	3.1×10^{-1}	1.0×10^{-5}

The brightness of BL Lac can change by ~ 400% within two days at optical frequencies (at radio frequencies the time scale of variations of BL Lac, as that of other AGNs, is slower, of the order of a few weeks, thus showing a greater extension of the radio structure).

However, the lower limit of the time scale of variations seems to exist, and does not seem to be conditioned by the time resolution power of the instruments. Thus, Miller and Gimsey (1978) have failed to detect significant changes in the optical brightness of BL Lac, 3C 66 A, Mrk 501 and 3C 371, (though using high time-resolution photometry) on time scales ranging from 20 s to a few hours. But day-to-day changes are present.

Generally, blasars vary more rapidly than quasars and vary more rapidly in polarization than in brightness. Optical polarization of most blasars can reach 10–20%, and similar values have been found in a few radio selected violently variable quasars, such as 3C 279 or 3C 345. However, as already indicated above, according to Stockman and Angel (1978), most quasars seem to be intrinsically unpolarized (less than 0.5% at least). Of course, rapid variations in polarization of some blasars do not exclude slow continuous variations, such as those observed in the radio polarization of A0 0235 + 164 during the outburst of 1975, when the plane of polarization had slowly rotated through 130° in several months.

Quasars are known to be variable in brightness on time scales of years and months (it seems that only the *continuum* varies: Baldwin *et al.* (1978)). However, 3C 273 varies in brightness by about 50% in a few months, and the amplitudes of variations tend to increase with increasing frequency.

The very small size of active galactic *nuclei* cannot be confirmed by Very Long Baseline Interferometry, since this *radio* technique can measure only the extension of the radio structures associated with the nuclei, and since from different evidences such radio structures are usually found to be much more extended than the nuclei.

However, the 'light travel argument' can be verified on comparing the VLBI measurements with the sizes deduced by this argument from the *radio variability* of some radio sources, and this verification seems very encouraging.

2.6. THE RESIDUAL PROBLEMS

We have seen above that, in the light of the most recent discoveries, the 'luminosity argument' against the cosmological interpretation of the quasar redshift has lost the greatest part of its value. As reported by Faulkner and Gaskell (1978) at a recent 'workshop' on AGNs at Lick Observatory, no doubts at all were raised over this interpretation. And, as they say, "the last nail in the non-cosmological redshift's coffin was hammered in by the work of A. Stockton".

Indeed, this astronomer (Stockton, 1978) has discovered what he calls an "almost incontrovertible evidence in favour of the cosmological nature of QSO redshifts, and that QSOs are more frequently associated with normal galaxies than had generally been assumed". He founds this statement on a survey of redshifts of a well-defined sample of galaxies near bright, relatively low-redshift QSOs. He discovered that in eight out of 27 fields studied by him, at least one galaxy had a redshift matching that of the corresponding quasar. In particular the redshift of one galaxy 75 arc seconds off the quasar 3C 273 has a redshift identical to the (already considerable) redshift ($z = 0.158$) of the emission lines of 3C 273. The 'blue' absolute magnitude of this galaxy is of the order of (−20.3).

Unfortunately, all this does not solve all of the 'deep mysteries' of AGNs. Many aspects of AGNs are still problematical and not much is yet satisfactorily explained: the physical nature of AGNs and the physical mechanisms involved in their energy production still pose many puzzles.

Of course, as emphasized by Rees (1978) the acceleration of relativistic particles is known to be widespread in astrophysical objects. They certainly occur in solar flares, in shock-waves in interplanetary space, in supernova remnants and in pulsars.

However, the *concentration* of an enormous *power* in a relatively *very small region*, indicated by the 'variability size' of AGNs (the '*power density*') sets an extremely difficult problem.

Theories that have been proposed hitherto: a series of supernova explosions, accretion onto a supermassive (10^7–$10^8 M_\odot$) black hole, a slowly rotating supermassive 'spinar', etc., all encounter very hard difficulties. However, it seems more and more clear that a gravitational source of energy is quite possible and that, for the moment, it is not yet absolutely necessary to speculate about other, unknown and 'revolutionary' sources of energy.

Absorption lines systems in many quasars also represent a tantalizing puzzle. Indeed, as reported, for instance, by Boroson *et al.* (1978) in the spectrum of PKS 0237 −23, one of the best studied (and most poorly understood) absorption line quasar, more than 200 absorption lines have been observed in the tiny range from 3700 to 4300 Å, of which about 90 lines remain unidentified.

One still does not know whether these absorption lines are due to intrinsic or intervening matter. If the absorption lines of an object such as PKS 0237 −23 are caused by intervening objects (galaxies, clusters of galaxies, etc.) then the observational results imply that they must be nonuniformly distributed on a large scale and must have high densities. On the other hand, intrinsic ('ejection') hypothesis leads to an implausibly large mass and energy outflow from the AGN. Only a single case is known of an absorption redshift system that is unambigously due to the presence of a galaxy: the case of the quasar 4C 32.33 and of the galaxy NGC 3067, separated by 1.9 arc minutes (Peterson, 1978).

Even more puzzling is the fact that, in a few quasars, the redshift of some systems of absorption lines is larger than the redshift of the system of emission lines of the same object. The considerable debate about the nature of absorption lines is far from being closed. The solution of some of the hitherto unsolved difficulties, could perhaps be solved through a re-introduction of Einstein's cosmological constant Λ because this could modify the theoretical density of the intervening galaxies (Canuto and Owen, 1978).

Still another difficulty concerning the AGNs comes from the discovery by H. Arp of the so-called *'abnormal redshifts'*. According to this astronomer some groups of galaxies 'almost certainly' situated at the same distance because of some (questionable and contested) symptoms of 'interaction' and nevertheless displaying different redshifts, must be interpreted as a possibility of (until now unexplained) 'intrinsic', non-cosmological redshifts, or at least, of non-cosmological components in the observed redshifts. (At any rate, non-cosmological components are permitted in the conventional interpretation, if instead of being 'fixed' in comoving coordinates the sources are supposed to possess a 'proper motion'.)

Thus, for instance, in a recent paper (Arp, 1978) a luminous emission line compact object of redshift $z = 0.043$ which seems to be situated slightly in front of the galaxy NGC 1199, is considered by Arp to be approximately at the same distance as the galaxy of redshift $z = 0.0086$. Hence, again, Arp's conclusion that most of the redshift of the 'compact object' is of an origin other than cosmological.

However, most astronomers seem to be convinced today that anomalous redshifts 'simply do not exist' (Rees, 1978) and that a simple alignment is interpreted by Arp as real proximity. The conventional view is called 'cosmological hypothesis' by those who still believe, as G. Burbidge, that some sort of anomalous redshifts (though absent in Seyferts and blasars) represent an essential part of the high redshift observed in the quasars analogous to those listed at the end of Table 13.2.

More precisely, most astronomers reject the hypothesis that 'anomalous redshifts', even if sometimes real, cannot be explained by conventional physical laws and that they indicate the necessity of some fundamentally new physics.

A slight hope for some sort of 'new physics' was recently provided by a revival of the old theory of 'tired light' by Pecker *et al.* (1973) and more recent papers by the same authors. In this theory, the redshifts, or more precisely the non-cosmological components of the observed redshifts, are supposed to result from a *photon–photon interaction* which would be possible if photons had a non-vanishing mass of the order of 10^{-48} g, a hypothesis difficult to verify in a terrestrial laboratory.

This theory has not received a general acceptance. Its difficulties are well discussed by Peebles (1971) in connection with the difficulties of the old theory of 'tired light'. However, in their attempts to find evidences against the theory of the expansion of universe the propounders of the photon–photon interaction have discovered some interesting indications of a 'non-isotropy' of the value of H_0, which may have a bearing on the 'Cosmological Principle'.

2.7. TOWARDS A NEW CLASSIFICATION OF ACTIVE GALATIC NUCLEI?

It results from the discussion presented above that it is not possible today to consider the nuclei of Seyfert 1 galaxies, the quasars and the blasars as entirely different classes of astronomical objects, defined for the most by purely apparent characteristics, such as the absence of a 'host galaxy' around quasars or a 'featureless' spectrum of blasars.

The unified concept of active galactic nuclei obviously represents major progress in the field of extragalactic astronomy. But on the other hand, a new subdivision of AGNs is not yet available because their intrinsic properties are not yet sufficiently known and, last but not least, because of a *different overlapping* of many properties in *different individual objects*. (We have already encountered a similar difficulty when attempting to classify the galactic X-ray sources.)

Meanwhile, one is condemned to the uncomfortable uncertainty of nomenclature and of definitions in the field of AGNs, which we have attempted to clarify above.

3. Clusters of Galaxies

3.1. GENERAL PROPERTIES OF CLUSTERS OF GALAXIES CONSIDERED FROM THE POINT OF VIEW OF OBSERVATIONAL COSMOLOGY

Clusters of galaxies have been classified by Abell (1958) in several ways.

In particular, they have been classified according to the (apparent) magnitudes of their brightest members. The corresponding groups called the '*distance groups*' are given in the Table 13.4.

TABLE 13.4
Abell's 'distance groups' (from Abell, 1958)

Distance group	No. 1	No. 2	No. 3	No. 4	No. 5	No. 6	No. 7
Magnitude range	13.3–14.0	14.1–14.8	14.9–15.6	15.7–16.4	16.5–17.2	17.3–18.0	Over 18.0

They also have been classified by Abell in the '*richness groups*'. These groups are defined by 'counts'. These counts refer to the number of galaxies counted in a cluster that are not more that two magnitudes fainter than the third brightest member. They are given in Table 13.5.

TABLE 13.5
Abell's 'richness groups' (from Abell, 1958)

Richness group	No. 0	No. 1	No. 2	No. 3	No. 4	No. 5
Counts	30–49	50–79	80–129	130–199	200–299	300 or more

Many rich clusters contain a 'central' supergiant galaxy, such as M 87 in the Virgo cluster (we put 'central' in inverted commas because of a frequent lack of spherical symmetry). This supergiant galaxy is generally surrounded by an extended halo of faint stars (and is often a strong radio source, with an AGN in its core).

Each rich cluster has three major components:

(a) Galaxies (including the centrally located one);
(b) Hot gas;
(c) Stars of the halo surrounding the centrally located galaxy.

The galaxies of the cluster describe different orbits around the center of mass of the cluster. However, encounters between galaxies result in an equipartition of kinetic energies among them. As in a gas, galaxies which are more massive than the average move more slowly than the average. Thus, the speed of a massive galaxy can become too small to remain on its original orbit, and it falls towards the cluster's center of mass.

Statistical mechanics applied to clusters of galaxies show that given the 'velocity dispersion' of its members, a well defined mass inside the cluster is necessary to prevent its 'dispersion' (or, which amounts to the same, to 'keep it gravitationally bound').

All clusters have a 'missing mass': the *observed* amount of mass is quite generally insufficient to keep the cluster gravitationally bound, with respect to the observed dispersion of velocities.

The mass of the observed galaxies in most rich clusters is about 10 times too small to keep them bound. In some clusters, such as the Perseus cluster, the missing mass problem is even more severe (by a factor of about 40).

This problem is much more disturbing than the similar problem of the mass missing for the 'closure' of the universe. Indeed, the clusters are *certainly* non 'dispersed' and *must* be bound in some way, whereas we cannot be certain that the universe *must* be 'closed', even if this would be more satisfactory aesthetically (see Subsection 5.2).

Of course, the missing mass in clusters of galaxies might be hidden in quadrillons of low mass (and consequently of low luminosity) stars partly attached to the halo of the 'central' galaxy, and invisible at the great distance of clusters.

However, most recent observations of clusters of galaxies with X-ray sensitive equipment (from satellites) have suggested the possibility that the missing mass is hidden in the *hot gas* mentioned above.

Indeed, the X-brightness (and consequently the X-luminosity L_x) of clusters considered as X-ray sources (it is generally impossible to know whether the source is *the cluster as a whole* or whether one observes one or a few *discrete* sources inside the cluster) is relatively high and corresponds to relativiely considerable values of L_x (see Table 13.6).

Table 13.6 shows that X-ray emission is a common characteristic of rich clusters of galaxies, and qualitatively corresponds to the suggestion that some 'hidden mass' in the form of hot ionized gas might be the source of this emission. However, quantitatively it is not yet certain that the amount of mass in the hot gas is sufficient to bind different clusters. For more details see Gorenstein and Tucker (1978).

4. Superclusters of Galaxies?

Many years ago Abell (1958, 1961) and DeVaucouleurs (1958) suggested independently, from their optical observations of clusters of galaxies, that these clusters might be physically associated into '*superclusters*'.

Presently some astronomers (Kalinkov, 1977) state that "there is now evidence

TABLE 13.6

The values of L_x (2–10 keV)/L_G for a few clusters of galaxies and for the giant radio galaxy Cyg A (which lies in the center of a rich cluster in Cygnus). The values followed by the note (a) (from Gursky and Schwartz, 1977) are relative to the 2–7 keV band. All other data from Culhane (1977)

Usual name	X-ray name	d_{clas} (Mpc)	$\frac{L_x}{L_G}$	'Central' galaxy	Velocity dispersion (3-dim. km s^{-1})	Size (Mpc)
Virgo cluster	3U 1228 + 12	22	0.2	M 87	≈ 1100	0.1
Centaurus cluster	3U 1247 − 41	68	1	NGC 4696	≈ 1200	0.3
A 2199 cluster	3U 1639 + 40	168	6.5(a)	NGC 6166	≈ 1500	–
Perseus cluster	3U 0316 + 41	110	22.5	NGC 1275	≈ 2400	0.5
A 401 cluster	3U 0254 + 13	452	32.5(a)			
A 2142 cluster	3U 1555 + 27	455(b)	75			
Cyg A (c)	3U 1957 + 40	337	30(a)			

Notes:
(a) A value relative to the 2–7 keV band.
(b) Estimated distance.
(c) Radio source, Radio galaxy in a rich cluster.

demonstrating beyond doubt the existence of second-order clusters, i.e. of superclusters, having characteristic size of about 50 Mpc".

Other astronomers think that the case for the existence of superclusters is not quite convincing, but that "it does seem to be improving" (Gorenstein and Tucker, 1978). Indeed, Murray *et al.* (1978) (Giacconi's group), have concluded from an analysis of X-ray observations made with the Uhuru satellite, that several unidentified sources were in parts of the sky where clusters of galaxies seemed to be grouped in superclusters. Hence, the statement that they have found evidence for the existence of a new class of X-ray sources: the superclusters. They published a Table of such X-ray sources candidates for an identification with 21 superclusters, and proposed to identify three of these X-ray sources with groups of galaxies considered by them as superclusters.

Their main conclusion was that "if the X-rays are produced by thermal bremsstrahlung in analogy with normal clusters, then in most cases there is significantly more mass contained in the X-ray emitting gas than in the clusters of galaxies which make up the supercluster".

The possible existence of hot intracluster gas was immediately interpreted, with much enthusiasm, as representing a considerable fraction of the 'missing mass' necessary to close the universe.

Unfortunately, soon after the preliminary announcement of the results reported above, a group of astronomers working with the OSO-8 satellite, scanned directly over the position of 4U 0135 − 11, one of the three X-ray sources proposed by Murray *et al.* (1978) for identification with superclusters.

The result of these new observations (made with an equipment six times more sensitive than that of Uhuru) published in November 1977 (*Nature*, **270**, 158) was that 4U 0134 − 11 was "either variable or *spurious*" (and more probably spurious than variable). At any rate, a variability of the X-ray emission on time scale of a few years, implying a source diameter of about 1 pc (see Table 13.3) would rule out a supercluster identification.

Therefore, for the time being (November 1978), it is not yet possible to be quite sure that the 'new class of X-ray sources' of Murray *et al.*, (1978) really provides the missing mass necessary to close the universe.

5. General Comments on the Confrontration of Cosmological Theories with Observational Data

The chief purpose of the present Chapter was to describe the main objects (quasars, blasars, clusters of galaxies, etc.) used (with more or less positive results) in the 'tests' of the cosmological theories exposed in Chapters 10 and 12.

However, we cannot leave our reader without a few brief indications concerning some of these tests and without some general comments on the confrontation of cosmological theories with observational data.

Several publications: Peebles (1971); Misner *et al.* (1973); Tinsley (1977); the Proceedings of IAU Symposium 63 (1974) and especially Gott *et al.* (1974), will greatly facilitate and reduce our task since we can refer the reader to these, more advanced, works for all details missing in our deliberately brief account.

A first justification for this brevity is that the progress in this field is so rapid that, at any rate, any detailed account would become out of date, by the time of printing our textbook.

A deeper justification is, however, that, as in other domains of astrophysics, most cosmological *observations* are usually much more clearly presented in review papers and monographs than the corresponding theoretical concepts.

One can divide, in a first rough approximation, the problems set by the confrontation of cosmological theories with observational data in two main classes:

(1) The problems which can be considered as almost definitively solved.
(2) The problems which are far from being solved.

The problems of the first category are those relative to the question of the reality of a '*hot big bang*'. Indeed, the overwhelming majority of cosmologists definitely favours the idea of a passage of the universe, in the very remote past, through a very condensed and hot state.

However, considering the objections raised by Arp (1974), Omnès and Puget (1974) and some other 'unorthodox cosmologists', against the 'hot big bang' model, it seems that, provisionally, one should consider this problem only as *almost* settled.

As to the problems of the second class, one can mention the following (very exciting) questions:

(I) Is the universe 'oscillating' and the space curved and closed (bound)? (Of course, as will be explained below, in Gott *et al.*'s (1974) opinion this problem probably belongs to the first class and the answer is negative!)

(II) Can we consider the cosmological principle as really confirmed by observations of the clusters of galaxies? In other words, can we neglect the tendency of these clusters to present (contrary to Sandage's opinion, exposed in Chapter 10, Subsection 2.4) a slightly non-homogeneous distribution (with an 'anisotropic' H_0)?

(III) Can we answer the question: How an initially homogeneous and isotropic universe could generate galaxies and clusters of galaxies?

(IV) Is it true (an idea initially suggested by Alfvèn and developed by Omnès) that about one half of all galaxies consists of anti-matter?

(V) Do the 'black holes' really exist, and what is their role in the dynamical equilibrium (or disequilibrium) of the universe? (We have seen in Chapter 9 that the study of galactic X-ray sources provides a beginning of an answer to this question.)

More generally, and to some extent combining all these problems, we have the problem concerning the 'initial singularity' in the neighbourhood of the big bang. How should we understand the theoretical concepts of a quasi infinite density, temperature and curvature near the 'origin' . . . ?

As stated by Zel'dovich (1974) "for many decades, the attitude of theoreticians towards this singularity was one of strong dislike to this unwanted child born from the marriage of general relativity with observation (no pills existing at the time of Einstein and Friedmann!). Now many of us are very happy to investigate the implications of all these infinities. Direct observation is out of question on this scale. More important are the links suggested by various theories which tie singularity problems to observational features."

The cosmological problems look more and more like a kind of 'astrophysical palaeontology', investigating the state of the universe in the remote past. We still wait an answer to a question such as: was the expansion initially well ordered, with only small departures from strict homogeneity and a strict isotropy? Or was the 'initial state' of the universe strongly turbulent and chaotic?

As emphasized by Zel'dovich, we thus arrive to a domain of research where the answer is strongly biased by subconscious attitudes of investigators concerning such things as order, chaos, antimatter, etc.

Fortunately, observational progresses help the cosmologists to extract the objective truth from a more and more stodgy magma of subjective individual prejudices. On the other hand, unfortunately many observations are still too often affected by the 'rush to publication'.

The late Russian physicist Arzymovich used to say that "there is nothing worse than doubtful theories confirmed by doubtful experiments".

5.1. The Problem of the Cosmic Relic Radiation

One of the fundamental observational data concerning the remote past of the universe is provided by the so-called relic 2.7 K radiation discovered by the 1978 Nobel Prize winners Penzias and Wilson.

According to Blair (1974) one of the main conclusions of these very difficult observations is that it is not yet possible to state that the observed spectrum is that of a black body at 2.7 K.

However, a great progress was accomplished in this domain in the following direction. Contrary to previous measures, not confirmed by more recent ones, the observations are compatible with the idea that the spectrum do correspond to the spectrum of a black body at 2.7 K.

As stated by Blair (1974): "There is a tendency to say that recent results lend support to the notion of a 2.7 K background; . . . This is true in the sense that these results are compatible with a 2.7 K background, while the earlier results weren't. But, in fact, the measurements give very little quantitative information about the submillimeter background radiation. At wavelengths less than 1 mm, all one has from these measurements is upper limits to the flux."

On the other hand, the extraordinary *isotropy* of this radiation (at large scale, of the order of 10°, which integrates the small scale fluctuations) seems to be confirmed. As stated by Partridge (1974): "It seems that we must face the possibility that we have in the past overestimated the magnitude of the cosmological anisotropy in the microwave background".

Besides, Partridge (1974) duly insists upon the right meaning of these concepts of isotropy or anisotropy of the 'cosmological radiation'. If the Earth were not revolving round the Sun, if the solar system were not revolving around the center of our Galaxy, and, finally, if our Galaxy were not revolving around an eventual center of a 'local supercluster of galaxies', etc. the cosmological radiation should be perfectly isotropic (in a non-rotating universe!). However, already because of all the motions mentioned above, a violetshift affects the radiation towards which we are moving.

The diurnal rotation of the Earth round its axis introduces every day a succession of different directions (since most radio telescopes are not steerable), hence a space-period of 360° for the observed radiation if one assumes the existence of an intrinsic anisotropy of the field. However, the observed intensity of the cosmic radiation does not vary more than by about 0.005 K around 2.7 K, in 24 hours!

5.2. The Problem of the Validity of the Cosmological Principle

According to Heller *et al.* (1974) one should consider in the cosmological principle separately:

(1) The Cosmological Principle for the Substratum of matter-energy.
(2) The Cosmological Principle for the Geometry, used, for instance, in the construction of the Robertson–Walker $(ds)^2$.

The 'Cosmological Principle for the Substratum' (CPS) postulates the homogeneity and the isotropy of the 'cosmic fluid', but it is necessary to define an universal frame of reference before speaking about this homogeneity and isotropy. The existence of such a reference system represents a hypothesis very much like the concept of the ether in classical electrodynamics. Thus one is led to consider a kind of cosmological '*neo-ether*'.

Since the cosmological radiation is generally interpreted as a 'relic' of remote past phases of evolution of the universe, one concludes that at that remote epoch the CPS was valid. However, since at that remote past the galaxies were not yet formed, this suggests that the inhomogeneities represented by galaxies should not affect the application of the CPS.

Concerning the existence of symmetries postulated by the Robertson–Walker metric, other geometries are presently known, so that the Cosmological Principle for Geometry (CPG) is less compulsory, and could be later submitted to observational tests.

5.3. THE PROBLEM OF THE VALUES OF Ω AND OF H_0: (CLOSED OR OPEN UNIVERSE?)

5.3.1. *Preliminary Conclusions*

If one assumes that at all epochs accessible to direct observation the universe obeys the cosmological principle, one is led to confront the observations with one of the models of general relativity which (if we postulate a zero cosmological constant Λ) is completely specified (in its present matter-dominated phase) by only two parameters, for instance H_0 and Ω (see Chapter 10).

We have already shown (see Chapter 12, Section 2) that if $\Omega > 1$ the universe is oscillating and *closed* (bound), whereas it is '*open*' (unbound) and monotonically expanding if $\Omega \leqslant 1$.

According to Tammann (1974), the present tendency is to ascribe H_0 a value near $55\ \mathrm{km\,s^{-1}\,Mpc^{-1}}$, but individual methods for determining H_0 give disparate values ranging from as low as 40 to as high as $110\ \mathrm{km\,s^{-1}\,Mpc^{-1}}$, and the value of 55 is far from being universally accepted.

As to the 'observational' value of Ω it is still uncertain. However, as pointed out by Gott *et al.* (1974); "Although there are loopholes in every case, the strongest arguments taken together point to an open universe, with

$$\Omega \approx 0.06 \pm 0.02. \tag{4}$$

It is remarkable that the best constraints come from very local data (age of elements in meteorites, interstellar deuterium, dynamics of nearby aggregates of galaxies)", not from observations of AGNs at large redshifts.

However, one must emphasize that direct determinations of H_0 and Ω are still subject to many statistical and systematic uncertainties: intrinsic luminosities of AGNs in the interpretation of the observations of the variation of brightness as a function of z, intrinsic diameters of radio sources and of clusters of galaxies in the interpretation of the variation of the angular diameter as a function of z – all reasons for which we omit to discuss the corresponding observational data.

The final result of Gott *et al.* (1974) is that

$$0.05 \leqslant \Omega \leqslant 0.09, \tag{5}$$

and

$$49 \leqslant H_0 \leqslant 65\ \mathrm{km\,s^{-1}\,Mpc^{-1}}. \tag{5'}$$

At any rate, they conclude that: "The objections to a closed universe are formidable but not fatal; a clear verdict is unfortunately not yet in, but the mood of the jury is perhaps becoming perceptible",

Let us now present briefly the method used by Gott *et al.* (1974) to reach this conclusion.

5.3.2. *Gott et al.'s Method*

We have shown (Chapter 10, Subsections 5.8 and 5.9) that the 'age' t_0 of the Friedmann universe is a definite function $t_0 = f(\Omega)/H_0$ of Ω and of H_0.

The determination of the age of the elements and age of stars makes use of long-lived 'nucleochronometers' Th^{232} and Re^{187}. On the other hand, the ages of globular clusters are currently estimated as 8 and 18 billion years. Consistent and generous limits to the age of the universe from these evidences are

$$8 < t_0 < 18 \text{ billion years.} \tag{6}$$

The theoretical relation $t_0 = f(\Omega)/H_0$ can be represented for a given value of t_0 by a line (iso $- t_0$) in the plane (Ω, H_0), (see Figure 13.2).

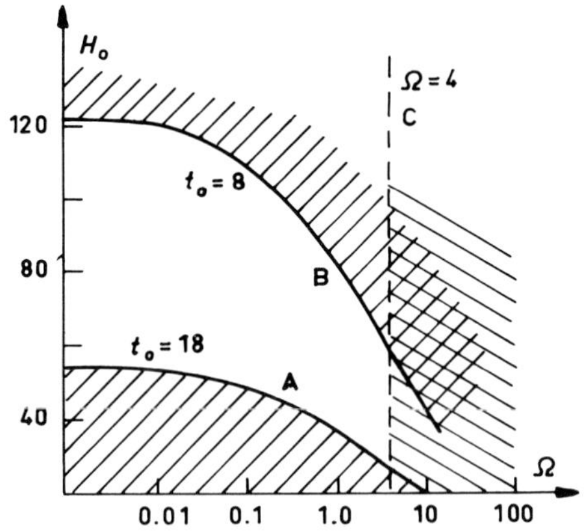

Fig. 13.2. The values of Ω and of H_0 excluded (cross-hatched areas) by the 'ages of the elements' and the 'ages of globular clusters'. [From Gott *et al.*, 1974.]

The two curved lines A and B in Figure 13.2 correspond to $t_0 = 8$ and $t_0 = 18$ billion years respectively. The cross-hatched areas below A and above B correspond to the domains of the plane (Ω, H_0) excluded by the inequalities (6).

On the other hand, and on assuming that all galaxies have the same luminosity $L(t_e)$, the value of Ω can, in principle, be deduced from the general relation (150) or from the approximate relation (159) of Ch. 12 between the observed integrated brightness $B(t_0)$ and the observed redshift z, applied to the observable AGNs.

However, statistical uncertainties concerning $L(t_e)$, already pointed out above, and systematic photometric corrections involved in this determination are so large that Gott *et al.* (1974) feel safe in concluding only that from this evidence $\Omega < 4.0$. This limit is shown by the vertical line C in Figure 13.2, and the cross-hatched area to the right of C.

We also know that one can attempt to obtain $\Omega = \rho_0/\rho_c$ from the 'observed masses' (actually derived from extrapolations concerning mass-to-light ratios) of astronomical objects composing the universe (galaxies, clusters of galaxies, etc.). However, such a direct determination can yield only a lower limit Ω^* to Ω, since it does not take into account any invisible or undetectable mass. Thus, for instance, Ω^* includes only all matter *within* clusters, but not the mass of an eventual intercluster gas or dust, and it does not take into account an eventual intergalactic medium or black holes.

Three independent determinations of Ω^* yield:

$$\Omega^* \approx 0.05 \pm 0.01. \tag{7}$$

This would permit to trace in Figure 13.2 a vertical line $\Omega = 0.05$, and to cross-hatch the domain at the left of this line, but we avoid doing so in order to not surcharge the figure without any benefit.

Finally, the theoretical determination of the mass fraction of deuterium synthesized in a homogeneous, isotropic universe with the present relic radiation temperature 2.7 K (the 'standard big bang') depends very strongly on the present density ρ_0 of the universe.

The details of these very specialized investigations in nucleosynthesis are quite outside the level of our introductory general textbook. The result, when one includes the possibility of 'nonstandard big bangs', is

$$\rho_0 < 2 \times 10^{-30} \mathrm{g\,cm^{-3}}. \tag{8}$$

Since, according to Equations (47) and (49) of Ch. 10, we have

$$\rho_0 = \rho_c \, \Omega = \frac{3H_0^2}{8\pi G} \, \Omega, \tag{9}$$

these relations give a new excluded domain in the plain (Ω, H_0), cross-hatched above D in Figure 13.3 (separated from Figure 13.2 in order to avoid a too overcharged picture).

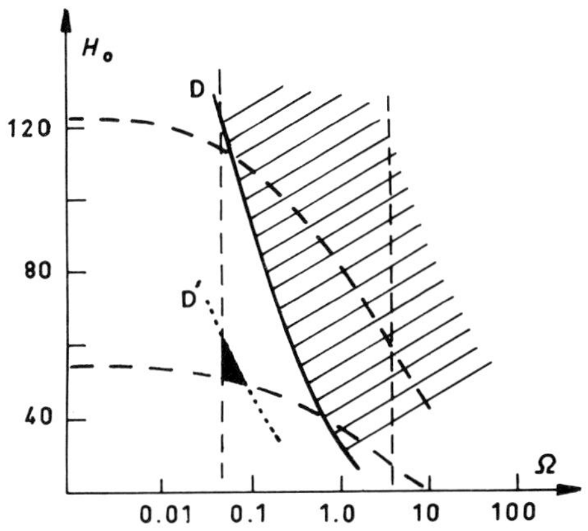

Fig. 13.3. The values of Ω and of H_0 permitted by the combination of all criteria (see the text): the black curvilinear triangle. [From Gott *et al.*, 1974.]

In Figure 13.3, the limits already introduced in Figure 13.2 are recalled by dotted lines, but the cross-hatchings of Figure 13.2 are not reproduced for the sake of clarity.

We see that the permitted region shrinks more and more. However, it becomes even more tiny if one takes into account the observed abundances of deuterium and hydrogen, which lowers ρ_0 to a value lower than 0.4×10^{-30}.

Hence a line D′ indicated by small 'dense' points and a new 'excluded region' to the right of D′ (which is not cross-hatched for more clarity).

Finally, one is left, as the only permitted region in the (Ω, H_0) plane, with only the very small curvilinear triangle corresponding roughly to inequalities (5) and (5′).

The reader who wants a more complete information about the 'standard big bang' is referred to the excellent summary by Weinberg (1973, pp. 528–570) (the Thermal History of the Early Universe and the Problem of the Helium Synthesis).

As conclusion let us quote, once more, Gott *et al.* (1974): "*A variety of arguments strongly suggests that the density of the universe is no more than a tenth of the value* [$\Omega = 1$] *required for closure. Loopholes in this reasoning may exist, but if so, they are primordial and invisible, or perhaps just black.*"

5.4. SOME MORE RECENT RESULTS

An important recent development in the problem of 'closure' of the Friedmann universe has been the discovery and employment of the 'Baldwin effect' (Baldwin *et al.*, 1978), i.e. of the strong correlation between continuum luminosity and emission line equivalent width of a particular subset of radio quasars with a flat radio spectrum.

This provides a 'luminosity calibration' for the considered quasars, from the measure of the widths of their emission lines, in a way somewhat similar to spectroscopic criteria of the luminosity of stars.

The knowledge of individual luminosities, brightnesses and redshifts of a sample of quasars allows one to construct a 'Hubble diagram' and to look for the best fit of the relations established in Chapter 12, Section 8 for different values of $q_0 = \Omega/2$.

The result is quite different from that of Gott *et al.* (1974) since Baldwin and his collaborators (1978) find a value of Ω well over 1. But this result provokes a lively controversy and it is not excluded that some hidden 'evolutionary effects' affect this treatment.

On the other hand, recent applications of the 'angular diameter-redshift' test for radio quasars with large redshifts (Hooley *et al.*, 1978) have shown that observations are consistent with the hypothesis that the overall physical sizes of the radio structures of radio quasars with the highest radio luminosity *do not change with cosmological epoch*.

Thus it seems, that Gott *et al.*'s conclusions cannot be considered as 'the last word' on the problem. Recent data (and theories with a non-zero cosmological constant) seem to open new perspectives.

As stated by Tinsley (1977) 'even with a strong preference for simplicity, we must not forget that the basic assumptions of the allowed class of models may be too restrictive or even incorrect.'

Moreover, all presently feasible tests for Ω, in the frame of Friedmann models, are so extremely sensitive to the evolution of intrinsic AGN properties that they provide too weak constraints on this parameter.

Finally, it is possible that theories more sophisticated that the conventional General Relativistic approach presented in the preceding Chapters will be soon constructed and found to better represent the observational data.

Appendix. A Dictionary of Abbreviations in the Field of Extragalactic Objects

(See also Table A1 in the Appendix to Chapter 9)

A	Abell's catalogue [1958, *Astrophys. J. Suppl.* 3, 211]
AO	Arecibo Observatory

AGN	Active Galactic Nucleus
BCS	Catalog of QSOs [1977, *Astrophys. J. Suppl.* **33**, 113]
BLRG	Broad Line Radio Galaxy
BSO	Blue Stellar Object
BL Lacs	BL Laçertae objects ('Blasars')
Kron.	Globular cluster [1956, *Publ. A.S.P.* **68**, 125]
LB	An object in *Luyten and Sandage* Catalog, (1966). Minneapolis, U.S.A.
M	Messier (Catalog)
Mk	Markarian's Catalog (Burakan Observatory, U.S.S.R.)
Mkn	Markarian's Catalog
MR	M = MIT, R = Ricker – Catalogue
NGC	New General Catalog, (1890), Dreyer, J. L. E.
NLRG	Narrow Line Radio Galaxy
Terzan.	Globular Cluster [1971, *Astron. Astrophys.* **12**, 477]
PHL	PHL Catalog [1967, *Astrophys. J.* **148**, 769]
PKS	Parkes Catalog of Radio Sources
QSG	Quasi Stellar Galaxy [Old name of 'radio quiet' quasars]
QSO	Quasi Stellar Object (Quasar)
QSS	A radio QSO (A radio quasar)
OSQ	Optically Selected Quasar
Zw	Zwicky, F., *et al.* [1961–1968, Catalog of Galaxies and Clusters of Galaxies, in 6 Vols. Pasadena, Cal. Inst. of Technology, U.S.A.]

References

1. WORKS CITED IN THE TEXT

(Works that can be used for further study are marked by an asterisk.)

Abel, G.: 1958, *Astrophys. J. Suppl.* **31**(3), 211.
Abell, G.: 1961, *Astron. J.* **66**, 607.
Apparao, K., *et al.*: 1978, *Nature* **273**, 450.
Arp, H.: 1974, In *Proceedings of IAU Symposium* **63**, 61, Reidel, Dordrecht.
Arp, H.: 1978, *Astrophys. J.* **220**, 401.
Baldwin, J. A. *et al.*: 1978, *Nature* **273**, 431.
Blair, A. G.: 1974, In *Proceedings of IAU Symposium* **63**, 144, Reidel, Dordrecht.
Boroson, T. *et al.*: *Astrophys. J.* **220**, 772.
*Burbidge, G. R. *et al.*: 1977, *Astrophys. J. Suppl.* **33**, 113.
Canuto, V. and Owen, J.: 1978, *Astrophys. J.* **225**, 79.
Culhane, L. J.: 1977, In *Highlights of Astronomy* **I**(4), 293, Reidel, Dordrecht.
De Vaucouleurs, G.: 1958, *Astrophys. J.* **63**, 253.
Edge, D. O. *et al.*: 1960, *Mem. Roy. Astron. Soc.* **68**, 37.
Fairall, A. P.: 1977, *Mon. Not. Roy. Astron. Soc.* **180**, 391.
Faulkner, J. and Gaskell, M.: 1978, *Nature* **275**, 91.
*Gorenstein, P. and Tucker, W.: 1978, *Sc. Amer.* **239**(5), 98.
Gott, J. R. *et al.*: 1974, *Astrophys. J.* **194**, 543.
Greenstein, J. and Matthews, T. A.: 1963, *Nature* **197**, 1041.
*Gursky, H. and Schwartz, D. A.: 1977, *Ann. Rev. Astron. Astrophys.* **15**, 541.
Hazard, C. *et al.*: 1963, *Nature* **197**, 1037.
Heller, M. *et al.*: 1974, In *Proceedings of IAU Symposium* **63**, 3, Reidel, Dordrecht.
Hooley, A. *et al.*: 1978, *Mon. Not. Roy. Astron. Soc.* **182**, 127.

Jauncey, D. L. *et al.*: 1978, *Astrophys. J.* (*Letters*) **223**, L1.
Kalinkov, M.: 1977, In *Highlights of Astronomy* I(4), 297, Reidel, Dordrecht.
Kristian, J.: 1973, *Astrophys. J.* (*Letters*) **179**, L61.
Margon, B. and Kwitter, K. B.: 1978, *Astrophys. J.* (*Letters*) **224**, L43.
Maza, J. *et al.*: 1978, *Astrophys. J.* **224**, 368.
Miller, H. R. and McGimsey Ben Q.: 1978, *Astrophys. J.* **220**, 19.
Miller, J. S. and Hawley, S. A.: 1977, *Astrophys. J.* (*Letters*) **212**, L47.
*Misner, C. W. *et al.*: 1973, *Gravitation*, Freeman, San Francisco.
Murray, S. S. *et al.*: 1978, *Astrophys. J.* (*Letters*) **219**, L89.
Mushotzky, R. F. *et al.*: 1978, *Astrophys. J.* (*Letters*) **226**, L65.
Omnès, R. and Puget, J. L.: 1974, In *Proceedings of IAU Symposium* **63**, 335, Reidel, Dordrecht.
Oke, J. B.: 1963, *Nature* **197**, 1040.
Partridge, R. B.: 1974, In *Proceedings of IAU Symposium* **63**, 157, Reidel, Dordrecht.
Pecker, J. C. *et al.*: 1973, *Nature* **241**, 338.
*Peebles, P. J. E.: 1971, *Physical Cosmology*, Princeton Univ. Press, Princeton.
Peterson, B. M.: 1978, *Astrophys. J.* **223**, 740.
*Proceedings of IAU Symposium **63**, (1974), *Confrontation of Cosmological Theories with Observational Data*, Reidel, Dordrecht.
Rees, M.: 1978, *New Scientist* **80**, 188.
Sandage, A.: 1961, *Sky and Telesc.* **21**, 148.
Sargent, W. L. *et al.*: 1978, *Astrophys. J.* **221**, 731.
Shull, J. M. and McCray, R.: 1978, *Astrophys. J.* (*Letters*) **223**, L5.
Schmidt, M.: 1963, *Nature* **197**, 1040.
Stockman, H. S. and Angel, J. R. P.: 1978, *Astrophys. J.* (*Letters*) **220**, L67.
Stockton, A.: 1978, *Nature* **274**, 342.
Swanenburg, B. N. *et al.*: 1978, *Nature* **275**, 298.
Tammann, G. A.: 1974, In *Proceedings of IAU Symposium* **63**, 47, Reidel, Dordrecht.
*Tinsley, B. M.: 1977, *Physics Today* **30**(6), 32.
Wampler, E. J. *et al.*: 1975, *Astrophys. J.* **198**, 49.
West, R. M. *et al.*: 1978, *Astron. Astrophys.* (*Letters*) **62**, L13.
Young, P. J. *et al.*: 1978, *Astrophys. J.* **221**, 721.
*Weedman, D. W.: 1977, *Ann. Rev. Astron. Astrophys.* **15**, 69.
Zel'dovich, Ya. B.: 1974, In *Proceedings of IAU Symposium* **63**, p. IX, Reidel, Dordrecht.
Zwicky, F. *et al.*, 1961–1968, *Catalogue of Galaxies and Clusters of Galaxies*, California Inst. of Technology, Pasadena.

2. REFERENCES PARTLY USED IN THE TEXT AND WHICH CAN BE USED FOR FURTHER STUDY

(See first the works cited in 1 marked by an asterisk.)

Abbreviations: *ApJ* = *Astrophys. J.*; *MN* = *Mon. Not. Roy. Astron. Soc.*; *AA* = *Astron. Astrophys.*; *Nat* = *Nature*; *Sc* = *Science*; *Sc Amer* = *Scient. Amer.*; *ST* = *Sky and Telescope*; *New Sc* = *New Scientist*; *ARAA* = *Ann. Rev. Astron. Astrophys.*; *IAUC* = *Int. Astr. Union Circ.*; *Rech* = *La Recherche* (French).

2.1. REVIEW ARTICLES (GENERAL)

1974, *Sc Amer*, **230** (4), 67. 1978, *Nat*, **272**, 599; **274**, 16; **274**, 419; **275**, 419; *MN*, **182**, 23P; *ST*, **55**, 107; **56**, 490; **56**, 499; *Sc*, **200**, 1031; *Sc Amer*, **237** (2), 32; *New Sc*, **80**, 284; *Rech*, 94, 944.

2.2. ACTIVE GALACTIC NUCLEI (AGNs)

2.2.1 Unified concept of AGN

1976, *ApJ*, **208**, 30; *Nat*, **263**, 279.

2.2.2. Individual AGNs (Entries in alphabetical order of the Catalogues)

3C 273: 1977, *ApJ*, **212**, 22; 1978, *Nat*, **274**, 16.

Fairall 9: 1978, *Nat*, **271**, 334; *ST*, **55**, 459.
M 87: 1978, *ApJ*, **219**, 408.
Markarian 132: 1978, *ApJ*, **223**, 758.
Markarian 421: 1978, *ApJ*, **222**, L3; *MN*, **182**, 489; *IAUC*, No. 3212, No. 3224.
Markarian 501: 1978, *ApJ*, **224**, L103.
NAB 0137 – 01: 1977, *ApJ*, **211**, L5.
NGC 4151: 1977, *Nat*, **270**, 319; 1978, *ApJ*, **221**, L7; **224**, 375.
NGC 5128(Cen A): 1978, *ApJ*, **219**, L81; **219**, 836; *MN*, **182**, 661.
NGC 6251: 1978, *Nat*, **272**, 131.
PKS 0349 – 14: 1977, *MN*, **181**, 435.
PKS 0548 – 322 (BL Lac): 1978, *IAUC*, No. 3261.
PKS 1402 + 044: 1978, *ApJ*, **222**, L81.
QSO 1418 + 54: 1978, *ApJ*, **223**, L67.
4U 0241 + 61: 1978, *Nat*, **274**, 16.
III Zw 2: 1978, *ApJ*, **222**, L91.

2.2.3. Seyferts (as X-ray sources)
1978, *ApJ*, **223**, 74; *Nat*, **271**, 334; *MN*, **183**, 129; **184**, 1P; *AA*, **65**, 115; **68**, L15.

2.2.4. N Galaxies (as X-ray sources)
1978, *Nat*, **275**, 624.

2.2.5. Search and identification procedures
1976, *Nat*, **263**, 372; 1978, *ApJ*, **221**, 468; **222**, 40; **222**, L81; **223**, 1.

2.2.6. Optical morphology
1978, *ApJ*, **223**, L67; *MN*, **182**, 361.

2.2.7. Radio morphology
1978, *Nat*, **272**, 131; *MN*, **184**, 335.

2.2.8. Selection effects
1978, *Nat*, **273**, 130.

2.2.9. Luminosities
1977, *ApJ*, **214**, 679; 1978, *Nat*, **273**, 438; **276**, 163.

2.2.10. Colours
1974, *ApJ*, **190**, 509.

2.2.11. Optical spectra
1977, *ApJ*, **211**, L5; 1978, *ApJ*, **223**, 364.

2.2.12. Absorption lines in quasars
1978, *ApJ*, **224**, 344.

2.2.13. Radio spectra
1978, *ApJ*, **222**, L3.

2.2.14. X-ray spectra
1976, *ARAA*, **14**, 173; 1978, *ApJ*, **224**, L103.

2.2.15. Theoretical models and physical mechanisms
1977, *Nat*, **265**, 225; 1978, *MN*, **182**, 537; **182**, 639; **184**, 87.

2.2.16. VLBI sizes (radio morphology)
1977, *ApJ*, **211**, 658.

2.2.17. Statisitical data: Space distribution
1976, *Publ. A.S.P.*, 88, 665; 1978, *ApJ*, **220**, 372; **220**, L1.

2.2.18. Anomalous redshifts
1978, *ApJ*, **219**, 367.

2.2.19. Cosmological hypothesis
1978, *MN*, **182**, 181.

2.3. CLUSTERS OF GALAXIES
1978, *ApJ*, **219**, 795; **220**, 8; **221**, 34; **221**, L43; **221**, 745; **222**, L85; **223**, 37; **223**, 185; **224**, 1; **224**, 718; **225**, 21; *MN*, **184**, 783.

2.4. SUPERCLUSTERS
1977, *ApJ*, **215**, 717; *ST*, **54**, 105.

2.5. BLACK HOLES
1978, *ApJ*, **221**, 721; **221**, 731; **222**, 667; **222**, 976; **225**, 687; *Nat*, **272**, 739; *MN*, **184**, 721; *ST*, **55**, 113.

2.6. MISCELLANEOUS COSMOLOGICAL OBSERVATIONS AND THEORIES

2.6.1. Observational tests
1978, *ApJ*, **221**, 383; *MN*, **182**, 27; **182**, 127; *AA*, **66**, 249; **68**, 353.

2.6.2. The missing mass problem
1977, *ST*, **53**, 253; 1978, *Nat*, **273**, 431; **273**, 519; *New Sc*, **78**, 668.

2.6.3. The neo-ether
1978, *New Sc*, 77, 298.

2.6.4. Cosmic radiation at 2.7 K
1978, *Nat*, **271**, 426; *Sc Amer*, **238**(5), 64; *New Sc*, **80**, 262; *Phys. Today*, **31**(1), 17.

2.6.5. The cosmological constant Λ
1978, *Phys. Today*, **30**(6), 32.

2.6.6. The origin of the universe
1978, *Nat*, **271**, 506.

2.6.7. Quiescent cosmology
1978, *Nat*, **272**, 211.

2.6.8. The value of H_0
1978, *Nat*, **274**, 449.

2.6.9. Unorthodox views in cosmology
1978, *MN*, **183**, 727; **183**, 749; **184**, 439; *AA*, **68**, 131.

2.6.10. Theoretical models
1978, *MN*, **182**, 537; **182**, 617.

TABLE OF VALUES

Value and units	*Item*	*Symbol or abbreviation*
Astronomical		
3.16×10^{7} s	1 year	yr
3.09×10^{18} cm	1 parsec	pc
1.99×10^{33} g	The mass of the Sun	$M_{\odot}$
3.83×10^{33} erg s^{-1}	The luminosity of the Sun	$L_{\odot}$
6.96×10^{10} cm	The radius of the Sun	$R_{\odot}$
30 kpc	Approximate diameter of our Galaxy	D_G
$1.4 \times 10^{11} M_{\odot}$	Approximate mass of our Galaxy	M_G
4×10^{43} erg s^{-1}	Approximate luminosity of our Galaxy	L_G
75 km s^{-1} Mpc^{-1}	Provisional value of the Hubble constant	H_0
2×10^{-31} g cm^{-3}	Provisional value of the mean mass density of the universe	ρ_0
10^{-26} W m^{-2} Hz^{-1}	1 Jansky = Unit of flux density	Jy
Physical		
6.67×10^{-8} c.g.s.	Gravitational constant	G
3.00×10^{10} cm s^{-1}	Speed of light in vacuum	c
6.63×10^{-27} erg s	Planck's constant	h
1.38×10^{-16} erg deg^{-1}	Boltzmann constant	k
1.66×10^{-24} g	Approximate mass of a baryon	m_B
6.02×10^{23} mole^{-1}	Avogadro number	N_A
9.11×10^{-28} g	Electron rest mass	m_e
1.67×10^{-24} g	Proton rest mass	m_p
5.29×10^{-9} cm	Bohr radius of the ground state of H	a_0
1.60×10^{-12} erg	Energy associated with 1 eV	eV
4.13×10^{-18} keV	Energy of a photon whose frequency is 1	Hz
4.80×10^{-10} e.s.u.	Electronic charge	e

INDEX

BY THE SAME AUTHOR

Basic Method in Transfer Problems
First edition: Oxford University Press (1952), London, (Out of print).
Second edition: Dover Publications Inc. (1963), New York, (Out of print).

La Recherche Scientifique (With J. C. Kourganoff)
First edition: Presses Universitaires de France (1958), Paris.
Fourth edition: Presses Universitaires de France (1971), Paris.
(Translated into Spanish, Italian, Japanese, Portuguese, and Greek).

Astronomie Fondamentale Elémentaire
Masson (1961), Paris, (Out of print).

Initiation à la Théorie de la Relativité
Presses Universitaires de France (1964), Paris, (Out of print).
(Translated into Russian, Spanish, and Polish. The author possesses a mimeographed translation into English).

Introduction to the General Theory of Particle Transfer
Gordon and Breach, 1969, New York.
(Translated into French).

Exercices d'Initiation Rapide au Russe Scientifique (With Mrs R. Kourganoff)
Dunod (1969), Paris.

Introduction to the Physics of Stellar Interiors
Reidel (1973), Dordrecht, Holland.
(Translated from the French).

La Face Cachée de l'Université
Presses Universitaires de France (1972), Paris.
(Translated into Spanish and Portuguese).

GEOPHYSICS AND ASTROPHYSICS MONOGRAPHS

AN INTERNATIONAL SERIES OF FUNDAMENTAL TEXTBOOKS

Editor:

BILLY M. McCORMAC (Lockheed Palo Alto Research Laboratory)

Editorial Board:

R. GRANT ATHAY (High Altitude Observatory, Boulder)
W. S. BROECKER (Lamont-Doherty Geological Observatory, New York)
P. J. COLEMAN, Jr. (University of California, Los Angeles)
G. T. CSANADY (Woods Hole Oceanographic Institution, Mass.)
D. M. HUNTEN (University of Arizona, Tucson)
C. DE JAGER (the Astronomical Institute at Utrecht, Utrecht)
J. KLECZEK (Czechoslovak Academy of Sciences, Ondřejov)
R. LÜST (President Max-Planck-Gesellschaft zur Förderung der Wissenschaften, München)
R. E. MUNN (University of Toronto, Toronto)
Z. ŠVESTKA (The Astronomical Institute at Utrecht, Utrecht)
G. WEILL (Institute d'Astrophysique, Paris)

1. R. Grant Athay, *Radiation Transport in Spectral Lines.*
2. J. Coulomb, *Sea Floor Spreading and Continental Drift.*
3. G. T. Csanady, *Turbulent Diffusion in the Environment.*
4. F. E. Roach and Janet L. Gordon, *The Light of the Night Sky.*
5. H. Alfvén and G. Arrhenius, *Structure and Evolutionary History of the Solar System.*
6. J. Iribarne and W. Godson, *Atmospheric Thermodynamics.*
7. Z. Kopal, *The Moon in the Post-Apollo Era*
8. Z. Švestka, *Solar Flares.*
9. A. Vallance Jones, *Aurora.*
10. C.-J. Allègre and G. Michard, *Introduction to Geochemistry.*
11. J. Kleczek, *The Universe.*
12. E. Tandberg-Hanssen, *Solar Prominences.*
13. A. Giraud and M. Petit, *Ionospheric Techniques and Phenomena.*
14. E. L. Chupp, *Gamma Ray Astronomy.*
15. W. D. Heintz, *Double Stars.*
16. A. Krüger, *Introduction to Solar Radio Astronomy and Radio Physics.*
17. V. Kourganoff, *Introduction to Advanced Astrophysics.*
18. J. Audouze and S. Vauclair, *An Introduction to Nuclear Astrophysics.*